D0603232

Life Science Automation
Fundamentals and Applications

For a list of related Artech House titles,
please see the back of this book.

Life Science Automation Fundamentals and Applications

Mingjun Zhang
Bradley Nelson
Robin Felder

Editors

Library of Congress Cataloging-in-Publication Data
A catalog record for this book is available from the U.S. Library of Congress.

British Library Cataloguing in Publication Data
A catalogue record for this book is available from the British Library.

Cover design by Igor Valdman

© 2007 ARTECH HOUSE, INC.
685 Canton Street
Norwood, MA 02062

ISBN-13: 978-1-59693-105-3

Contents

CHAPTER 10

DNA and Protein Microarray Fabrication Automation   279

PART IV

System Integration   303

CHAPTER 11

Automation of Nucleic Acid Extraction and Real-Time PCR:
A New Era for Molecular Diagnostics   305

Preface

Life science automation is a fast growing field since it is bringing unprecedented levels of productivity, safety, and efficiency to laboratories and other life science establishments. To the best of our knowledge, this is the first comprehensive book on life science automation. Like many others, we hope it can provide readers with both fundamentals and advanced applications of the subject. We have tried to cover most of the representative subjects related to life science automation. The focus is on fundamental sciences and engineering principles behind various life science automation systems and their applications. This book is intended to serve not only students, but also life science professionals who have yet to learn of the countless benefits of automating.

The book consists of five parts. The first part presents introduction to nucleic acids and clinical laboratory testing, basic analytical chemistry, and basic healthcare delivery methods. The second part offers overviews of fundamental engineering principles for life science automation ranging from principles of human-machine interfaces and interactions, fundamentals of microscopy and machine vision, to control mechanisms and robotics for life science automation. The goals of the first two parts are to introduce basic science and engineering principles for life science automation. The third and fourth parts present various applications of the life science automation systems in biotechnology, pharmaceuticals, and healthcare. The goals are to showcase how the fundamental science and engineering principles presented in earlier parts may be applied to solve practical life science automation problems. The fifth part discusses current active research areas and future directions for life science automation.

This book cannot provide a comprehensive view of this vast and rapidly expanding field. Due to space and time constraints, many excellent advancements have not been collected. We look forward to having these published in a future edition. As always, any comments to improve the book are highly appreciated.

We thank all the contributors' excellent and dedicated work, as well as the technical editor's and the reviewers' excellent help in making this book available.

Mingjun Zhang
Bradley Nelson
Robin Felder
Editors
Palo Alto, California
July 2007

PART I

Life Science Basis for Automation

CHAPTER 1

Introduction to Nucleic Acids and Clinical Laboratory Testing

Mark A. Williams[†]

On April 2, 1953, a single-page paper submitted to the journal *Nature* by Francis Crick and James Watson identified the structure for deoxyribose nucleic acid, or DNA [1]. In it, they described some of the basic features of the molecule. DNA is constructed from four basic building blocks called nucleotides; these, in turn, are connected in a particular sequence with an alternating sugar-phosphate backbone that makes up a molecule of DNA. Finally, two DNA molecules are then entwined in a double helix structure to form a DNA fiber. This proposed structure immediately suggested an answer to the question of the time: what was the material that allowed cells to pass on genetic information from generation to generation? This was a concept that was not lost on the authors who wrote in this paper a spectacularly simple, single-sentence summation:

> It has not escaped our notice that the specific pairing we have postulated immediately suggests a possible copying mechanism for the genetic material.

This paper introduced a new branch of biologic science that had ramifications for understanding cellular function, genetics, medicine, and ultimately application for human health. This chapter, in part, is an extension of that original paper, bringing together the fundamentals of studying DNA and automation. This area of study will be integral to providing faster and more reliable techniques to investigate causes of disease and strategies for maintaining health and providing tailored therapy to individuals rather than using population-based methods to treat medical conditions.

In the first part of this chapter, the physical characteristics of DNA and its sister molecule ribonucleic acid (RNA) along with the proteins that are the mechanisms for their duplication will be discussed. I will not emphasize the in vivo aspects of nucleic acids or how they are manipulated in the cell, but rather the in vitro aspects that are more applicable to this subject. So, there must be a discussion of not only the molecules themselves but also how they interact with the test environment both

---

†   University of Virginia, Department of Pathology

physically and chemically and with other commonly used inert materials used in manufacturing. This will establish a basis for engineers and give insight into how this amazing polymer should be handled so that it may be studied or used for diagnostic purposes. The last part of the chapter will focus on the larger picture of why we test DNA and RNA and how this plays into medical decision making and provides a statistical basis for this process. It is hoped that this latter section will inspire engineers to create novel technologies to facilitate more accurate, rapid, and reliable ways to measure DNA.

1.1 Basics of Nucleic Acid Structure

1.1.1 General Principles

DNA and RNA are polymers assembled from basic units called nucleotides that each are a combination of a circularized five carbon sugar (a ribose ring) that is attached to both a pyrimidine or purine molecule and a phosphate group (Figure 1.1). Collectively, the pyrimidines and purines are called bases, and the result of attachment of a base to a ribose sugar is called a nucleoside. Nucleotides result when a phosphate group has been attached to the 5′ carbon of the sugar. The carbons of the ribose sugar are numbered in a clockwise fashion starting with the carbon attached to the base, or 1′, and ending with the carbon attached to the phosphate group. The purine bases that are found in both DNA and RNA include adenine and guanine. The pyrimidine bases found in nucleic acids include cytosine, thymine, and uracil. Cytosine is commonly found in both DNA and RNA, thymine is found exclusively in DNA, and uracil is found exclusively used in RNA sequences (Figure 1.2). Bases are linked into polymers through the phosphate group attached to the 5′ carbon of the first nucleotide to the hydroxyl group attached to the 3′ carbon of the next nucleotide in the sequence. The sequential linking of bases constitutes the primary structure of the nucleic acid.

The 5′ to 3′ linkage ensures linearity and directionality to the molecule so that DNA is always a nonbranching polymer and one end is always the starting point. Furthermore, 5′ to 3′ is the direction of duplication and the direction of

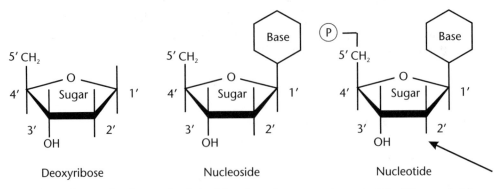

Figure 1.1 Progression from a circularized 5-carbon ribose sugar to nucleotide. Either pyrimidine or purine bases are attached to the 1′ carbon. The difference between deoxyribonucleic acid and ribonucleic acid is the addition of a hydroxyl group at the 2′ carbon (arrow).

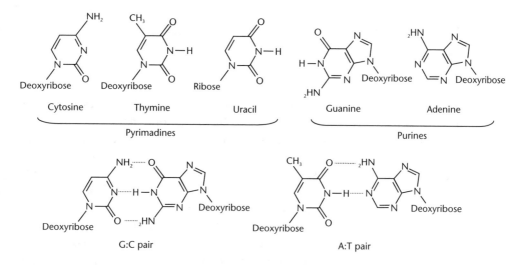

Figure 1.2   These are the most common biochemical conformations of the nucleic acid bases allowing standard Watson-Crick base pairings. Occasionally, conformational changes of the bases allow non-Watson-Crick base pairing due to different H-bonding patterns.

transcription to RNA that in turn is read 5′ to 3′ in the translation from nucleic acid to polypeptide polymer during protein synthesis. The two strands of DNA exist in an antiparallel fashion (Figure 1.3); that is to say that in relation to each other, one molecule of DNA is oriented 5′ to 3′, and the other strand is oriented in the opposite direction, 3′ to 5′.

As you can see, reading the noncoding strand from 5′ to 3′ would give a completely different sequence of bases than the coding strand read 5′ to 3′. It is the 5′ to 3′ strand that contains the sequence used for cell function, while the other strand is used specifically for the purpose of duplication. When communicating nucleic acid sequences, it is customary to list only the single letter abbreviations of the bases with the beginning of the sequence corresponding to the 5′ end on the left as if reading a sentence in English.

1.1.2   In Vitro Aspects of DNA

There are two primary features of DNA structure: an alternating sugar-phosphate backbone and a linear sequence of bases. The backbone has a significant negative charge along its length due to the phosphodiester bonds which link the nucleotides together. This negative charge makes this portion of the molecule very hydrophilic. The bases, on the other hand, tend to be more hydrophobic. Both of these characteristics help promote the conformation of the double stranded DNA fiber (dsDNA) in aqueous solution where the bases tend to collect on the "inside" of the fiber with the backbone facing the "outside" of the fiber. The result is the familiar double stranded helix that is the secondary structure of DNA (Figure 1.4).

The bases have unique molecular structures that only allow certain conformations within the molecule. Consequently, only specific pairs of bases form between the two DNA strands. In DNA, adenine (A) will always pair with thymine (T), and guanine (G) will always pair with cytosine (C). This rigid scheme is the basis for

Noncoding strand   3'—TTAATCGCTAGCGCGATATCGATGCTGA—5'
Coding strand   5'—AATTAGCGATCGCGCTATAGCTACGACT—3'

Figure 1.3   Antiparallel DNA strands.

consistent duplication of the primary structure of DNA. The bases are each linked to its partner by hydrogen bonds (a type of chemical bond that is fairly weak especially in comparison to the covalent phosphodiester bond that links nucleotides together). Additionally, G:C pairs tend to form three hydrogen bonds while A:T pairs tend to form two hydrogen bonds. This differential between the paired bases has important ramifications for the local structure of any one portion of the DNA molecule. Those sections of DNA that have a higher G:C content—and thus a lower proportion of A:T pairs—require more energy to break the two strands apart in comparison to those regions with higher A:T content (Figure 1.5). The formation of these hydrogen bonds, even though relatively weak, when multiplied by millions of base pairs contributes significantly to the stability of the double helix structure under conditions found in the cell nucleus. When these conditions are duplicated in vitro, manipulation of DNA becomes dependent on the ability to disrupt the central hydrophobic core of the DNA fiber and the ability to modify the environment surrounding the molecule and its association with the negatively charged DNA backbone.

DNA can be disrupted by introducing heat into a test system to form single stranded DNA (ssDNA) in a process called denaturation or melting. Once the reaction is cooled, the DNA molecules form once again into the more thermodynamically stable dsDNA conformation. This process of renaturation or annealing is a second order reaction where two molecules come together to form the double helix H. The rate of this reaction is dependent on the relative concentrations of each of the molecule strands A and B. As varying strands come together, they attempt to form DNA fibers and either dissociate because of poor initial complementary base pairing or begin to anneal faster than the strands can dissociate. Those molecules with the greatest compatibility are more likely to move beyond the initial base pair linkages and bond along their entire length. Thus, if particular compatible ssDNA molecules are present in greater amounts—for instance, if a particular region of the genome had multiple copies of a particular gene—then one would expect this particular construct to anneal faster than others present in the reaction mix. Knowing this, in combination with probe techniques, a quantitative estimate of the number of copies of a particular gene can be generated.

Increasing the temperature of a reaction is usually the primary means for denaturing DNA fibers to individual molecules. However, other physical properties that can be manipulated include pH and salt concentration present in the reaction solution. These tend to affect the backbone of the DNA molecule. Hydrogen bonding can be modified by adding compounds that specifically disrupt hydrogen bonding. When temperature is used to denature dsDNA, the point at which 50% dsDNA becomes ssDNA is called the melting temperature or T_m (Figure 1.5). These changes can then be measured using parameters such as intrinsic viscosity, absorbance, optical density, and sedimentation coefficients, or the use of certain probes with attached fluorochromes that coexcite only when annealed in close proximity to each other.

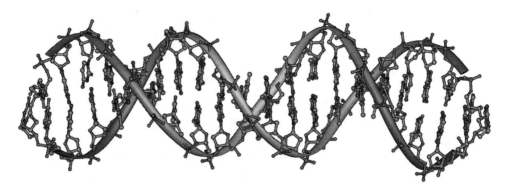

Figure 1.4   The secondary structure of DNA.

While the most stable form of DNA is the double stranded conformation under physiologic conditions, the two sugar-phosphate backbones do produce a repelling force that tends to counteract the various other forces holding the DNA fiber together. T_m is increased as the salt concentration of the solution is increased. This occurs because the electrostatic forces repelling the two DNA strands are lessened due to masking by positively charged ions in solution such as sodium (Table 1.1). Conversely, as the salt concentration is decreased, T_m also decreases since the two chains are exposed to each other and therefore more likely to repel each other. Under in vitro conditions this usually means a >0.2M salt concentration in order to allow single-stranded DNA (ssDNA) to approach one another. Similarly, there are compounds that increase the solubility of weakly polar organic molecules in water like those found in nucleotides. These agents tend to disrupt the base inter-actions in double-stranded DNA (dsDNA) and promote disruption of the

Table 1.1   Properties Affecting T_m

	Effect on Structure	*Effect on T_m*	*Mechanism*
Decreased salt concentration of reaction solution	Promotes formation of ssDNA	Decreased	Decreased masking of negative charge of sugar-phosphate backbone
Increased salt content of reaction solution	Promotes formation of dsDNA	Increased	Increased masking of negative charge of sugar-phosphate back-bone
Formamide, formaldehyde, and ethylene glycol	Promotes formation of ssDNA	Decreased	All promote solubility of weakly polar organic molecules in aqueous solution
Increased G:C content	No effect	Increased	Greater amount of H-bonding
Decreased G:C content	No effect	Decreased	Lesser amount of H-bonding
Longer length of DNA in base pairs	Greater flexibility and conformation possibilities	Increased	Greater amount of H-bonding
Shorter length of DNA in base pairs	Comparatively rigid structure	Decreased	Lesser amount of H-bonding

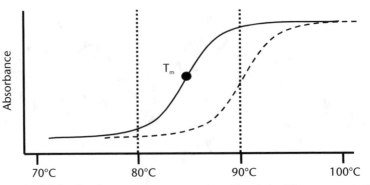

Figure 1.5 Graph of light absorbance versus temperature. The dashed line represents DNA with a higher G:C content compared to dsDNA with higher A:T content (solid line). T_m marks the halfway point of the transition between dsDNA and ssDNA

DNA fiber. In this case, the concentration of these types of agents and T_m are inversely related.

The shorter the sequence, the more likely a properly paired double helix conformation will result since the number of pairing possibilities are decreased. Furthermore, shorter sequences are "stiffer" in that they do not have as many conformational possibilities in the single stranded form. In other words, shorter lengths will *zip up* faster. On the other hand, not as much energy is required to dissociate shorter sequences of DNA, so they will dissociate at lower T_m.

As ssDNA increases in length, it acquires the ability to bend and fold on itself. Increased length also allows for the possibility that several homologous sequences are present in different parts of the molecule. Thus, as ssDNA becomes longer, homologous areas can anneal not only in proper sequence but with any of these homologous regions in a random fashion, creating odd molecular shapes such as hairpins or loops. The ssDNA may also conform in such a way as to allow intrastrand binding where a single DNA molecule loops around to bind with itself. When these random base-pairings occur, they have intervening unpaired bases that lead to instability of the molecule that manifests as a lower T_m.

Testing strategies often use differential T_m amongst varying stands of DNA to identify sequence variations. For example, herpes simplex virus comes in two major varieties, type I and type II. Most of the viral genome between these two variants has similar homology except in certain small regions. A probe consisting of a small piece of DNA can be designed such that it is entirely homologous to one of the viral types. This same probe does have the ability to bind to the other viral type; however, since it does not have complete homology to the other type, base-pairing along its entire length will not be complete. DNA fibers formed between molecules that are incompletely homologous will have lower T_m in comparison to those with complete homology. This information can be easily detected revealing not only the presence of viral DNA, but also which variant of the virus is present.

To summarize, the formation of dsDNA in vitro is dependent on the sequence characteristics of the two molecules, the salt and pH concentration of the reaction solution, and the concentration of ssDNA. This last feature contributes significantly to the rate of reaction, while the first two parameters determine whether the reaction will actually occur.

1.1.3   Unidirectional Flow of Genetic Information: The Central Dogma

The primary structure is the sequence of bases that forms the genetic code that in turn stores all the information required for cellular metabolism, cell differentiation, cell division, and ultimately generation of a new organism. Generally, this is the information being sought in nucleic acid testing. At its simplest, the DNA sequence is broken into blocks called genes that each code for a particular protein. During transcription from DNA to RNA, the entire DNA sequence in a gene is transcribed into a single stranded RNA molecule. In most nonbacterial genomes, the raw DNA sequence of a gene contains regions that do not code for the eventual protein. These regions are called introns and are excised from the RNA transcript in a process called splicing that leaves only coding regions or exons in the resultant RNA molecule (see Figure 1.6). Once modified, the much shortened RNA is then translated into protein after it has made its way to the cell cytoplasm from the cell nucleus. The RNA base sequence is read in a series of codons that consists of three bases in sequence. Codons are each associated with a specific amino acid that in turn are used to construct proteins, and they are analogous to nucleotides in the DNA molecule. Once constructed, proteins are used in many ways by the cell to perform various functions. The key point is that errors in the genetic sequence of a gene translate directly into errors of protein construction that may affect protein behavior and ultimately may have downstream effects on cell function. Those errors, also called mutations, which do not have an effect on protein generation or function, are called silent mutations. All together, transcription from DNA to RNA and then translation to protein constitutes the central dogma of molecular biology.

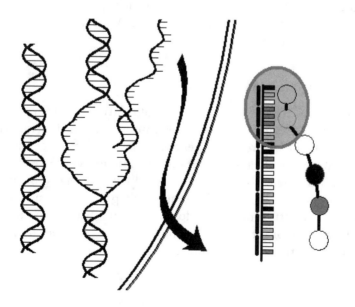

Figure 1.6   The flow of genetic information is unidirectional from transcription of DNA to RNA in the nucleus to translation of nucleic acid "text" into proteins out in the cytoplasm of the cell. The linear sequence of proteins is not reverse translated into RNA. With some exceptions, notably in viruses, reverse transcription of RNA into DNA usually does not occur in eukaryotic cells. However, reverse transcription utilizing viral proteins and RNA templates can be a useful tool in molecular diagnostic testing.

1.2 Manipulation of DNA Under Testing Conditions

1.2.1 Extraction

For the purpose of in vitro testing, the nucleic acids must be freed from the confines of cells in a patient sample. Double stranded DNA does not exist in the cell nucleus on its own as a free molecule. Left to itself, dsDNA is prone to breakage, rearrangement of the linear sequence, or loss of genetic material. One would expect that events such as these would be detrimental to cell function since the effect would be to introduce a considerable amount of mutations. Consequently, organisms have developed highly conserved proteins that help to ensure the structural integrity of DNA. That is to say, a histone protein extracted from a human cell would have a similar sequence as those taken from a frog or a dog. All of these intimately associated proteins along with all the other proteinaceous and phospholipid material including the cell membrane must be broken down and the DNA removed in a process called extraction.

Extraction provides the raw material for DNA analysis; in fact, it makes the entire human genome available for examination. There is no test available currently that can study the entire human genome at once; however, this is usually not necessary. Most often, DNA testing is used to answer narrow questions concerning certain diseases or congenital conditions. Thus, only small parts of the entire genome are of interest, such as certain genes, particular mutation hot spots, or known regions of variability from person to person. One may have already guessed that the ability of a test to resolve a small portion of DNA is dependent on the amount of the sequence present in the reaction solution. The larger the amount of DNA present, the greater the signal that is produced and subsequently the more it is likely to be detected. Many times, the sought after DNA target is in very small amounts requiring amplification.

1.2.1.1 Testing for RNA

The main purpose of RNA is as a carrier of information from point to point in the cell, and to act as regulatory step in the translational machinery. Thus, in order for the cell to maintain control of the genetic signal, RNA messages must be removed so that the signal is not always being translated into protein. An enzyme that performs this function is called RNAase and is tasked with cutting RNA into nonfunctional pieces. As it happens, RNAase is ubiquitously present and makes RNA manipulation much more difficult. Hence, close attention must be paid to sample handling where the activity of these enzymes must be halted in order to preserve the target integrity. This activity must then be inhibited or removed for the remainder of the test.

1.2.2 Amplification

Prior to 1983, DNA amplification required growing large numbers of organisms and then extracting their DNA in bulk. The disadvantage of this was that although you were getting the same genome from millions of cells, each copy was not guaranteed to be the same. In 1983, Kary Mullis received the Nobel Prize in Chemistry for developing the polymerase chain reaction (PCR). This process makes possible amplification of a specific target from a small amount of DNA such that only the targeted DNA is amplified at the end of the reaction. Many of the molecular-based tests for nucleic acids rely on amplification and without it, advances in nucleic acid testing could not be possible.

1.2.2.1   The Polymerase Chain Reaction

PCR is relatively simple. It requires a sample of DNA (template DNA), DNA primers that are chosen specifically to bracket the region of interest, an optimized aqueous reaction solution, DNA polymerase, ACGT nucleotide triphosphates, and the ability to raise and lower the temperature of the reaction chamber. The reaction is begun by heating the mixture so that the sample DNA melts all to ssDNA. The reaction is cooled. At this point, the ssDNA could anneal back together with its originally paired sister molecule; primers, however, are also present in the reaction chamber and are considerably shorter than the sample DNA. Since annealing occurs faster for shorter sequences, the primers anneal to the target sequence faster than the longer sister DNA molecules. The likelihood of primer binding is further enhanced by ensuring a high concentration of primer in the reaction mixture relative to the template DNA. The primers are the means by which DNA polymerase attaches to DNA and begins making the polymer and are required in both in vitro and in vivo settings. Without them, DNA polymerase would never attach to ssDNA and begin synthesis. Once primer binding occurs, DNA polymerase begins to extend the molecule.

Originally, the PCR required constant attention because of the inability of human DNA polymerase to remain functional under the high temperature requirements of the reaction. Once the reaction was heated, human DNA polymerase became irreversibly denatured and thus nonfunctional after one cycle, requiring repeated addition of DNA polymerase to replenish the functional protein. Since then, DNA polymerases (e.g., Taq polymerase) from certain thermophilic bacteria that often live in temperatures up to 110°C are used in the reaction. The advantage is that the reaction can cycle without having to add DNA polymerase after raising the reaction temperature. The disadvantage is that Taq polymerase does not have the ability to self-correct errors when it synthesizes DNA as does human DNA polymerase.

The base pair mismatch rate is approximately once every 10,000 base pairs. This effectively limits the possible length of DNA for testing since the longer the sequence, the more likely errors can be introduced. Moreover, if the error is introduced early enough, the error is also amplified throughout the entire reaction. The length of DNA in base pairs that PCR can easily amplify is in the range of 2,000 to 3,000 base pairs. With longer sequences, the polymerase has a tendency to fall off the molecule or cannot completely synthesize the molecule due to a lack of time before the next temperature cycle begins. Once DNA synthesis is completed, the polymerase simply falls off the end of the DNA molecule and the cycle is repeated. When the temperature is again increased, the dsDNA separates, the reaction is cooled and another round of synthesis occurs (Figure 1.7). This process is repeated through 25 to 30 cycles, increasing the targeted amount exponentially until the primers and nucleotides in the reaction are utilized.

1.2.2.2 The Importance of Primers

Primer development becomes important in modifying the sensitivity and specificity of the results. As primer length increases, so does the specificity of the sequence, and thus there is a greater likelihood that the region of interest will be targeted. On the other hand, shorter probe sequences may have more options for binding along the DNA molecule, but may not bind in the right place since the sequence is less specific for the correct region. The optimum primer length that balances sensitivity and specificity has been established between 10 to 20 bases.

DNA primers are not simply compatible antiparallel DNA sequences since they are used at differing points along the template DNA. Using the coding and noncoding example above, a sample set of primers for each of the strands would be as shown in Figure 1.8.

Two things to notice right away are that: (1) when placed in parallel fashion, the sequences are different from 5′ to 3′; and (2) there is roughly equal amount of A:T and G:C pairings between the two primers.

A higher T_m is required for DNA with greater G:C content. In the above example, primers should be able to anneal and melt at close to the same T_m, otherwise, if one primer has greater G:C content compared to the other primer, one primer will be favored over the other and inefficient PCR will result. As a general rule, the T_m difference between primers should not exceed >5°C and T_m for the amplification product should not exceed the T_m of the primers by >10°C. One might surmise then each PCR attempting to amplify differing sections of the genome would have different primers and therefore different thermodynamic conditions. Designing the actual

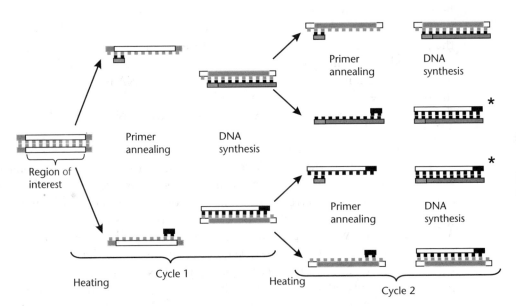

Figure 1.7 The first two cycles of the PCR reaction. Notice as the reaction cycles increase, the amplification product designated with * eventually increases in amount and greatly outnumbers the template DNA after only a few cycles.

biochemical aspects of a nucleic acid test is not likely within the purview of an engineer; however, the design of the analyzer, especially one that is intended to perform several tests in this category, must have enough flexibility to allow adjustment for differing T_m among primer sets.

1.2.2.3   A Major Stipulation for PCR

With its ability to amplify very small amounts of DNA, the PCR is a very sensitive method under the correct conditions (see Table 1.2). However, this feature can be detrimental if each new test run is not started with absolutely clean materials. In a modern molecular pathology laboratory, great lengths are taken to ensure samples do not contaminate each other in order to avoid nonspecific or misleading results due to amplification of a contaminant. Such precautions can include single use materials, different clothing for workers during each step of the process and unidirectional flow of work with separate rooms for each step of the process. For example, amplification of DNA from microorganisms may be helpful for initial diagnosis of an infection. The advantage would be a faster turn around time in comparison to traditional methods used to establish a diagnosis, such as bacterial culture. Cross-contamination of patient and control samples is a foreseeable risk in PCR instruments due to the enormous amount of final product produced. This error in testing is called carryover, where a reagent or a particular patient sample is inadvertently included in another test reaction as a result of incomplete cleaning of materials used for processing the test or as a result of splash, spray or aerosolization of test materials into another test reaction chamber. In particular, one can see that carryover of positive control DNA from a microorganism will subsequently cause positive results in whatever test was contaminated. This phenomenon must be taken into account when designing these types of analyzers. Contamination may be controlled by rigorous design of disposable pipetting tips, control of the creation of aerosols, and by rigorous cleaning of all contact surfaces.

1.2.3   Detection

1.2.3.1   Separation

As mentioned above, the physical properties of the sugar phosphate backbone make the DNA molecule amenable to manipulation by both size and negative charge. DNA can be separated by placing sample DNA in aqueous solution housed in a matrix gel and applying a charge to the solution bathing the gel. Due to its relatively negative charge, DNA will move toward the positive node. Over time, fragments of DNA will travel a certain distance that is proportional to both size and charge; thus, variably sized pieces of DNA can be separated from each other and visualized.

3′—TTAATCGCTAGCGCGATATCGATGCTGA—5′ → 5′—AATTAGCG—3′
5′—AATTAGCGATCGCGCTATAGCTACGACT—3′ → 5′—AGTCGTAG—3′

Figure 1.8   A sample set of primers for each of the strands.

This process is affected by the chemical composition of the gel matrix. The gel matrix is a cross-linked web of molecules through which the aqueous solution and DNA can travel. By modifying the number of cross-linking molecules, the "holes" can be made larger or smaller and thus allow for faster or slower progression of DNA fragments through the gel. Other characteristics that can be adjusted to fine tune the resolution of an assay include the path length of the gel, the time that charge is applied, and the amount of voltage. Once separated, DNA is usually compared to a *size ladder* of known fragment lengths by direct visualization (Figure 1.9).

1.2.3.2   Restriction Enzymes

The Nobel Prize was awarded to three investigators in 1978 for discovering restriction enzymes, which made possible genetic manipulation of DNA using recombinant technology. These enzymes are used by bacterial organisms to cleave double stranded DNA at certain recognition sites as a defense against bacterial viruses. The recognition sites consist of a particular nucleic acid sequence that is unique to each type of enzyme. Many of these enzymes produce an asymmetric cleavage of the DNA fiber with several bases left unpaired on one end rather than a clean break. These sites tend to occur at the same location in the genome sequence between individuals (Figure 1.10). This *sticky end* can be recombined with another fragment of DNA that has also been similarly cleaved, resulting in a new DNA sequence. However, recombining DNA is usually not the basis for investigations in the work-up of medical disease. Instead, situations are sought were a particular enzyme is not able to cleave a portion of DNA, often in the case of a mutation that alters a recognition site for an enzyme that would otherwise normally be present. For instance, a mutational analysis using restriction enzymes might entail amplifying a region of DNA where mutations occur that also include recognition sites. Once the DNA sample has been subjected to the action of the enzyme, particular lengths of DNA should remain. Any mutations such as additions or deletions of base pairs should alter the length of these fragments allowing for detection by gel separation.

1.2.3.3   Hybridization

Hybridization is an important concept in molecular diagnostics and has been obliquely described above concerning the structure of DNA. Defined, hybridization

Table 1.2   Various PCR Methods

Amplification Technique	Description
RT-PCR	RNA is used as a template to amplify DNA. Used to characterize active gene expression in a cell.
Nested PCR	Used to increase specificity of amplification with the use of two primer sets with one set closer to each other in the sequence than the other set.
Quantitative real-time PCR	Used to measure the amount of amplified material as the test reaction occurs using fluorescent probes.
Multiplex PCR	Several genes are assayed at once in the same reaction using multiple primer sets.
Asymmetric PCR	Used to preferentially amplify one of the DNA template strands.

is the production of a double stranded nucleic acid fiber due to complementary base pairing; however, it is not limited to DNA. Both DNA and RNA can combine together in either homologous pairs (i.e., DNA-DNA or RNA-RNA) or heterologous pairs. These various hybridization possibilities can be manipulated utilizing changes in the reaction environment to control for stringency, or the limit of how many base pair mismatches that are allowed between two nucleic acid molecules. Detecting characteristics of nucleic acids becomes a function of designing a nucleic acid probe that specifically binds to a region of interest and a method for detecting this hybridization.

1.2.3.4   Types of Probes and Methods for Detection

The first probe modality was the use of radioactive labels attached to probes that produced a detectable signal that could be recorded by scintillation or radiography. These types of probes, by their nature, may be hazardous to laboratory workers if handled improperly, and cause considerable expense as a waste product. Other, nonradioactive modalities include fluorescent labels and indirect affinity labeling. Affinity labeling is based on two proteins, avidin and biotin, that have a very high affinity for binding to each other. By using these two proteins, great flexibility in the type of detection method can result since these proteins can be attached to many different types of reporter molecules. For instance, avidin can be bound to an enzyme that can either convert a substrate producing a chemical change to the solution such as a color difference or it may utilize the same reaction that occurs in a firefly tail and create light. Biotin, on the other hand, may be attached to a surface or another molecule of interest, allowing the biotin-avidin pair to attach and provide a selective signal (Figure 1.11). Both of these modalities are easily detectable and can be quantified.

Hybridization can occur in two types of phases, either liquid or solid. In liquid phase based hybridization, both probe and target interact in solution, increasing the

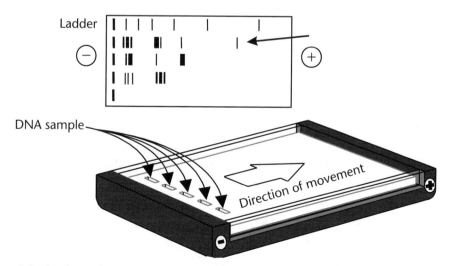

Figure 1.9   An electrophoresis apparatus showing how DNA is loaded into the matrix gel and the direction of movement. A sample result is shown. Smaller DNA fragments travel faster and farther than larger fragments (arrow).

likelihood of interaction. The advantages to this method are that it is a relatively simpler method and it is amenable to quantification. Disadvantages include: (1) the inability to determine size of the successful hybridization product, usually only a single target can be assayed at once, and (2) low levels of target versus high levels of a cross-reactive DNA sequence may produce similar test results. Solid phase hybridization usually involves fixation of a probe to a solid surface and bathing this construct in sample nucleic acid. One method uses multiple probes that are fixed to the surface, allowing for multiple targets to be assayed at one time, called a dot/blot assay. A major drawback that is similar to liquid based hybridization is the inability to determine size; however, both this type of solid phase hybridization and liquid phase hybridization can tolerate some degradation of the original sample. A southern blot is a solid phase method used to determine both a specific sequence and its size (an example of one is seen in Figure 1.10) where DNA is separated, fixed onto a solid surface, and then bathed in probe. A similar method called the northern blot is the same in all respects except it is an assay used for RNA molecules.

1.3   Statistics and Test Utilization Used in Medical Decision Making

There are three main classifications of questions that are asked in medical decision making: those that pertain to diagnosis, those that pertain to treatment, and those

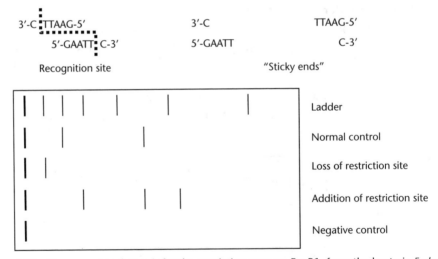

Figure 1.10   The sequence shown is for the restriction enzyme EcoR1, from the bacteria *Escherichia coli*. The region of interest has one restriction site in the normal control resulting in two fragments. No cleavage occurs, thus resulting in one single DNA strand, when loss of the site occurs and an additional fragment is seen when a recognition site is added to the sequence.

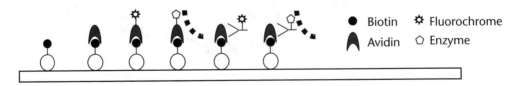

Figure 1.11   Steps of the test.

that pertain to prognosis. In other words, what is going on, what do we do about it, and what is going to happen in the future? As engineers, the concepts of statistical analysis are probably very familiar and are similarly utilized in medical applications. In comparison to systems of manufacturing or designing tests for industrial applications, the variability that is present in biologic systems is quite large not only from person to person, but also within an individual at different points in time. Likewise, the ability to freely test any hypothesis is more readily undertaken when the object of study is a widget as opposed to a human being. Subsequently, much of the working knowledge that is utilized in medicine is either empirically or population based, producing many discussions about likelihoods and percentages concerning diagnosis, treatment, or prognosis. On the other hand, when considering these prospects on an individual basis, these questions have binary solutions: the disease is there or not there; the treatment is applied or not applied; either the patient will respond or not will not respond; the patient will recover or not recover. Unfortunately, the best way to manage this patient by patient, is to apply population-based studies to the individual. Given 100 people, based on clinical data, some event will occur to the 50th person at a certain point in time—we just cannot know who that individual will be.

In order to put medical testing into perspective, the entire process of physician and patient interaction should be discussed. The process of medical decision making entails first the interview and subsequent direct observation of the patient. Collectively, this is called the history and physical examination (H&P). Most of the time, given enough clinical experience, the physician can answer all three question modalities based on the H&P. There are some disease processes that are so well characterized that further medical testing is unnecessary because the H&P is able to reach a significant level of data to support a high pretest probability or what many physicians would call a high clinical suspicion. Many times, the H&P is not definitive but has raised the pretest probability of certain condition or set of conditions to a significant likelihood. One could consider the H&P as a test in itself; it is still used heavily in modern medicine because of its ability to produce a reasonable pretest probability for nearly every human affliction. What is a reasonable pretest probability? It is near the point wear a flip of a coin could resolve the answer, or in other words, the point where the likelihood of a condition being true is approximately 50%. Any less, and the results from subsequent testing, as we will see, become unreliable and any greater, the need for subsequent testing is superfluous.

1.3.1   Sensitivity and Specificity

When the possible conditions of disease, either present or not present, are placed on one axis and test results of either positive or negative are placed on the other, four categories result (Table 1.3). Category 1 is the condition where the disease is present and the test is positive indicating a true-positive result. Category 2 represents a condition where a false-positive result occurs; Category 3, a false-negative; and Category 4 a true-negative. Dividing the true-positive results by the total number of those with the disease condition determines the clinical sensitivity of the test. A test with high sensitivity is more likely to be positive in all those patients with the condition; however, this type of test casts a wide net and may include some individuals

Table 1.3 Test Results

	Disease Present	*Disease Absent*	
Positive test	Category 1: TP	Category 1: FP	PPV: TP/(TP + FP)
	Category 1: FN	Category 1: TN	NPV: TN/(TN + FN)
	Sensitivity TP/(TP + FN)	Specificity TN/(FP + TN)	Total population

TP: true positive; FP: false positive; FN: false negative; TN: true negative; PPV: positive predictive value; NPV: negative predictive value

who do not have the condition. In other words, when this test is negative, the likelihood of the disease condition being present is very small. This type of test is used in medical decision making to rule out a condition and thus is useful as a screening test. Dividing the number of true-negative results by the total number of individuals who do not have the condition renders the specificity. A positive result for a test with a high specificity usually rules in a condition and is often used as a confirmation of a screening test. By the nature of these conditions, as a test becomes more sensitive, the less specific it becomes and vice versa, and often the sensitivity and specificity are experimentally established to maximize one or the other or both depending on how the test will be used. The experimental conditions are often carefully selected populations and may have certain pretest probabilities. Many times for the purpose of establishing sensitivity, a population with the condition is used, while a normal population of healthy individuals is used to determine specificity.

1.3.2 Predictive Values

Going back to Table 1.3, when the true-positives are divided by the total amount of individuals with a positive test this represents the likelihood that a positive result reflects a true disease condition or the positive predictive value. Similarly, true-negatives divided by the total number of individuals with negative tests give the negative predictive value or the probability that a negative test reflects absence of disease. These values, unlike sensitivity and specificity, are subject to variation as pretest probability or prevalence of disease in a population changes as shown in Tables 1.4 through 1.7 which hold data for a hypothetical screening test.

Table 1.4 Profile of a Screening Test: Experimental

	D+	*D−*	*Totals*
Test +	782	36	818; PPV = 0.956
Test −	68	114	182; NPV = 0.626
Totals	850	150	1,000
	SN = 0.92	SP = 0.76	

Table 1.5 Profile of a Screening Test: Disease with 2% Prevalence

	D+	*D−*	*Totals*
Test +	182	235	253; PPV = 0.055
Test −	2	745	747; NPV = 0.997
Totals	20	980	1,000
	SN = 0.92	SP = 0.76	

Table 1.6   Profile of a Screening Test: 50% Pretest Probability of Disease

	D+	D−	Totals
Test +	460	120	580; PPV = 0.793
Test −	40	380	420; NPV = 0.904
Totals	500	500	1,000

SN = 0.92 SP = 0.76

Table 1.7   Profile of a Screening Test: 90% Pretest Probability of Disease

	D+	D−	Totals
Test +	828	24	852; PPV = 0.972
Test −	72	76	148; NPV = 0.513
Totals	900	100	1,000

SN = 0.92 SP = 0.76

Note for Tables 1.4–1.7:   After a screening test is designed and a prototype has been produced, it must be assayed in an experimental population. Table 1.4 represents data from an original study used to establish sensitivity and specificity of the new test. Individuals are determined to be positive for a condition if the gold standard test is positive. Usually the gold standard is a method that is the best representation of the true state of the condition but is difficult to implement in a clinical setting because of complexity or expense. The sensitivity and specificity for this test has been established as 92% and 76%, respectively. This condition has a prevalence of 2% in the population; notice the relative amounts of false-positives and true-positives making the positive predictive value (PPV) very low where there is only a 5% chance that a positive result means the disease is present. However, the negative predictive value (NPV) is very high, which means that if the test is negative, there is a nearly 100% chance that the condition is absent. PPV and NPV only approach values found under experimental conditions when the likelihood of disease approaches those set in the experiment. PPV becomes 100% if all individuals in the population have the condition, while NPV becomes 0%.

These tables are showing graphically what was stated previously about how pretest probability affects the reliability of test results. As you can see, as the prevalence of a condition increases in a population, the more likely a positive test result represents the truth. Conversely, screening tests will more likely result in false-positives when applied to the general population.

1.3.3   Preanalytical Versus Analytical Stages of Testing

In order to achieve results on an everyday basis that are similar to the original experimental conditions, all parameters for the test must be equivalent to that setting. It is tempting to consider the only issue to be solved is just the design of the test apparatus. However, the analyzer does not exist in a vacuum and the result issued by the analyzer is greatly affected by not only the analytical stage of the test but the preanalytical phase as well.

The preanalytical phase includes everything pertaining to the test prior to the actual testing procedure. Thus, the preanalytical stage begins with the order from a physician and encompasses everything from the forms used for ordering the request to specimen storage and processing prior to introduction into the analyzer. Design of the analyzer must take this into account and may encompass some aspects of the preanalytical stage and may include the hardware and software interface between the analyzer and the hospital information system, the design of the collection devices, the amount of time allowed between procurement of the specimen and analysis, or centrifugation parameters for different sources of material.

Table 1.8 Steps of Preanalytical Versus Analytical Stages of Testing

Elements Contributing to the Preanalytical Stage of Testing	Elements Contributing to the Analytical Stage of Testing
Test requisition	Reagents
Preparation of the patient	Storage
Specimen collection techniques	Water use
Collection devices and containers	Measurement of mass
Body fluids and tissues	Measurement of volume
Precentrifugation/ storage	Temperature control
Centrifugation/ storage	Evaporation of specimen
Calibration of equipment	Extraction
	Mixing
	Detection of results
	Calculation
	Choice of significant figures
	Report formatting
	pH
	Buffer solutions

Incorporation of these aspects may be very extensive, but may allow for a less complex design of the analyzer. On the other hand, in order to allow more flexibility at the preanalytic stage (i.e., the ability to accept various sized collection containers or variable temperatures of samples), the design of the analyzer becomes necessarily more complex.

Each step of this process (see Table 1.8) can introduce variation over and above that originally established under experimental conditions for the purpose of establishing sensitivity and specificity. Incorporating design elements that can help limit this variation will help increase the accuracy and precision of the analyzer.

1.4 Summary

This chapter has introduced basic concepts concerning nucleic acids and some of the main physical characteristics that are readily manipulated in order to study these amazing molecules. Designing a "black box" analyzer for nucleic acid testing will require incorporating practically every advanced technology we have available today, from software and electronic design to optics to biochemistry. Furthermore, designing analyzers for medical testing will require an understanding of how they fit into the context of decision making in medicine and its impact on that process.

Automated nucleic acid testing has the most potential of all medical testing for maximizing both sensitivity and specificity to the greatest degree. On the one hand, sensitivity can be raised significantly thanks to the polymerase chain reaction where minute amounts of target signal can be amplified enormously. On the other hand, the potential to create a highly specific test can be achieved through the appropriate design of primers and probe material. As an engineer, you will be integral in crafting these tools and designing methods that will help bring about more precise, more reliable, and more rapid answers to our most complex questions.

Reference

[1]   Watson, J. D., and F. H. C. Crick, "A Structure for Deoxyribose Nucleic Acid," *Nature*, Vol. 171, 1953, p. 737.

Basic Analytical Chemistry for Engineers

Kerstin Thurow,[†] Stefanie Hagemann,[†] and Norbert Stoll[‡]

This chapter introduces basic analytical principles for life science automation. Introductions on separation methods (chromatographic methods), bioanalytical methods, and physical measurement principles are presented. Future challenges for the development of suitable analytical methods for applications in life sciences are presented at the end of the chapter.

2.1 Introduction

Analytical chemistry deals with the development of methods, equipment, and strategies for analyzing the qualitative and quantitative composition of chemical or biological moieties, their spatial structures and dynamic behaviors. Analytical chemistry has become a cross-sectional discipline that plays a major role in the progress of science, technology, and medicine. It is used as much in the development of technological products as in foodstuffs and pharmaceuticals. Analytical methods are used for clinical testing, monitoring, and control in pharmaceuticals, drinking water, and waste water as well as in the determination of heavy metals and other environmental contaminants.

The development of high-performance methods in spectroscopy and chromatography in the middle of the twentieth century introduced major areas of physics, measuring technology, information technology, material sciences, as well as biology and genetic engineering, into analytical chemistry.

Selecting a suitable analytical process requires clear and definite goals. The following aspects must be taken into account:

1. What specimen needs to be analyzed? Apart from pure substances, possible examination objects may include material samples, environmental samples, or even biological matrices. Depending on the type of examination object or matrix, decisions must be made on the sampling and sample preparation processes to be applied.

2. What needs to be analyzed in the examination object? Depending on the specific issue involved, a variety of analytical measuring methods may be

† Center for Life Science Automation, Rostock
‡ Institute for Automation, University of Rostock

applied. Specifically in the field of physical measuring technology, a distinction is drawn between element and structure-selective processes. Apart from that, bioanalytical methods may also be applied for analytical tasks in the biological context.

3. What purpose should the analysis serve? One important issue is the accuracy of the process to be used, which will depend on whether the examination results are to be used as general prospective data (screening) or in clinical contexts for diagnostic purposes.

This chapter will summarize the processes currently applied in analytical chemistry. Our aim is to familiarize engineers involved in life sciences with the general principles of bioanalytical and physical measuring technology.

2.2 Chromatographic Separation Methods

2.2.1 General Principles

Chromatographic processes are based on the distribution of substances to be separated between stationary and mobile phases [1, 2]. Chromatographic processes are categorized according to aggregate states in the two phases, separation process, or type of technology used. Gas-chromatographic and liquid-chromatographic processes have become the most commonly used methods due to their power and instrumentation simplicity.

Depending on the physicochemical properties of the substances and temperatures involved, dynamic equilibrium is established, leading to substance separation. Coefficient K_x is used to express the tendency of compounds to remain in the mobile or stationary phase:

$$K_x = \frac{C_{stat}}{C_{mob}} \tag{2.1}$$

where C_{stat} is the concentration of substance X in the stationary phase, and C_{mob} is the concentration of substance X in the mobile phase.

Or capacity factor k' is used:

$$k' = \frac{n_{stat}}{n_{mob}} \tag{2.2}$$

where n_{stat} is molar number of substance X in the stationary phase, and n_{mob} is molar number of substance X in the stationary phase.

Chromatographic process selectivity is dependent on the specific distribution coefficients of the individual substances and the retention times involved. The retention time of a substance is constant under identical chromatographic conditions (type of separation column, composition of the mobile phase, flow rate, temperature, and so on), and can therefore be referred to in identifying a compound in a mixture.

The number of theoretical plates, N_{th}, and length l of the column are used to measure the efficiency of the chromatographic system. The theory of plates divides the stationary phase into separate steps that are taken to be theoretical separation steps. The higher the number of theoretical plates, the higher the resolution of the system. The value can be calculated as follows:

$$N_{th} = 5.54 \left(\frac{t_R}{w_h} \right)^2 \tag{2.3}$$

where t_R is retention time of the test component, and w_h is peak width at half peak height.

More often, the height equivalent to a theoretical plate (HETP) is used for characterization, which can be calculated from the quotient between the length of the column, l, and the number of theoretical plates, N_{th}.

$$H.E.T.P. = \frac{l}{N_{th}} \tag{2.4}$$

The lower this figure is, the higher the chromatographic resolution of the system.

The number of theoretical plates depends on the flow rate in the mobile phase. Apart from that, various diffusion processes act against compound separation in the chromatographic system, leading to remixing and band broadening; this affects chromatographic separation efficiency. The Van Deemter formula describes the influence of these parameters:

$$H.E.T.P. = A + \frac{B}{v} + Cv \tag{2.5}$$

where A, B, and C are coefficients in turbulent diffusion, longitudinal diffusion, and mass transfer, respectively, and v is flow rate.

Similar physicochemical properties can lead to similar retention times for different substances, which leads to overlap in chromatographic peaks. Specific or selective detection systems have to be used in these cases for a clear identification of the substances contained.

2.2.2 Gas Chromatographic Methods

Gas chromatography is currently one of the most widely used processes for separating complex mixtures. A fundamental prerequisite is that the substances to be analyzed evaporate without decomposition.

A wide variety of detectors have been described for gas chromatography; the main devices used are flame ionization detectors and electron collectors. Mass spectrometry has enjoyed increasing importance since it can be used as a universal system for detecting a large number of substances. Table 2.1 summarizes the detectors available for gas-phase processes.

Sample injection is especially important in gas chromatographic processes [2]. Usually, the sample is injected using a hot split or splitless injection system. This

Table 2.1   Detectors for Gas-Phase Processes

Name	Type	Selective for	Detection Limit	Linear Range
Flame ionization detection (FID)	Selective	Ionization in air-hydrogen flames	5 pg C/s	10^7
Wallerstein's differential (WLD)	Universal	Heat conductivity	400 pg/ml	10^6
Electron capture detection (ECD)	Selective	Gas-phase electrophiles	0.1 pg Cl/s	10^4
Nitrogen phosphorus detection (NPD)	Selective	Nitrogen, phosphorus, heteroatoms	0.4 pg N/s 0.2 pg P/s	10^4
Electrolytic conductivity detector (ELCD)	Selective	Halogens, nitrogen, sulfur	0.5 pg Cl/s 2 pg S/s 4 pg N/s	10^6 10^4 10^4
Flame photometri detection (FPD)	Specific	Phosphorus, sulfur	20 pg S/s 0.9 pg P/s	10^3 10^4
Fourier transfor infrared detection (FTIRD)	Universal	Molecule resonance	1,000 pg in a strong absorber	10^3
Mass selective detection (MSD)	Universal	Mass	10 pg to 10 ng (SIM or SCAN)	10^5
Atomic emission detection (AED)	Universal	Wavelength	0.1 to 20 pg/s depending on the element concerned	10^4

Detection limits: pg X/s = picogram of specific element per second; pg/ml = picogram per milliliter; ng = nanogram.

process is limited to thermally stable compounds in its application. Suitable alternatives include on-column injectors as well as cold injection systems, which enable cold injection followed by sequential heating in the injection space. Cold injection reduces selective vaporization of volatile substances from the injection needle or valve when injecting fluid samples, where hot injection would discriminate against highly volatile substances. While sample volumes are limited to 1 to 3 μl depending on the solvent used in the above processes, large-volume injection technology enables sample injection of up to 1 ml. This can often eliminate the need for concentrating samples during sample preparation while allowing nonvolatile "contaminants" to be selectively held back in the injection system, thus preventing column contamination.

2.2.3   Liquid Chromatographic Methods

High-pressure liquid chromatography (HPLC) is another chromatographic separation method that is often used in substance analysis [1–3]. This method is especially used in separating nonvolatile, highly polarized, and ionic, as well as thermally unstable substances, and for analyzing compounds with high molecular weight. This method requires that all of the substances to be analyzed are soluble in the solvent used. It is important to optimize the method for the particular compounds to be tested, since variations in the chromatographic conditions can affect the efficiency of this method. Gradient elutions that correspond to temperature programming in gas chromatography are especially influential.

Various detectors have been described for the detection of compounds eluted from the column, and are summarized in Table 2.2 [4].

Additionally, the past few years have seen the development of several opportunities for adapting mass-spectrometry detectors to liquid chromatography (see Section 2.4.4). Specialized inlet and ionization techniques are required due to the high solvent volumes involved. Apart from particle beam (PB) and thermospray (TS), electrospray (ESI) and atmospheric pressure ionization (APCI) are the standard methods used today. The last two methods are based on electrical excitation of substances using high voltage potentials. The major advantage of these techniques lies in the low thermal stress on the molecules in the sample. HPLC/MS combinations are used in a variety of areas; apart from determining amino acids and peptides, a large number of applications in the analysis and determination of metals and metal complexes have been described [5, 6].

2.2.4   Electrophoresis

2.2.4.1   Introduction

Electrophoretic processes are based on the migration of ions inside an electric field. Apart from movements of ions with low molecular weight, movements of colloid particles, macro molecules, viruses, and whole cells can be used.

Table 2.2   Detectors for Liquid Phase Processes

Detector	Selective for	Detection Limit	Comments
UV detector (constant λ)	UV-active components	0.3 ng/ml	Adjustment by changing the filter
UV detector (variable λ)	UV-active components	0.3 ng/ml	Broadly applicable optimum in absorption maximum
Diode-array detector	UV-active components		Method development for UV spectra
Refraction index detector	Universal	0.7 μg/ml	No gradient elutions possible
Fluorescence detector	Fluorescent components	0.8 pg/ml	
Electrochemical detector	Components that can be oxidized or reduced	1 pg/ml	No gradient elutions possible
Conductivity detector	Ionic species	Depending on eluent	

The classical form of electrophoresis is used in a variety of ways in biochemical and biomolecular research for separating, isolating, and analyzing proteins and other biopolymers. A distinction is drawn between carrier-free and carrier electrophoresis (Figure 2.1).

Carrier-free electrophoresis uses the movement of ions in buffered solutions layered over the sample solution. A direct-current potential using platinum electrodes partly separates the sample solution, with the ions collecting at the border between the buffer and the sample solution. This process does not allow complete separation.

In *carrier electrophoresis*, the ions migrate onto a carrier, which may be paper or gel. The carrier is dipped into conductive electrolyte ending in buffered solutions that cover the electrodes. There is a direct voltage of more than 100V on the carrier. Migrating zones are formed for various particle types that can be evaluated by dying or photometry.

2.2.4.2 Capillary Electrophoresis

Capillary electrophoresis (CE) is a modern technique that has developed as a combination of various electrophoresis and chromatography methods, and enables rapid and efficient separation of dissolved and charged particles in an electric field [3, 7–9]. Introducing a mixture of various substances into the separation system and laying a potential difference results in separation of the compounds. How long this takes depends on migration speeds in the electric field. The separation principle applied in capillary zone electrophoresis (CZE) is mainly based on the varying magnitude of a charge of the species dissolved at a certain pH level. The migration of cations in the electric field produces an accompanying flow (the *electro-osmotic flow*, EOF). According to the corresponding migration speeds, cations, neutral molecules, and anions with differing charge and mass relationships can be separated (Figure 2.2).

While suppressing EOF, the following relationship applies for migration speed v:

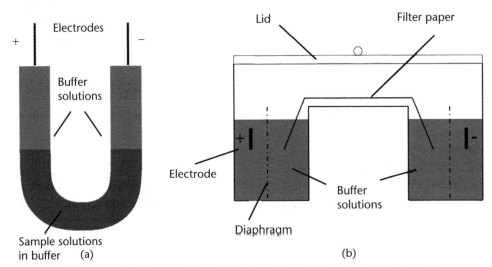

Figure 2.1 Equipment for (a) carrier-free electrophoresis, and (b) carrier electrophoresis on paper.

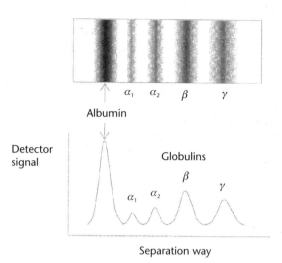

Figure 2.2   Electropherogram for serum proteins with photometric valuation. (*From:* [1]. © 1995 VCH Verlagsgesellschaft. Reprinted with permission.)

$$v = \mu_{EP} \cdot F = \frac{\mu_{EP} \cdot U}{l} \qquad (2.6)$$

where v is migration speed, μEP is electrophoretic mobility, F is field strength, U is voltage applied, and l is capillary length.

The migration time t is a result of the following equation:

$$t = \frac{l}{v} = \frac{l^2}{\mu_{EP} \cdot U} \qquad (2.7)$$

The efficiency of the separation thus depends more on the voltage applied than on capillary length. Taking EOF into account, the two following equations apply for migration speed v and migration time t:

$$v = \frac{\left(\mu_{E0} + \mu_{EP}\right) \cdot U}{l} \qquad (2.8)$$

$$t = \frac{l^2}{\left(\mu_{E0} + \mu_{EP}\right) \cdot U} \qquad (2.9)$$

The selectivity characteristics of electrophoretic methods are determined by the migration times of various substances. Under identical circumstances, the migration times of a certain compound are constant, thus enabling the identification of substances.

The parameters that mainly influence the result of separation by capillary electrophoresis are the solvent, buffer system, capillary length, and temperature as well as voltage used.

Selecting the correct buffer system is essential for optimal separation. The buffer selected should have a good buffer capacity in the target pH range, low absorption in the wavelength range set for the detector, and a low degree of mobility.

Heavy requirements are made on detection due to the small dimensions of the capillary and sample volumes. *UV detection* is by far the most frequently used technology, while *diode arrays* yield more wide-ranging information. Indirect UV detection (see Section 2.4.2.2) involves the addition of highly absorbent substances to the solvent, and it is used in applications for ions with small radii that often do not absorb radiation in the UV range.

At the moment, capillary electrophoresis represents an alternative method to HPLC due to often shorter measuring times. In particular, however, capillary electrophoresis can be used together with HPLC with regard to the information content of the results.

In the past few years, capillary electrophoresis has developed into a powerful technique due to its high separation efficiency, short analysis times, and low sample and electrolyte consumption. The various application areas in research and diagnostics include the determination of peptides and proteins, nucleic acids, and pharmaceutical and agrochemical products.

2.3   Bioanalytical Detection Methods

2.3.1   Protein Analysis

Biological processes center on proteins. Almost all molecular transformations that define the cellular metabolism involve protein catalysts. Proteins are made up of smaller units of polymer chains that consist of 20 different amino acid residues. The sequence of amino acids in the molecule indicates the chemical and physical properties of a protein and its effects on the organism.

A number of analytical methods have established themselves in protein analysis. Molar mass, amino acid sequence in a protein, tertiary conformation, and protein stability are among the parameters analyzed.

2.3.1.1   Processes for Determining Protein Mass

The molar mass of a protein is a fundamental characteristic often used in classifying proteins. Various processes can be used for determining molecular mass [10]:

- *Electrophoresis:* Electrophoresis is the simplest method for determining the molecular mass of proteins. The disadvantage of this process is that it only yields information on the molecular mass of the subunits that form the protein, and their stoichiometry in native proteins (see Section 2.2.4).
- *Gel filtration:* This process is used for determining the molecular mass in native proteins. To achieve a good result, nonspecific interactions between proteins and gel matrix have to be excluded, which can usually be achieved by selecting an appropriate buffer. This method is not very precise, and requires calibration with standard proteins of a known mass [10, 11].

- *Ultracentrifugation:* This process takes advantage of the fact that the speed with which particles sediment in a centrifuge depends on their mass. Ultracentrifuges can reach rotation speeds of up to 80,000 rpm, and centrifugal forces of above $600,000 \times g$ are possible. The sedimentation coefficient of a protein (s, sedimentation speed per unit centrifugal force) is usually expressed in units of 10^{-13}s. The relationship between molecular mass and sedimentation coefficient is nonlinear. Table 2.3 summarizes sedimentation coefficients for a selection of proteins.

- *Mass spectrometry:* Mass spectrometry has gained a great deal of importance in recent years in analytical determination of large biological macromolecules. In particular, the electrospray ionization (ESI) and matrix-assisted laser desorption (MALDI) methods have made a major contribution in this area (see Section 2.4.4). Currently, biological macromolecules with a molecular mass of up to several hundred kDa can be measured at an accuracy of 0.01% [10, 12, 13].

2.3.1.2 Processes for Quantitative Determination of Proteins

There is a variety of methods available for the quantitative determination of proteins [7, 10]:

- *Biuret method:* This method is based on reactions between substances containing several peptide compounds in the molecule and Cu^{2+} ions under alkaline conditions. The result is a violet-colored product that can be measured at a wavelength of 540 nm. (See Figure 2.3.)

Table 2.3 Sedimentation Coefficients in a Selection of Proteins

Protein	Molecular Mass (kD)	Sedimentation Coefficient
Lipase (milk)	6.7	1.14
Ribonuclease A	12.6	2.00
Cytochrome c	13.4	1.71
Myoglobin	16.9	2.04
α-chymotrypsin	21.6	2.4
Crotoxin	29.9	3.14
Concanavalin	42.5	3.5
Diphtheria toxin	70.4	4.6
Cytochrome oxidase	89.8	5.8
Lactate dehydrogenase	150	7.31
Catalase	222	11.2
Fibrinogen	340	7.63
Hemocyanine	612	19.5
Glutamate dehydrogenase	1,015	26.6
Yellow mosaic virus protein	3,013	48.8

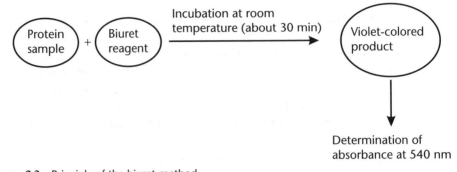

Figure 2.3   Principle of the biuret method.

- *Lowry method:* The sensitivity of the biuret method can be significantly enhanced by adding Folin-Ciocalteau reagent. The color reaction between peptide and Cu^{2+} is combined with reduction of phosphomolybdates and phosphotungstates by tyrosine, tryptophan, and cystein residues. The Lowry method is far more dependent on the type of protein to be determined. (See Figure 2.4.)
- *Bicinchoninic acid assay (BCA):* This method also uses Cu^+ formation and reaction with bicinchonates resulting in a violet-colored product, which can be measured at a wavelength of 562 nm. The sensitivity of this process is similar to that of the Lowry method, but is easier to apply in practice. (See Figure 2.5.)
- *Bradford assay:* This is the best-known and most frequently used process for the quantitative determination of proteins. The method is based on the formation of Coomassie brilliant blue G-250 on proteins under acidic conditions. This results in a shift in the absorption spectrum from 465 to 595 nm. The color reaction depends on the content of alkaline amino acids (especially arginine) and aromatic acid residues. (See Figure 2.6.)
- *Spectrophotometry methods:* Proteins contain chromophore groups that absorb both near and far UV radiation, and this can be used for determining

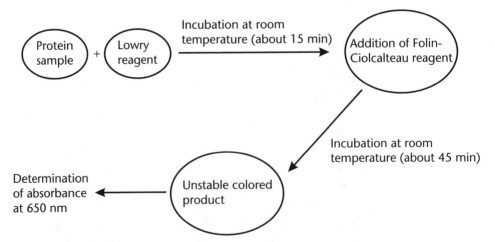

Figure 2.4   Principle of the Lowry method.

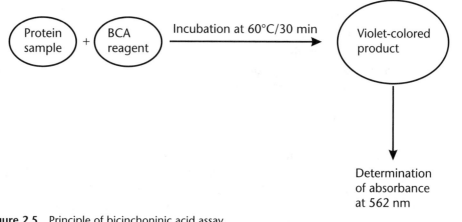

Figure 2.5   Principle of bicinchoninic acid assay.

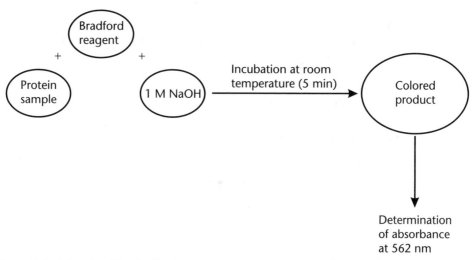

Figure 2.6   Principle of the Bradford assay.

protein concentrations. The advantage of this method is that the proteins are not destroyed.

2.3.1.3   Processes for Determining Amino Acid Sequences in Proteins

The first protein with a complete identified amino acid sequence was insulin hormone. The complete sequence was revealed in 1953. Today, amino acid sequences are known for hundreds of thousands of proteins. Information on amino acid sequence is necessary for the following reasons:

• Knowledge of the amino acid sequence in a protein is essential in determining its three-dimensional structure, and for understanding the molecular mechanisms by which the protein takes effect.

- Sequence comparisons within analogous proteins of different species reveal information on protein function as well as evolutionary background.
- Many genetic diseases are caused by mutations leading to amino acid exchange in a protein. Analysis of amino acid sequences can be useful in developing diagnostic tests and effective therapies.

There are four steps to sequencing a protein [11].

Preparation.   The complete amino acid sequence of a protein contains the sequences of each subunit. The number of different subunits is determined by end-group analysis (Figure 2.7). Each polypeptide chain has an N-terminal and a C-terminal residue. The number of chemically different polypeptides in a protein can be determined by identifying these end groups. Several methods can be used for determining the N-terminal:

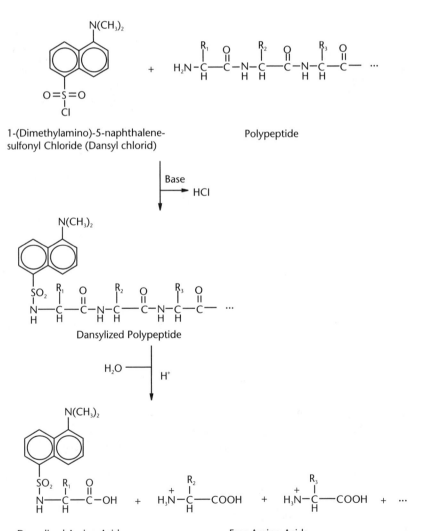

Figure 2.7   End group analysis of a protein by dansyl chloride reaction.

- Reaction between primary amines with 1-dimethylaminonaphthyl-5-sulphonylchloride (dansyl chloride) by forming dansylated polypeptides followed by acidic hydrolysis and liberation of the modified N-terminal residue;
- Identification using the first step in the Edman degradation process (q.v.);
- Separating the N-terminal residue using amino peptidases.

Carboxy peptidases can be used for identifying the C-terminal residue. These are enzymes that act as catalysts towards the hydrolytic separation of the C-terminal residue on a polypeptide. The amino acid liberated can then be isolated and identified.

Separating polypeptide chains connected by disulphides requires that the disulphide bonds between Cys residues be broken, which may be achieved by oxidation using performic acid or by reduction using mercaptans (Figure 2.8).

Splitting Polypeptides.   Polypeptides with more than 40 to 100 residues cannot be sequenced by Edman degradation (q.v.), and have to be broken down either

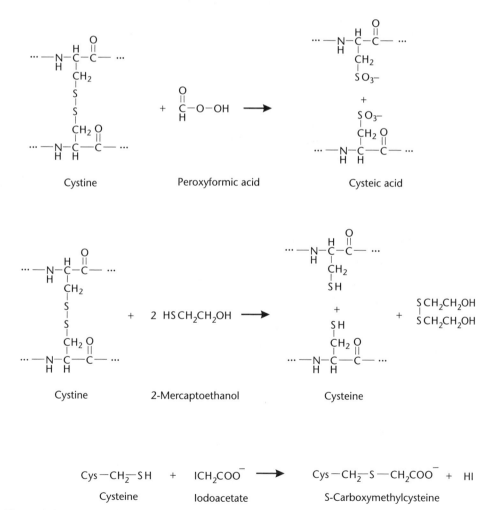

Figure 2.8 Severing disulphide bonds using performic acid and 2-mercaptoethanol.

chemically or using enzymes into specific fragments that are small enough to be sequenced. Various endopeptidases (enzymes that act as catalysts towards hydrolysis of internal peptide bonds) can be used for fragmenting these polypeptides. Trypsin digestion enzyme has high specificity and breaks peptide bonds at the C-terminal side (Figure 2.9). Other enzymes that can be used include chymotrypsin, elastase, thermolysin, pepsin, and endopeptidase V8.

Edman Degradation.   After isolating the peptide fragments that have been formed by specific fragmentation reaction, their amino acid sequence can be determined. This is usually achieved by repeated Edman degradation cycles. Phenylisothiocyanate (PITC) reacts with the N-terminal amino group of the peptide under slightly alkaline conditions to form a phenylthiocarbamoyl (PTC) adduct. This adduct is then treated with anhydrous trifluoroacetic acid, which separates the N-terminal residue as a thiazolinone derivate without hydrolyzing the other peptide bonds. This means that the N-terminal residue is liberated while the remaining polypeptide chain remains intact. The entire amino acid sequence of a polypeptide chain is then determined by repeating the Edman degradation cycle and identifying the amino acid liberated after each cycle [3]. (See Figure 2.10.)

The Edman degradation technique has now been automated, and it can be rapidly applied with low material utilization. Sequence analysis can be carried out on 5 to 10 pmol of a peptide since much less than 1 pmol of a PTH amino acid can be detected and identified.

Reconstructing the Protein Sequence.   After sequencing the individual peptide fragments, their sequence in the original polypeptide needs to be determined. This is achieved in a second protein separation process with the reagent of another specificity, which is followed by comparing amino acid sequences in overlapping peptide fragments.

2.3.2   DNA and RNA Analysis

Nucleic acids are macromolecules consisting of three building blocks in a molar ratio of 1:1:1. The building blocks are as follows:

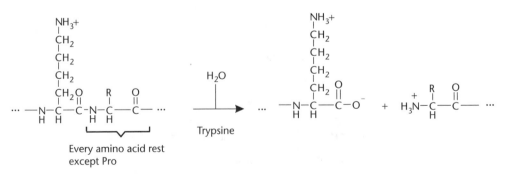

Figure 2.9   Breaking polypeptide bonds using trypsin.

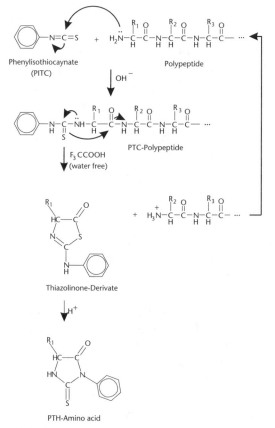

Figure 2.10   Edman degradation.

- N-containing, heterocyclic purine bases (adenine, guanine) and pyrimadine bases (cytosine, uracil in RNA);
- Pentoses (ribose or 2-desoxiribose);
- Phosphates.

The amounts of DNA and RNA in homogenates can be determined by color reaction between diphenylamine or orcinol and ribose. After isolating the DNA or RNA fraction, the quantities can also be determined according to phosphate content or base absorption.

2.3.2.1   Detection of DNA and RNA Sequences by Southern and Northern Blotting

Blotting techniques refer to the transfer of substances separated by electrophoresis onto a firm matrix, for example, cellulose acetate film. There are three main advantages to blotting technology, as follows:

- An identical and reproducible copy of the pattern is formed on the firm matrix, but in immobilized form.

- After the substances have been transferred onto the firm matrix, a wide-ranging variety of detection reactions can be applied in order to identify the substances, which would be difficult or impossible in separation gel.
- Multiblots, in which proteins may be transferred to multiple stacked membranes simultaneously, allow the investigator to examine the concentrations of many proteins simultaneously.

In *Southern blotting*, DNA fragments separated by electrophoresis are transferred onto a solid phase, usually a nitrocellulose membrane, by capillary forces. The DNA to be analyzed is treated with one or several restriction enzymes, and then separated up according to size by gel electrophoresis. The resulting separation pattern in the gel consists of the DNA fragments which are separated into single strands by alkalis and transferred onto a membrane (mostly nylon or nitrocellulose) and permanently fixated there. After that, the membrane is treated with a chemically or radioactively marked genetic probe. This probe consists of single-strand DNA complementing the sequence being assayed for. If this sequence is to be found anywhere on the membrane, the probe will form base pairs and will bind permanently into this area. All nonspecific bonds are then washed off. Detection takes place by laying the membrane onto X-ray film, photo paper, or a phosphor imaging plate.

Northern blotting is a variation on the Southern blotting technique for identifying certain RNA molecules in a mixture. The sample to be analyzed consists of RNA, but the technique is otherwise basically the same as Southern blotting. An example of where Northern blotting might be used would be to compare protein-encoding mRNA from a mutant organism with that of its "normal" counterpart.

The blotting techniques also include *Western blotting* (immunoblot or protein blot). A protein pattern separated by electrophoresis is transferred onto a solid phase. Western blotting serves to identify antibodies in body fluids. Three reaction steps are necessary, as follows:

1. Separation of a protein mixture into individual fractions by electrophoresis;
2. Transfer of the separated proteins onto a secondary carrier;
3. Visualization of the protein fraction on the secondary carrier. This is achieved by immunoreaction using specific antibodies. The immuno-complexes formed are then visualized by applying an enzyme-conjugated secondary antibody (conjugate) and substrate solution in a color reaction. The band pattern that results is then subjected to analysis.

Blotting technologies are used for genetic fingerprinting. This concept refers to the fragmentation of a substance, such as DNA, followed by two-dimensional separation by electrophoresis and chromatography for identification purposes. DNA fingerprinting has particular diagnostic importance in the detection of genetic defects (inherited diseases), forensics, and in prenatal diagnostics and paternity testing.

2.3.2.2 Detection of DNA and RNA Sequences by Polymerase Chain Reaction

Hardly any other invention has changed the face of biological science as rapidly and sweepingly as the polymerase chain reaction (PCR). This technique reproduces minimal amounts of DNA rapidly to such an extent that the tiniest DNA samples can be identified, analyzed, and otherwise utilized after PCR. The medical field has recognized the potential, with new applications appearing on a steady basis. Wherever genetics can reveal anything about the origin or course of a disease, PCR is the method of choice.

With PCR, immeasurably small amounts of DNA can be copied as often as needed for the resulting DNA to be analyzed by ordinary laboratory methods such as sequencing. In theory, one single DNA molecule is all that is needed; PCR is one of the most sensitive biological techniques in existence [13–19].

PCR serves towards reproducing DNA. This constant copying is achieved using certain proteins known as polymerases. These enzymes are capable of combining building blocks of DNA to long molecule strands. First, they need the individual DNA building blocks; that is, the nucleotide bases adenine, thymine, cytosine, and guanine (A, T, C, and G). Second, they only need a small piece of DNA to connect their own building blocks to; these are known as *primers*. Third, the polymerases need a *template* in the form of a longer DNA molecule, which serves as a kind of matrix to construct a new strand. Once all three of these conditions have been fulfilled, the enzyme creates exact copies of the original (see Figure 2.11).

An example of where this process is important is in cell reproduction, where DNA polymerases reproduce the genetic code of the organism. Apart from that, there are also RNA polymerases that construct molecule strands according to RNA building blocks; in particular, they construct mRNA, which are working copies of the genes. These enzymes can be used to reproduce any sequence of nucleic acids in PCR. Usually, the molecules produced are DNA strands. If RNA is to be copied, the RNA strands are usually converted to DNA using the *reverse transcriptase* enzyme first. This method is known as *reverse-transcription* PCR (RT-PCR). The reproduction that follows only needs a small part of the DNA range in question, which is used as the original for creating the corresponding primer. Using this primer, PCR can copy completely unknown DNA sequences. Genes are always accompanied by similar short sections, which can be used to reproduce completely unknown genes. This is an important replacement for what is referred to as molecular cloning, where DNA strands are copied in a time-consuming process involving bacteria or other host organisms. PCR reaches this goal faster, more easily, and more importantly, in vitro—in a test-tube. Additionally, PCR can use known areas and longer DNA molecules such as in chromosomes as a starting point before delving into unknown regions.

PCR is based on constantly repeating this three-step cycle (see Figure 2.12):

1. *Denaturation:* The reaction solution with the DNA molecules (which are to be copied), the polymerases (which actually copy the DNA), the primers (that serve as the starting DNA strand), and the nucleotides (which are connected to the primer) are heated up to 95°C. The DNA molecules are split into their two complementing strands, which is referred to as *denaturation*.

2. *Annealing:* By lowering the temperature to 55°C, the primer binds with the DNA in a process known as *hybridization*. This connection is only stable if the primer and DNA section complement each other, which means that base

pairs can form between them. The polymerases begin to bind more complementing nucleotides at these sites, thus strengthening the bond between the primers and DNA.

3. *Polymerization:* The temperature is now increased to 72°C. This is the ideal working temperature for the polymerases used, which bind other nucleotides to the DNA strands produced. Apart from that, the loose bindings between the primers and those DNA sections that do not completely complement the primer are broken.

Constantly repeating these three steps doubles the number of DNA molecules copied each time. After 20 cycles, one single DNA double helix is copied into around a million molecules.

Currently, the polymerases used normally originate from the bacterial species *Thermus aquaticus*, whose polymerases are stable at temperatures of 95°C, as required by PCR.

Today, PCR is used everywhere where the exact sequence of a DNA molecule needs to be determined. Applications include the genome project, exploring new genes, analyzing changes in inherited genetic code, and searching for targets. Current topics include detailed changes in the genome with effects such as vulnerability to disease and varying reactions of patients to medication. Additionally, genes may also serve as targets for new pharmaceutical products. PCR therefore plays a major role in this central area of pharmaceutical research.

If PCR is to be used for detecting a certain DNA segment alone, this is referred to qualitative PCR. Usually, the standard protocol is applied. Since only one single DNA molecule in a sample is theoretically necessary for detection, PCR is an extremely sensitive method. Many applications require that the originally available DNA molecules be quantified in the PCR reaction. There is a variety of approaches for counting the number of DNA copies to reach this value, even in the individual chain reaction steps. The middle or exponential phase of PCR is usually selected for this purpose, which is when the number of DNA templates almost doubles in each cycle. Most of the quantitative PCR methods used today are based on a recent development from 1992: a dye known as ethidium bromide (EtBr) that fluoresces in double-strand DNA when stimulated by light. The fluorescence observed reveals the amount of DNA formed at every stage in the chain reaction, which is why the processes known as real-time PCR. Parallel testing with the same DNA of a known amount provides a comparison curve under identical conditions; this is conditional upon the DNA being known. Quantitative PCR is used in pharmaceutical applications including research and evaluation of targets for new active agents [17].

2.3.3 Enzymatic Analysis

Catalysts synthesized by living organisms are known as *enzymes*, a term that comes from the Greek *en* for "inside" and *zyme* meaning "yeast." These are also referred to as biocatalysts. The first enzyme to be detected was pepsin, which is to be found in the stomach; the urease enzyme was isolated in 1926 [13, 18].

Enzymes are proteins or proteides with the relative molar mass ranging from 10,000 to 2 million and consist of amino acid chains connected by peptide chains.

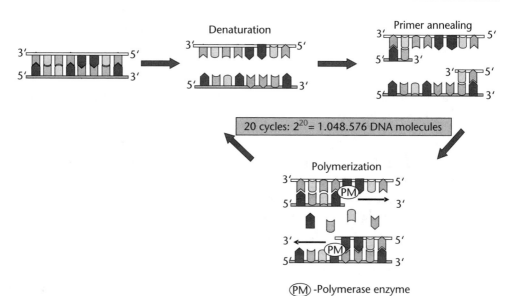

Figure 2.11 Principle of PCR.

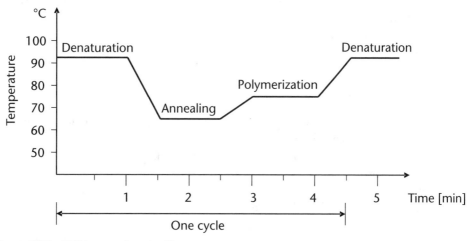

Figure 2.12 PCR temperature profile.

Many enzymes only take effect in the presence of a cofactor. In this process, the original protein, which is known as the apoenzyme, forms a holoenzyme after reacting with the cofactor.

The following chemicals may serve as cofactors:

- Metal ions such as Zn^{2+} in alcoholdehydrogenase and carboxypeptidase, Mn^{2+} in phosphortransferase, or iron ions in cytochromes;
- Coenzymes such as NAD^+, which transfers hydrogen groups and electrons, or coenzyme A, which transfers acyl groups. As an example, alcohol-

dehydrogenase together with its coenzyme, NADH, can reduce aldehyde to alcohol molecules. (See Figure 2.13.)

Enzymes possess a high degree of specificity, meaning that they each only act as a catalyst for a certain type of reaction. Enzymes are therefore categorized into the following six main groups:

1. Oxidoreductases act as catalysts in redox reactions.
2. Transferases catalyze the transfer of groups such as amino groups (transaminases).
3. Hydrolases catalyze hydrolytic separation of chemical bonds, such as in carbonic acid ester hydrolysis.
4. Lyases catalyze the breaking of chemical bonds in elimination and addition reactions on double bonds
5. Isomerases catalyze isomerization.
6. Ligases (synthases) catalyze production of chemical bonds by connecting together two substrates using ATP.

According to the type of substrate or the particular catalyst reaction concerned, the main groups are further divided into subgroups and sub-subgroups containing the individual enzymes.

On the other hand, the specificity of an enzyme can be set towards certain compounds. For example, urease only catalyzes hydrolysis in urea.

2.3.3.1 Enzyme Kinetics

Enzymes function as catalysts. With an enzyme present, the activation energy of a reaction is decreased, and the reaction takes place several times more rapidly than the same reaction without a catalyst [1, 13, 18].

The formal kinetics can be applied to enzymatic reactions. Enzymatic substrate saturation reactions are a special case—in a certain substrate concentration range, the reaction speed is independent of substrate concentration.

Substrate saturation is to be understood according to enzyme kinetics as developed by MICHAELIS and MENTEN, where enzyme E and substrate S form product P according to the following reaction equations:

$$E + S \underset{k_{-1}}{\overset{k_1}{\rightleftharpoons}} ES \qquad (2.10)$$

$$ES \xrightarrow{\ k_2\ } E + P \qquad (2.11)$$

where k_1, k_{-1}, k_2 are speed constants.

The enzyme-substrate complex is formed in a reversible reaction. The result of the product formation is that the enzyme is liberated again. Figure 2.14 shows the time course for concentrations of enzymatic reactions.

In order to use the MICHAELIS-MENTEN kinetics model for determining enzymes or substrates, a simple relationship is needed to describe the starting reaction speed depending on enzyme and substrate concentration.

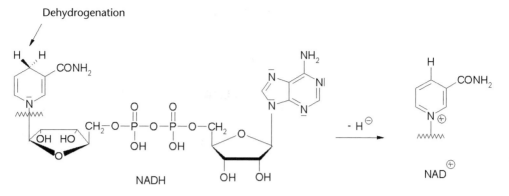

Figure 2.13   Oxidation of NADH to NAD⁺.

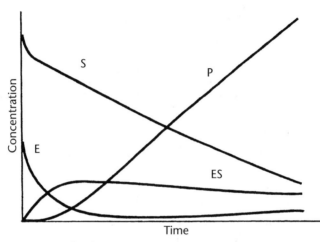

Figure 2.14   Concentration development by time in an enzymatic reaction.

The starting reaction speed, v_0, is a result of the reaction equation for forming the product:

$$v_0 = k_2 [ES] \tag{2.12}$$

where $[ES]$ is the concentration of the enzyme-substrate complex.

Since the concentration of the enzyme-substrate complex is not known, an attempt is made to express this concentration by enzyme and substrate concentration. The resulting equation applies for the formation of this complex:

$$\frac{d[ES]}{dt} = k_1 [E][S] \tag{2.13}$$

where $[E]$ is the concentration of the enzyme, and $[S]$ is the concentration of the substrate.

The complex decomposes as follows:

$$-\frac{d[ES]}{dt} = k_{-1}[ES] + k_2[ES] \tag{2.14}$$

On the condition that:

$$\frac{d[ES]}{dt} = 0 \tag{2.15}$$

The two equations can be combined as follows:

$$k_1[E][S] = k_{-1}[ES] + k_2[ES] \tag{2.16}$$

Taking into account the total concentration of enzyme c_E where

$$c_E = [E] + [ES] \tag{2.17}$$

the enzyme concentration can be replaced, and the equation can be simplified as follows:

$$k_1(c_E - [ES]) = k_{-1}[ES] + k_2[ES] \tag{2.18}$$

The equation can be rearranged to yield the MICHAELIS-MENTEN constant, K_M:

$$K_M = \frac{k_{-1} + k_2}{k_1} = \frac{[S](c_E - [ES])}{[ES]} \tag{2.19}$$

Table 2.4 shows examples for the MICHAELIS-MENTEN constant.

Enzymatic activity is measured in the international unit Katal (abbreviated to kat). One kat is the mass of an enzyme required for 1 mol of substrate to react per second under predefined reaction conditions. Since the activity of an enzyme is dependent on factors such as temperature, pH value, and substrate concentration, enzyme activity is normally measured at 30°C with the current experimental conditions also included.

2.3.3.2 Enzyme and Substrate Determination

An excess in substrate is used when analytically determining an enzyme. One typical example of enzyme determination is the analysis of glutamic-oxalacetic trans-

Table 2.4 Examples of Analytically Useful Enzymatic Reactions with their Michaelis-Menten Constants

Enzyme	Substrate	K_m, mM
Catalase	H_2O_2	25
Hexokinase	Glucose	0.15
Glutamate dehydrogenase	Glutamate	0.12

aminase (GOT). Of all of the transaminases in the human body, this enzyme has the highest concentration and is an important diagnostic parameter for amino acid metabolism. The enzymatic reaction is based on the following equation:

$$L - aspartate + \alpha - ketoglutarate \overset{GOT}{\rightleftharpoons} oxalacetate + L - glutamate$$

Enzymatic reactions are used to determine substrates in a variety of applications. Table 2.5 summarizes the classical substrates and the enzyme reactions used to analyze them.

Following enzymatic reactions often requires an additional indicator or other auxiliary reactions. The NAD coenzyme is particularly important as an indicator, and it is used in testing blood alcohol content amongst other applications. Detection is based on the various UV spectra of the two redox forms, NAD^+ und NADH (Figure 2.15).

Coupling enzymatic and auxiliary reactions mainly involves photometric detection principles. A typical example is the analysis of hydrogen peroxide in glucose oxidation in the presence of glucose oxidase using a dye such as o-dianisidine (Figure 2.16).

The oxidized diimine form of o-dianisidine is used as a dye, and it is measured at 460 nm. An indicator can also be used in connection with an auxiliary reaction. Substrates such as uric acid, ⟨–amylase, and creatine kinase can be genetically analyzed using the following reactions in combination:

- Enzyme reaction: Substrate + ATP $\overset{Enzym}{\rightleftharpoons}$ Substrate – Phosphate + ADP;

- Auxiliary reaction: ADP + Phosphoenolpyruvate $\rightleftharpoons$ Pyruvate + ATP;

- Indicator reaction: Pyruvate + NADH + H^+ $\rightleftharpoons$ Lactate + NAD^+.

ADP is adenosine-5-diphosphate, and ATP is adenosine-5-triphosphate.

Table 2.5   Substrate Analyses Based on Enzymatic Reactions

Substrate	*Enzyme Reaction*
Clinical	
Glucose	Glucose + O_2 + H_2O $\overset{GOT}{S}$ D-gluconolacetone + H_2O_2
Blood alcohol	Ethanol + NAD^+ $\overset{ADH}{S}$ Acetaldehyde + NADH
Industrial	
Penicillin	Penicillin + H_2O $\overset{Penicillinase}{S}$ Penicillinic acid
Catechol	Catechol + O_2 $\overset{Catechol\ 1,2\text{-}Oxygenase}{S}$ Dicarboxylic acid
Environmental	
Phenol	Phenol + O_2 $\overset{Polyphenoloxidase}{S}$ o-benzochinone

GOD: Glucose oxidase; ADH: Alcoholdehydrogenase.
Source: [1].

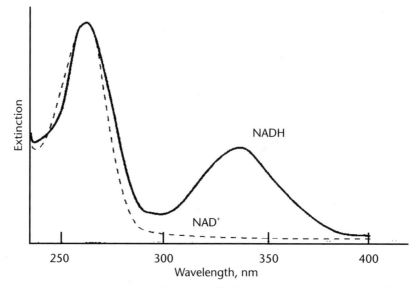

Figure 2.15   UV absorption spectra of NAD⁺ and NADH in aqueous solution. (*From:* [1]. © 1995 VCH Verlagsgesellschaft. Reprinted with permission.)

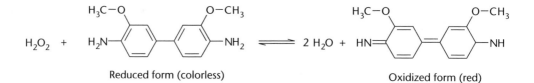

Reduced form (colorless)   Oxidized form (red)

Figure 2.16   Using o-dianisidine as an indicator.

NADH can also be used as an indicator since it has an absorption wavelength of 340 nm.

2.3.3.3   Examples of Enzymatic Analyses

Many pharmaceuticals and toxins function by blocking enzyme activity. In competitive inhibition, the inhibitor competes with the substrate for the same active site on the enzyme. In noncompetitive inhibition, the inhibitor binds to another active site on the enzyme, thus reducing its catalytic effect.

The following is a list of examples of enzymatic reactions.

Analysing Trypsin Activity.   Trypsin is a proteolytic enzyme that is formed in an inactive form, trypsinogen, in the pancreas. Responding to a hormonal signal, trypsinogen is secreted by the pancreas into the duodenum, where it is converted into trypsin by another enzyme (enteropeptidase). Trypsin plays a key role in the cascade of proteolytic activity in the reaction of the body to food ingested.

Specifically, trypsin catalyses the separation of peptide bonds from lysine and arginine residues, and is most effective in a slightly alkaline environment (pH 7.6 to

8) at body temperature. Trypsin activity is regulated by endogenous trypsin inhibitors, while also controlling the activation and catabolism of many intracellular and extracellular proteins.

The principle of determining trypsin activity consists of splitting N-benzoyl-DL-arginine-p-nitroanilide • HCl (BAPNA) under the influence of trypsin into a yellow dye (p-nitroaniline), which can be colorimetrically determined at a wavelength of 405 nm (see also Figure 2.17).

A substance that inhibits trypsin activity can be determined by colorimetric determination of the p-nitroaniline produced.

Analyzing COX/LOX Activity. Inflammation reactions are caused by prostaglandins, which are synthesized in various intermediate steps from phospholipids, parts of the cell membrane, and enzymatic reactions. Specific receptors in the target cells and organs respond to prostaglandins, representing the human body's response to nitrous oxides. Due to the essential importance of arachidonic acid in the process, the synthesis of prostaglandins from phospholipids is also referred to as the arachidonic acid metabolism. Apart from phospholipase, which catalyzes the formation of arachidonic acid from phospholipids, lipoxygenase (LOX) and, in particular, cyclooxygenase (COX) (also referred to as prostaglandin H-synthase) play a central role in forming these inflammation mediators. The cyclooxygenase (COX) enzyme is also of great pharmacological interest since most nonsteroidal anti-inflammatory drugs (NSAIDs) are based on COX inhibition.

Methods such as enzyme-linked immunosorbent assay (ELISA) (see Section 2.3.4.2) are used for analyzing catalytic COX activity; these techniques can be automated, and they are suitable for high-throughput screening.

The principle of the COX-2 inhibition test is based on the use of the enzyme's peroxidase activity. Peroxidase activity is determined colorimetrically by the formation of oxidized N,N,N′,N′-tetramethyl-p-phenylendiamine (TMPD) at 590 nm (Figure 2.18).

Lipoxygenases, enzymes that catalyze hydroperoxidation in fatty acids, have enjoyed increasing interest in pharmaceutical research as potential therapeutic approaches against inflammatory diseases and cancer.

The LOX assay is based on the same principle as the Cayman Lipoxygenase Inhibitor Screening Assay Kit. LOX enzyme activity is measured by colorimetrically

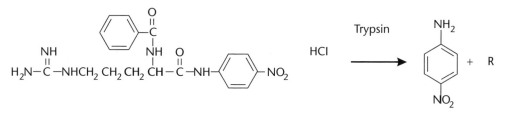

BAPNA   p-Nitroanilin

Figure 2.17 Enzymatic conversion of N-benzoyl-DL-arginine-p-nitroanilide • HCl (BAPNA) to p-nitroaniline.

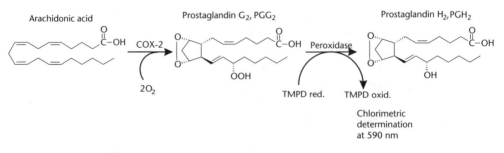

Figure 2.18   Principle of COX-2 activity analysis.

assessing the hydroperoxide produced in the conversion of fatty acids such as arachidonic or linolic acid at 500-nm wavelength (Figure 2.19).

Analyzing Kinase Activity.   The kinases are another class of modifiable molecules that have attracted attention due to the essential role they play in intracellular signal transduction. Many new kinase inhibitors have been identified, some of which are now undergoing clinical testing [19].

As an example, neuronal stem cells and dopaminergic neurons differentiate on the activity of special signal molecules, Wnt proteins, which are secreted by neighboring cells. The central element of this signal cascade is protein kinase GSK-3β (glycogen synthase kinase 3β), which possesses a phosphate residue without the signal transfer of Wnt. This activates GSK-3β and destabilizes (phosphorylates) β-catenine, and the Wnt target genes cannot be expressed. When Wnt binds to its receptor, LRP, the disheveled protein is activated, leading to deactivation of GSK-3β (dephosphorylation). As a result, β-catenine is no longer degraded, and it is in a position to induce target gene expression in the cell nucleus. Inhibition (deactivation) of GSK-3β thus leads to differentiation from neuronal stem cells to differentiated dopaminergic neurons.

Kinase activity detection takes place via the reaction with the residual ATP still present in the solution after the reaction with GSK-3β. ATP activates the luciferase enzyme, which then becomes luminescent (see Section 2.4.2.3). This can be recorded by a luminescent reader, where the light intensity is proportional to the residual

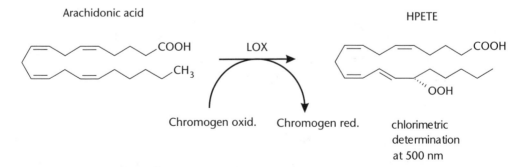

Figure 2.19   Principle of analyzing LOX activity.

amount of ATP present from the experiment. If the light measured is low, a large amount of ATP has been used up by the GSK-3β protein kinase (the test substance used therefore did not inhibit GSK-3β). If the light intensity is high, GSK-3β has been inhibited since no ATP has been converted (the test substance is an inhibitor).

2.3.4 Immunological Methods

Immunochemical methods provide another approach in making use of biochemical reactions. These methods are based on antigen-antibody reactions [1, 20].

The immune system will react when foreign substances are introduced into the organism. A specific substance is formed that is able to bind to the foreign substance and neutralize it. The foreign substance is referred to as an antigen, and the substance produced by the immune system is known as the antibody. The reaction between the antigen (Ag) and antibody (Ab) leads to the formation of an immune complex according to the equilibrium reaction shown in Figure 2.20.

The first use of immune reactions can be traced back to the detection of insulin in serum using insulin antibodies as early as in the 1950s.

The antibodies are proteins such as γ-globulins in the form of what are known as IgG globulins. Antigens are also compounds of higher molecular weight with relative molecular masses of above 1,000. While compounds of low molecular weight (haptenes) can also be determined by immunochemistry, they first need to be coupled to a carrier protein to form compounds of a higher molecular weight.

One particular aspect of immunochemical methods is that the formation of antibodies by the organism is the biological reaction that replaces the synthesis process otherwise carried out by the analytical reagents. This makes it possible to develop assays for badly definable analysis substances such as humines.

Detection principles that may be applied in immunochemical reactions include analysis of luminescence, enzymatic activity, or radioactivity. The most important processes include radioimmunoassay (RIA), or when using enzymatic detection, enzyme-linked immunosorbent assay (ELISA).

Table 2.6 summarizes immunochemical reactions by giving examples.

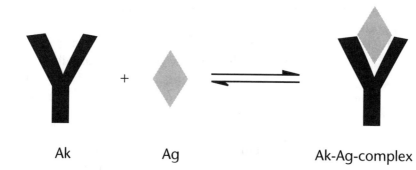

Ak   Ag   Ak-Ag-complex

Figure 2.20 Ag-Ab equilibrium reaction.

Table 2.6   Examples of Immunochemical Reactions

Substance for Analysis	Assay
Insulin (peptide hormone)	RIA
Penicillin G	RIA
Morphine	RIA
Trypsin (Protein)	RIA
Atrazine	ELISA
Benzene, toluene, xylene	ELISA
Benzo[a]pyrene	RIA
Cytokine (Il-6, Il-2, TNF, INF, and so forth)	ELISA
Immunoglobulins (such as IgE)	ELISA
Infection (virus, parasite, bacterial components, specific antibodies)	ELISA
Choriongonadotropin (hCG, pregnancy test)	ELISA

2.3.4.1   Radioimmunoassay

Radioimmunoassay combines the high specificity of an antigen-antibody reaction with the high detection power of radiochemical methods. Concentrations down to a few pg/ml can be detected using this method [1].

Radioimmunoassay is based on the competing complex formations of natural antigens (Ag) and a labeled antigen (Ag*) with the antibody (Ab). The competitive equilibrium in the subreactions can be expressed in the following equations:

$$Ag + Ak = AgAk \; ; \; K = \frac{[AgAk]}{[Ag][Ak]} \tag{2.20}$$

$$Ag^* + Ak = Ag^*Ak, k^* = \frac{[Ag^*Ak]}{[Ag^*][Ak]} \tag{2.21}$$

The conditions for the analytical use of competitive equilibrium are as follows:

1. Concentrations of labeled antigens and antibodies must remain constant throughout the analysis. The only variable allowed is the antigen concentration.
2. The number of labeled antigen molecules must be larger than the number of binding sites on the antibodies. The antiserum is usually diluted until around 50% of the marked antigens are bound in the absence of the antigen.
3. The effectiveness of the antigen must be identical in both the standard solution and in the samples to be analyzed.

The simultaneous equilibrium states allow the conclusion that the concentration of the labeled complexes formed, Ag*Ab, increases with decreasing concentration of the unmarked antigen.

To measure the activity of the complex, this is separated from the reaction mixture. The steps to be taken in radioimmunoassay are as follows:

1. Immobilizing the antibody;
2. Reaction between antibody and antigen from the sample and the labeled antigen also added;
3. Separation of the immune complex;
4. Measurement of radioactivity.

The isotopes ^{32}P, ^{35}S, and ^{3}H are very frequently used for the production of labeled antigens since these elements are present in a large number of substances to be analyzed, and also because they allow safety regulations to be upheld with ease. Apart from that, ^{125}I is also used for marking, as this gamma radiation emitter is available in 100% nuclide purity and has a relatively low radioactive intensity. The steps necessary for labeling marking with ^{125}I are as follows (Figure 2.21):

1. Isotope exchange if iodine is already present in the molecule, such as in thyroxin (T4) and triiodothyronine (T3);
2. Introduction of ^{125}I in molecule residues such as tyrosyl or histidine, such as in peptide hormones;

$$2\,^{125}I^- \xrightarrow{\text{Chloramine-T}} \,^{125}I_2 \longrightarrow \,^{125}I^+ + \,^{125}I^-$$

3. Introduction of the ^{125}I-labeled tyrosyl residue into the molecule.

2.3.4.2 Enzyme-Linked Immunosorbent Assay

Enzymes can be used in a somewhat less elaborate detection process than radioimmunoassay. Instead of radiochemically labeled antigen molecules, ELISA uses antigen molecules that are labeled by an enzyme. The result is a competitive enzyme-analyte conjugate bound to the antibody.

ELISA can be used to detect proteins, viruses, and also compounds with low molecular weight such as hormones, toxins, and pesticides in a sample (e.g., blood serum, milk, and urine). The process uses the property of a specific antibody to bind to the substance to be detected, the antigen. The antibody or antigen is first labeled with an enzyme. The reaction catalyzed by the enzyme services as evidence for the presence of that antigen. There are several ELISA techniques, with sandwich ELISA and competitive ELISA dominating the field.

The *sandwich ELISA* technique uses two antibodies that both bind specifically to the antigen in question. Antibodies against the antigen of interest are immobilized on the microtiter plate. After blocking the nonspecific binding sites, the

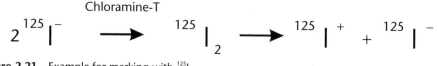

Figure 2.21 Example for marking with ^{125}I.

antigen is added. A second antibody, also specific for the antigen and coupled to the detecting enzyme, is added, followed by the enzyme substrate which initiates the detection reaction. (See Figure 2.22.)

The color intensity corresponds proportionately to the concentration of the chromagen formed, and therefore also to the concentration of the antigen to be analyzed in the sample.

Competitive immunoassay is also often used. Antibodies against the relevant antigen are immobilized by absorption in the wells of a microtiter plate. Antigen, both free and enzyme-conjugated, is added to the wells. After washing, enzyme-substrate solution is added to initiate the detection reaction. The amount of product formed depends on the ratio of free and enzyme-coupled antibody, which can be determined using a standardization curve [1, 21]. (See Figure 2.23.)

The detection limits for ELISA lie within the ng/l to μg/l range.

2.4 Physical Detection Methods

Apart from bioanalytical methods (see Section 2.3), physical principles and phenomena can also be used for testing for the existence of substances and identifying

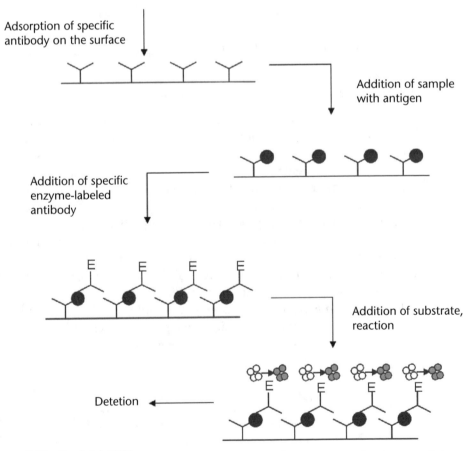

Figure 2.22 Sandwich ELISA.

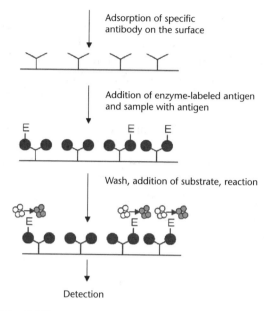

Figure 2.23   Competitive ELISA.

their structure. Spectroscopy is one of the most important physical detection methods used, and covers all methods that involve the interaction between electromagnetic radiation and matter.

Electromagnetic radiation can take several forms including light, heat, ultraviolet light, microwaves and radio waves, as well as gamma rays and X-rays.

Electromagnetic radiation can be described as having the character of both particles and waves. Under the weight aspect, properties such as wavelengths, frequencies, and velocities can be attributed to electromagnetic radiation. Absorption and emission properties are described from the particle aspect of E-M radiation.

The section will introduce the most important physical processes for detecting substances.

2.4.1   Atomic Spectroscopy

Atomic spectroscopy is based on absorption, emission, and fluorescence by atoms. Depending on the energy of the radiation stimulant, the valence electrons (stimulated by UV of visible light) or inner electrons (stimulation by X-rays or ionic radiation) are subject to spectroscopy [1, 22, 23].

Bohr's atomic model is used to explain the formation of linear spectra. Transitions between varying energy levels result from the absorption or release of energy. The following equation applies to the absorption or emission of radiation:

$$\Delta E = E_b - E_a = h * v = h\frac{c}{\lambda} \tag{2.22}$$

where ΔE is energy difference; E_a, E_b is energy at levels a and b; h is Planck's constant; λ is wavelength; and c is speed of light.

According to the dipole selection rule, only certain electron transitions lead to high-intensity lines in atomic spectroscopy. Each element forms characteristic lines due to its unique electron configuration in the electron shell. The lines contained in the spectrum, and their positions (wavelength, energy) and relative sizes, can be used for the unequivocal qualitative identification of an element, even when they overlap at various wavelengths.

If an atom is subjected to sufficient thermal energy, outer electrons can be ejected against the pull of the atomic nucleus. Depending on the processes detected, a distinction is drawn between *atomic absorption* (AAS) and *atomic emission* (AES). The number of possible emission lines is limited by Kirchhoff's law, as each substance can only absorb light at the same frequency at which it emits.

The intensity I of the lines observed in thermal equilibrium is determined by the Boltzmann distribution:

$$I = \frac{h\upsilon}{4\pi} W \cdot N_1 = \frac{h\upsilon}{4\pi} W \cdot N_0 \frac{g}{Z_s} \cdot e^{\frac{-\Delta E}{kT}} \tag{2.23}$$

where N_0 is number of particles in the ground state, N_1 is number of particles in the excited state, W is transition probability, and Z_s is total states.

Flames, arcs, sparks, and plasma can serve as excitation sources. Several elements can be subject to excitation at the same time in plasma spectroscopy, and the method can be used for multielement measuring. The linear measuring range for atomic emission spectroscopy ranges between 10^5 and 10^6, and main as well as trace components can be measured together while avoiding error-prone thinning procedures.

Atomic emission spectra show many lines. This means that various lines are available when analyzing for a particular element. However, this is also the main problem in emission spectroscopy, which is to be found in spectral disturbances. Eliminating spectral disturbances requires the use of an elaborate optical system for separating the line spectra. Element selectivity is only determined by the high resolution of the monochromator in atomic emission spectroscopy.

Recently, atomic emission spectroscopy has been used as a detection method in connection with gas chromatography.

2.4.2   Optical Molecule Spectroscopy

Optical molecule spectroscopy is an important process in revealing structures. This process is mainly distinguished by infrared spectroscopy, UV/VIS spectroscopy, and fluorescence spectroscopy.

2.4.2.1   IR Spectroscopy

The interaction between radiation of intensity I_0 and the sample to be analyzed causes oscillations in the beam due to absorption I_A, reflection I_R, and scatter I_S; the relationship between total approaching light and light that passes through, I, can be described as follows [22]:

$$I_0 = I + I_A + I_R + I_S \tag{2.24}$$

Methods using excitation in the infrared range are referred to as absorption methods. Infrared waves are usually characterized by the wave number detected, $\bar{v}$, such that:

$$\bar{v} = \frac{1}{\lambda} \tag{2.25}$$

where λ is wavelength of light absorbed.

The freedom of movement of a molecule within three-dimensional space is described by translation, rotation, and resonance. The fundamental resonance of a molecule is normal resonance, which refers to independent resonance without energy being converted. Excitation of resonance in a molecule by infrared radiation is requires a change in dipole moment.

Infrared spectroscopy (IR) is a typical structurally selective method that delivers information on the type of bonds between the individual elements in a compound [22, 23]. The aims of spectrum analysis may vary. In some cases, the identity of the bond is to be identified by its infrared spectrum. Comparing the spectrum from an unknown compound with the spectra listed in a library of spectra is sufficient for identification. In other cases, the aim may be to confirm a suspected structure, or to select the formula of a molecule from various possibilities. Finally, infrared spectroscopy may be used to identify the structure of a completely unknown substance. Identifying the structure of a molecule may involve characteristic group resonance patterns as listed by example in Tables 2.7 and 2.8.

2.4.2.2   UV/VIS Spectroscopy

In the UV and visible light range, the outer electrons in the sample molecules are raised to an excited state, which leads to absorption [1, 22, 23]. Certain selection rules apply in the electron transitions in *UV/VIS spectroscopy*, which are

Table 2.7   Wave Numbers of Typical Valence Bond Resonance Patterns

Resonance	Substance Classification	Number of Waves per cm
CH valence bond resonance	Aliphatic	2,800–3,000
C-H valence bond resonance	Unsaturated bonds	3,000–3,100
C=C valence bond resonance	Olefins	1,630–1,680
C=C valence bond resonance	Aromatics	1,500–1,600
C=C valence bond resonance	Alkynes	2,100–2,260
C-H vertical resonance	Olefins and aromatics	700–1,000

Table 2.8 Typical Wave Numbers in Functional Groups

Functional Group	Wave Number per cm	Intensity
O-H	3,200–3,650	Variable
N-H	3,300–3,500	Medium
C-O	1,050–1,300	Strong
C=O	1,690–1,760	Strong
NO_2	1,300–1,370	Strong

determined by electron spin, changing symmetry during the transition, and changes permitted in the resonance quantum numbers (Frank Condon principle).

$\sigma \rightarrow \sigma^*$ transitions are the most energetic electron transitions, and are mostly observed in vacuum UV in saturated hydrocarbons. The absorption wavelengths range between 125 and 190 nm.

$n \rightarrow \sigma^*$ electron transitions absorb electromagnetic radiation in wavelengths of up to 230 nm. The actual wavelength depends on the binding state in the molecule and electronegativity of the heteroatoms.

$n \rightarrow \pi^*$ and $\pi \rightarrow \pi^*$ transitions are the most important electron transitions in UV/VIS spectroscopy. These are characterized by longer wavelengths and the highest intensities (transition probabilities). Table 2.9 summarizes typical resonance patterns.

UV spectroscopy is currently mainly used as a detection principle in chromatography. UV/VIS spectroscopy is often conditional upon the introduction of chromophore groups into the compound to be analyzed.

2.4.2.3 Fluorescence Spectroscopy

Molecular fluorescence, phosphorescence, and chemoluminescence, collectively termed luminescence, constitute another method of visual spectroscopy.

Table 2.9 Maximum Wavelengths and Maximum Extinctions in Compounds in Functional Groups

Compound/ Group	λ_{max}, nm	ε_{max}, l/mol/cm	Transition
H_2O	167	1,480	$n \rightarrow \sigma^*$
CH_3Cl	173	200	$n \rightarrow \sigma^*$
CH_3OH	184	150	$n \rightarrow \sigma^*$
CH_3NH_2	215	600	$n \rightarrow \sigma^*$
$(CH_3)_3N$	227	900	$n \rightarrow \sigma^*$
C=O	280	20	$n \rightarrow \sigma^*$
C=S	500	10	$n \rightarrow \sigma^*$
C=N-	240	159	$n \rightarrow \sigma^*$
-N=N-	350	12.6	$n \rightarrow \sigma^*$
-N=O	660	20	$n \rightarrow \sigma^*$

Luminescence generally refers to the emission of light after excitation. Apart from electron excitation, molecules can be stimulated by chemical or mechanical energy (chemoluminescence, bioluminescence, radioluminescence, mechanoluminescence, thermoluminescence, electroluminescence). Many molecules show photoluminescence after exposure to UV or visible light.

After being stimulated to the first resonance state, $S1$, for periods in the picosecond to femtosecond range, the molecule relaxes via $S1$ into its ground state, $v0$, in deactivation without electromagnetic radiation emission, or crosses over to the triplet state via "intersystem crossing." The molecule may fluoresce from the $S1$ state or phosphoresce from the $T1$ state into its ground state, or deactivate without electromagnetic radiation emission [1, 22–24].

Fluorescence refers to the emission of light from the first singlet excitation state to the ground state of the molecule. This process is allowed by quantum mechanics, and is therefore extremely fast, taking place in nanoseconds.

Fluorescence spectroscopy is always used where slight differences in the electron structure in the molecules are to be detected, or where major changes take place around the electrons in the molecule (solvate shell, inclusion in membranes, and so on). Fluorescence spectroscopy is an extremely sensitive method where even slight changes in surrounding conditions can result in a significant change in spectra. Another advantage of fluorescence spectroscopy is that it is the only method that can visually separate substance mixtures by using varying excitation frequencies.

Phosphorescence refers to the transition between the molecule's triplet excitation state to its singlet ground state. This process is forbidden in quantum mechanics, and therefore occurs in time periods of microseconds to seconds.

In accordance with the Franck Condon principle, absorption and emission transitions always occur from the corresponding ground resonance state.

Table 2.10 shows some examples of fluorescence excitation and emission characteristics of selected compounds.

Chemoluminescence light is observed where the electrons in a chemical species are raised to a state of excitation during a reaction, then to release radiation after returning to their ground state. In biological systems, this kind of radiation emission is observed as bioluminescence.

Table 2.10   Examples of Fluorescence Excitation and Emission Characteristics

Molecule	Excitation max.	Fluorescence max.	Absorption max.	Comment
Adenine	377	425	260	Phosphate buffer at pH 7
Cytosine	355	382	267	Phosphate buffer at pH 7
Thymine	309	402	264	Phosphate buffer at pH 7
Guanine	365	420	275	Phosphate buffer at pH 8
Tryptophane	278	353	280	Phosphate buffer at pH 7
Tyrosine	274	303	274	Phosphate buffer at pH 7
Phenylalanine	257	282	257	Phosphate buffer at pH 7

2.4.3 Nuclear Magnetic Resonance Spectroscopy

2.4.3.1 Introduction

Nuclear magnetic resonance (NMR) spectroscopy is one of the most important instrumental analytical methods of today. This method is a fixed component in the repertoire of modern analytics due to its ability to deliver clear structural information without interference.

The effect is based on the interaction of electromagnetic radiation in the radio frequency range—typically 300 to 900 MHz—with atomic nuclei (nuclear spin) as occurs when strong external magnetic fields are present. The most important NMR-active nuclei are ^1H and ^{13}C since they appear in most molecules. Other important nuclei are ^{19}F, ^{31}P, ^{17}O, ^{15}N, and some metals. Due to the longevity of the excitation states, the spectra have high resolution. Electron streams in the molecular surroundings of the nuclei in question influence the local magnetic field, thus leading to spectral shifts. Apart from that, fine splitting occurs due to interactions with neighboring nuclear spin (spin coupling), which is characteristic for the arrangement of the nuclei in the molecule. A host of chemical and physical information can be gleaned from NMR spectra, such as the arrangement of functional groups in the molecule and their spatial orientation. Given the appropriate circumstances, the intensities of NMR signals are also proportional to the number of nuclei observed, thus enabling quantitative analysis at the same time [1, 24].

2.4.3.2 Physical Principles

The NMR effect is based on atomic nuclei with the nuclear spin (quantum spin number $I \neq 0$). Depending on the composition of protons and neutrons, an isotope will either be NMR-active or not. The two most important isotopes of hydrogen ^1H ($I = \frac{1}{2}$) and ^2H ($I = 1$) are NMR-active. Of all of the carbon isotopes, only ^{13}C has nuclear-magnetic characteristics ($I = \frac{1}{2}$). The ^{12}C carbon isotope cannot be analyzed using NMR spectroscopy.

Due to the moving charge, the spinning impulse L of the nucleus gives rise to a magnetic spin moment, μ. A powerful, constant magnetic field, B_0, provided by a cryomagnet leads to alignment of the magnetic spin moments towards and against B_0, with those in field direction having a slight energy advantage. The two states show a very weak energy difference, enabling spectroscopic excitation in the radio frequency range.

$$\Delta E = h \cdot v = 2\mu_z B_0 = \lambda \cdot h \cdot B_0 \qquad (2.26)$$

where h is Planck's constant and $\gamma = \mu/L$ – gyromagnetic ratio.

At $B_0 = 9.4T$, the resonance frequency is 400 MHz. Since the thermal energy acts strongly against complete alignment of the magnetic nuclear moments, the population difference between the ground and excitation states is only a few ppm, which sets a certain limit to the sensitivity of NMR spectroscopy.

2.4.4 Mass Spectrometry

2.4.4.1 General Principle of Mass Spectrometry

Mass spectrometry (MS) [1, 3, 22, 25–27] is a method that delivers both element-selective and structure-selective information. In contrast to the atomic and molecular spectroscopy methods above, no radiation effects are measured; instead, this method focuses on the molecule and fragment ions that result from reactions in ionization and separation processes due to the energy situation in the molecule concerned in order to identify the molecule structure.

The principle of mass spectrometry is based on the separation and measurement of molecule and fragment ions with respect to mass and intensity information. There is no direct determination of mass; rather, the mass-to-charge ratio, m/z. Ions can be influenced by electrical and magnetic fields, which enables mass separation for analytical purposes (see Figure 2.24).

Since ions are displaced charged particles, separation can take place in a variety of ways, such as the dependency of speed on mass, or in combination with the deflection of displaced charged particles. One possibility is deflection in a *magnetic sector field*, where the deflection radius of the ions is a result of the equalization between Lorentz and centrifugal forces. The physical principles applied yield the mass spectrometer equation for systems involving magnetic separation, as follows:

$$M \approx \frac{m}{z} = \frac{B^2 r_B^2}{2U_a} \tag{2.27}$$

where M is molar mass, m is mass, z is charge number, B is magnetic field strength, r_b is magnetic field radius, and U_a is acceleration tension.

An *electrical sector* provides another way of influencing ions; in the analyzer, the ions are scattered into a cylindrical electrical sector according to their kinetic energy-charge ratio, where ions with the same acceleration potential are not mass-separated in the conventional sense. The electrical sector is usually used in connection with magnetic mass separators as the energy-focusing element in a double-focusing arrangement (see Figure 2.25).

The theory of *quadrupole mass spectrometry* was developed by Paul and Steinwedel in 1953. This process reached a breakthrough with the possibility of manufacturing high-precision rods with hyperbolic cross-sections, which led to increased resolution capacity. In quadrupole MS, the ions are separated by deflecting the ions formed in a periodical electrical field. In a preset, adjustable voltage, only particles with a certain mass succeed in completing the longitudinal path between the rods and the detector. This method is therefore based on the mass-dependent resonance frequency of the moving ions with high-frequency alternating electrical fields for mass separation [22].

Quadrupole focusing has the following advantages over other processes:

- The equipment needed is relatively simple since only an electrical field with moderate voltage and frequencies of a few hundred kilohertz is needed.

- The ionic trajectories are largely independent of the initial conditions, meaning that higher intensities can be reached and ionic sources with higher velocity homogeneity can be used.
- Determining ions in a certain mass (at known charge) is simple.

Another process for separating ions is *time-of-flight mass spectrometry*. Ions are separated according to the time they take to pass through a predefined distance. Since the time taken increases with increasing mass, this process is suitable for separating and determining masses [28].

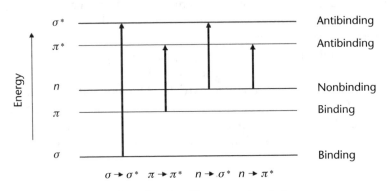

Figure 2.24   Electron transitions in molecules with n, σ, and π electrons.

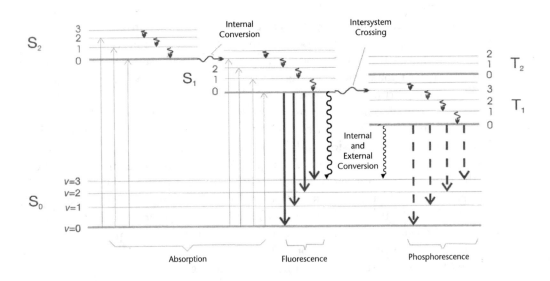

Figure 2.25   Jablonski term diagram.

2.4.4.2   Influence of Ionization Principles on the Selectivity of Mass Spectrometry

Selecting an ionization type is essential for the mass spectrum achieved, and therefore the information yielded serving selective identification of compounds.

Electron impact ionization (EI) is the method most commonly used. In this process, the vaporized compounds are ionized on impact with electrons. The high energy levels of the ions (standardized to 70 eV) not only ionize the molecule, but also induce various fragmentation processes. The characteristic key fragment ions in the mass spectra indicate that certain structural elements are present. There are often no molecular ions left in the spectra revealed by EI, and the molecular mass has to be determined by other means. This especially applies to complex mass spectra.

The ionization types most commonly used include chemical ionization, field absorption and ionization, fast atom bombardment, and so on.

In *chemical ionization* (CI), the vaporized compounds are ionized using a gas (methane, isobutene, and so forth); the energy applied to the molecule is far lower than in electron impact ionization, and fragmentation reactions are suppressed.

Compared to the ionization processes where molecules are ionized by bombardment with particles, *field ionization* (FI) and *field desorption* (FD) are "soft ionization" processes that only transfer minimum energy onto the compounds during the ionization process, and fragmentation hardly occurs. These methods are especially suitable for substances that have very unstable bonds in the molecule and are thermally unstable. The molecule is ionized by exposure to a strong electrical field. The potential wall is deformed to the extent that the electrons penetrate to the outside, and a positively charged ion remains.

The past years have seen various specialized processes developed for examining liquid-phase substances such as those with high solvent concentrations as in *high-performance liquid chromatography* (HPLC) and *Cation exchange* (CE). *Electrospray ionization* (ESI) enables the separation of substance molecules and solvents by drying and spraying with nitrogen [29]. The substance is then ionized using a capillary voltage. In more recent times, this method has mainly been used for analyzing peptides, proteins, hormones, and other compounds occurring in nature, but is also suitable for analyzing ionic metal compounds with high mass numbers.

2.4.4.3   Selectivity by Characteristic Isotope Patterns

Element-selective information in mass spectra is mainly obtained from the characteristic isotope patterns. Most elements occur in nature as various isotopes in varying amounts. The intensity distribution for n atoms in element A with isotopes x_a and y_a can be calculated using the following equation:

$$(ixa + iya)^n = i_{xa}^n + i_{xa}^{n-1} * i_{ya} + \frac{1}{2}n * (n-1)i_{xa}^{n-2} * i_{ya}^2 \cdots \qquad (2.28)$$

This method is not appropriate for monoisotopic elements such as fluorine, sodium, aluminum, phosphorous, manganese, cobalt, arsenic, rhodium, and so on.

2.4.4.4 Selectivity by Exact Mass Determination Using High-Resolution Mass Spectrometry

Results from high-resolution mass spectrometry (HRMS) are especially important, as these yield more information from mass determination compared to the lower-resolution methods, and therefore additional information for determining sum formulae for element combinations.

The distinction between these methods is based on the results that the processes used in mass determination return. While lower resolutions limit the results to the determination of integer masses in the individual lines of the spectrum, exact masses are measured in high-resolution mass spectroscopy. High-resolution mass spectrometers with a resolution above 10,000 are usually used for determining exact mass.

High-resolution mass spectra can be obtained in processes such as mass *peak-to-peak gap measurement* or *peak matching*, as shown in Figure 2.26.

Mass determination using peak-to-peak gap measurement is based on measuring the gaps between ions of known mass. The mass scale is determined according to a suitable internal reference. Interpolation between masses of several reference signals yield the mass of the sample, which may take place by linear interpretation or, as applicable, interpolation processes of a higher order.

In peak matching, the signals from the ion to be measured are adjusted to those of a comparison ion of a known mass (reference) by varying the acceleration tension to the same trajectory radius in the magnetic deflection field.

According to the mass spectroscopy equation, there is an inverse relationship between mass and acceleration tension in sample and reference signals for ions with a simple charge in the presence of constant magnetic induction, B.

$$m_x : m_{ref} = U_{ref} : U_x \qquad (2.29)$$

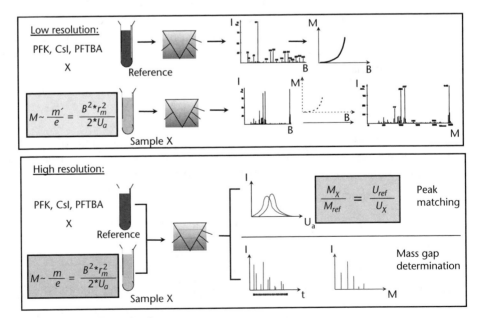

Figure 2.26 Mass determination process.

where m_x is probe signal mass, m_{ref} is reference signal mass, U_x is acceleration tension (sample), and U_{ref} is acceleration voltage (reference).

The unknown mass of the sample signal can be calculated using the familiar ratio $U_{ref} : U_x$ and the predefined mass of the reference signal R_{ref}. The peak matching process has a high level of accuracy with errors in the lower millimass range. The measuring process can be automated, but this is a complex process.

2.4.4.5 Selectivity by Determining Fragmentation Paths Using MS/MS Analysis

Another way of increasing mass spectrometry sensitivity is the use of MS/MS analysis. First, a conventional ionization process is initiated. Individually interesting ions are selectively broken into a neutral fragment and an ion with lower mass by unimolecular or collision-activated excitation in a field-free space. This process can be used to obtain information relevant to structure [30].

2.5 Future Challenges

The main challenges to analytical chemistry today are in the following areas:

- *Miniaturization:* With life sciences undergoing increasing automation, there is also a constant tendency towards smaller samples for analysis. The development of processes to deal with samples constantly decreasing in size while maintaining constant or increasing sensitivity must keep pace.
- *Decreasing analysis time:* Modern high-throughput screening series require that an increasing number of samples per time unit be processed. This may be reached by heavy parallelization (such as in visual reading systems) or the development of more rapid analytical processes based on the familiar classical principles of physics.
- *Online analytics:* Sample-taking, preparation, and transport represent a particular characteristic of substance measurement. Contrary to classical inline sensor equipment, *offline sensors* with substance transport to or through the sensor system concerned have generally dominated in substance-oriented measuring technology. This may give rise to delays and inactive periods in measuring analysis, which can affect the real-time aspect while following process data. For this reason, there is a major drive towards developing online sensors, inline sensors, or equivalent sensor systems. Systems of this type are already available for some applications [31].

References

[1] Otto, M., *Analytische Chemie*, Weinheim: VCH Verlagsgesellschaft, 1995.

[2] Böcker, J., *Chromatographie*, Würzburg: Vogel Buchverlag, 1997.

[3] Walsh, G., *Biopharmaceuticals: Biochemistry and Biotechnology*, New York: John Wiley & Sons, 1998.

[4] Patonay, G., *HPLC Detection Newer Methods*, New York: VCH Publishers, 1992.

[5]   Niessen, W. M. A., and R. D. Voyksner, *Current Practice of Liquid Chromatography-Mass Spectrometry*, Amsterdam: Elsevier Science, 1998.

[6]   Lee, M. S., *LC/MS Applications in Drug Development*, New York: John Wiley & Sons, 2002.

[7]   Mikkelsen, S. R., and E. Corton, *Bioanalytical Chemistry*, New York: John Wiley & Sons, 2004.

[8]   Khaledi, M. G., *High-Performance Capillary Electrophoresis: Theory, Techniques, and Applications*, New York: John Wiley & Sons, 1998.

[9]   Baker, D. R., *Capillary Electrophoresis: Techniques in Analytical Chemistry*, New York: John Wiley & Sons, 1995.

[10] Pingoud, P., et al., *Biochemical Methods: A Concise Guide for Students and Researchers*, Weinheim: Wiley-VCH Verlag GmbH & Co KGaA, 2002.

[11] Voet, D., J. G. Voet, and C. W. Pratt, *Fundamentals of Biochemistry*, New York: John Wiley & Sons, 2002.

[12] Chapman, J. R., *Protein and Peptide Analysis by Mass Spectrometry*, Totowa, NJ: Humana Press, 1996.

[13] Bommarius, A. S., and B. R. Riebel, *Biocatalysis: Fundamentals and Applications*, WILEY-VCH Verlag GmbH & Co KGaA, Weinheim, 2006.

[14] Bustin, S. A., *A–Z of Quantitative PCR*, IUL Biotechnology, No. 5, IUL Biotechnology Series, La Jolla, CA: International University Line, 2004.

[15] Mullis, K. B., and F. A. Faloona, "Specific Synthesis of DNA In Vitro Via Polymerase-Catalyzed Chain Reaction," *Methods Enzymol*, Vol. 155, 1987, pp. 335–350.

[16] Higuchi, R., et al., "Simultaneous Amplification and Detection of Specific DNA Sequences," *Bio/Technology*, Vol. 10, 1992, pp. 413–417.

[17] Wilfingseder, D., and H. Stoiber, "Quantifizierung von PCR-Produktmengen Durch Real-Time PCR-Verfahren," *Antibiotika Monitor*, Heft 12, 2002.

[18] Reymond, J.-L., *Enzyme Assays*, Weinheim: Wiley-VCH Verlag GmbH & Co KGaA, 2006.

[19] Beveridge, M., et al., "Detection of p56(Ick) Kinase Activity Using Scintillation Proximity Assay in 384-Well Format and Imaging Proximity Assay in 384- and 1536-Well Format," *J. Biomol. Screen*, Vol. 5, 2000, pp. 205–212.

[20] Bladon, C. M., *Pharmaceutical Chemistry: Therapeutic Aspects of Biomolecules*, New York: John Wiley & Sons, 2002.

[21] Crowther, J. R., *ELISA: Theory and Practice*, Methods in Molecular Biology, Vol. 42, Totowa, NJ: Humana Press, 1996.

[22] Böcker, J., *Spektroskopie*, Würzburg: Vogel Buchverlag, 1997.

[23] Ingle, J. D., and Jr. S. R. Crouch, *Spectrochemical Analysis*, Upper Saddle River, NJ: Prentice-Hall, 1988.

[24] McLennan, F., and B. R. Kowalski, (eds.), *Process Analytical Chemistry*, New York: Blackie Academic & Professional, 1995.

[25] Hoffmann, E. D., and V. Stroobant, *Mass Spectrometry: Principles and Applications*, New York: John Wiley & Sons, 2002.

[26] Herbert, C., *Mass Spectrometry Basics*, Boca Raton, FL: CRC Press, 2003.

[27] McLafferty, F. W., and F. Turecek, *Interpretation of Mass Spectra*, 4th ed., Mill Valley, CA: University Science Books, 1993.

[28] Cotter, R. J., *Time-of-Flight Mass Spectrometry: Instrumentation and Application in Biological Research*, Washington, D.C.: American Chemical Society, 1997.

[29] Cole, R. B., *Electrospray Ionization Mass Spectrometry*, New York: John Wiley & Sons, 1997.

[30] Busch, K. L., G. L. Glish, and S. A. McLuckley, *Mass Spectrometry/Mass Spectrometry: Techniques and Applications of Tandem Mass Spectrometry*, New York: VCH Publishers, 1988.

[31] Profos, P., and T. Pfeifer, *Handbuch der industriellen Messtechnik*, München Wien: R. Oldenbourg Verlag, 1992.

CHAPTER 3

Basic Health Care Delivery for Engineers

Lanis L. Hicks and Gordon D. Brown[†]

3.1 Introduction

The modern health care system is an integration of complex technologies and human touch in the delivery of services. As a result of services provided by various permutations of technology and personal interactions, the health care system emerges as one of the most complex systems in existence. Clearly, the functioning of this complex system is difficult to understand, model, and analyze. However, a thorough understanding of the structural complexity and operational performance of the health care system is paramount to strategically improving the system's outcomes.

Often touted as the "best" in the world or as the "worst" in the industrialized world [1–3], the U.S. health care system undoubtedly lies somewhere between these two extremes. While there are many things about the U.S. health care system that are outstanding (e.g., highly trained professionals, well-equipped institutions, innovations, and many choices), there are other aspects that need improvement (e.g., high costs, uninsured populations, access, safety, inequities, disparities, and quality). Improvements in the current system are necessary to provide American citizens with the highest possible level of health. The United States can no longer afford (health wise, socially, or economically) to maintain the status quo of its current health care system [4]; improvements in the efficiency and effectiveness of the health care system must be made. Thus, to reengineer the U.S. health care system, an understanding of the current health care system, and how it functions, is needed.

This chapter discusses the U.S. health care system and its major subsystems and functions. How decisions are made in the health care system and the consequences of those decisions are presented. The types and roles of health care professionals, institutions, and technologies in the production function are examined, as are the changes occurring in the structure of clinical decision making and value migration in the health system. In addition, explanations of how different support systems operate and their implications for the performance of the system are provided, as are the role of evaluation in health services and the consequences of the increased evaluation requirements on the development and diffusion of technology in health care.

†   Department of Health Management and Informatics, School of Medicine, University of Missouri

3.2 The Health Care System: A Holistic Perspective

The health delivery system can be viewed from a system theoretical perspective as a set of highly interactive, dynamic functions represented as a patterned set of relationships. The health system is relatively simple in concept, but complex in structure and function. Functional details determine the appropriateness of automation technology and its potential for being adopted and utilized by the system. The U.S. health system can be considered, at the macro level, as an amalgamation of health systems with very different structures, making meaningful generalizations about the system difficult.

The U.S. health care system is a mixture of private and public delivery entities, where government-run organizations provide services directly to certain segments of the population and government financing provides access to services for other segments of the population. The government's involvement in providing care has a substantial impact, not only on the availability and accessibility of care, but also on the overall performance of the system, and on the adoption and diffusion of technology. Four major governmental providers of health care are the Department of Veteran Affairs, the military, the Indian Health Service, and the Public Health Service. Each makes considerable contributions to the provision of services in the health care system, and each impacts significantly on the adoption of innovative practices in health care (see Figure 3.1).

Some health services delivery systems are more centralized in structure and authority and, therefore, more mechanistic in nature. Examples of such systems are the Veterans Administration and the armed services. The Indian Health System has, historically, been centralized, but is undergoing a decentralization strategy with greater local adaptation and concomitant variation. The private medical care system consists of very autonomous and decentralized corporations, staffed by physicians with considerable professional autonomy. This system is sometimes characterized as

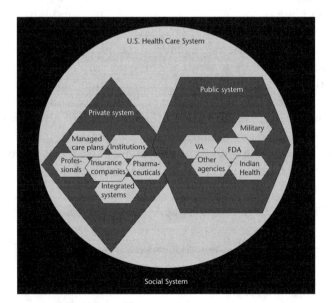

Figure 3.1 Components of the U.S. health care system.

a nonsystem, because it is not centralized with a single formal structure and common set of standards, rules, and policies. While it lacks the mechanistic nature characteristic of bureaucratic structures of nationalized health systems, it still has all of the characteristics and behaviors of a system.

Although the U.S. health system interacts with the larger social system in a purposeful manner, its fundamental purpose, defined by its relationship with society, is not always easily understood or viewed with common agreement. Defining and placing a value on outcomes (clinical, social, emotional, and economic) is one of the greatest challenges of health professionals, executives, and policy makers. If the health of a population is a paramount goal of an egalitarian society, issues of equity and access for all members have great importance. In such a society, the health and welfare of the entire population would be considered as the measure of the status of society itself. It would be the genotypic function of society [5]. In contrast, some systems may value health services according to their contribution to maintaining a healthy work force. Health would thus be considered an input into production and productivity, and gross national product would be the value measure to society. Improving the health of the current or future work force may become a social priority. Economists, politicians, and ethicists debate what value *should* be assigned to health care; but ultimately, each society assigns its own attitudes, beliefs, and values to a health care system. Knowing the value a society assigns to health care is critical in determining how the system will be structured and how it will operate.

At the micro systems level, the health system can be viewed as a process of individual patients interacting with the health system receiving services. The system works best for acute illnesses where the patient encounters the system, receives a service, and leaves the system. If a patient has a chronic disease that has to be managed over time, and typically, may involve many health professionals, the system does not work as well. The breakdown of the system is in its attempt to coordinate services and integrate clinical information across the entire clinical process. The relationship between patient and clinician, as they interact in the clinical process, needs to be well understood by those automating this process.

Historically, clinical decision making has been structured around individual physicians and other health professionals interacting with individual patients. This *illness model* clearly defines the roles of both patient and clinician. There have been significant levels of patient autonomy in seeking professional consultation when and where the patient perceived it was needed, and with the freedom to choose the source of care and individual professional desired. When a patient assumes the *sick role*, he or she is relieved from certain normal social duties, such as going to work or caring for the family. In addition, under this privileged position of the sick role, there are specific responsibilities of the patient, namely, that he/she will seek medical care and follow the orders of a qualified medical practitioner. If competent medical treatment was not sought and followed, negative sanctions against the sick role might be imposed, and the individual might lose privileges. For example, the individual might be perceived as a lingerer or hypochondriac, and expected to resume his/her normal role as a worker, parent, friend, and so on.

The sick role is a complex social dynamic within the community, and its definition depends on the culture of the family and a community [6]. This complex social relationship transcends the scientific definition of the disease and its treatment.

Physicians, and other health professionals, must be perceived as part of the complex social culture of illness and treatment. Traditionally, the physician-patient relationship has been defined by a complex process of transference, where the patient who sought treatment submitted passively to an otherwise forbidden invasion of their personal privacy, giving physical access to the clinician, as well as answering questions of a very personal nature. The physician-patient relationship became a highly respected and individualized social relationship [7]. There was, and still is, considerable opposition to altering this historical relationship, with interventions that violate the autonomy of the patient and physician in carrying out the diagnostic and therapeutic process. Attempts to automate elements of the clinical process, or to introduce provider-integrated information, pose dramatic changes to this physician-patient dynamic. These changes are occurring but involve more than an understanding of the technology on which they are based. They require an understanding of the culture and values in both the micro clinical systems, as well as the social and cultural context in which services are provided. These micro and macro processes are discussed within the context of different delivery systems within the U.S health system.

3.2.1 The Department of Veterans Affairs

The medical care system of the Department of Veterans Affairs (VA) is the largest integrated health care delivery system in the United States. In addition to the direct provision of care, the VA has responsibility for training future health care professionals and conducting a substantial amount of medical research. The VA system has undergone major restructuring of its management system in recent years. The current management system focuses on data-driven processes to ensure the "delivery of the right care in the right place at the right time." The VA has adopted a performance measurement system to ensure actions are aligned with the goals of the organization [8]. With the implementation of this restructuring, the VA has become a national leader in the provision of a coordinated continuum of care.

The VA system's data management strategy is key to system development and performance. This coordinated data management system provides a single access point for information, facilitating intrasystem performance. Other factors facilitating this transformation are the establishment of telecommunications structures, which enhance the ability of the total organization to integrate technology, and the implementation of innovative management techniques. All these procedural and technological transformations serve to improve the quality of care delivered to veterans. As a result of these changes, the current VA system is a national leader in the use of integrated health information systems and provides significant influence throughout the health care system through its contributions to teaching, research, and service innovations.

3.2.2 The Military Health Services System

The Military Health Services System is a large health care system that provides medical services to active and retired members of the Armed Services, their dependents, and survivors. It is a highly structured system, with a strong focus on integration of services supported by an integrated health information network. In 2005, the U.S. military unveiled its Internet-based electronic medical record (EMR), which globally transfers information. It is designed to be able to interface with the VA system in the future. Part of the system is an interactive field electronic data-collection device by which the medical record of the recipient can be updated at any location that the military health services system is accessed. The interactive nature of this information system is intended to reduce errors, improve quality of care, and lower costs [9]. Collectively, these new technologies provide the infrastructure for a seamless delivery of medical care on a global scale.

The Department of Defense (DoD), through the Military Medical Services System, is a leader in the development and adoption of telemedicine services for use in remote areas where access to medical services is limited. The use of telemedicine technologies enables the provision of patient-centered, cost-effective, and real-time care to members of the Armed Forces, their dependents, and retirees. Such technologies allow specialty services to be provided to facilitate the improved management of complex medical cases, provide routine consultations, and to reduce the need for airlift medical evacuations, reducing lost duty time for active personnel.

Another key innovation in military medicine is the use of medical logistics to improve the efficiency and effectiveness of the medical supply chain. The goal of this process is to use the best business practices for information and distribution management and to develop a standard automated information system. The military system is well positioned to play a key role in the development of such a system, given its expertise and synergy in medical logistics, information technology, and research capabilities. The need to be able to provide urgent and complex medical care under demanding operational environments creates pressures for the system to become a leader in the development of technologies and systems to improve the care delivered [10]. Military nuclear, biological, and chemical defense research plays a critical role in developing new treatments for medical problems, which can then be transferred to the civilian health care system.

3.2.3   Indian Health Service

The Indian Health Service (IHS) was created to assess and improve the health status of American Indians and Alaskan Natives (AI/AN). The agency attempts to ensure that "comprehensive, culturally acceptable personal and public health services are available and accessible to AI/NA people" [11]. Under recent self-determination provisions, tribal decisions have resulted in a more decentralized national health care system. Services delivered have also changed, with telemedicine connections being used to bring services to remote areas, and modern technologies and information systems implemented to support the system. Tribal leaders are taking sophisticated and active roles in advocating and planning for their own communities, although the system still faces serious shortages of professionals and financial resources.

As tribal governments take on more responsibilities for the health of their population, the partnership with the IHS continues to be an important element of that development. With many of the services being provided to AI/AN populations in some of the most remote and poorest areas of the country, the use of technology and automation plays an important role in the improvement of the health of communities, and not just the medical needs of individuals within those communities.

3.2.4 Public Health Service

The U.S. Public Health Service (PHS) is the largest public health program in the world. While the Public Health Service has currently moved away from the direct provision of personal health care, except for very limited amounts, it has major influence and control over the direction of the health care system and the health of the population [12]. The various operating divisions of the PHS provide a wide range of grants and contracts, conduct extensive amounts of internal and collaborative research, and provide regulatory oversight of all public and private health providers. Thus, the PHS continues to have major influence over the development, introduction, and diffusion of new technologies and automation within the health care system.

There are eight agencies within the public health service: (1) Agency for Health Care Research and Quality (AHRQ), (2) Agency for Toxic Substances and Diseases Registry (ATSDR), (3) Centers for Disease Control and Prevention (CDC), (4) the Food and Drug Administration (FDA), (5) the Health Resources and Services Administration (HRSA), (6) The Indian Health Services (HIS), (7) the National Institutes of Health (NIH), which encompasses 27 institutes and centers, and (8) the Substance Abuse and Mental Health Services Administration (SAMHSA). In addition, the Office of Public Health Preparedness (OPHP) and the Office of Public Health and Sciences (OPHS) are located within the Office of the Secretary of the Department of Health and Human Services. The Office of Surgeon General is located within OPHS. Familiarity with the activities on each of the government-run organizations will be invaluable in understanding the future direction of the U.S. health care system.

3.3 Health Care Subsystems

The health system can be viewed as having a number of subsystems, including production, support, guidance, and adaptation. The production system, or health services delivery system, includes health professionals and such organizations as clinics, hospitals, nursing homes, and increasingly, home care. The structure of the systems describes how clinical and health services are organized, owned, and relate to one another, and how health services are delivered. The support function includes financing, professionals, education, and institutions (see Figure 3.2).

Automation technology (such as robotics and diagnostic tools) will be considered in this chapter as a production technology and viewed as an existing resource that can be applied to support health services delivery. The guidance function includes public policies and such activities as accreditation, licensure, and regulations. The guidance function largely determines how rapidly and in what areas automation technology is diffused. Diffusion of health technology is as much a political issue as it is a technical one. The adaptation function includes long-term planning, as well as research and development. This discussion focuses primarily on the

Figure 3.2   Structure of the U.S. health care system.

production and support functions of the system, with an integration of the guidance and adaptation functions and their influence on the other system components.

Clearly, the most promising areas for the application of automation technology are in biological and biomedical research. Biomedical research provides rich opportunities for the application of automation technology, as it involves very large databases, standardized processes, and complex computational requirements. The understanding gained from the development of automation applications in biomedical research includes testing this technology for direct application to health services delivery. The interdisciplinary development of new science links basic scientists with computational scientists to carry out life sciences research. The application of basic science to clinical areas through *translational research* links basic science, clinical science, and computational science, providing an interdisciplinary foundation for the application and integration of positive research findings in health care. A further extension of this research program is in health services, exploring how the synthesis of clinical and computational technology will serve as a catalyst for health systems change. This area of study can appropriately be termed *transformational research*, which integrates engineering, information science, behavioral science, and organizational theorists [13]. Automation research should not be limited to addressing computational issues in biomedical research, but also should explore the potential of automation across a broad range of disciplines and applications. This chapter focuses on the applications of automation in health services delivery.

The interrelationships of the subsystems of the health services delivery system determine the structure and function of the system. The health system in the United States is complex, in that a new technology or policy is assessed by its impact on quality, cost, access, patient satisfaction, and by its effect on traditional professional roles and organizational structures.

The physiology of the health system is described by the function of the various subsystems and the relationships between and among them. Subsystems can be delineated and analyzed from a classical systems model perspective. A small change in one might cause a major change in the function of the overall system, while a major change in one might cause only a small change in the overall system (homeostatic interrelatedness). For example, one subsystem might either facilitate or block acceptance of a new technology, such as automation, even if the technology is technically sound and supported by good scientific evidence. Technology acceptance may be blocked, even if the new technology has demonstrated its effectiveness in improving quality of health services or to be cost-effective because of its impact on the status quo of a particular subsystem.

Innovations introduced into the health system might be rejected for a number of reasons, including financial, political, strategic, legal, or cultural. Regulatory approval, for example, of many applications have to be received from the FDA, the federal agency with responsibility for approving all pharmaceuticals and other medical devices before they can be released to physicians and other health professionals for use. Approval requires a long period of clinical trials to demonstrate the efficacy and safety of all medical products. Automated processes, such as surgical robots, have to be approved by the FDA, based on extensive clinical trials. Final approval might include restricted use of the device to those applications that have been tested or that have demonstrated safe use for a defined range of applications.

To change a culture, there must be shifts in beliefs, traditions, values, perceptions, and feelings held by health professionals, patients, and families. Changing the structure of clinical decision making and clinical processes for health professionals might seem easy in the face of good scientific evidence, but it is complex and encounters considerable change resistance. In general, individuals do not like change. As a result, new systems need to be designed to integrate into existing cultural beliefs, norms, values, and behaviors rather than attempting to change the underlying structure. The anatomy and physiology of the health system are explored as a means of understanding the potential effect of automation, as well as sources of resistance.

3.4 Evidence-Based Decision Making in Health Care Delivery

Decision making in health care delivery can be categorized according to how decisions are structured, including their interdependency. The application of automation technologies in the life sciences is based on assumptions about the structure of the decisions they are intended to automate. Decisions can be considered as mechanistic in nature if they follow a given order, are predictable, lend themselves to precise measurement, and are deterministically interdependent [14]. These decisions structures characterize much of biomedical research and many aspects of clinical science. They also characterize many of the business decisions in the health system, such as payroll, billing, and supplies management. Traditionally, business functions have tended to be more mechanistic, or deterministic, than clinical functions, and have become more automated. In general, the health system, however, has lagged behind other sectors in adopting automation due to the nature of clinical work and the individualized nature of health services for patients. To some degree, this lag may be the result of cost-based reimbursement funding models that enabled the industry to perpetuate current practices without transformational change. In this chapter, the concentration is on the clinical side of the health services delivery enterprise and not the business side, although business applications are rich with potential.

The clinical function makes up the core of the production function, and is generally much more complex and less mechanistic in nature. Within the clinical function, the structure of the decision processes is more variable and customized. Mechanistic decision processes lend themselves much easier to automation. This does not imply that such applications are simple, but rather recognizes the range of complexity in clinical decision making. The application of automation technologies will, accordingly, vary depending on the structure of the decision processes. In many clinical areas, measurement of outcomes and processes is less precise and not entirely predictable, requiring creativity within the system to explore innovative decision-making behaviors. For example, surgical procedures tend to be more standard and repetitive than the treatment of chronic diseases, such as type II diabetes. In surgical services, procedures are more precise, variables more known and controlled, and tasks more repetitive. Not all variation in surgical procedures carried out by a surgeon is good, however. Some variations might be the result of errors or lack of precision and may cause harm to the patient. The application of robotics to surgery has demonstrated the precision and portability of this technology in clinical care. It has demonstrated evidence, in some settings, of greater three-dimensional spatial accuracy, greater reliability, and more precision, consequently resulting in better outcomes [15].

3.4.1   The Structure of Clinical Decision Making

Decision making has been recognized as a step in a complex problem-solving process, not as a single point of making a choice. There have been many representations of this process, but most include some aspect of problem identification, analysis, selection, implementation, and feedback. Feedback is critical for individual or organization evaluation, including organizational learning [16] (see Figure 3.3).

The development and application of automation technology should appropriately consider all stages of problem solving, as this is the context within which both clinicians and health organizations view clinical decision making. Improved problem solving can result from improvement in any or in all steps in the problem solving process. It is important to consider how decision structures might differ at each stage of the problem solving process, and how automation might assist within a single stage or for the overall process.

Different levels of automation of decision making have been identified in the literature: information acquisition, information analysis, decision selection, and action implementation [17]. Automation can be applied across the range of these decision processes, but must be specific to the type of decision that is being made. The use of automation in clinical decision making will be more easily accepted at the information acquisition and analysis end of the spectrum, but may also include

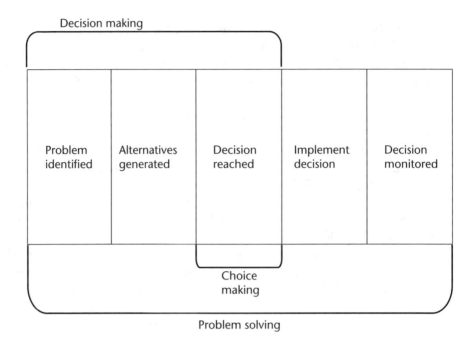

Figure 3.3   Scopes of choice making, decision making, and problem solving.

decision recommendations. The application of automation technology to clinical decisions at the point of action implementation is the point where the automated device is directly in control of the clinical service being provided. Automating decisions at the action implementation stage will be much more difficult, technically and politically, although that is the threshold currently being challenged by automation, and probably where considerable potential lies.

Some automation implementation, such as robotics, is at the action implementation stage. Thus, it will need to pass much more stringent approval processes, such as by the FDA. If the automation devices are made passive, such that they are always under the direct control of the surgeon or other provider, there will be fewer invasions into the clinical decision process, and, therefore, less resisted by providers and less external review by public regulatory agencies. The importance of the regulatory and political subsystems in the application and use of automation thus depends on how it is used in the clinical process. Although approval will be more complex, many promising applications of this new technology are self-directing, where they are monitored but not controlled by the provider. The development of such devices opens increased possibilities of researchers claiming intellectual property rights from such developments and engaging in increased innovation entrepreneurship.

3.4.2   Automation Applied to Types of Clinical Decision Making

The application of automation technology to clinical decision making requires an understanding of the outcomes of clinical decisions. Automation has the potential of improving the memory capacity brought to decision making and of increasing the total object field that is being perceived (see Figure 3.4). Decision making quality can be enhanced, if information in the object field can be expanded and systematically presented in the decision making process. Such information might include systematically presenting all relevant information about a patient's genetic code, current and past physical, social, and mental conditions, as well as current symptom otology. Expanding the object field increases the amount of relevant information that is perceived by the decision maker. Decision making can also be enhanced by drawing on the total memory content of what is known about this disease, based on the totality of collected scientific evidence. This expands the evoked set, assuring all scientific evidence is brought to the point of clinical decision making [18]. Automation has the potential of bringing vast amounts of information about the clinical condition of a patient and matching it with the accumulated knowledge on that condition from the best scientific evidence. If the knowledge from either of these processes can be increased and/or processed more rapidly at the point of decision making, clinical decisions and, potentially, health outcomes can be improved.

Where clinical decisions can be standardized, drawing on clear scientific evidence applied to well-defined and accepted clinical processes, automation through advanced information technology has considerable potential. More complex clinical tasks present greater potential benefit from automation [19]. Automation is not a matter of simply applying technology to all clinical decisions, but one of understanding the nature of the clinical task and clinical decision making process, and applying technology selectively in areas in which automation can assist the process. Clinical procedures, in general, have considerable variability because of variability in disease processes themselves, or due to individual patient conditions and

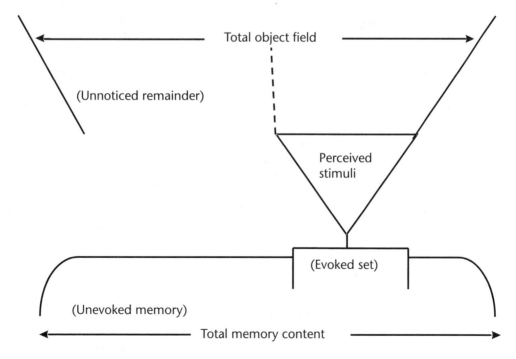

Figure 3.4 Object field of decision making. (*From:* [18]. ©1970 McGraw-Hill, Inc. Reprinted with permision.)

preferences. Variability requires increased judgments by clinicians and other health professionals that might be assisted by, but not replaced by, automated processes.

These qualities are consistent with the assumptions of complex adaptive systems. Some decisions might be well structured and lend themselves to automation, while other processes are less structured and continue to depend on human involvement. The latter is probably most characteristic of clinical decision making, particularly decisions made in specialties other than surgery or other areas of procedural medicine, where conditions are more unstable, emergent, or less defined at presentation. It is worth noting that reimbursement approaches in the health field classify clinical services as either procedural or cognitive. Automation has broad potential application to both procedural and cognitive medicine, but will be applied and utilized differently.

Most clinical areas in medicine can be classified as cognitive medicine. The use of automation technology in cognitive medicine inherently assumes the involvement of clinicians in the decision process. This changes the context of decision making and how automated technologies are developed and deployed. It changes the unit of analysis from the automated technology per se, to one of interactions between people and machines. This interaction design extends the model of decision making from decisions and tasks, discussed above, to include the cognitive attributes of humans. This approach to decision making focuses on the potential of automation in reducing cognitive effort by the clinician.

The use of automation to assist clinicians and other health professionals in decision making consists first of understanding the nature of the clinical process and the structure of tasks that make up the process. This approach to automation focuses on improved decision making by task complexity reduction. Sintchenko and Coiera [19] have identified five steps in this process: (1) selection of the domain and relevant tasks; (2) evaluation of the knowledge complexity for tasks selected; (3) identification of cognitively demanding tasks; (4) assessment of unaided and aided effort requirements for this task accomplishment; and (5) selection of computation tools to achieve this complexity reduction. Where decision aids can provide reduction in cognitive effort, the speed and accuracy of decisions can be improved, and a basis for learning and embedding knowledge in the organization is increased. This knowledge has value for the organization, with an increase in learning, and in transcending individual clinicians working within the organization. The design of interaction decision support systems is more applicable to the majority of clinical decision making. It is also less invasive in the clinical decision process and will be professionally and politically more acceptable. Clinicians will have less resistance to change if speed and accuracy can be improved and clinician time and effort are not increased.

3.4.3 Automation Applied to the Clinical Decision Process

Clinical decisions are increasingly being evaluated as part of the clinical process, and not simply at the discrete point of each individual clinical encounter. This chapter discusses the decision process in two contexts. First, as discussed above, it is a systematic sequence of steps individual clinicians use to identify clinical problems, determine courses of treatment, monitor outcomes, and confirm the appropriateness of the diagnosis and course of treatment (learning). In this section, the decision process is viewed as the sequencing of individual clinical decisions that transcend the entire clinical process. Increased attention is being given to clinical processes that transcend a single clinician, a single location, and/or a single point in time. Historically, activities within the clinical tasks were coordinated by hand-offs from one clinician or health professional to another. The means of coordinating tasks across the clinical work process was through the standardization of skills of each profession [20]. Each profession knew the specialized competencies and practice privileges of other health professionals, and work was coordinated by bringing the appropriate individuals into a clinical case through consultations or referrals.

The structure of clinical processes has considerable potential for automation as a means of structuring clinical tasks, as they transcend individual health professionals, departments, and institutions. The coordination of tasks across the clinical process assumes there can be some degree of standardization of the clinical process itself. The application of this technology will require a profound transformation of the health system itself, as health organizations have, traditionally, protected the decision making autonomy of individual clinicians.

Standardized clinical processes, referred to as clinical protocols or clinical pathways, are considered to consist of the standardization of the clinical process as a means of coordinating work. These concepts are similar to the critical pathway used in industrial engineering, although the term "critical" is generally not the preferred terminology in medicine, due to possible misinterpretation by patients and families regarding the seriousness of the condition or level of care required.

The standardization of clinical processes supports the application and adoption of advanced information technology, which supports data mining to analyze patterns of clinical care. Information technology will facilitate the development of standardized protocols, and bring evidence-based solutions to the point of clinical decision making. One example of this is the use of automated patient education after the individual is diagnosed with a chronic disease. The increase in the rate of chronic diseases within society also enhances this development, because the clinical process is much more complex. Complexity results from the cyclical clinical processes of chronic illnesses, and it transcends individual clinicians and institutions.

The management of clinical decisions that transcend individual clinicians, departments, institutions, and systems inherently requires direct involvement of the organization in the clinical process. Clinical decisions were, historically, made totally by clinicians, relatively free from organizational interference. The organization carefully protected the decision making autonomy of individual clinicians. The new era of health care, however, is a collaborative transformational information exchange, which has been driven by the implementation and utilization of technology.

Clinical protocols involve the coordination of work of multiple health professionals and also the direct involvement of the organization. In his classic work *The Social Transformation of American Medicine*, Paul Starr concluded with a chapter on "The Coming of the Corporation." In that chapter, he characterized the coming of the corporation as change in the type of ownership, horizontal integration, diversification, vertical integration, and industry concentration [21]. All of these predictions have proven true. However, the corporation has gone beyond the size and complexity of health organizations to become directly involved in restructuring the clinical process. The restructuring of the clinical process presents both an opportunity and challenge to automation.

With health organizations now directly involved in clinical decision making, it is appropriate to have a thorough understanding of how work processes are structured in organizations as well as of human dynamics in various work settings. Organizational structure reflects how tasks, jobs, and the work process are configured, how they are changed, and how workers perform in an increasingly knowledge-based system. These qualities of the organization thus become part of the clinical process, and automation technologies need to address these organizational issues, as well as the technical biomedical and clinical issues. The traditional business processes of personnel, finance, logistics, legal, and management that functioned separately from the clinical process now become an integral part of it. As discussed earlier, health organizations have progressed further in automating business processes than clinical processes. While this is true, business processes in health care have been developed around their own logic, including their unique information systems. These systems must now be redesigned around the logic of the clinical function [22].

The development and implementation of clinical protocols have been slow to evolve, in part because they have been considered a technical or clinical issue outside the context of the organization. Research has demonstrated that the organization is the appropriate unit of analysis for this research, and that automation technology must include organizational knowledge as well as clinical knowledge. Quaglini et al. [23] have developed a useful analysis, in which they recognized three essential components for guideline-based care flow systems, including knowledge representation, model simulation, and implementation within the organization. In it, the automation model structures the clinical tasks to be performed and the organizational requirements to perform them, including resource allocation. These systems are not static, but dynamic and knowledge based. The application of automation technologies to structure and to support complex clinical processes offer an immense opportunity for improving the quality of clinical outcomes and the efficiency of health care delivery.

3.5 Care Providers

The U.S. health care system consists of a complex mixture of health care professionals and health care organizations. Direct-care personnel in the health care industry range from the most highly trained professionals to on-the-job trained technicians. Health care organizations range from large, sophisticated integrated health delivery systems employing thousands of individuals to private offices with only two or three people employed. While the health care industry is, currently, very labor-intensive, technology may make it less labor-intensive in the future.

3.5.1 Health Professionals

The U.S. health system consists of a production system that has, at its core, the health professions, such as allopathic physicians (MDs) and doctors of osteopathy (DOs), nurses, and a wide range of allied health professions, such as pharmacists, physical therapists, occupational therapists, medical technologists, and social workers. According to the U.S. Department of Labor, Bureau of Labor Statistics [24], there are 17 health diagnosis and treating professions and another 15 health technologist and technician professions, each with its own recognized body of knowledge, selection and training requirements, and set of recognized skills and area of practice. Most of these professions are now trained in university settings, although the training requirements and approaches are heavily controlled by the professions.

It is not possible to understand the function of the health system without understanding the role of the health professions in society. The social mandate of the health professions is to best serve the public [25]. The tendency of the professions is also to protect and advance the profession, including increasing the size and stature of the profession by such means as higher educational requirements and regulatory control over entry into the profession. Historically, many of the professions started with on-the-job training requirements, culminating with a written examination and licensure. They have migrated to 2-year college programs, baccalaureate degree programs, master's programs, and now many are considering doctoral programs as the minimum for research, teaching, and practice in the profession. The case can be made that increasing educational requirements reflect the public mandate for quality and service, but one can also make the case that it increases cost and concentration of resources in major medical centers.

The level of training and competence of health professionals is reinforced through accreditation of educational programs by the professions, licensure boards of the state, and certification by professional societies. Because health care services are so technical in nature and there are potentially harmful consequences of incompetent providers and products, a substantial amount of emphasis is placed on protecting the consumer. Licensure, accreditation, certification, and continuing education requirements provide to the public assurance of a level of competency and quality, but they also perpetuate established boundaries of professional practice and traditional practices that are sometimes hard to change.

Replacing a highly professionalized labor force with automation, for example, will likely meet with resistance that is far beyond that which would be encountered in manufacturing or in other service industries. This resistance may block the introduction and widespread use of automation, no matter what evidence existed on cost-effectiveness and quality improvement. Resistance by the professions could be considered to be a hindrance to the diffusion of automation, but the responsibility of health professions in society is to protect individuals from harm and unproven technology. It is the professionals who are held accountable by society and not the machine.

3.5.2 Organizational Providers

Clinical care has been provided through a range of institutional partners, such as hospitals for acute, in-patient care; rehabilitation facilities for chronically impaired patients; clinics for ambulatory patients; nursing homes for chronic care patients; and hospice care for terminally ill patients. Health institutions have been primarily focused on the business function of organizations, including financial, human resources, legal, managerial, and clinical support services. Historically, health organizations have been independent in their structure and strategy. Hospitals have their roots in the private, nonprofit sector, many with a strong religious heritage. The social mandate of these nonprofit institutions is to serve the community, including the medically indigent and other underserved populations. The operation of these institutions is under increasing stress, due to such factors as the continuing rapid expansion of basic sciences knowledge (such as the human genome project), increasing medical technology (such as automation), the increasing demand for services due to the aging of the population, and changes in patient preferences and desire for greater involvement in their care [26].

Ambulatory care settings, including outpatient clinics, physician, dentist, and other health professional offices, laboratories, and other service organizations, are typically small, for-profit entities. These facilities range from a private practitioner's office that employs only one additional assistant, to large group practices employing hundreds of diverse workers. The diversity of organizations in this sector makes generalizations difficult, and it also means that there are large variations in the sophistication of technologies and information systems employed in this group, a group that while accounting for over three-fourths of all health care establishments, employs only about one-third of the health care workforce [24]. However, many of these organizations may become prime targets for medical technologies that impact the way in which services are delivered to their clients and patients.

Clinics are primarily for-profit institutions, ranging from a large number of small, two or three-physician clinics, to large complex clinics with hundreds of physicians. Physicians may function as independent practitioners within clinics or be employed by the clinic with a salary, and, typically, some bonus arrangement. The latter are referred to as staff model clinics. Other models of clinic structure include group practices in which physicians are joint owners, network models that link group practices together, and management service organizations that manage groups and allow physicians to maintain their independent clinical practices. The organizational environment within which clinical practices are carried out, and the strategy for the adoption of automation technology, is dependent on the ownership and structure of the clinic.

There is an increasing consolidation of hospitals and clinics into large regional or national integrated delivery systems. This consolidation has been the result of increased pressures on hospitals and clinics for economies of scale in the purchase of expensive technology, access to capital, competitive power in the market, and to some degree, branding. Branding is a concept whereby an institution makes known its identity and builds its reputation for quality that characterizes the institution and everyone who works in it [27]. Institutions, such as Mayo Clinic, M.D. Anderson, and Cleveland Clinic, serve as examples of large integrated systems with a strong brand name.

Large integrated systems are structured as formal organizations and consist of many different levels of care, such as hospitals, clinics, and, frequently, nursing homes, and home care. However, this integration has not enabled these institutions to provide well-integrated clinical services. The reasons for this are that these organizations were consolidated primarily for business, not clinical, reasons, and they have concentrated on building stronger business units. This focus has included purchases of expensive equipment, such as electronic medical records and advanced information technology. There has been a lag in the integration of clinical care within these institutions, due in part to the continued dominance of the professions for clinical decision making and the management of the clinical process.

3.5.3 Automated Devices: Control and Oversight

One of the areas in which the federal government, through the FDA, retains considerable control and influence over the health care system is the market for medical devices and pharmaceuticals. The FDA has the responsibility for protecting the health of the population by "assuring the safety, efficacy, and security of human and veterinary drugs, biological products, medical devices, the nation's food supply, cosmetics, and products that emit radiation" [28]. The items that are overseen by the FDA account for about 25% of all expenditures in the United States. Before a product under the responsibility of the FDA can come to market, it must be reviewed and approved by the agency. A major difference between drugs and devices relates to their stability. When a drug is created, its chemical formula cannot be modified without it becoming a different drug. Devices, on the other hand, are typically developed incrementally and most will change or improve over time.

A medical device is defined as [29]:

an instrument, apparatus, implement, machine, contrivance, implant, in vitro reagent, or other similar related article, including the component parts, or accessory, which is:

- Recognized in the official national formulary, or the United States Pharmacopoeia, or any supplement to them,
- Intended for use in the diagnosis of disease or other conditions, or in the cure, mitigation, treatment, or prevention of disease, in man or other animals, or
- Intended to affect the structure or any function of the body of man or other animals, and which does not achieve any of its primary intended purposes through chemical action within or on the body of man or other animals and which is not dependent on being metabolized for the achievement of any of its primary intended purposes.

In addition, certain electronic radiation emitting products that have medical applications or claims are also considered to be medical devices. Medical devices are distinguished from drugs, whose primary intended use is achieved through chemical action being metabolized by the body.

Medical devices are categorized into three major classes, with regulatory control increasing with each class. Classification of the device depends upon the intended use of the device and upon indications for use. It also depends upon the risk the device has for patients and/or the users of the device. Class I devices involve the least amount of risks, while class III devices involve the greatest amount of risks.

In determining the reasonable safety and effectiveness of the medical devices, the FDA evaluates the problems that can arise from their use. These problems include such things as mechanical failure, faulty device and poor manufacturing quality, adverse effects from materials implanted in the body, improper maintenance/specifications, user errors, compromised sterility/shelf life, and electromagnetic interference among devices. The regulations do recognize the dynamic process of device development and, therefore, do allow for unregulated incremental improvements in a prototype device.

Recently, attention has been focused on increasing the role of the FDA in protecting public safety after a product is on the market. Focus is on postmarketing safety by monitoring devices to identify device problems whose failure could cause adverse health effects or deaths. Efforts are also underway to develop unique methods to identify medical devices and to provide better information in patient records about their medical devices, and create an electronic system for reporting adverse events for medical devices. The difficulty in the regulatory approach is to not make the process so detailed and labor intensive that valuable devices are delayed from reaching the market in a timely manner, but to ensure that the devices that do reach the market are safe and effective.

The FDA also has the responsibility for ensuring that both brand name and generic drugs work correctly and that the health benefits derived from them outweigh any known health risk. The FDA also closely regulates over-the-counter (OTC) drugs to ensure that they are safe, effective, and properly labeled. The process required to bring a drug to market its long and complicated. The drug evaluation process of the FDA scrutinizes everything from the design of clinical trials, to the severity of side effects, to the conditions under which the drug is manufactured. It is recognized that all drugs involve risk; therefore, the focus is on determining if the benefits of the drug outweigh the side effects and risks associated with it. After the drug comes to market, surveillance continues to identify risks that come to light when the drug is used in larger numbers of patients who are different from those studied before approval was granted [30]. If problems are identified, the FDA takes action to notify the public and providers, change the drug's label, or even remove the product from the market.

3.5.4 Value Migration in the Health System

The health system is complex and dynamic in its operation and is constantly in the process of transformation. Like most social systems, the health system is slow to transform, in part due to its decentralized structure and the internal checks and balances that resist change. Christiansen [31] has suggested that the health system cannot be responsibly changed from the inside, due to the many self-serving checks and balances, and that change needs to be imposed from external business forces. Others have suggested using external governmental forces to bring about transformative change. While the current system may be slow to adopt rapid, transformational change, it does continue to transform itself in a more evolutionary manner from within.

Bopp and Brown [32] have described the long-term transformation of the U.S. health system in terms of a value migration. Value in the health system has traditionally been placed in highly specialized hospital care provided by medical specialists. The hospital became the center of the health care universe and was the focus for all the support systems. For example, the financing of health care services through health insurance initially required covered services to be limited to inpatient care.

These services were the more high tech and high cost, so this system was quite ratio-nal in its approach. The initial health insurance plans were started by hospitals through the Blue Cross insurance programs. High tech meant institutionalized care, which was related to high costs and high risks for citizens. The hospital-based pri-vate insurance industry was the solution and brought the system into equilibrium. Although new technology has changed how and where care can be given, with patients in some markets moving to predominantly ambulatory care centers, the hospital-based inpatient care is still the dominant orientation of the industry, and new technology is frequently assessed in terms of how well it fits into this structure.

In the 1970s, a new form of financing was promoted, where providers were paid for services based on capitation payments. A capitation form of payment meant that providers were prepaid an amount for a defined population and were motivated to keep that population out of high-cost facilities. This form of payment was strongly advocated by private industry and the federal government as a means of controlling costs by reducing inpatient services. The application of this form of payment to Medicare services provided a major motivation to shift from hospital care to outpa-tient care. Value was thus shifted to outpatient clinics. Hospitals responded by con-tracting, purchasing, or initiating their own network of ambulatory care clinics, creating networks of integrated delivery systems (IDS), which became a new organi-zational form. This was a natural organizational response by hospitals, as the net-work of ambulatory clinics was viewed as an effective way of feeding the hospital system. This response did cause a shift in the value chain and a migration to ambula-tory care. Instead of free-standing clinics, these organizations became part of larger systems and were managed by these systems.

Services provided through ambulatory care centers included both acute short-term services, such as same-day surgery, as well as services for chronic care patients who did not need to be treated in inpatient settings. The impact of chronic care patients on ambulatory care was magnified by the fact that the elderly, who make up the greatest percentage of chronic care, have become the most rapidly growing segment of the population. Between 2000 and 2010, the population is projected to grow from 35.0 to 40.2 million, and continue growing to 71.5 million by 2030. It is projected, that by 2010, the 65 and older population will comprise 13.2% of the population, and by 2030 it will comprise 19.6% and continue increasing to 20.9% by 2050. While the number of retirees is expected to double by 2030, the workforce is expected to increase only 18% by that time [33].

Chronic care services include not only acute services provided through ambula-tory care clinics, but medical and psychosocial services provided through home care, hospice care, and a range of community social service agencies. For children with chronic diseases, such as asthma, this potentially involves schools as well. Tra-ditionally, these services were considered low tech and high-touch services, allow-ing little automation, but requiring high human resource investments from a range of health and social service workers. This assumption is now being challenged by automation technology, allowing more chronic care services to be provided outside the traditional methods. Remote sensing in home care, nursing homes, and hospice services are examples of how automation is allowing more and better services to be provide in settings outside of traditional acute care settings [34].

Telemedicine is another example of how technology is impacting both acute and chronic care. Initial telemedicine applications consisted of limited applications through specialized audio-visual facilities that allowed visual and audio linkages between traditional professional and institutional settings. These linkages included physician-to-physician or physician-to-hospital models. With the Internet and automation technology, telemedicine has been redefined as e-health applications that potentially include all physicians and increasingly involve patients.

This e-health technology is causing a shift from institution-centric information systems to patient-centric systems, and a shift in the role of patients as subservient participants in the care process to that of coproducers of the health care experience [35]. Information and automation technology empower consumers and changes the nature of the organizational landscape in health care. It also changes the market for technology from a limited number of institutional purchasers to millions of individual purchasers. A similar market phenomenon occurred in the telephone industry, when the population moved from home telephones to cell phones, with a corresponding explosion in the market for new technology packaged in new ways [36].

The migration of value from high-tech hospital, institutional care to community-based, home and personal systems has been enabled by automation and advanced information technology. The structure of the health system is changing from one characterized by organizational ownership to a structure based on information networks. Value resides not only in the heads of medical and nursing specialists in centralized institutions, such as hospitals, but is embedded in distributed information systems in the form of knowledge. Knowledge systems are greatly empowered by automation that enables patients to self-monitor and self-treat conditions that were formerly the purview of specialized health professionals. The patient continues to take a greater role (e.g., responsibility) in the clinical process, however. Automation has also enabled services to be provided with increased precision, thus improving quality. One example of this is the new, easy-to-use machines with remote monitoring that have been developed for home hemodialysis, increasing access to home hemodialysis and improving outcomes of patients selecting this treatment modality. The impact on costs can be measured in terms of both the actual cost of treatment, as well as reduced cost to the patient or consumer due to less costly travel, time away from work, cost of caring for children while treatment is being provided, and emotional cost of being in unfamiliar settings. Part of the challenge to enabling this form of care to achieve its full potential in the system is to change the structure of the services and the payment mechanisms.

3.6   The Mandate for Improved Clinical Outcomes

There is increasing pressure on U.S. health organizations to become more accountable for clinical outcomes, including quality and patient safety. This trend became a movement with the release of the report, *To Err Is Human* [37], on patient safety by the Institute of Medicine in 2000. This report estimated that U.S. hospitals caused the death of up to 98,000 patients per year, due to accidents and injuries not related to the disease for which the patients were hospitalized. This landmark report and subsequent reports have accelerated the trend toward accountability for clinical outcomes by hospitals and health organizations. If hospitals are to be held account-able for clinical outcomes, they must also become accountable for clinical processes, the historic domain of the health professions.

Increased accountability for clinical outcomes and processes has drawn on a body of science that is well developed and applied to manufacturing industries and other service industries. This science draws heavily on industrial engineering and operations research to improve work flows and processes. Deming's work [38] on total quality management (TQM) has led to the development of such analytical management practices as Six Sigma, and while it is widely discussed in health orga-nizations, so far the application is more limited. In part, this is due to the fact that clinical decisions and clinical processes are much more complex than manufactur-ing processes, with major challenges for measurement, standardization, and auto-mation. In part, the slow pace is also due to resistance by health professionals and organizations to apply this new science to an area of decision making that has, tra-ditionally, been viewed as an autonomous relationship between a patient and a health professional.

The Institute of Medicine Report [37] revealed that the high rate of errors expe-rienced in the health field were the result of transfers among health professionals and locations over the course of the treatment process. Errors were not the result of the failure by individual health professionals due to poor training or lack of com-mitment. This fact had been known for a number of years, but the IOM reports served as a national call to arms to examine critically the manner in which health services were provided and to improve the processes for delivering them. For the first time, the health system started to explore the concept of quality improvement and the work of Deming and Juran. The notion of Six Sigma quality was a standard far out of reach in most health organizations. Decision makers in these organiza-tions, however, did start to examine critically the process by which health services were delivered.

Many of the traditional beliefs and assumptions in the health system are increasingly being challenged. One assumption has involved the inability of the health system to measure and compare clinical outcomes. If each diagnosis and treatment were very individualized and customized, comparisons across popula-tions would not be possible. The measurement and comparison of clinical treat-ments has, increasingly, been carried out and reported. The work by Wennberg, reported in the *Dartmouth Atlas* [39], chronicles a disturbing variation in different rates of treatment that cannot be explained by differences in disease and disease severity. These findings increasingly put pressure on health organizations to exam-ine treatment rates and clinical outcomes. As health organizations are increasingly being held accountable for health outcomes, they must also start to explore clinical treatment processes. The examination of clinical process-outcome relationships ushers the world of Deming into the health system, by looking at the evidence on process-outcome relationships and initiating process improvement strategies. The health organization is thus being thrust into structuring and managing the clinical processes and, therefore, increasingly invading the domain of the patient-professional relationship.

3.7   The Support Function of the Health System

In addition to the complex mix of production functions in the health care system by providers, the operation of the health care system is impacted tremendously by numerous supporting functions. As briefly discussed earlier, one of these supporting functions is the way in which health care is financed in the United States and the implications of the financing mechanisms on the operations of the system. Another is the health care educational system, which was briefly introduced in connection with the role of health professionals in the system. The way in which the development and diffusion of technology in health care occur also has a substantial impact on the operations of the system.

3.7.1   Financing the Health Care System

Expenditures on health care have been growing at a rapid rate the past 40 years, and the rate of growth is expected to continue to outpace economic growth in the foreseeable future. In 1965, the year before the implementation of the major governmental programs of Medicare and Medicaid, the United States spent $41.6 billion on health care, an amount equal to $204 per person, or 5.9% of the gross domestic product (GDP). By 2005, expenditures had increased to more than $2 trillion, an amount equal to $6,683 per person, or 16.2% of the GDP. Unless substantive changes are made in the health care system, it is projected that health care expenditures will reach more than $4 trillion in 2015, an amount equal to $12,347 per person, or 20.0% of the GDP (see Figure 3.5 for trends in health expenditures as percent of GDP) [40].

As health care continues to absorb a larger share of the GDP, there will be fewer resources available for other sectors of the economy. A concern is that as health care consumes a larger share of total output, the overall well-being of the population may begin to suffer, as more and more other valuable services are forgone in order to

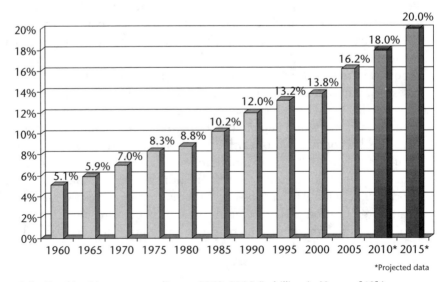

*Projected data

Figure 3.5   Total health care expenditures, 1960–2015 (in billions). (*Source:* [40].)

assign scarce resources to the health care sector. It is important that as new technologies, new drugs, and new medical procedures are introduced into the health care system their relative value be established. While spending an increasing proportion of total resources on health care is not necessarily inappropriate, it is important to be able to demonstrate the value to society contributed by those expenditures of limited resources.

The way in which we pay for the complex mixture of health care services in the United States is also complicated. Just as the provision of care is a combination of public and private entities, so are the payment mechanisms for those services. Unlike most markets in which consumers pay directly for the goods and services they consume, insurance (public and private) is the source of payment for most health care services today. The widespread use of health insurance increases the complexity of the operations of the health care system. Since a third party is now paying a significant share of the price of the services, the direct link between the supplier and the purchaser is removed, creating different incentives for the production and consumption of health care. The rising costs of health care are increasingly putting pressure on the system to modify the financing mechanisms.

While many industrialized countries have implemented predominantly public financing systems for their health care industry, the United States continues to support a system that relies on both private and public financing mechanisms. For the employed population and their dependents, the primary financing mechanism is employer-based private insurance. For the population aged 65 and older, and the disabled and those with end-stage renal disease, the federal government, through the Medicare program, is the primary financing mechanism. For categorically eligible low-income individuals, the primary source is the state-federal partnership Medicaid program. And, as discussed earlier, specific populations have health care services financed by other federal programs. For a growing number of individuals in the United States, who have neither public nor private health insurance, their health care services must be financed out of personal resources or through charitable sources. In 2005, 15.9% of the population did not have insurance coverage [41]. Figure 3.6 illustrates the major sources of financing for the $2 trillion health care expenditures in the United States in 2005.

The majority of private health insurance is obtained through place of employment, with the employers paying a substantial amount of the insurance premiums for the worker, and often making contributions to the premium of the dependents. As costs continue to increase, employers are shifting a higher percentage of premiums to the employee, or discontinuing offering insurance benefits altogether. As it becomes more difficult for insurance companies to pass their premium costs on to employers, they scrutinize more carefully the types of services they cover with their policies. This increases the need for entities trying to introduce technologies to demonstrate the value added of their products.

The billing procedures by which providers are paid for their services are diverse and complex. While providers typically establish the amount they will *charge* for a particular service, most third-party payers set the *rate* that they will pay providers for that service through a variety of mechanisms. For professional services, the most common way of setting the rate to be paid to providers is through the development of a fee schedule. A fee schedule is simply a list of services that will be covered by the

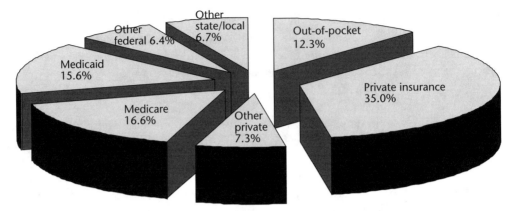

Figure 3.6   Sources of funding for health care services, 2005.

plan with the corresponding fee that insurers are willing to pay providers for that service. Typically, this fee schedule is established for each unit of service provided, although the fee schedule may be applied to a designated bundle of services.

Under a fee-for-service system, providers are paid for each individual distinct unit of care provided. Since providers are paid for each service delivered, the incentive is to provide more units of services in order to increase reimbursement. Since insured consumers are not responsible to a substantial portion of the bill, they do not have a strong financial incentive to limit the number of units of service consumed or to be too concerned about the price of the service. However, as the ability of public and private insurers to simply pass on rising costs is curtailed, more attention is focused on changing the incentives within the system. The act of changing incentives places more risk on the providers (e.g., for investment and production decisions), impacting the capital expenditures necessary for the integration and diffusion of medical technologies.

Historically, fee schedules were established based on the usual, customary, and reasonable charges in local areas. In 1992, Medicare established a national fee schedule for more than 7,000 covered services using the resource-based relative value scale (RBRVS) method [42]. The RBRVS system encompasses three components: physician work or time, physician practice expenses, and professional liability insurance expenses. The system reflects an effort to provide uniform definitions of physician's work. Medicare also implemented a system of monitoring the quantity of services supplied by providers and making adjustments in the fee schedule to prevent professionals from simply increasing the quantity of services supplied to increase revenues. While the system was designed for Medicare providers, many Medicaid programs, private insurance companies, and managed care organizations use variations of the system to determine their fee schedules. The implementation of the RBRVS system has created incentives for professionals to evaluate the efficiency with which they produce services, which, in turn, may have an impact on the types of new technologies adopted.

When Medicare was first implemented in 1966, it paid hospitals on a cost-plus basis. Under this reimbursement structure, hospitals informed Medicare what it had cost them to provide services to Medicare patients during the year, and Medicare then paid them that amount plus an additional 6% for reinvestment into the organization. Shortly after implementation, Medicare dropped the "plus" component and simply reimbursed retrospectively determined costs. Under this type of system of being reimbursed for all costs incurred in the provision of care, hospitals had no incentive to be cost efficient, or to carefully evaluate efficiency before adopting technologies into the organization. As a result, costs of hospital care rose rapidly as hospitals expanded facilities, equipment, and services.

In 1983, Medicare adopted a prospective payment system based on diagnosis-related groups (DRGs). DRGs reflect groups of medically similar patients with stable patterns of resource use. Each DRG group is clinically coherent and has a similar amount of hospital costs associated with the care provided to patients in that group. Instead of retrospectively paying a per diem rate for each day a patient was in the hospital, the prospective DRG-based system establishes an all-inclusive amount prior to provision of care that it will pay for the entire stay of the patient. Again, private insurance companies, managed care, and other payers have followed the lead of Medicare and adapted the prospective system for their purposes. Under the DRG-based system, hospitals are now placed at financial risk for the provision of care in their facilities, and therefore have an incentive to provide services in a cost-effective way and to evaluate both the costs and the benefits of new technologies before they are adopted. Because of prospective payment systems, hospitals evaluate more closely the financial implications of new technologies before acquiring them for the organization.

Another financial method that many managed care organizations have adopted is capitation, especially for paying physicians under their plans. Capitation is simply a fixed amount of money paid to the provider for each individual that is enrolled in that provider's panel, and is usually expressed in terms of a set fee per member per month (pmpm). Under a capitation plan, the provider assumes financial risk for a set of predefined services provided to the members of his or her panel. If fewer or lower costs are incurred in the provision of care than the predetermined amount, then the provider keeps the difference. If the members cost more, then the provider loses money. Typically, providers need an adequate sized panel for the law of averages to work. In addition, the plan may use a stop-loss feature under which the provider is responsible for cost up to some maximum amount for an individual patient, and then all additional costs revert back to the managed care plan. Or, the provider may purchase re-insurance, under which costs above the designated amount are transferred to the new insurance company. Under a capitated system, providers, again, are more likely to consider both cost and benefits associated with adopting a new technology, because they now are at financial risk for the results of their decisions. This shifting of financial risk to the providers has implications for the types and amounts of technologies that will be adopted in the health care system.

A number of reimbursement strategies have recently been implemented in efforts to improve quality of care, reduce errors, and increase patient safety. One of these strategies is the pay-for-performance demonstration project implemented by Medicare. Other insurance companies and health care organizations are also implementing variations of pay-for-performance strategies. Under these various pay-for-performance strategies, providers are rewarded with small price differentials for complying with accepted medical standards of practice, or processes of care. These price differentials are intended to change behavior of providers by rewarding better performance. A critical factor in the implementation of effective pay-for-performance programs is the selection of appropriate measures to align incentives and rewards across systems.

Many managed care organization have incorporated various pay-for-performance incentives into their contracts with physicians. Typically, these incentives provide a financial reward to physicians for meeting quality targets involving such factors as: patient satisfaction, access to care (e.g., waiting time necessary to get an appointment), emergency room utilization, and certain preventive care measures (e.g., mammogram screenings, immunization rates, and cancer screening rates). When providers meet the targets for these events, then they receive a bonus payment. In order to meet the targets, providers often will need to change the methods by which they practice medicine and develop a perspective and practice, since additional data are often needed to track progress towards meeting targeted goals. Generally, increased data collection will require additional computerized registries to track patients and generate timely reports. There is an increased focus on improving quality, reducing errors, and increasing patient safety within the realm of costs constraints. Thus, information technologies and other technologies for improving processes and efficiency of care delivery will continue to evolve.

3.7.2 Development and Diffusion of Medical Technology

Another support function in the health care system is the development and diffusion of medical technology. Medical technology encompasses translational research (the practical application of biomedical research to clinical practice) and transformational research (the application of systems science to the delivery of health care). As discussed earlier, health care costs continue to increase rapidly in the United States, consuming an ever-increasing share of the GDP. A significant factor contributing to these rising costs is technology—devices, drugs, treatment, and practice techniques. While technologies have increased longevity, reduced disabilities, eliminated certain diseases, and stabilized chronic conditions, such advances contribute significantly to the costs of the health care system. It is estimated that increasing use of health care services accounted for about one-half of the increases in real, per capita health care spending, and technology is a major contributor to the increase in use [43].

Medical technologies take many different forms, from simple instruments to sophisticated machines, to different ways of performing a procedure, to new biologicals and pharmaceuticals, to organizational and support systems for delivering care. Technologies are used in the provision of patient care in hospitals, outpatient clinics, offices, laboratories, nursing homes, homes, and so on. Various types of medical technologies that have been developed over the years resulted from research and development (R&D) within the health care field, and through adaptations of the results in research and development and other fields. These changes in technology have affected significantly the way in which health care services are delivered and the outcomes that can be achieved with the medical intervention.

Technologies are developed in response to problems that currently do not have satisfactory solutions. As problems or issues are identified, efforts are made to develop technologies that are simple, compatible, and can be used effectively in real-world settings. As these technologies are developed, it is important to demonstrate both their technical effectiveness as well as their cost-effectiveness in delivering care. Once technologies have been developed, then the technology must be transferred to the health delivery system. After the value of the technology to the industry has been demonstrated, then that technology can be diffused throughout the industry.

As technologies are developed, their safety must be demonstrated. A major controversy occurs in deciding who should be subjected to the experimental evaluations of the technology. A number of restrictions were imposed to protect human subjects. However, if the new technologies offer potential life-saving opportunities, then who should be offered the opportunity to be an experimental subject? Other serious ethical and moral issues arise with gene mapping of humans, genetic cloning, and stem cell research. While these areas offer substantial potential benefits for health care, they also present ethical dilemmas [44]. For example, the development of the technology to perform in vitro fertilization, whose application results in spare embryos, then raises the question regarding what to do with these extra embryos. Another example involves the use of life-support technologies; who should make the decisions and when should such decisions be made regarding the continuation of the mechanical support when chance of recovery is minimal, or expectations for quality of life are low. Decisions surrounding these issues have major implications for the development and diffusion of technologies in health care.

Information technologies have already begun to transform how health care services are delivered. Clinical information systems, such as electronic medical records (EMRs) and clinical decision support systems, support the delivery of patient care by organizing, processing, and retrieving information needed to make appropriate medical decisions. When implemented and used appropriately, these systems can increase efficiency and reduce medical errors. Administrative information systems, such as billing systems, can interface with the clinical information systems to reduce coding errors and to improve efficiency in the financial, materials management, and administrative activities of the system. Decision support systems provide tools to enhance managerial decision making and improve the functioning of the health care system [45]. Investments in these types of information systems have lagged behind in the health care industry, but pressures are increasing to speed the diffusion of such technologies in health care. The premium that the U.S. society places on high technology in health care (the technological imperative) is also putting pressure on the system for more rapid diffusion of information technologies in the health care system.

In many instances, there is at least a perceived positive relationship between the sophistication of the technology and the quality of care delivered. When these technologies enable quicker and more accurate diagnoses of medical conditions, provide new or quicker medical cures, reduce the likelihood of medical errors or inappropriate care being delivered, or reduce the invasiveness of medical interventions, outcomes are improved. Not only can many of these medical technologies reduce the risks associated with the medical intervention, they can also reduced recovery time after the intervention. Some of the new imaging technologies, for example, dramatically reduce or eliminate the need for exploratory surgery, greatly improving outcomes and the quality of medical care. Medical technologies can also improve the quality of life for patients by enabling individuals with certain medical conditions to return to a normal life or to overcome their disabilities [46]. As the proliferation of medical technologies occur, however, it is also important to consider that not all technologies equate to higher quality.

The adoption of technology in health care has, historically, occurred differently than it has in other industries. In most industries, technologies are adopted only when they have a positive impact on the production efficiency of the organization; organizations adopt technologies that enable them to produce more output with the same inputs or to produce the same output with fewer inputs—increasing efficiency in the production process. In health care, technologies have been adopted without careful consideration of the impact of the technology on the method or cost of producing health care services. In health care, technology adoption decisions have tended to increase cost rather than to decrease cost.

As health care costs continue to rise, however, increasing attention is being given to assessing the relative costs and benefits of the new technologies. Technology assessment is "the process of examining and reporting properties of a medical technology used in health care, such as safety, effectiveness, feasibility, indications for use, cost and cost effectiveness, as well as social, economic, and ethical consequences, whether intended or unintended" [47]. The goal of technology assessment is to determine what works in the practice of medicine and how cost-effective the solution is. The results of the assessment should strengthen the scientific basis and provide evidence for decision making and improve the process of delivering effective and efficient health care.

In the performance of the technology assessment, it is important to consider the role of the technology. Is the technology intended to improve morbidity or mortality, to improve quality of life, or to improve the economic performance of the organization? When judging the value of any new technology, it is important to consider the importance of the changes in the outcome measured, in terms of both short and long-term health implications [48]. Increased emphasis will be placed on the evidence-based approach for comparisons between old and new technologies. The adoption of any new technology will have to pass the test of technical efficacy, acceptance by clinicians, organizational cost-effectiveness, and health system regulation.

Automation embedded within the existing structure of clinical decision making will be much easier to implement and use, because it does not alter the structure of the clinical decision making process. The acquisition and maintenance of devices, historically, fell to the organization and could be paid for by passing the costs through to insurance companies, and thus the government and private corporations. The acquisition of automated devices did not have to pass a very rigorous cost-effectiveness test within the organization. The issue was not one of costeffectiveness, but rather, cost shifting to insurers. This finance pass-through strategy enabled the health system to become very "high tech" in nature, with the inherent condition of being a very high-cost system.

Automation technology must demonstrate its cost-effectiveness to be adopted by hospitals and clinics but must also be considered within the realities of insurance reimbursement schemes. The technical efficacy of a device may well be demonstrated in clinical trials, but it may still be rejected by clinicians, by hospitals and clinics, or by third-party insurers. This constitutes at least three levels of acceptance that must be considered in a diffusion strategy. The government might be considered as a fourth level of one considers regulatory approval by agencies such as the FDA. There are many forces within the health system that serve to facilitate or block innovation, and many times they are not based on technical performance. They may be based entirely on tradition, or power, or for other reasons, creating an inability of health leaders to bring about innovation in the system. An understanding of the complexity of health system organizations and professionals will assist in developing a diffusion strategy.

3.8   Conclusions

This chapter provides an overview of the health care system as a set of complex, highly interactive and dynamic functions. It is the functional details of the health care system that determine the appropriateness of automation technology and its potential for being adopted and utilized by the health care system.

The U.S. health care system is complex in structure and function. It is an amalgamation of public and private structures and functions, making generalizations and standardization difficult. This chapter has focused on the production function of the health care system. Discussions focus on the impact of production on the overall performance of the system and on the adoption and diffusion of technology. Within the system, the impact of a new technology is assessed by its impact on quality, cost, access, patient satisfaction, and by its effect on traditional professional roles and organization structure. One subsystem or function may either facilitate or block acceptance of a new technology, such as automation, irrespective of its technical soundness or scientific evidence. The interconnected relationships among subsystems must be clearly understood if automation is to be effectively developed and diffused in the health care system.

References

[1]   Schuster, M., E. McGlynn, and R. Brook, "How Good Is the Quality of Health Care in the United States?" *Milbank Quarterly*, Vol. 76, No. 4, 1998, pp. 517–563.

[2]   Starfield, B., "Is US Health Really the Best in the World?" *JAMA*, Vol. 284, No. 4, July 26, 2000, pp. 483–485.

[3]   Oberland, J., "The US Health Care System: On a Road to Nowhere?" *Canadian Medical Association Journal*, Vol. 167, No. 2, July 23, 2002, pp. 163–168.

[4]   Davis, K., C. Schoen, and S. Schoenbaum, "A 2020 Vision for American Health Care," *Archives of Internal Medicine*, Vol. 160, No. 22, December 11/25, 2000, pp. 3357–3362.

[5]   Katz, D., and R. L. Kahn, "Growth of Organizational Structures and Subsystems," in *The Social Psychology of Organizations*, New York: John Wiley & Sons, 1966, pp. 110–111.

[6]   Parsons, T., *The Social System*, Glencoe, IL: The Free Press, 1951.

[7]   Jellinek, M., "Change in Traditional Relationship: Erosion of Patient Trust in Large Medical Centers," *Hasting Center Report*, Vol. 6, No. 3, June 1976, pp. 16–19.

[8]   Doebbeling, B. N., A. F. Chou, and W. M. Tierney, "Priorities and Strategies for the Implementation of Integrated Informatics and Communications Technology to Improve Evidence-Based Practice," *Journal of General Internal Medicine*, Vol. 21, Suppl. 2, February 2006, pp. S50–S57.

[9]　Schoomaker, E. B., "Medical Innovator: Delivering Innovative Health Care Solutions," *Military Medical Technology, Online Edition*, http://www.military-medical-technology.com/print_article.cfm?DocID=1581, visited August 29, 2006.

[10]　Hetz, S. P., "Introduction to Military Medicine: A Brief Overview," *Surgical Clinics of North America*, Vol. 86, No. 3, June 2006, pp. 675–688.

[11]　Indian Health Service, *The First 50 Years of the Indian Health Service: Caring and Curing*, U.S. Department of Health and Human Services, 2006.

[12]　Grosse, S. D., et al., "From Public Health Emergency to Public Health Service: The Implications of Evolving Criteria for Newborn Screening Panels," *Pediatrics*, Vol. 117, No. 3, May 2006, pp. 923–929.

[13]　Reid, P. P., et al., (eds.), *Building a Better Delivery System: A New Engineering/Health Care Partnership*, Washington, D.C.: The National Academies Press, 2005.

[14]　Burns, T., and G. M. Stalken, *The Management of Innovation*, rev. ed., Oxford, U.K.: Oxford University Press, 1994.

[15]　Buckingham, R. A., and R. O. Buckingham, "Robots in Operating Theatres," *BMJ*, Vol. 311, No. 7018, December 2, 1995, pp. 1479–1482.

[16]　Huber, G. P., *Managerial Decision Making*, Glenview, IL: Scott Foresman and Company, 1980.

[17]　Parasuraman, R., T. Sheridan, and C. Wickens, "A Model for Types and Levels of Human Interaction with Automation," *IEEE Transactions on Systems, Man, & Cybernetics, Part A: Systems & Humans*, Vol. 30, No. 3, May 2000, pp. 286–297.

[18]　Shull, Jr., F. A., A. L. Delbecq, and L. L. Cummings, *Organizational Decision Making*, New York: McGraw-Hill, 1970, p. 77.

[19]　Sintchenko, V., and E. Coiera, "Which Clinical Decisions Benefit from Automation? A Task Complexity Approach," *International Journal of Medical Informatics*, Vol. 70, Nos. 2–3, July 2003, pp. 309–316.

[20]　Mintzberg, H., *The Structuring of Organizations*, Upper Saddle River, NJ: Prentice-Hall, 1979.

[21]　Starr, P., *The Social Transformation of American Medicine*, New York: Basic Books, 1982, p. 427.

[22]　Brown, G., and T. Stone, "Information Strategy Related to Enterprise and Organizational Strategies," *Strategic Management of Information Systems in Healthcare*, Health Administration Press, 2005, pp. 31–50.

[23]　Quaglini, S., et al., "Flexible Guideline-Based Patient Care Flow Systems," *Artificial Intelligence in Medicine*, Vol. 22, No. 1, April 2001, pp. 65–80.

[24]　Bureau of Labor Statistics, *Occupational Outlook Handbook, 2006–07 Edition*, U.S. Department of Health and Human Services, 2006, http://www.bls.gov/oco/home.htm, visited September 19, 2006.

[25]　Friedson, E., "The Development of Administrative Accountability in Health Services," *American Behavioral Scientist*, Vol. 19, No. 3, January–February 1976, pp. 286–298.

[26]　Bodenheimer, T., "High and Rising Health Care Costs, Part 1: Seeking an Explanation," *Annals of Internal Medicine*, Vol. 142, No. 10, May 17, 2005, pp. 847–854.

[27]　Miller, D. W., "A Core Strategy—Developing a Brand: A Health Care Organization That Builds Strong Brand Loyalty Will Ensure Its Position for the Future," *Health Forum Journal*, Vol. 44, No. 6, January–February 2001, pp. 36–38.

[28]　Phillips, S. J., and R. S. Phillips, "Devices and the Food and Drug Administration," *Artificial Organs*, Vol. 29, No. 5, 2005, pp. 363–365.

[29]　Food and Drug Administration, "Getting to Market with a Medical Device," http://www.fda.gov/cdrh/devadvice/3122.html, visited September 8, 2006.

[30]　Center for Drug Evaluation and Research, *2005 Report to the Nation: Improving Public Health Through Human Drugs*, U.S. Department of Health and Human Services, 2005.

[31] Christiansen, C., *The Innovator's Dilemma: When New Technologies Cause Great Firms to Fall*, Boston, MA: Harvard Business School Press, 1997.

[32] Bopp, K., and G. Brown, "Aligning Information Strategy and Business and Clinical Strategies: Information as a Strategic Asset," *Strategic Management of Information Systems in Healthcare*, 2005, pp. 121–148.

[33] Federal Interagency Forum on Agency-Related Statistics, *Older Americans Update 2006: Key Indicators of Well-Being*, May 2006.

[34] Demiris, G., S. M. Speedie, and L. L. Hicks, "Assessment of Patients' Acceptance of and Satisfaction with Teledermatology," *Journal of Medical Systems*, Vol. 28, No. 6, December 2004, pp. 575–579.

[35] Brown, G., K. Bopp, and S. Boren, "Assessing Communications Effectiveness in Meeting Corporate Goals of Public Health Organizations," *Journal of Health & Human Services Administration*, Vol. 28, No. 2, 2005, pp. 159–188.

[36] Schopp, L. H., et al., "Design of a Peer-to-Peer Telerehabilitation Model," *Telemedicine Journal and e-Health*, Vol. 16, No. 2, 2004, pp. 243–251.

[37] Kohn, L., J. Corrigan, and M. Donaldson, (eds.), *To Err Is Human: Building a Safer Health System*, Washington, D.C.: National Academies Press, 2000.

[38] Deming, W. E., *The New Economics of Industry, Government, Education*, 2nd ed., Cambridge, MA: MIT Press, 2000.

[39] Wennberg, J. E., Dartmouth Medical School, Center for the Evaluative Clinical Sciences, *The Dartmouth Atlas of Health Care in the United States*, American Hospital Association, 1996.

[40] Centers for Medicare & Medicaid Services, "National Health Care Expenditures Projections: 2005–2015," http://www.cms.hhs.gov/NationalHealthExpendData/downloads/proj2005.pdf, visited August 26, 2006.

[41] DeNavas-Walt, C., R. D. Proctor, and C. H. Lee, *Income, Poverty, and Health Insurance Coverage in the US: 2005*, Current Population Reports P60-231, Washington, D.C.: U.S. Government Printing Office, 2005.

[42] Johnson, S. E., and W. P. Newton, "Resource-Based Relative Value Units: A Primer for Academic Family Physicians," *Family Medicine,* Vol. 34, No. 3, March 2002, pp. 172–176.

[43] Goldman, D. P., and E.A. McGlynn, *US Health Care: Facts About Cost, Access, and Quality*, RAND Corporation, 2005.

[44] Califf, R. M., "Issues Facing Clinical Trials of the Future," *Journal of Internal Medicine,* Vol. 254, 2003, pp. 426–431.

[45] Austin, T., et al., "A Prototype Computer Decision Support System for the Management of Asthma," *Journal of Medical Systems,* Vol. 20, No. 1, February 1996, pp. 45–55.

[46] Hatcher, M., and I. Heetebry, "Information Technology in the Future of Health Care," *Journal of Medical Systems*, Vol. 28, No. 6, December 2004, pp. 673–688.

[47] Committee for Evaluating Medical Technologies in Medical Use, *Assessing Medical Technologies*, Washington, D.C.: National Academies Press, 1985.

[48] Fleisher, L. A., S. Mantha, and M. F. Roizen, "Medical Technology Assessment: An Overview," *Anesthesia and Analgesia*, Vol. 87, No. 6, December 1, 1998, pp. 1271–1282.

PART II

Engineering Basis for Life Science Automation

CHAPTER 4

Principles of Human-Machine Interfaces and Interactions

Marcia K. O'Malley[†]

This chapter introduces basic principles of human-machine interfaces and human-machine interactions, including issues of levels of autonomy, teaming, human performance enhancement, and shared control between machine and human operator. Specific challenges that face life sciences and micro-nano applications are given.

4.1 Introduction

Human-machine interaction (HMI) can be generally defined by drawing upon related literature on human-computer interaction (HCI) and human-robot interaction (HRI). Indeed, the body of research on HCI is much broader than that for HRI or HMI individually, and current challenges in HCI can be extended and applied to human-machine interactions for discussion as it applies to life science automation.

The discipline of HCI is generally considered to address the design, evaluation, and implementation of human-operated interactive computing systems, and the study of related topics concerning such systems. In this context, there can be multiple humans and/or machines considered. The Association for Computing Machinery's (ACM) Special Interest Group on Computer-Human Interaction (SIGCHI) lists several areas of primary study, including joint performance of tasks by humans and machines, communication structures between entities, capabilities of humans to use and learn to use such machines, interface design and fabrication concerns, specification, design, and implementation processes, and design trade-offs. Given this broad range of topics which are of concern to researchers in the field, there are clearly multiple relevant disciplines involved in HCI research, including computer science, psychology, sociology and anthropology, and industrial design.

HRI specifies the computer as a robotic device, and the majority of the field deals with industrial robots, mobile robots, or personal or professional service robots [1]. When we shift away from the terminology "human-computer" interface and seek to specify the human-robot or human-machine interface, the manifestation of the machine (as opposed to software) brings into the fold the engineer. In turn, the designer of such interfaces must consider the additional aspects of machine

---

†   Rice University

design, ergonomics, safety, interaction modality, and other concerns. HRI is typically distinguished from HCI and HMI in that it concerns systems with complex, dynamic control systems, systems that exhibit autonomy and cognition, and which operate in changing, real-world environments [2]. When considering the principles of human-machine interaction as they apply to life science automation, we note the absence (typically) of interaction with systems that exhibit cognition. Also, systems for the life sciences are typically operating in controlled, laboratory environments, thus further distinguishing HMI from HRI. The remainder of this chapter will focus only on those relevant aspects of human-machine interaction and human-machine interfaces that are pertinent to life science automation, and will specifically omit discussion of cognitive architectures in human-robot interaction and mobile robotic systems.

4.2 Fundamentals of Human-Machine Interaction

There are four primary elements of all human-machine interfaces, as summarized by Degani and Heymann [3]. These include the machine's behavior, the operational goals or task specification, the model that user has about the machine's behavior (referred to as the user model), and the user interface (through which the user obtains information about the machine's state and responses). They state that these four elements must be suitably matched in order to insure correct and reliable user-machine interaction. Their work describes an evaluation methodology to determine if an interface provides to the human sufficient information about the machine so that the task can be successfully and unambiguously completed. In order to carry out such evaluation, a set of assumptions are made including the existence of an underlying model of machine behavior, machine behavior that is unambiguous and deterministic, specified operational requirements for the system, and formal representation of the user model gleaned through training materials and documentation.

Human-machine interfaces can be problematic, as summarized by Woods et al. [4]. Specifically, the interface to operator may provide inadequate information about the current state of the machine, there may be perplexing interactions among machine components that are not clear to the operator, and the operator may have an incomplete user model of the machine's behavior. These problems may limit the operator's ability to anticipate future behavior of the machine, leading to confusion on the part of the operator, and the potential for errors [3].

Hoc provides a discussion of failures in human-machine systems and focuses on four primary types of failure that result from both the design of the interface, and the cooperation between man and machine [5]. Specifically cited are loss of expertise, complacency, trust and self-confidence, and loss of adaptability. Loss of expertise is often attributed to the operator taking a passive role [6] and is a clear trade-off when low-level functions such as decision implementation, or high-level functions such as decision making are automated [7]. Complacency arises when automation performs high level functions, and operators take solutions presented by automation systems for granted and without question [5]. When an operator can choose between manual or automatic operation of a system, trust and self-confidence in the machine are an important factor in the utilization of the automation [8–10]. Finally, loss of adaptability of the human-machine system, primarily due to limited feedback to the human operator during automated decision making, leads to a syndrome of human out of the loop and makes it difficult for the operator to regain manual control of the system [11, 12].

With these definitions of human-machine interactions in mind, the remainder of this section will provide a brief introduction to the categories of robotics and machines used in life science automation. In addition, some background on design of automation systems, performance measures, human-machine teaming, and communication between humans and machines is presented.

4.2.1   Robotics and Machines for Life Science Automation

Using the categories of robotics given in the introduction (industrial robotics, professional service robotics, and personal service robotics), we can discuss scenarios that are relevant to life science automation. Typically, an industrial robot is considered to have three elements: manipulation of the physical environment (pick and place), computer control, and operation in industrial (controlled) settings [1]. It is reasonable to extrapolate these elements to a life science environment such as a controlled laboratory setting. This application crosses over into the realm of professional service robots. Indeed, robotic manipulators are being used in chemical and biological laboratories to handle and transport samples with high speed and precision, and automation in life science applications is increasing dramatically [13–15]. Later chapters of this book will address specific examples and case studies. The reader is encouraged to explore Chapman's review of lab automation, which focuses on specific examples such as automated mapping of the detailed structure of a protein, cell culture systems, sample storage and retrieval, and information management systems [13].

4.2.2   Design of Automation Systems

Automation is defined by Parasuraman and Riley as, "a device or system that accomplishes (partially or fully) a function that was previously, or conceivably could be, carried out (partially or fully) by a human operator" [11]. When considering automation for life sciences, one must first consider the degree to which the task will be automated, the subtasks of the procedure to be automated, communication between the human operator and machine, and more traditional HCI issues (usability, human factors, reliability, and safety). What to automate is typically determined by identifying functions that humans do not wish to perform, or cannot perform as accurately or reliably as machines.

Clear in the literature is a need to maintain a human in the loop, even when incorporating automation into a given application [16]. Humans are able to make well-formed decisions even with an absence of complete or correct information. Additionally, humans possess retention skills that aid in problem solving [17]. These abilities establish confidence in the reliability of an automation system, since the human operator can resume manual control when the environment changes, or when subsystems fail or encounter unexpected states.

In discussing automation, and in particular human interaction with automated systems, it is helpful to categorize the degree and nature of the automation. A number of autonomy scales have been proposed in the literature. A common taxonomy is based on ten levels of autonomy (LOA) and is summarized in Figure 4.1.

This scale was combined with a four-stage model of human information processing [19–21] and proposed as a revised taxonomy for automation [18]. These four stages include acquisition and sensory processing; analysis and perception/working memory; decision making; and response and action selection. The model for types and levels of automation is depicted in Figure 4.2.

For the acquisition stage of information processing, automation can be in the form of sensors that scan or observe the environment and provide data to the operator. In the most basic of terms, this can mean incorporating sensors into a system and relaying that data to a human operator or to the next automation stage. Acquisition can also involve the organization of incoming information according to some

HIGH 10. The computer decides everything, acts autonomously, ignoring the human.
 9. Informs the human only if it, the computer, decides to.
 8. Informs the human only if asks, or
 7. Executes automatically, then necessarily informs the human, and
 6. Allows the human a restricted time to veto before automatic execution, or
 5. Executes that suggestion if the human approves, or
 4. Suggest one alternative.
 3. Narrows the selection down to a few, or
 2. The computer offers a complete set of decision/action alternatives, or
LOW   1. The computer offers no assistance: Human must take all decisions and actions.

Figure 4.1   Levels of automation of decision and action selection. (*After:* [18].)

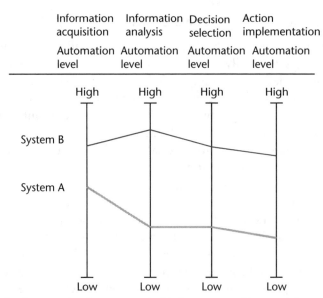

Figure 4.2   Levels of automation incorporated with four stages of human information processing. (*After:* [18].)

predetermined criteria, and then indicating subsets of the data or information for the operator.

The analysis stage involves cognitive functions such as working memory and processes of inference drawing. This stage can be automated in a basic sense by applied extrapolation or prediction to incoming data, or in a more advanced way by integrating information that has been collected into a single value or piece of information for the operator.

Decision making is automated via methods that select among several alternatives. One approach is the incorporation of conditional logic operations that select from a number of predefined alternatives. As emphasized by Parasuraman and coauthors [18], these automated decision making systems make explicit or implicit assumptions about the value or cost of a given outcome, which can be uncertain.

Finally, automation of action is in the form of machine execution, replacing actions performed by the human operator. Such actions can be manual physical tasks, such as the sort and staple options on many photocopier machines.

Alternative definitions and taxonomies of levels of automation are presented in [22–25], including the introduction of adaptive automation (also termed adjustable autonomy), where the user or the machine can initiate changes in the level of automation. An application of adjustable autonomy for micro assembly is discussed in [26], where human in the loop operation is desirable due to the complexities of the microscale environment, but some degree of automation of the machine improves overall system performance.

Based on the LOA scale presented here, Parasuraman et al. propose a framework for automation design that can be applied to systems for life science automation, shown in Figure 4.3. As represented in the figure, automation can be applied to any of the four stages of human information processing (acquisition, analysis, decision, action), and more than one stage can be automated in any given system. After deciding which stage(s) will be automated, the level of automation that should be applied must be determined. Inherent in this decision are many trade-offs, and the process will likely be iterative. Any specified level of automation for a given stage should be thoroughly evaluated to determine the primary and secondary consequences of the automation. Primary criteria are focused on direct impact on human performance, while secondary criteria include reliability, costs, ease of system integration, efficiency/safety trade-offs, and many more.

Note that any manifestation of automation can be adaptive, where the level and/or type of automation could vary depending on situational demands during operational use. Such adaptability can be automated itself, or controlled by the human operator.

Human-centered automation systems should be designed such that the human is not left with a fragmented or difficult job. Specifically, task allocation between the human and the automation should be such that the performance of the team exceeds that of the individuals. Key points to consider are assurance that the human can monitor the system, that the operator receives sufficient feedback on the state of the system, and that the behavior of the automation system is predictable in that the human is able to form a reliable mental model of the machine's operation [6, 19].

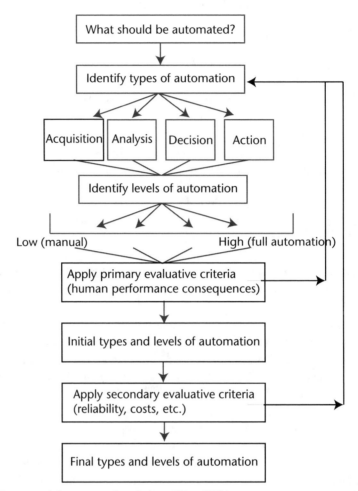

Figure 4.3 Framework for automation design. (*After:* [18].)

4.2.3 Performance of Human-Machine Systems

When automating tasks or systems that were formerly controlled manually, human activity is not simply replaced by machine, but instead human activity is changed in potentially unintended and unanticipated ways [11]. As a result, the automation system presents new and unique coordination demands on the human operator [27]. Given these new demands on the human operator, studies of human performance with human-machine and human-automation interaction systems are increasing. Indeed, these performance criteria can be used to determine the degree and type of automation that is suitable for the system.

Typically performance is measured in the resulting human-machine interface system, comparing task performance to the equivalent manual system. It is expected and desirable that both human and system performance are enhanced by automation, but the degree of performance enhancement is highly dependent on the types and levels of automation that are incorporated in the man-machine interface system. Human interaction with automation has been studied from a number of perspectives, including theoretical analyses, laboratory experiments, modeling, simulation, field studies, and analysis of real-world systems and accidents.

Four primary performance measures are presented here, as they are used in the automation design framework presented by Parasuraman et al.: mental workload, situation awareness, complacency, and skill degradation [18]. Mental workload reduction is a clear benefit of automation systems, since reducing operator workload (be it cognitive or even physical) can free up the operator to accomplish other tasks simultaneously. However, if the automation system is difficult to initiate, or if the operator is required to manually provide significant amounts of data to the system for automated analysis and decision making, the intended outcome of mental workload reduction can actually result in increasing load. Situation awareness is another performance measure. If the automation system is handling decision making or is updating an environment model, the human operator may be unaware of such updates compared to if they make such changes themselves. This lack of awareness can affect the ability of the human operator to form a sufficient mental model of the automation system's processes, increasing the probability of errors should the human operator need to regain manual control or override of the system. Third, if the automation is not reliable, the operator may become complacent and may not be aware of errors in the automated processes, basing future actions on incorrect outcomes. Finally, when decision making processes are automated, cognitive skills of the human operator can be degraded as they are no longer processing data, analyzing potential actions, and making such decisions themselves.

In addition to these primary performance criteria, based on the performance of the human and machine system, there are a number of secondary criteria which can be used to determine the utility of automation for a particular subtask or system. These secondary measures include, among others, the reliability of automation and the cost of decision/action outcomes. Reliability is often measured in probabilistic terms (mean time to failure, or raw values normalized to one), and there are consequences of poor reliability in terms of operator trust of the system, leading to underutilization of the automation. In such cases, other benefits, in terms of the primary performance measures, are lost. In terms of cost, factors such as the risk involved in the system, and the potential benefit or harm that could come to the human operator or those in the vicinity of the automated system, must clearly be considered. Such considerations can influence not only the selection of subtasks to automate, but the level of automation that is implemented.

A number of other performance measures are available for review in the literature, including neglect tolerance (a measure of how a system's current task effectiveness degrades over time when the system is neglected by the user), task effectiveness (how well the operator-automation system accomplishes the task), robot attention demand (the fraction of total task time that the human operator must attend to a given automated system), fan-out (the number of automated tasks that the human operator can simultaneously control), and interaction effort (the amount of time required to interact with the automated system) [28]. A study of the effects of automation on reliability is presented in [29]. Additional studies of human performance with automation systems are presented in [6, 12, 30], with a focus on situation awareness and workload. Leveson and Palmer studied error reduction via automation design [31].

These numerous performance criteria serve as a guiding principle in the iterative human-machine and human-automation system design process, and selection of performance criteria will vary according to the application and desired outcomes.

4.2.4 Human-Machine Teaming

When considering teaming between humans and machines, there are two primary topics for discussion: the architecture of the team, and task allocation. Team architectures are focused on how to organize teams of humans and machines, including the optimal number of human and machine team members. Research on teaming architectures seeks to identify situations which require various structures, for example authoritarian, hierarchical, or democratic [32].

In work by Scholtz, the role of the human is proposed as a defining aspect of human-machine teaming, with potential roles including supervisor, operator, teammate, mechanic/programmer, and bystander [33]. The supervisor is responsible for monitoring and controlling the overall situation and has sole responsibility for changing the larger goals and/or intentions of the robot or machine system, while the operator is able to modify internal software or models when the robot or system behavior is not acceptable. The mechanic deals primarily with the physical interventions with the system, but must be able to determine if the interaction has the desired effect on system behavior. The peer serves as a teammate to the robot or machine, and can give commands within the larger goal/intention framework defined by the supervisor. Finally, the bystander has a subset of actions available for interacting with the system. Scholtz's categorization of potential roles for the human are based on work by Kidd, who notes that human skill is always required in robotic systems, and that the goal of the designer should be to support and enhance human skill rather than substitute the robot for the human [34]. Specifically, he argues for developing robotics and human-machine systems such that humans can be more productive and efficient, for example by using machines for routine or dangerous tasks. Indeed, studies in both human-computer and human-machine interaction have demonstrated that performance of complex tasks improves when the system is designed to support rather than eliminate the human [35, 36]. In her work, Scholtz focuses on situational awareness as a measure of system effectiveness [33], recalling the three levels of situational awareness defined by Ensdley [37]. This taxonomy defines level 1 as perception of cues, level 2 as the ability to comprehend or integrate information and determine relevance to specific user goals, and level 3 as the ability to forecast future events and dynamics based on the perception and comprehension of the current system state.

In contrast to studies of team architectures, research on task allocation seeks to balance the skills of human and machine for a given task. In applications such as teleoperation, it is straightforward to utilize the controllability of the robotic manipulator to damp out tremor from the human operator, while in other applications the proper allocation of tasks may not be so straightforward. Research also seeks to determine how to build systems with dynamic task allocation, perhaps based on human workload. One study has investigated the efficiency of human-robot interactions for shared-control teleoperation with adjustable robot autonomy in order to reduce human workload, with promising results [38]. Through experiments, they showed that a shared control scheme is more tolerant to neglect, and results correlated well with measures of workload and ease of use. In another study, Marble et al. examined how the human operator was able to work with the machine at varying levels of autonomy (related to task allocation), with a focus on situation awareness and task performance, for a search and rescue task [39]. They found that participants were able to maintain awareness of the completeness of the task in all modes of autonomy, regardless of user experience with robotic systems. Additionally, they found that operators differed in their ability to function with higher levels of autonomy, with some users fighting the robot for control. Finally, they found that performance in the shared mode benefited from practice, indicating that as robot autonomy increases in a system, there may be an increased need for training of the human operator so that they may understand how and why the robot or machine will exhibit various behaviors.

When developing a human-machine interface for life science automation, therefore, the roles of human and machine must be clearly defined in the context of the task. Once roles are defined, then the degree of autonomy of each component can be specified, and tasks can be allocated to human and machine in a manner that exploits the strengths of each participant. It should be noted that safety is of great import in human-machine collaborative environments, especially as the human works in close proximity to the machine. The literature addresses safety in human-robot collaborative settings, which can be extrapolated to human-machine interactions for the life sciences. For example, Heinszmann and Zelinsky have proposed a control algorithm for robot manipulators, such as those that may be used for automation of pick and place tasks, which results in predictable and understandable robot actions, and limited forces between the robotic device and its environment [40]. In related work, Kulic and Croft have derived a measure of danger during human-robot interactions that can be explicitly computed based on impact force during potential collisions between the human and robot. Motion strategies that minimize the danger index have been developed and demonstrated both in simulation and experimentally [41, 42]. They extend their work to incorporate measurement of physiological signals to determine human response to robot motions, with decreased anxiety when safe planning is carried out for high speed operation of the robot [43].

4.2.5   Communication

Communication in human-machine systems can be categorized as direct or mediated [32]. Direct human-machine communication is accomplished via speech, vision, gesture, or remote operation of manipulators (teleoperation). Mediated human-robot communication is accomplished via virtual environments, graphical user interfaces, and collaborative software agents.

Teleoperation, or remote control via robotic manipulators, remains a primary mode of interaction in human-machine systems. Such systems can be multimodal, reflecting visual, auditory, and/or haptic feedback from the remote site to the human operator. These systems have been quite successful in a number of applications such as space robotics and surgical robotics, but they can also be expensive and limited in scope. Challenges arise such as limited degree-of-freedom of control, limited feedback to the operator, increased cognitive demand on the operator, and the potential for communication delays between master and slave. Other direct means for communicating with machines and systems include speech, gestures, facial expressions, body posture, and written gestures [44, 45]. Often, however, inclusion of such communication modalities between humans and machines are limited due to scientific challenges in speech recognition, natural language processing, and computer vision algorithms.

As a result of the challenges of direct communication in human-machine interfaces, we are often relegated to use physical interaction and interfaces, which can include computer terminal interfaces, touch screens, physical haptic interfaces, and other input devices. For automation applications, often the physical interface is used not only to operate but also to program the machine to carry out the desired tasks. One common approach is programming by demonstration [46]. The goal of programming by demonstration is to simplify complex machine programming by allowing the human operator to carry out the specified task as desired, and then recreate the actions with the machine. This can be done either within the real environment or in a simulated environment, and sometimes with the use of synthetic aids [47].

4.3   Current Research

Given the background on human-machine interactions discussed in the previous section, some current research trends in the field will now be presented. Specifically, this section will focus on physical haptic interaction between human and machine. In the context of the topics presented thus far, recent research advances in haptic human-machine interaction will be presented. Focus areas include performance specifications for haptic devices, human-machine teaming architectures, and human performance enhancement via haptics.

4.3.1   HMI Interaction Via Haptic Devices

One consideration in human-machine interfaces is the design requirements for the physical device. The proper design of any machine requires a well-defined set of performance specifications. The requirements for device design become even more important when the human operator is physically coupled to the interface, and when haptic (force) feedback is provided to the operator. Although much work has been accomplished in the field in general (see, for example, the surveys [48, 49]), hardware specifications for haptic interfaces that relate machine parameters to human perceptual performance are notably absent. The absence of such specifications is most likely because haptic interface performance specifications must consider issues of human perception, which is complex in nature and difficult to assess quantitatively. With the recent introduction of several commercially oriented haptic devices and applications, the need for a set of specifications to guide the cost-optimal design of haptic devices is that much more pronounced.

The vast majority of the research literature related to this topic has generally either focused on quantitative measures of human factors, measures of machine performance independent of human perception, or the effects of software on the haptic perception of virtual environments. Regarding the first area, psychophysical experiments conducted by several research groups have quantified several haptic perception characteristics, such as pressure perception, position resolution, stiffness, force output range, and force output resolution (for example, [50–54]). Since these experiments did not involve haptic interface equipment, however, they were not able to create a direct link between machine performance and human perception during haptic task performance.

Within the second area of research, optimal machine performance has been characterized in the literature, yet these measures are typically disparate from human perceptual measures. When designing high-performance equipment, designers seek to build a device with characteristics such as high force bandwidth, high force dynamic range, and low apparent mass [55, 56]. These are typically qualitative specifications, however, since the designers have little reference information regarding the quantitative effects of these machine parameters on the performance of humans with regard to perception in a haptically simulated environment. Several researchers have incorporated human sensory and motor capability as a prescription for design specifications of a haptic interface [57–59]. Such measures are logical, though indirectly related to haptic perception and most likely quite conservative for common haptic tasks. Colgate and Brown offer qualitative suggestions for haptic machine design that are conducive to the stable simulation of high impedances [60]. Though simulation of a high impedance is a useful and logical performance objective for a haptic device, the objective is not directly based upon measurements of human perception.

Finally, researchers have studied the effects of software on the haptic perception of virtual environments (for example, [61–63]), yet these experiments did not address the relationships between haptic interface hardware design and haptic perception. Recent work has addressed the relationship between haptic interface hardware and human perception, and in particular measures the effects of varying virtual environment force and virtual surface stiffness in a simulated environment on human perceptual capabilities in a haptic environment [64–66]. Virtual surface stiffness is of interest as a machine parameter because hardware selections, including position sensors and computers, can limit achievable virtual surface stiffnesses. A good discussion of the relationship between hardware and achievable surface stiffness is given in [60].

In one study [64], identification, detection, and discrimination tests were performed to characterize the effect of maximum endpoint force on the haptic perception of detail. Results indicate that haptic interface hardware may be capable of conveying significant perceptual information to the user at fairly low levels of force feedback (3N to 4N). While higher levels of force output in a haptic simulation may improve the simulation in terms of perceived realism, the results of these experiments indicate that high levels of force feedback are not required to reach maximum information transfer for most aspects of the haptic display of detail.

In a similar study [65], identification, detection, and discrimination tests were performed to characterize the effect of virtual surface stiffness on haptic perception of detail in a simulated environment. Results indicate that haptic interface hardware may be capable of conveying significant perceptual information to the user at low to moderate levels of simulated surface stiffness (approximately 400 N/m when virtual damping is also present) for gross stylus-type perceptual tasks.

In a follow-up study [67], experiments were conducted to compare human perceptual performance in a real environment to performance in a simulated environment for two perception tasks: size identification and size discrimination. Findings indicate that performance of size identification tasks with haptic interface hardware with reasonable maximum force output can approach performance in real environments, but falls short when virtual surface stiffness is limited. For size discrimination tasks, performance in simulated environments was consistently lower than performance in a comparable real environment. Interestingly, significant variations in the fidelity of the haptic simulation do not appear to significantly alter the ability of a subject to identify or discriminate between the types of simulated objects described herein.

These findings can be extrapolated to teleoperation environments which may be more common in life science applications, where forces at the slave manipulator are reflected to the human operator, and they give some insight into the requirements on the master manipulator's force reflection capabilities. To insure good performance of size identification tasks, designers of haptic interfaces should first aim to create simulated environments with high virtual surface stiffness, and should treat maximum force output of the haptic device as a secondary design goal, since limited force output had an insignificant effect on performance when compared to performance in a real environment. For size discrimination tasks, designers of haptic devices should aim to reach recommended minimum levels of maximum force output *and* virtual surface stiffness (3N and approximately 470 N/m, respectively) to insure acceptable performance, but should note that this performance will never reach the level that can be attained in a comparable real environment.

4.3.2 Teamwork

While teaming of humans and machines can strictly refer to humans working in close proximity to mobile robots or industrial automation, teaming of human and teleoperated robotic agents is also a feasible architecture for task completion, and one that has clear application for life science applications where the existence of a human in the loop is of great importance, yet the capabilities of the machine or robotic device should be exploited. In one study, a simplified, hypothetical extra-vehicular activity (EVA) assembly task featuring human-robot teaming was simulated with hardware-in-the-loop to study the human-robot interaction problem [68]. The task was purposefully designed to require more than two hands and, therefore, multiple agents so that meaningful interactions could take place. A long structural beam, too awkward for one agent to handle alone, was inserted into a fixed socket and pinned in place. In the experiment, the number of information channels and types of communication were varied to study the effects of such means on performance of the assembly task. Three communication modes were compared with and without force feedback provided via a haptic device. The three modes included force feedback to the teleoperator via a visual display, visual feedback of force combined with verbal cueing from the worksite human to the teleoperator, and visual feedback of force combined with verbal and gestural cueing from the worksite human to the teleoperator.

The assembly team consisted of one robot and three humans. One human, the coworker, is collocated with the robot at the worksite, while the other two, the teleoperator and the monitor, are placed in different remote locations. Performance metrics for the assembly task included task success, task completion time, maximum contact force/torque, and cumulative linear/angular impulse. Task success describes the degree to which a team was able to meet all task objectives. Task completion time reflects how efficiently resources were used in accomplishing the task. Maximum contact force/torque quantifies the risk of hardware failure or damage due to excessive momentary peak loads at the beam-socket interface. Cumulative linear/angular impulse quantifies the risk of hardware failure or damage due to excessive wear and tear as a result of extended contact at the beam-socket interface [69].

The most significant result of this experiment is the comparison of maximum contact force in the beam receptacle across pairs and feedback modes. In the case of no force feedback, where the teleoperator was limited to only a visual display of the forces and torques in Robonaut's arm, peak forces ranged between 40N and 110N. As additional feedback modes were added, such as verbal cues and gesturing, peak forces tended to decrease. In fact, in the case where visual force information, verbal cues, and gestures were all employed, peak forces were roughly half that of the other nonforce feedback trials. When the teleoperator was provided with force feedback via a haptic device, peak forces were quite consistent and ranged between 30N and 50N. Standard errors were much smaller for the force feedback case. This is a significant result due to the fact that large forces in the receptacle are transferred to the robot during constrained motion and contact, leading to larger loads on the hardware. When the teleoperator had kinesthetic information regarding the contact forces, a significant reduction in peak forces was observed, regardless of the other methods of communication between teammates. Differences in the roles played by each subject (task leader or teleoperator) were insignificant for this comparison.

For applications in micro- or nanomanipulation, or in a laboratory setting, it will be beneficial to provide multiple modalities of feedback to the human operator, especially haptic feedback of the forces at the slave environment via the operator interface, in order to minimize damage to the biological sample.

4.3.3   Robots for Performance

Virtual environment (VE) technology offers a promising means of enhancing human performance and training humans for motor skill acquisition. Computationally mediated training has many potential advantages over physical training like lower risk and cost, and better data collection and evaluation. Training in VE aims to transfer what is learned in the simulated environment to the equivalent real-world task. Virtual training can be designed either to provide a *virtual practice medium* that matches the targeted physical medium as closely as possible, or to behave as a *virtual assistance* to improve training effectiveness by providing additional feedback in ways that are possibly not realizable in the physical world.

Most forms of interaction with computerized simulations involve only visual and auditory information. However, it is shown that the addition of haptic feedback to virtual environment simulations provides benefits over visual/auditory-only displays via reduced learning times, improved task performance quality, increased dexterity, and increased feelings of realism and presence [69–74].

To exploit performance enhancement and training capabilities of virtual environments with haptic feedback, various virtual assistance paradigms have been proposed. These training paradigms are inspired by various motor learning theories and are realized through different assistance schemes such as promoting more practice, demonstrating a strategy, augmenting feedback error, and reducing feedback error.

Among these methods, the most common form of haptic assist is achieved through the introduction of forbidden zones in the workspace via so called *virtual fixtures* [75]. Virtual fixtures are analogous to the use of training wheels when riding a bicycle, or a ruler when drawing straight lines. These virtual fixtures have been shown to significantly improve task performance in virtual environments [76, 77]. However, since the feedback provided by virtual fixtures is independent from the dynamics of the system to be learned, and because this feedback becomes available intermittently only to prevent large errors, from the perspective of training, virtual fixtures provide nothing more than a safer medium for practice. The assistance provided by virtual fixtures is not aimed to assist the mechanism of learning, but is designed merely to facilitate safer practice. Learning still takes place through *virtual practice*.

Another form of virtual trainer is motivated through teaching by demonstration. In these *record and play* strategies [78–82], the dynamics of an expert are recorded while performing the task and these dynamics are played backed to the novice to assist learning. In this kind of assist, the novice is not actively involved in the task during training. Once the preferred strategy to achieve the task has been played back a couple of times, the novice is allowed to practice to mimic the demonstrated dynamics. This paradigm does not account for the differences due to user-specific dynamics, and also prevents the novice from forming their own strategies.

Patton et al. [83] propose to train reaching movements by generating custom force fields designed to drive subjects to adopt to a prechosen trajectory. This strategy is based on aftereffects of adaptation and aims to alter the feedforward command in the central nervous system. However, this approach is not effective for long-term training since the aftereffects tends to wash out after relatively short periods.

In [84], Todorov et al. utilize error augmentation strategies to speed up human motor learning of a dynamic task. By amplifying the instantaneous error, modified dynamics are displayed to the user to promote faster convergence of error-based adaptation mechanisms. Capitalizing on a form of assistance not realizable in the physical world, this technique resulted in significant increases in learning rates. The limitation of this technique lies in its applicability to complex tasks since augmenting the error in these cases can significantly degrade performance, rendering successful task completion infeasible.

Finally, error reduction has been implemented through a *shared controller* for performance enhancement and training [85–87]. O'Malley and colleagues have proposed shared control as an active assistance paradigm where the feedback is provided by a controller, which is dependent upon the system states, as depicted in Figure 4.4. By dictating the type and level of active control between the computer and the human on the virtual system's dynamics, shared control constitutes the most general form of virtual assistance or training. Virtual fixtures, record and play strategies, and transient dynamics amplification are all encompassed as special cases of shared control since these paradigms can easily be realized through shared controllers of specific structures. Shared control has been shown to improve task performance in both physical and virtual environments [88, 89]. Other shared controllers have been proposed for training in minimally invasive surgery [90]. However, effects of these controllers on training is yet to be studied. Finally, the authors' implementation of error reduction with a shared control architecture is shown to improve performance of the task as well as affecting motor skill acquisition through improved retention from one training session to the next compared to practice without assistance [85].

These shared control techniques can be used to create intelligent teleoperation systems for life science applications that go beyond reducing operator tremor. Forbidden regions of the workspace can be established, along with guidance forces to carry out various tests such as analysis of material properties at the micro- and nanoscale. When extrapolated to the training scenario, standard laboratory practices could be incorporated into training protocols for young researchers who must manipulate small-scale structures in difficult environments. Such cues can all be conveyed via the haptic operator interface.

4.4   Future Issues

A number of key research directions have been proposed as the result of a DARPA/NSF study on human-robot interaction, which directly apply to human-machine

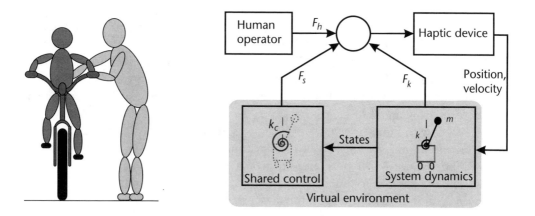

Figure 4.4   Schematic representation of shared controller, where an expert collaborates with the operator. Block diagram representation shows implementation of shared controller with a virtual environment displayed via a haptic interface. The shared controller acts on the states of the system to be controlled.

interface issues in life science automation [32]. These include studies of human intervention with different levels of autonomy; interaction modalities that can be used in various physical environments; and the development of roles for robots and humans within teams. In this spirit, the following section highlights several unique challenges to the field of human-machine interfaces and interactions applied to life science automation and micro- and nanoapplications.

4.4.1   Challenges for Applications in Life Science Automation

A significant challenge for human-machine interaction in the life sciences is the degree of dependency on individual applications that exists. While some procedures such as pick and place tasks may be quite standard and could be completed with a common machine hardware platform, often the laboratory environment presents design constraints for the system, resulting in a need for customized hardware for each application of interest. Tasks may require a wide range of degree of freedom from a manipulator to be completed, or may require a range of end-effector tool options. Additionally, tasks may lend themselves to different interaction modalities, be they haptic, audio, or visual. The geometry of the workspace, and the task to be completed, will define sensor resolution requirements (for controlled positioning tasks) and actuator requirements (for meeting speed and load carrying specifications). Due to the variations in environment and task specification, all-inclusive solutions are unlikely. Therefore, designers of human-machine interfaces for life sciences must consider the issues of what tasks should be automated, how tasks should be allocated between human and machine, performance requirements for the collaborative system, and teaming architectures presented earlier in the chapter to ensure a successful solution.

4.4.2   Challenges for Micro- and Nanoscale Applications

Over the last decade, considerable interest has been generated in building and manipulating nanoscale structures and objects. Experiments have been conducted to interact with nanoparticles, molecules, DNA, and viruses [91–94], measurement of mechanical properties of carbon nanotubes [95], and bottom-up nanoassembly [96, 97]. Nanomanipulation refers to the use of external forces for controlled positioning or assembly of nanoscale objects in two or three dimensions through cutting, drilling, twisting, bending, pick and place, push and pull kind of tasks [98]. Figure 4.5 depicts some basic mechanical manipulation tasks that can be performed at nanoscale using an AFM cantilever [99].

Due to limitations in current understanding of nano-scale phenomenon, nanomanipulation systems are typically implemented using a telerobotic setup [100–103]. Figure 4.6 shows the setup of a proposed nanomanipulation system. The human operator commands a slave robot through the master robotic interface. During manipulation, the operator may be provided force or visual feedback, from the environment, or both. Visual feedback is useful for locating the objects of interest, whereas haptic feedback plays an important role in contact and depth perception.

Several research efforts have focused on the development of scanning probe microscopes (SPMs) for nanomanipulation [93, 102–104]. These systems are

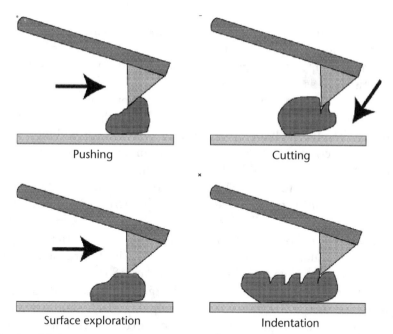

Figure 4.5   Possible nanomanipulation tasks using an AFM cantilever: pushing, cutting, surface exploration, and indentation. (*After:* [99].)

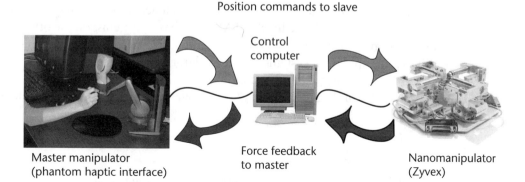

Figure 4.6   Typical nanomanipulation setup. A human operator commands the slave nanomanipulator. Force feedback to the operator may or may not be provided.

generally restricted to two dimensions with a very limited third dimension. Using a scanning tunneling microscope (STM) probe, manipulation of atoms or molecules can be achieved by applying voltage pulses between the probe and the surface of the sample. This was first achieved by Eigler and Schweitzer in 1990 [104]. Sitti and Hashimoto [102] present the design of an AFM-based telerobotic nano-manipulation system. They successfully positioned latex particles with 242- and 484-nm radii on Si substrates, with 30-nm accuracy, using an AFM cantilever as a manipulator. They adopt a two-stage manipulation strategy. First, the image of the particles is obtained using AFM tapping mode. Then the particle is pushed by moving the substrate with a constant velocity. A virtual reality interface including 3D projection display and force feedback for SPM-based manipulation, known as the nanoManipulator, is presented in [93, 103]. In SPM-based nanomanipulation systems, such as these, there is no real-time visual feedback from the environment. Following their experience with the nanoManipulator, Guthold et al. [93] reported: "Force feedback has proved essential to finding the right spot to start a modification, finding the path along which to modify, and providing a subtler touch than would be permitted by the standard scan-modify-scan experiment cycle." Hence, haptic feedback is critical in these nanomanipulation systems.

As compared to the SPM-based nanomanipulation systems, scanning electron microscopy (SEM) or transmission electron microscopy (TEM) based systems provide real-time visual feedback from the nanoscale environment. Dong et al. present a 16 degree of freedom (DOF) nanorobotic manipulator that operates inside a field emission scanning electron microscope (FESEM). Their system supports up to four end-effectors and has a workspace of 18 mm × 18 mm × 18 mm. The relatively large workspace, multiple end-effectors, and large number of DOFs allow complex operations with such a 3D nanorobotic manipulator. The authors report nanorobotic manipulation to be more effective in constructing complex nanostructures than self-assembly and SPM-based systems. These systems, however, are restricted to operate in vacuum.

A comparison of SPM-based systems with SEM/TEM-based systems is presented in Table 4.1. The primary limitation of SPM-based systems is their inability to image and manipulate simultaneously. Due to drift and hysteresis, the position of tip relative to the sample may change over time. Hence, during manipulation, the user relies only upon haptic feedback. SEM/TEM-based systems, on the other hand, provide real-time visual feedback but present challenges in force sensing. The nanorobotic manipulation systems provide a larger workspace for operation than SPM-based systems, but cannot match the position resolution of SPMs. AFMs can also operate in liquids, making them particularly suitable for biological applications.

Table 4.1 SPM Versus SEM/TEM-Based Nanomanipulation Systems

System Description	SPM-Based Nanomanipulators Modified SPM (AFM/STM) System	Nanorobotic Manipulation Systems Robotic System Operates Within SEM/TEM
Number of probes	1	Multiple independent probes
Image resolution	1 nm	~ 5 nm
Position accuracy	1 nm	5–10 nm
Visual feedback during manipulation	No	Yes
Haptic feedback during manipulation	Yes	Some systems
Force/position sensing	Yes	Some systems
Biological sample manipulation	AFM-based systems	No

Haptic feedback has been reported to be critical for nanomanipulation [93]. This is especially true for SPM-based systems, where no real-time visual feedback is available during manipulation. Nanorobotic manipulators that operate inside SEM/TEMs can also benefit from incorporation of haptic feedback as it helps improve sensation of contact. Nanoscale objects could be fragile, and force feedback helps the user apply controlled forces for manipulation.

Various methods of force feedback have been used for SPM-based systems. An AFM cantilever can be used as a nanoscale force sensor. Using an optical detection system, coupled normal and frictional forces can be sensed inside an AFM [100]. Piezoresistive force sensors can be incorporated into the AFM probe during fabrication to provide a compact manipulation system [101, 105]. The STM-based nanomanipulator presented by Taylor et al. [103] provides a method of feedback of surface topography during manipulation using virtual springs. In later work [93], the authors report that regions of high and low friction can also be represented as high or low regions topologically.

Force sensing in nanorobotic systems that operate inside SEM/TEMs is more challenging. Arai et al. present a carbon nanotube–based pico-Newton level force sensor [106]. The nanotubes are attached to an AFM cantilever and their deformation is measured from SEM images. We are not aware, however, of any use of this information for force feedback. The range of this carbon nanotube–based sensor is limited to pico-Newton levels. In addition, the nanotube may not be the ideal end-effector for general purpose nanomanipulation due to its high length-diameter ratio. Hence, there is a need for improved force sensors for such systems.

Gupta, Patoglu, and O'Malley present a vision-based force sensing scheme for a nanorobotic manipulation system that operates inside an SEM [107], shown in Figure 4.7. An AFM cantilever is used as the end-effector and is visually tracked to

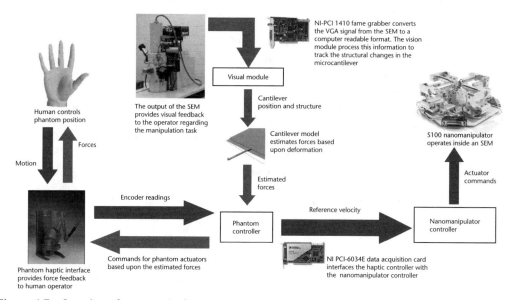

Figure 4.7   Overview of nanomanipulation system proposed by Gupta et al. [107].

measure its deformation. The work is motivated from similar work by Greminger and Nelson [108], who use visual template matching for tracking of a microcantilever to sense nano-Newton forces at the micro level. Their approach of template matching, however, is not suitable for our application of nano-manipulation due to variable occlusion of the cantilever and loss of coherence in consecutive frames, due to the nature of implementation of the magnification functionality in the environmental SEM (ESEM). Hence, a global search strategy is proposed that can be used for force sensing directly or to find a suitable point of initialization for subsequent template matching.

As described, many of the challenges in human-machine interaction at the nanoscale, and extended to the microscale, are based on the issues of sensing and actuation of micro- and nanoscale manipulators. Additionally, the decisions of what modalities and data to render to the human operator must be made with the understanding that forces of interaction at these scales differ from macroscale physics, which humans are more familiar dealing with.

References

[1]   Thrun, S., "Toward a Framework for Human-Robot Interaction," *Human-Computer Interaction*, Vol. 19, Nos. 1–2, 2004, pp. 9–24.

[2]   Fong, T., C. Thorpe, and C. Bauer, "Collaboration, Dialogue, and Human Robot Interaction," *10th International Symposium on Robotics Research*, Lorne, Victoria, Australia, 2001.

[3]   Degani, A., and M. Heymann, "Formal Verification of Human-Automation Interaction," *Human Factors*, Vol. 44, No. 1, 2002, pp. 28–43.

[4]   Woods, D. D., N. Sarter, and C. Billings, "Automation Surprises," in *Handbook of Human Factors and Ergonomics*, G. Salvendy, (ed.), New York: John Wiley & Sons, 1997, pp. 1926–1943.

[5]   Hoc, J.-M., "From Human-Machine Interaction to Human-Machine Cooperation," *Ergonomics*, Vol. 43, No. 7, 2000, pp. 833–843.

[6]   Endsley, M. R., "Automation and Situation Awareness," in *Automation and Human Performance: Theory and Applications*, R. Parasuraman and M. Mouloua, (eds.), Mahwah, NJ: Lawrence Erlbaum, 1996, pp. 163–181.

[7]   Bainbridge, L., "Ironies of Automation," in *New Technology and Human Error*, J. Rasmussen, K. D. Duncan, and J. Leplat, (eds.), New York: John Wiley & Sons, 1987, pp. 271–284.

[8]   Lee, J., and N. Moray, "Trust, Control Strategies, and Allocation of Function in Human-Machine Systems," *Ergonomics*, Vol. 35, 1992, pp. 1243–1270.

[9]   Lee, J., and N. Moray, "Trust, Self-Confidence, and Operators' Adaptation to Automation," *International Journal of Human-Computer Studies*, Vol. 40, 1994, pp. 153–184.

[10]   Muir, B. M., "Trust Between Humans and Machines, and the Design of Decision Aids," in *Cognitive Engineering in Complex Dynamic Worlds*, E. Hollnagel, G. Mancini, and D. D. Woods, (eds.), New York: Academic Press, 1988, pp. 71–84.

[11]   Parasuraman, R., and V. Riley, "Humans and Automation: Use, Misuse, Disuse, Abuse," *Human Factors*, Vol. 39, No. 2, 1997, pp. 230–253.

[12]   Endsley, M. R., and D. B. Kaber, "Level of Automation Effects on Performance, Situation Awareness and Workload in a Dynamic Control Task," *Ergonomics*, Vol. 42, No. 3, 1999, pp. 462–492.

[13]   Chapman, T., "Lab Automation and Robotics: Automation on the Move," *Nature*, Vol. 421, No. 6, 2003, pp. 661–666.

[14] Meldrum, D., "Automation for Genomics, Part One: Preparation for Sequencing 10.1101/gr.101400," *Genome Res.*, Vol. 10, No. 8, 2000, pp. 1081–1092.

[15] Meldrum, D., "Automation for Genomics, Part Two: Sequencers, Microarrays, and Future Trends 10.1101/gr.157400," *Genome Res.*, Vol. 10, No. 9, 2000, pp. 1288–1303.

[16] Ruff, H. A., S. Narayanan, and M. H. Draper, "Human Interaction with Levels of Automation and Decision-Aid Fidelity in the Supervisory Control of Multiple Simulated Unmanned Air Vehicles," *Presence*, Vol. 11, No. 4, 2002, pp. 335–351.

[17] Dunkler, O., et al., "The Effectiveness of Supervisory Control Strategies in Scheduling Flexible Manufacturing Systems," *IEEE Transactions on Systems, Man and Cybernetics*, Vol. 18, No. 2, 1988, pp. 223–237.

[18] Parasuraman, R., T. B. Sheridan, and C. D. Wickens, "A Model for Types and Levels of Human Interaction with Automation," *IEEE Transactions on Systems, Man & Cybernetics, Part A (Systems & Humans)*, Vol. 30, No. 3, 2000, pp. 286–297.

[19] Billings, C., *Aviation Automation: The Search for a Human-Centered Approach*, Mahwah, NJ: Lawrence Erlbaum, 1997.

[20] Lee, J. D., and T. F. Sanquist, "Maritime Automation," in *Automation and Human Performance: Theory and Applications*, R. Parasuraman and M. Mouloua, (eds.), Mahwah, NJ: Lawrence Erlbaum, 1996, pp. 365–384.

[21] Sheridan, T. B., "Rumination on Automation, 1998," *Annual Reviews in Control*, Vol. 25, 2001, pp. 89–97.

[22] Kaber, D. B., and M. R. Endsley, "The Effects of Level of Automation and Adaptive Automation on Human Performance, Situation Awareness and Workload in a Dynamic Control Task," *Theoretical Issues in Ergonomics Science*, Vol. 5, No. 2, 2004, pp. 113–153.

[23] Kaber, D. B., and M. R. Endsley, "Out-of-the-Loop Performance Problems and the Use of Intermediate Levels of Automation for Improved Control System Functioning and Safety," *Process Safety Progress*, Vol. 16, No. 3, 1997, pp. 126–131.

[24] Yanco, H. A., and J. Drury, "Classifying Human-Robot Interaction: An Updated Taxonomy," *2004 IEEE International Conference on Systems, Man and Cybernetics, SMC 2004*, The Hague, the Netherlands, October 10–13, 2004.

[25] Scerri, P., D. V. Pynadath, and M. Tambe, "Towards Adjustable Autonomy for the Real World," *Journal of Artificial Intelligence Research*, Vol. 17, 2002, pp. 171–228.

[26] Ferreira, A., "Strategies of Human-Robot Interaction for Automatic Microassembly," *2003 IEEE International Conference on Robotics and Automation*, Taipei, Taiwan, September 14–19, 2003.

[27] Woods, D. D., "Decomposing Automation: Apparent Simplicity, Real Complexity," in *Automation and Human Performance: Theory and Applications*, R. Parasuraman and M. Mouloua, (eds.), Mahwah, NJ: Lawrence Erlbaum, 1996, pp. 1–16.

[28] Goodrich, M. A., and D. R. Olsen, Jr., "Seven Principles of Efficient Human Robot Interaction," *SMC '03 Conference Proceedings. 2003 IEEE International Conference on Systems, Man and Cybernetics*, Washington, D.C., October 5–8, 2003.

[29] Parasuraman, R., and M. Mouloua, (eds.), *Automation and Human Performance: Theories and Applications*, Mahwah, NJ: Lawrence Erlbaum, 1996.

[30] Lockhart, J. M., et al., "Automation and Supervisory Control: A Perspective on Human Performance, Training, and Performance Aiding," *Proceedings of the 37th Annual Meeting of the Human Factors and Ergonomics Society*, Seattle, WA, October 11–15, 1993.

[31] Leveson, N. G., and E. Palmer, "Designing Automation to Reduce Operator Errors," *1997 IEEE International Conference on Systems, Man, and Cybernetics. Computational Cybernetics and Simulation*, Orlando, FL, October 12–15, 1997.

[32] Burke, J. L., et al., "Final Report for the DARPA/NSF Interdisciplinary Study on Human-Robot Interaction," *IEEE Transactions on Systems, Man and Cybernetics, Part C (Applications and Reviews)*, Vol. 34, No. 2, 2004, pp. 103–112.

[33] Scholtz, J., "Theory and Evaluation of Human Robot Interactions," *36th Hawaii International Conference on Systems Sciences*, Big Island, HI, January 6–9, 2003.

[34] Kidd, P. T., "Design of Human-Centered Robotic Systems," in *Human Robot Interaction*, M. Rahimi and W. Karwowski, (eds.), London, U.K.: Taylor and Francis, 1992, pp. 225–241.

[35] Abbott, K. A., S. M. Slotte, and D. K. Stimson, *Federal Aviation Administration Human Factors Team Report on the Interfaces Between Flightcrews and Modern Flight Deck Systems*, Federal Aviation Administration: Washington, D.C., 1996, pp. D1–D3.

[36] Espinosa, J. A., et al., "Coming to the Wrong Decision Quickly: Why Awareness Tools Must Be Matched with Appropriate Tasks," *Human Factors in Computer Systems, Computer-Human Interaction (CHI) Conference*, 2000.

[37] Endsley, M. R., "Direct Measurement of Situation Awareness: Validity and Use of SAGAT," in *Situation Awareness Analysis and Measurement*, M. R. Endsley and D. J. Garland, (eds.), Mahwah, NJ: Lawrence Erlbaum, 2000.

[38] Crandall, J. W., and M. A. Goodrich, "Characterizing Efficiency of Human Robot Interaction: A Case Study of Shared-Control Teleoperation," *2002 IEEE/RSJ International Conference on Intelligent Robots and Systems*, Lausanne, Switzerland, September 30–October 4, 2002.

[39] Marble, J. L., et al., "Evaluation of Supervisory Vs. Peer-Peer Interaction with Human-Robot Teams," *Proceedings of the 37th Annual Hawaii International Conference on System Sciences*, Big Island, HI, January 5–8, 2004.

[40] Heinzmann, J., and A. Zelinsky, "Safe-Control Paradigm for Human-Robot Interaction," *Journal of Intelligent and Robotic Systems: Theory and Applications*, Vol. 25, No. 4, 1999, pp. 295–310.

[41] Kulic, D. and E. A. Croft, "Real-Time Safety for Human-Robot Interaction," *2005 12th International Conference on Advanced Robotics*, Seattle, WA, July 17–20, 2005.

[42] Kulic, D. and E.A. Croft, "Safe Planning for Human-Robot Interaction," *Journal of Robotic Systems*, Vol. 22, No. 7, 2005, pp. 383–396.

[43] Kulic, D. and E. Croft, "Anxiety Detection During Human-Robot Interaction," *2005 IEEE/RSJ International Conference on Intelligent Robots and Systems*, Edmonton, Alta., Canada, August 2–6, 2005.

[44] Perzanowski, D., et al., "Building a Multimodal Human-Robot Interface," *IEEE Intelligent Systems* [see also IEEE Intelligent Systems and Their Applications], Vol. 16, No. 1, 2001, pp. 16–21.

[45] Stiefelhagen, R., et al., "Natural Human-Robot Interaction Using Speech, Head Pose and Gestures," *2004 IEEE/RSJ International Conference on Intelligent Robots and Systems (IROS)*, Sendai, Japan, September 28–October 2, 2004.

[46] Cypher, A., et al., (eds.), *Watch What I Do: Programming by Demonstration*, Cambridge, MA: MIT Press, 1993.

[47] Aleotti, J., S. Caselli, and M. Reggiani, "Evaluation of Virtual Fixtures for a Robot Programming by Demonstration Interface: Human-Robot Interaction," *IEEE Transactions on Systems, Man, and Cybernetics Part A: Systems and Humans*, Vol. 35, No. 4, 2005, pp. 536–545.

[48] Burdea, G. C., *Force and Touch Feedback for Virtual Reality*, New York: John Wiley & Sons, 1996.

[49] Srinivasan, M. A., "Haptic Interfaces," in *Virtual Reality: Scientific and Technological Challenges*, N. I. Durlach and A. S. Mavor, (eds.), Washington, D.C.: National Academies Press, 1994.

[50] Pang, K. D., H. Z. Tan, and N. I. Durlach, "Manual Discrimination of Force Using Active Finger Motion," *Perception and Psychophysics*, Vol. 49, No. 6, 1991, pp. 531–540.

[51] Tan, H. Z., et al., "Human Factors for the Design of Force-Reflecting Haptic Interfaces," *ASME Dyn. Sys. and Control Div.*, 1994.

[52] Durlach, N. I., et al., "Manual Discrimination and Identification of Length by the Finger-Span Method," *Perception and Psychophysics*, Vol. 46, 1989, pp. 29–38.

[53] Beauregard, G. L., M. A. Srinivasan, and N. I. Durlach, "The Manual Resolution of Viscos-ity and Mass," *ASME Dyn. Sys. and Control Div.*, 1995.

[54] Jones, L. A., "Matching Forces: Constant Errors and Differential Thresholds," *Perception*, Vol. 18, No. 5, 1989, pp. 681–687.

[55] Ellis, R., O. Ismaeil, and M. Lipsett, "Design and Evaluation of a High-Performance Proto-type Planar Haptic Interface," *Advances in Robotics, Mechatronics, and Haptic Interfaces*, 1993.

[56] Brooks, T. L., "Telerobotic Response Requirements," *IEEE International Conference Pro-ceedings on Systems, Man and Cybernetics*, 1990.

[57] Adelstein, B. D., and M. J. Rosen, "Design and Implementation of a Force Reflecting Manipulandum for Manual Control Research," in *Advances in Robotics*, H. Kazerooni, (ed.), New York: ASME, 1992, pp. 1–12.

[58] Lee, C. D., D. A. Lawrence, and L. Y. Pao, "A High-Bandwidth Force-Controlled Haptic Interface," *ASME Dynamic Systems and Control Division*, 2000.

[59] Gupta, A., and M. K. O'Malley, "Design of a Haptic Arm Exoskeleton for Training and Rehabilitation," *IEEE/ASME Transactions on Mechatronics*, Vol. 11, No. 3, 2006, pp. 280–289.

[60] Colgate, J. E., and J. M. Brown, "Factors Affecting the Z-Width of a Haptic Display," *Pro-ceedings of the 1994 IEEE International Conference on Robotics and Automation*, 1994.

[61] Millman, P. A., and J. E. Colgate, "Effects of Non-Uniform Environment Damping on Haptic Perception and Performance of Aimed Movements," *ASME Dyn. Sys. and Control Div.*, 1995.

[62] Rosenberg, L. B., and B. D. Adelstein, "Perceptual Decomposition of Virtual Haptic Sur-faces," *IEEE Symposium on Research Frontiers in Virtual Reality*, San Jose, CA, 1993.

[63] Morgenbesser, H. B., and M. A. Srinivasan, "Force Shading for Haptic Shape Perception," *ASME Dyn. Sys. and Control Division*, 1996.

[64] O'Malley, M., and M. Goldfarb, "The Effect of Force Saturation on the Haptic Perception Of Detail," *IEEE/ASME Transactions on Mechatronics*, Vol. 7, No. 3, 2002, pp. 280–288.

[65] O'Malley, M. K., and M. Goldfarb, "The Effect of Virtual Surface Stiffness on the Haptic Perception of Detail," *IEEE/ASME Transactions on Mechatronics*, Vol. 9, No. 2, 2004, pp. 448–454.

[66] Hale, K. S., and K. M. Stanney, "Deriving Haptic Design Guidelines from Human Physio-logical, Psychophysical, and Neurological Foundations," *IEEE Computer Graphics and Applications*, Vol. 24, No. 2, 2004, pp. 33–39.

[67] O'Malley, M. K., and M. Goldfarb, "On the Ability of Humans to Haptically Identify and Discriminate Real and Simulated Objects," *Presence*, Vol. 14, No. 3, 2005, pp. 366–376.

[68] O'Malley, M. K., and R. O. Ambrose, "Haptic Feedback Applications for Robonaut," *Industrial Robot*, Vol. 30, No. 6, 2003, pp. 531–542.

[69] Williams, L. E. P., et al., "Kinesthetic and Visual Force Display for Telerobotics," *2002 IEEE International Conference on Robotics and Automation*, Washington, D.C., May 11–15, 2002.

[70] Massimino, M. J., and T. B. Sheridan, "Teleoperator Performance with Varying Force and Visual Feedback," *Human Factors*, Vol. 36, No. 1, 1994, pp. 145–157.

[71] Meech, J. F., and A. E. Solomonides, "User Requirements When Interacting with Virtual Objects," *IEE Colloquium (Digest) Proceedings of the IEE Colloquium on Virtual Reality —User Issues*, March 25, 1996, pp. 1–3.

[72] O'Malley, M. K., et al., "Simulated Bilateral Teleoperation of Robonaut," *AIAA Space 2003*, Long Beach, CA, 2003.

[73] Adams, R. J., D. Klowden, and B. Hannaford, "Virtual Training for a Manual Assembly Task," *Haptics-e*, Vol. 2, No. 2, 2001.

[74] Richard, P., and P. Coiffet, "Human Perceptual Issues in Virtual Environments: Sensory Substitution and Information Redundancy," *Proceedings of the 4th IEEE International Workshop on Robot and Human Communication, RO-MAN'95*, Tokyo, 1995.

[75] Rosenberg, L. B., "Virtual Fixtures: Perceptual Tools for Telerobotic Manipulation," *1993 IEEE Annual Virtual Reality International Symposium*, Seattle, WA, September 18–22, 1993.

[76] Haanpaa, D. P., and G. P. Roston, "Advanced Haptic System for Improving Man-Machine Interfaces," *Computers & Graphics*, Vol. 21, No. 4, 1997, pp. 443–449.

[77] Bettini, A., et al. "Vision Assisted Control for Manipulation Using Virtual Fixtures: Experiments at Macro and Micro Scales," *2002 IEEE International Conference on Robotics and Automation*, Washington, D.C., May 11–15, 2002.

[78] Gillespie, R. B., et al., "Virtual Teacher," *Proceedings of the 1998 ASME International Mechanical Engineering Congress and Exposition*, Anaheim, CA, November 15–20, 1998.

[79] Yokokohji, Y., et al., "Toward Machine Mediated Training of Motor Skills—Skill Transfer from Human to Human Via Virtual Environment," *Proceedings of the 1996 5th IEEE International Workshop on Robot and Human Communication, RO-MAN*, Tsukuba, Japan, November 11–14, 1996.

[80] Kikuuwe, R., and T. Yoshikawa. "Haptic Display Device with Fingertip Presser for Motion/Force Teaching to Human," *2001 IEEE International Conference on Robotics and Automation (ICRA)*, Seoul, May 21–26, 2001.

[81] Henmi, K., and T. Yoshikawa, "Virtual Lesson and Its Application to Virtual Calligraphy System," *Proceedings of the 1998 IEEE International Conference on Robotics and Automation*, Part 2 (of 4), Leuven, Belgium, May 16–20, 1998.

[82] Feygin, D., M. Keehner, and R. Tendick, "Haptic Guidance: Experimental Evaluation of a Haptic Training Method for a Perceptual Motor Skill," *Proceedings of the 10th Symposium on Haptic Interfaces for Virtual Environment and Teleoperator Systems, HAPTICS 2002*, 2002.

[83] Patton, J. L., and F. A. Mussa-Ivaldi, "Robot-Assisted Adaptive Training: Custom Force Fields for Teaching Movement Patterns," *IEEE Transactions on Biomedical Engineering*, Vol. 51, No. 4, 2004, pp. 636–646.

[84] Todorov, E., P. Shadmehr, and E. Bizzi, "Augmented Feedback Presented in a Virtual Environment Accelerates Learning of a Difficult Motor Task," *Journal of Motor Behavior*, Vol. 29, No. 2, 1997, pp. 147–158.

[85] O'Malley, M. K., et al., "Shared Control in Haptic Systems for Performance Enhancement and Training," *Journal of Dynamic Systems, Measurement and Control, Transactions of the ASME*, Vol. 128, No. 1, 2006, pp. 75–85.

[86] O'Malley, M. K., and A. Gupta, "Skill Transfer in a Simulated Underactuated Dynamic Task," *Proceedings of the 12th IEEE International Workshop on Robot and Human Interactive Communication*, Millbrae, CA, October 31–November 2, 2003.

[87] O'Malley, M. K., and A. Gupta, "Passive and Active Assistance for Human Performance of a Simulated Underactuated Dynamic Task," *Proceedings of the 11th Symposium on Haptic Interfaces for Virtual Environment and Teleoperator Systems*, Los Angeles, CA, March 22–23, 2003.

[88] Yoneda, M., et al., "Assistance System for Crane Operation with Haptic Display—Operational Assistance to Suppress Round Payload Swing," *IEEE International Conference on Robotics and Automation*, 1999, pp. 2924–2929.

[89] Griffiths, P. G., and R. B. Gillespie, "Sharing Control Between Humans and Automation Using Haptic Interface: Primary and Secondary Task Performance Benefits," *Human Factors*, Vol. 47, No. 3, 2005, pp. 574–590.

[90] Nudehi, S. S., R. Mukherjee, and M. Ghodoussi, "A Shared-Control Approach to Haptic Interface Design for Minimally Invasive Telesurgical Training," *IEEE Transactions on Control Systems Technology*, Vol. 13, No. 4, 2005, pp. 588–592.

[91] Resch, R., et al., "Manipulation of Nanoparticles Using Dynamic Force Microscopy: Simulation and Experiments," *Applied Physics A: Materials Science & Processing*, Vol. 67, No. 3, 1998, pp. 265–271.

[92] Guthold, M., et al., "Identification and Modification of Molecular Structures with the NanoManipulator," *Journal of Molecular Graphics and Modelling*, Vol. 17, 1999, pp. 188–197.

[93] Guthold, M., et al., "Controlled Manipulation of Molecular Samples with the Nano-Manipulator," *IEEE/ASME Transactions on Mechatronics*, Vol. 5, No. 2, 2000, pp. 189–198.

[94] Guthold, M., et al., "Quantitative Manipulation of DNA and Viruses with the Nanomanipulator Scanning Force Microscope," *Surface and Interface Analysis*, Vol. 27, Nos. 5–6, 1999, pp. 437–443.

[95] Nakajima, M., F. Arai, and T. Fukuda, "In Situ Measurement of Young's Modulus of Carbon Nanotubes Inside a TEM Through a Hybrid Nanorobotic Manipulation System," *IEEE Transactions on Nanotechnology*, Vol. 5, No. 3, 2006, pp. 243–248.

[96] Fukuda, T., F. Arai, and L. Dong, "Assembly of Nanodevices with Carbon Nanotubes Through Nanorobotic Manipulations," *Proceedings of the IEEE*, Vol. 91, No. 11, 2001, pp. 1803–1818.

[97] Requicha, A. A. G., et al., "Nanorobotic Assembly of Two-Dimensional Structures," *Proceedings of 1998 IEEE International Conference on Robotics and Automation*, 1998.

[98] Sitti, M., and H. Hashimoto, "Teleoperated Nano Scale Object Manipulation," *Recent Advances on Mechatronics*, 1999, pp. 322–335.

[99] Sitti, M., "Survey of Nanomanipulation Systems," *Proceedings of the 2001 1st IEEE Conference on Nanotechnology*, 2001.

[100] Sitti, M., et al., "Development of a Scaled Teleoperation System for Nano Scale Interaction and Manipulation," *Proceedings of 2001 IEEE International Conference on Robotics and Automation*, 2001.

[101] Sitti, M., and H. Hashimoto, "Two-Dimensional Fine Particle Positioning Under Optical Miscroscope Using a Piezoresistive Cantilever as Manipulator," *Journal of Microelectronics*, Vol. 1, No. 1, 2000, pp. 25–48.

[102] Sitti, M., and H. Hashimoto, "Controlled Pushing of Nanoparticles: Modeling and Experiments," *IEEE/ASME Transactions on Mechatronics*, Vol. 5, No. 2, 2000, pp. 199–211.

[103] Taylor II, R. M., et al., "The Nanomanipulator: A Virtual-Reality Interface for a Scanning Tunneling Microscope," *Proceedings of SIGGRAPH Computer Graphics*, 1993.

[104] Eigler, D. M., and E. K. Schweitzer, "Positioning Single Atoms with Scanning Tunneling Microscope," *Nature*, Vol. 344, 1990, pp. 524–526.

[105] Tortonese, M., R. C. Barrett, and C. F. Quate, "Atomic Resolution with an Atomic Force Microscope Using Piezoresistive Detection," *Applied Physics Letters*, Vol. 62, 1993, pp. 834–836.

[106] Arai, F., et al., "Pico-Newton Order Force Measurement Using a Calibrated Carbon Nanotube Probe by Electromechanical Resonance," *Proceedings of IEEE International Conference on Robotics and Automation*, 2003.

[107] Gupta, A., V. Patoglu, and M. K. O'Malley, "Vision Based Force Sensing for Nanorobotics Manipulation," *IMECE 2006 ASME International Mechanical Engineering Congress and Exposition*, Chicago, IL, 2006.

[108] Greminger, M. A., and B. J. Nelson, "Vision-Based Force Measurement," *Pattern Analysis and Machine Intelligence*, Vol. 26, No. 3, 2004, pp. 290–298.

CHAPTER 5

Fundamentals of Microscopy and Machine Vision

Ge Yang[†] and Bradley J. Nelson[‡]

Microscopy is a basic tool of life science and is widely used to visualize and measure the structure and dynamics of biological systems. Microscopy automation is essential to achieving the high efficiency, reliability, and accuracy required by studies of complex biological systems and is an important element of many life science automation solutions. Implementation of microscopy automation is made through the application of machine vision techniques, which automate the image collection and, more importantly, the image analysis and understanding processes of microscopy. This chapter introduces the fundamentals of microscopy and machine vision, focusing on concepts, theories, and methods that are essential to their integration. Representative applications of machine vision in microscopy automation are presented. Future developments and challenges are discussed [1–63].

5.1 Introduction

Microscopy has always been a basic tool of life science for visualizing the structure and dynamics of biological systems [1, 42]. High-level computational image analysis and understanding significantly expand the roles of microscopy and transform it into a powerful tool for acquiring high-resolution quantitative structural and dynamic measurements [34, 53, 56]. Because of the important roles of microscopy, its automation is required as part of many life science automation solutions for maximal efficiency, reliability, accuracy, and minimal bias and information loss in image data analysis.

Implementation of microscopy automation is made through the application of machine vision techniques, which partially or completely automate the processes of image collection, analysis, and understanding [15, 51]. The goal of this chapter is to introduce the fundamentals of microscopy and machine vision. Although different microscopy modalities are used in life science [49], the focus is exclusively on light

† Laboratory for Computational Cell Biology, Department of Cell Biology, The Scripps Research Institute, La Jolla, California
‡ Institute of Robotics and Intelligent Systems, Swiss Federal Institute of Technology Zurich (ETH)

microscopy, the dominant choice of technique for live imaging and the primary microscopy application for machine vision.[1]

To fully understand the integration of machine vision with microscopy, a brief historical review is necessary. The origination of contemporary microscopy techniques can be traced back at least to the early 1900s [49]. Development of novel microscopes and related labeling, imaging, and data processing techniques played an important role in the development of life science, which in turn provided a primary driving force for advances in microscopy [45, 53]. In comparison, machine vision started in the 1960s primarily as an application of computer vision to automation of industrial manufacturing processes [15, 51]. Machine vision differs from computer vision in that it is more specialized: computer vision aims to develop theoretical frameworks and computational techniques for general image understanding [57], whereas machine vision focuses on industrial automation. Another basic difference is that, unlike computer vision, machine vision often includes hardware development as an integral part of its solutions [15, 51]. Despite such differences, computer vision is always a critical source of concepts, theories, and methods in the development of machine vision techniques.

Driven by the integration of electronic video devices with microscope systems and the development of digital computers, convergence of machine vision with microscopy started in the late 1960s to early 1970s under the concept of *automated microscopy*, with the goal of using computers for pattern recognition purposes [24]. Successful applications in automated blood cell counting and chromosome classification were developed. From the late 1970s until the early 1990s, application of machine vision in microscopy automation grew steadily but was limited primarily to low-level image processing rather than high-level image understanding [24]. In the meantime, important advances were made in development of machine vision techniques for industrial automation, while the fundamental limitations of traditional light microscopy became apparent, especially its lack of selectivity, sensitivity, and molecular details in its readouts.

Since the late 1990s, integration of machine vision with microscopy for automation has started to accelerate significantly due to the synergy of several factors. First, fluorescence microscopy revolutionized light microscopy by overcoming its basic limitations [23, 24, 29, 39]. Automated microscope systems dramatically improved the efficiency of image data collection. Massive computation power became more easily available, making advanced machine vision techniques accessible to more applications. Last, but most importantly, developments in life science—especially in areas such as functional genomics, proteomics, drug screening and testing, and system biology—make it essential to study complex biological systems at large scales, even over entire genomes or proteomes. Such studies depend on microscopy for high-resolution structural and dynamic measurements and require automation for maximal efficiency, reliability, and accuracy. Today, life science applications, especially high-throughput/high-content screening applications, routinely generate large volume of image data [35, 37, 40, 46]. Although image data management can now be efficiently handled using databases, image data analysis remains a major performance bottleneck and a fundamental challenge to machine vision.

---

1.   In the rest of this chapter, *microscopy* and *light microscopy* are used interchangeably.

Application of machine vision in microscopy automation is not limited to large volume image data analysis or high-throughput/high-content studies. The role of machine vision in minimizing bias and information loss through automated image data analysis must be emphasized. With the development of new labeling and imaging techniques and the widespread use of quantitative analysis and modeling in life science, machine vision becomes even more important in providing quantitative measurements for not only high efficiency but also high reliability and accuracy [22, 34, 58].

This chapter aims to introduce the fundamentals of microscopy and machine vision to microscopy automation practitioners. Because of the breadth of both fields, we have to limit our introduction to basic concepts and principles rather than technical details. It is expected that readers interested in a specific topic will use the references provided as a starting point to find related literature. The rest of this chapter is organized as follows: we start with introducing the fundamentals of microscopy because the physical properties of microscopy image formation impose fundamental constraints on machine vision techniques. Then we introduce several machine vision techniques that are selected based on their importance to microscopy automation, including *autofocusing, image registration, feature tracking*, and *segmentation*. A representative application of machine vision in microscopy automation is analyzed in detail. Future developments and challenges are discussed.

5.2 Fundamentals of Light Microscopy

We start with introducing the basic structure of optical microscopes from a user's perspective. Then the two fundamental issues in microscopy, namely contrast generation and resolution configuration, are addressed. Different contrast generation techniques are reviewed. Resolution configuration is discussed with a summary of basic metrics to characterize the performance of microscopes. Next, specialized three-dimensional microscopy techniques are introduced. Several important practical issues in microscopy are discussed subsequently. Finally, the fundamental constraints imposed by microscope optics on image formation are summarized. Although geometrical and physical optics are not covered, knowledge of these subjects is valuable to understanding microscopy and to customizing microscope systems for specific applications [8, 44, 49].

5.2.1 Basic Structure of Optical Microscopes

To achieve optimal image quality in microscopy automation, it is essential to understand the structure of light microscopes and methods to correctly align and adjust their components [24, 39, 50]. Depending on whether its objective lens is on the same side as the specimen, an optical microscope can be either upright or inverted (an inverted microscope is shown in Figure 5.1). Structurally, an optical microscope is composed of a mechanical base and various optical components that form its *optical path*. The mechanical base provides mechanical supports to the optical components, facilitates their alignment, adjustment, and exchange, enables their coupling with other external functional units, and suppresses vibration. The optical path of a microscope consists mainly of the following components [24, 50] (Figure 5.1):

1. *Illumination source:* This provides illumination light at required wavelengths. For illumination over wide continuous wavelength spectra, sources such as tungsten-halogen lamps, arc-discharge lamps, or light-emitting diodes (LEDs) can be used. For illumination within narrow spectrum bands, wavelength filters need to be added. For high-intensity illumination within very narrow spectrum bands used in confocal microscopy, laser sources are required [24, 39].

2. *Illumination filter:* This conditions light emissions from the illumination source to meet specific requirements in terms of wavelength, phase, and polarization.

3. *Condenser:* This gathers and focuses light from the illumination source so that a light cone can be formed to provide uniform illumination over the field of view.

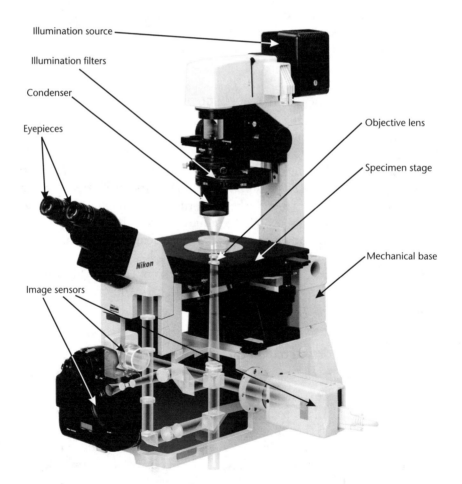

Figure 5.1 Optical path of an inverted microscope. Major components along the optical path of a Nikon TE-2000 inverted microscope are shown. (Image courtesy of Nikon. Annotations have been added by the authors.)

4. *Specimen stage:* This holds the specimen and controls its position and orientation within the field of view. Stage movement along the optical axis (often called the *z*-axis) is used for focusing or three-dimensional scanning.

5. *Objective lens:* This collects light emission from the specimen and is the most critical component in determining the resolution, contrast, magnification, intensity, and color fidelity of the image formed. A revolving nosepiece is used to hold multiple objective lenses for convenient switching.

6. *Image filter:* This further transforms passing light in terms of factors such as wavelength, phase, and polarization, to satisfy specific imaging requirements before the final image is projected onto an image sensor.

7. *Image sensor:* This collects influx of photons and converts optical signals into digital image signals for transmission and storage [39].

8. *Eyepiece:* This is also called an ocular. As an alternative to image sensors, they provide further magnification of the intermediate image of the specimen for direct view by human operators.

All these components must be precisely aligned and adjusted to achieve optimal balance between image resolution, contrast, and intensity [39, 50]. Contemporary microscope design provides motorized control of almost all these components, making it possible to automate microscope operation using computer control. Excellent illustrations of these components with detailed introduction to light microscopy can also be found at the *Optical Microscopy Primer* Web site, http://micro.magnet.fsu.ed/primer.

5.2.2 Contrast Generation in Microscopy

The fundamental role of any microscope is to provide adequate contrast and resolution [24]. Without adequate contrast, it would be impossible to differentiate between the specimen and its background in images. Among the many contrast generation methods developed for life science, fluorescence microscopy is the most important and useful. In practice, selection of contrast generation methods, including related labeling techniques, is application-specific as it depends on the geometrical, physical, biochemical, and even genetic properties of the specimen [47]. Correspondingly, it is often necessary and feasible to optimize machine vision technique for a specific contrast generation method. In the following, commonly used contrast generation techniques are classified and reviewed. There is overlap between these classifications as they are made from different perspectives.

5.2.2.1 Transmitted Light and Reflected Light

Depending on whether the optical path of a microscope passes through the specimen, its illumination can be categorized as either *transmitted light* or *reflected light*. Transmitted light illumination (also called transillumination) is often used in imaging thin specimens, where contrast is generated by specimen staining or refraction at object boundaries (see Figure 5.1). Kohler illumination is required for all transmitted light settings [24, 50]. However, contrast under transmitted light often is poor since many biological specimens are transparent.

Reflected light illumination is commonly used to image thick and opaque specimens such as tissues and embryos. Several variants have also been developed for life science. An example is total internal reflection fluorescence (TIRF) microscopy, which is widely used to image thin cell-substrate interfaces or cell membranes. The penetration depth of imaging into the specimen is generally below 200 nm [2].

5.2.2.2 Brightfield and Darkfield

Under conventional illumination with axial transmitted light, the specimen appears dark against a bright background. This contrast configuration is called *brightfield*. It can be reversed by using a special condenser so that only the light scattered by the specimen can enter the objective lens. The corresponding configuration is called *darkfield* as the specimen appears bright against a dark background. Darkfield is particularly useful in imaging small particles or thin and narrow objects [24].

5.2.2.3 Phase Contrast

Phased-based contrast generation relies on the difference in refractive indices between the specimen and its background [Figure 5.2(a)]. When a light beam passes through the specimen, it undergoes a phase shift (hence called the deviated beam) as compared to the light waves passing the background (the undeviated beam). These two types of waves are recombined using a special optical component called the phase plate. Contrast is generated depending on whether the interference between these two waves is additive or subtractive. Phase contrast provides an improvement over transmitted light. However, its application is also limited due to its low resolution, limited contrast, and generation of artificial shadows [Figure 5.2(a)].

Differential interference contrast (DIC) provides an important improvement over conventional phase contrast techniques by using a complete beam splitting and recombining scheme [39, 49]. Incident illumination is first split into two orthogonally polarized beams using a beam-splitting prism. After these two beams pass through the specimen, they are recombined using another prism. Final images are formed after further conditioning using an analyzer. DIC is capable of generating high-contrast images of very thin specimens. However, like conventional phase contrast, DIC can also suffer from generation of artificial shadows.

5.2.2.4 Fluorescence Microscopy

Fluorescence microscopy is the most important and the most extensively used contrast generation technique in contemporary light microscopy [Figure 5.2(b–d)]. It utilizes the fluorescence properties of fluorophores such as organic dyes, fluorescent proteins, or quantum dots. A common property of these markers is that they absorb excitation irradiation within a narrow wavelength band and emit light within another narrow band at a longer wavelength [23, 29]. Fluorescence microscopy has a few important advantages over the contrast generation techniques introduced previously. First of all, it is highly sensitive and can even be used for detection of single molecules [39]. It can provide high resolution spatial and temporal readouts of the structure and dynamics of living organisms. Second, it is highly specific.

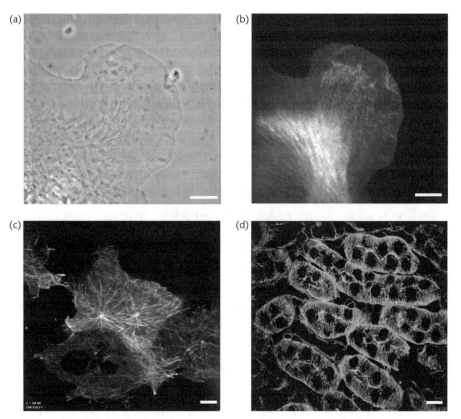

Figure 5.2   Phase and fluorescence contrast generation. (a) The leading edge of a PtK1 cell under phase contrast. There are many particle-like shadows due to backscattering. (b) The same region with actin network labeled using X-rhodamine, an organic dye. Notice the improvement in specificity and contrast of (b) over (a). Bar: 10 μm. (c) EB3 (a microtubule plus end binding protein) labeled with green fluorescent protein (GFP) in vero cells. The firework-like patterns are generated by the highly specific labeling of microtubule plus ends. Bar: 5 μm. (d) Spatial colocalization of different proteins inside a mouse kidney section. Bar: 5 μm. (a) is taken using a normal inverted microscope under phase contrast. (b), (c), and (d) are taken using different spinning disk confocal microscopes.

Fluorophores can be conjugated to antibodies for highly specific labeling of targeted proteins. Third, it is suitable for living imaging. In particular, cloning of fluorescent proteins makes it possible to label almost any proteins for live imaging with high specificity [23] [Figure 5.2(c)]. Finally, since different fluorophores can be differentiated by their excitation and emission wavelengths, multiple markers can be used to study the complex interaction of proteins within living organisms [Figure 5.2(d)]. All these properties make fluorescence microscopy ideal for quantitative study.

An important related concept is epi-illumination, the dominant illumination setting for fluorescence microscopy. Under this configuration, a dichromatic mirror is used to direct the illumination beam into the objective lens, which is also used as a condenser. Epi-illumination has several advantages over transillumination. In particular, the condenser and the objective lens are always perfectly aligned under epi-illumination so that alignment errors are eliminated [24, 39].

Details of several more specialized but less commonly used contrast generation techniques can be found in [24, 39].

5.2.3 Resolution Configuration and Related Performance Metrics

In addition to contrast generation, another fundamental issue in microscopy is to configure microscopes for adequate resolution, which determines their capability to resolve fine details. Microscope resolution can be defined in different ways based on different definitions of when details are considered resolved. The most commonly used is the Rayleigh limit, which is based on a definition of the minimal distance between two neighboring but resolvable points (Figure 5.3) and can be computed as

$$D = \frac{0.61 \cdot \lambda}{NA} \tag{5.1}$$

where D is the resolution, λ is the illumination wavelength, and NA is the numerical aperture of the objective lens used. According to this definition, microscope resolution is determined by both the illumination (excitation) wavelength[2] and the numerical aperture (NA) of the objective lens. In practice, however, the illumination

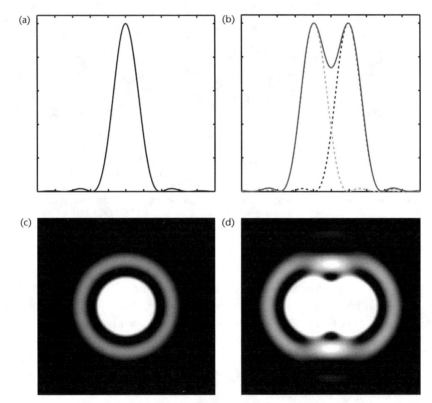

Figure 5.3 Airy pattern and Rayleigh limit. (a) shows the theoretical response of a point feature in microscope image plane, often referred to as the Airy pattern. (b) The Rayleigh limit is based on the notion that the closest resolvable distance is reached when the maximum of one Airy pattern coincides with the first minimum of its neighboring Airy pattern. The corresponding 2D image for (a) and (b) are shown in (c) and (d), respectively. Contrast has been adjusted to make higher order maxima visible. Rayleigh criterion is intuitive but does not consider the actual statistical nature of the photon collection process in imaging [43].

---

2.   The visible light spectrum used by fluorescence microscopy is approximately from 380 to 780 nm.

wavelength is often fixed as it is determined by the fluorescent markers selected. Consequently, choosing objective lenses with higher NA becomes the only feasible way to achieve higher spatial resolution. An important fact about light microscopy is that under this criterion, the spatial resolution limit of optical microscopes is approximately 200 nm. Breaking this limit is an important research topic in contemporary light microscopy [43].

The importance of microscope resolution configuration resides in that choosing a high NA objective lens for high resolution will also influence many other aspects of microscope image formation. To understand this issue, several additional microscope performance metrics are introduced in the following. Understanding these metrics is essential to understanding the basic constraints in microscopy image formation, which must be considered in development of related machine vision techniques.

5.2.3.1   Field of View

The field of view (FOV) of a microscope determines the size of the region that can be imaged. It can be computed using the following equation:

$$FOV = \frac{FN}{M} \tag{5.2}$$

where M is the optical magnification, and the field number FN is the diameter of the image field measured at the intermediate image plane of the microscope (usually in millimeters). In order to actually achieve the resolution determined by the wavelength and the NA, high optical magnification is required because the spatial sampling in the image plane must adequately match the resolution by satisfying the Shannon sampling theorem [39]. But higher magnification comes at the expense of a smaller field of view. In terms of imaging, this means that only a smaller and more limited region of the specimen can be observed.

A common misconception about optical magnification is that it also determines resolution. However, provided that the spatial sampling requirement is satisfied, no new details can be resolved by merely increasing the size of objects in images.

5.2.3.2   Depth of Field

The depth of field of a microscope is the distance between the closest and farthest in-focus position along the optical axis. It determines the thickness of the imaging region (also called the *optical section*) within the specimen. The total depth of field d_{total} of a microscope system depends on its optical configuration and can be computed using the following equation:

$$d_{total} = \frac{\lambda n}{NA^2} + \frac{ne}{M \cdot NA} \tag{5.3}$$

where n is the refractive index of the objective lens, and e is the smallest resolvable distance of the image detector [24]. Choosing higher NA for higher resolution will cause smaller depth of field. In terms of imaging, this means that regions outside of the optical section will be blurred in images and will generate undesirable background image signals. Addressing this issue is critical to 3D microscopy. Related techniques are introduced in the next section.

5.2.3.3 Light Collection Power (Image Intensity)

The power of a microscope to collect light, which determines the intensity of the image formed, is determined by its numerical aperture and optical magnification. For transillumination, this relation is given by

$$I \propto \frac{NA^2}{M^2} \tag{5.4}$$

For epi-illumination in fluorescence microscopy, this relation is given by

$$I \propto \frac{NA^4}{M^2} \tag{5.5}$$

Under the same magnification, increasing NA will significantly increase image intensity, especially for fluorescence microscopy. On the other hand, under the same NA, image brightness decreases rapidly as magnification is increased for higher spatial sampling. In terms of imaging, lower image intensity can cause a lower signal-to-noise ratio (SNR), which can create significant difficulties to subsequence image analysis and understanding.

5.2.4 Three-Dimensional Microscopy

Since the structure and dynamics of living organisms are inherently three dimensional, 3D microscopy is particularly important to life science. Although the introduction to contrast generation techniques and the discussion on resolution configuration are applicable to microscopy at both 2D and 3D, more specialized techniques are required for 3D microscopy.

As explained previously, a basic property of high-resolution microscopy is that only a thin optical section of the specimen is perfectly in focus. This provides a very useful tool to image a highly localized and specific region of the specimen. To achieve 3D microscopy, this optical sectioning capability must be combined with scanning at multiple depths along the optical axis of the microscope so that the 3D structure of the specimen can be reconstructed. The main technical challenge, however, is to minimize undesirable artifacts generated by blurred, out-of-focus features in each optical section. Commonly used techniques to address this issue include:

1. *Deconvolution microscopy:* Deconvolution microscopy is essentially a computational technique. Since the microscope imaging process can be described in spatial domain as the convolution with a point spread function *PSF* (*x,y*) [see also Figure 5.3(a, c)], the resulting image in each optical section is the convolution of *PSF* (*x,y*) with the actual artifact-free but unknown image *I* (*x,y*). The basic idea of deconvolution microscopy is to reverse this process by deconvolution so that an estimation of *I* (*x,y*) can be

obtained. This technique has become reasonably mature, with many commercially available software packages providing different algorithms for this purpose. A detailed review is given in [50].

2. *Scanning confocal microscopy:* Despite it effectiveness, the performance of deconvolution microscopy is fundamentally limited by the information loss in the imaging process and the influence of artifacts. Scanning confocal microscopy provides an optical solution to reduce the undesirable contribution of blurred, out-of-focus features [Figure 5.2(b–d)]. Under conventional microscopy configuration, the entire field of view is imaged simultaneously, a configuration called *widefield* microscopy. The configuration for confocal microscopy is significantly different. Under the commonly used point scanning confocal mode, a focused laser beam is used to scan over the specimen so that only a highly localized in-focus region of the specimen is imaged at each instant. An exit pinhole is used to further reduce the contribution of light from the neighboring out-of-focus regions, so that only light from the scanning spot can pass through and be collected. Other implementations using line scanning or spinning disk scanning are also commercially available [39].

3. *Two-photon confocal microscopy:* An important limitation of conventional confocal microscopy is its excessive illumination: Although only the light emission from a small in-focus region is collected for imaging at each moment, a much wider region is *illuminated* by the scanning beam. This imposes a basic constraint on imaging due to photobleaching and photodamage, which will be introduced in the next section. The two-photon technique provides a solution to this problem by ensuring that only the region in focus is irradiated, and thus further reduces the contribution of out-of-focus components. An important advantage of two-photon microscopy is that it can image deep (up to hundreds of microns) into the specimen because of its longer excitation wavelength. More details can be found in [16, 17, 39].

5.2.5   Several Practical Issues in Microscopy

5.2.5.1   Photobleaching and Photodamage in Fluorescence Microscopy

Because of the importance of fluorescence microscopy in contemporary light microscopy, it is essential to understand the concepts of photobleaching and photodamage. A basic property of fluorophores is that they emit photons under excitation. However, such light emission capability is not static. Under constant irradiation of photons, fluorophores will gradually lose such capability. This phenomenon is called *photobleaching*. In terms of imaging, photobleaching causes a gradual decrease in image intensity and image contrast. The change of image intensity over time can be described using the following equation:

$$I(t) = I_0 e^{-\lambda t} \qquad\qquad (5.6)$$

where I_0 denotes the initial average image intensity, and the parameter λ is called the photobleaching rate. This intensity change often must be explicitly addressed in related machine vision techniques. On the other hand, it has been exploited to study protein colocalization and dynamic turnover in cells by actively inducing photobleaching [28]. Another related concept is *photodamage*. Although the exact mechanism is not well understood, it is known that long exposure to illumination induces generation of free toxic radicals that cause cell death. This imposes a basic constraint on the duration of imaging and the frequency of exposure.

5.2.5.2 Special Requirements of Live Imaging

The ability to image specimens under live conditions is critical to life science. However, several practical issues must be addressed in implementation. The most important requirement is to ensure specimen viability, which requires not only the control of photobleaching and photodamage but also the control of environmental factors such as temperature and culture medium condition. Live imaging creates several problems for machine vision techniques. For example, imaging condition often fluctuates significantly due to specimen dynamics and photobleaching. Machine vision techniques that are robust against such fluctuations are required. As another example, specimen position drift is common, which must be canceled using image alignment techniques before accurate measurements can be obtained [39].

5.2.6 Discussion

We can now summarize the basic properties of image formation under high resolution microscopy that are required by life science applications. Because of the use of high numerical aperture and high magnification to resolve and sample fine features, microscopy image formation is subject to the following constraints:

- Small field of view;
- Small depth of field;
- Strong dependence of image intensity on microscope configuration;
- Significant fluctuation in imaging conditions.

These basic constraints differentiate image formation under microscope optics from that under macroscale optics, which is assumed in many industrial automation applications. Consequently, machine vision techniques developed for macroscale applications may not be directly applicable to microscopy. Instead, customization of machine vision techniques often is required to address these constraints. This will be the guideline for the introduction of machine vision techniques in the next part of this chapter. In general, it is important for practitioners of machine vision techniques to recognize these constraints and the importance of practical experience with microscopy.

5.3 Fundamentals of Machine Vision

This section introduces several machine vision techniques that are selected based on their importance to microscopy automation, including *autofocusing, image alignment, feature tracking*, and *image segmentation*. Emphasis is placed on introducing basic concepts and principles. Comprehensive introductions to machine vision techniques are given in [15, 51, 52, 57].

5.3.1 Autofocusing

Autofocusing (automated focusing) is a basic requirement of microscopy automation since microscopes must be precisely focused to achieve optimal image quality. Autofocusing techniques utilize a basic property of microscopic optics that in-focus features appear "sharper" compared to out-of-focus features because of better preservation of high spatial frequency details [27]. The amount of high frequency details within a given region of an image can be quantified in different ways using different focus measures. These measures are useful not only for autofocusing but also for many other machine vision techniques because they provide depth information [36].

To qualify as a focus measure, the function designed to quantify high frequency details must have only one global maximum (i.e., unimodal) and must increase monotonically as the selected region of interest approaches its in-focus position and decrease monotonically as the region departs from its in-focus position [54, 55]. Therefore, with these measures, the focusing process is essentially a search for their maximum [54]. Due to the influence of image noise, these conditions may not be perfectly satisfied. However, nonoptimal local maxima in focus measures often can be tolerated since autofocusing usually uses global search strategies to find the global maximum [27].

Among the many focus measures available, the following three spatial domain operators—namely, sum-of-image-variance M_{IV} (5.7), sum-of-gradient-magnitude M_{GS} (5.8), and sum-of-modified-Laplacian M_{SML} (5.9)—have been reported to be the most effective based on extensive experimental results [55]:

$$M_{IV} = \frac{1}{|E|} \iint_E \left(I(x,y) - \mu_I \right)^2 dxdy \tag{5.7}$$

$$M_{GS} = \frac{1}{|E|} \iint_E \|\nabla I(x,y)\|^2 dxdy \tag{5.8}$$

$$M_{SML} = \frac{1}{|E|} \iint_E \left(\left| \frac{\partial^2 I}{\partial x^2} \right| + \left| \frac{\partial^2 I}{\partial y^2} \right| \right)^2 dxdy \tag{5.9}$$

where $I(x,y)$ denotes the original image, E denotes the selected region in which the focus measure is computed, and μ_I denotes the average image intensity within E. Implementation and performance evaluation details are given in [27, 36, 54, 55].

On the other hand, focus measures can also be defined on various image transform domains. For example, we have proposed the following measure in wavelet transform domain

$$M_{wt}^{i} = \frac{1}{wl}\left[\sum_{(i,j)\in E_{LH}} \left(W_{LH}(i,j) - \mu_{LH}\right)^{i} + \sum_{(i,j)\in E_{HL}} \left(W_{HL}(i,j) - \mu_{HL}\right)^{i} + \sum_{(i,j)\in E_{HH}} \left(W_{HH}(i,j) - \mu_{HH}\right)^{i} \right]$$

(5.10)

where W and μ denote the wavelet transformation of the original image and its average within subbands LH, HL, HH, and w and l denote the width and length of the region on which the measure is computed, respectively. This operator effectively utilizes the small depth-of-field property of microscopy images and often performs better than spatial domain measures. Implementation and performance evaluation details are given in [60, 61].

In practice, several performance metrics must be considered in development, comparison, or selection of focus measures:

- *Accuracy:* Focus measures must provide an accurate indication of in-focus positions.
- *Selectivity:* Contributions of out-of-focus features must be sufficiently suppressed in focus measures.
- *Robustness:* Under noise perturbations, focus measures must remain unimodal and provide satisfactory accuracy and selectivity.

Detailed mathematical quantification of these metrics is given in [60].

Several practical issues also must be addressed in implementation of focus measures, especially where to apply these operators and how large the operator window must be. Placement of a focus measure depends on the structure and geometry of the specimen and should be determined through active selection of regions of interest. The size of a focus measure window depends on the inherent scale of images. If the window size is too small, the focus measure may become unstable (i.e., non-unimodal) under the influence of noise. On the other hand, if the window size is too large, the focus measure may lose accuracy and selectivity. Detailed discussion of these issues with examples is given in [60].

Although machine vision hardware is beyond the scope of this chapter, it should be pointed out that execution of autofocusing requires a motorized position control system to adjust the distance between the specimen and the microscope objective lens. Recently, several commercial microscope systems added functions to precisely maintain predefined in-focus positions through feedback control. Such functions are particularly useful for living imaging as they can compensate for the inherent position drift of microscopes over long periods of imaging.

5.3.2 Image Alignment

Image alignment, also referred to as image registration, is a basic technique in machine vision. Its goal is to establish the correspondence between two images taken at different times or using different modalities [63]. An important application of

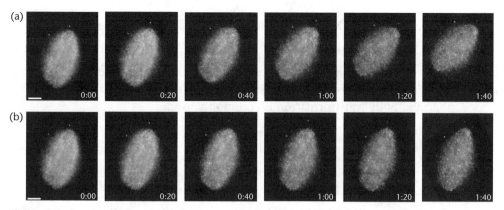

Figure 5.4 Image alignment. (a) shows a series of unaligned widefield images of a *Xenopus* egg extract spindle. (b) shows the image series after alignment using a correlation maximization algorithm. The alignment algorithm can accommodate translation, rotation and changes in spindle shape. Bar: 10 μm [9].

image alignment in microscopy automation is to correct specimen position drifts, a common problem in live imaging. For example, cell biology studies often require tracking of fluorescently labeled protein molecules inside cells to understand their dynamics. However, cell position drifts must be cancelled using image alignment to avoid bias in tracking results (Figure 5.4). Another important application of image alignment is to establish correspondence between images taken using multiple modalities. For example, simultaneous use of light microscopy and electron microscopy to image the same specimen is a useful technique to break the resolution limit of light microscopy. Image alignment is required to locate the same region within images from the two microscopy modalities.

A wide variety of image alignment techniques are available [63]. In particular, medical image analysis, which is another application field of computer vision, provides an important source of image alignment techniques, especially for deformable object images and 3D images. Many medical image registration algorithms have been implemented in the open-source ITK[3] software package. A comprehensive survey of medical image registration techniques is given in [31].

Depending on whether explicit image feature detection is required, image alignment techniques can be classified as either feature-based or region-based [63].

5.3.2.1   Feature-Based Image Alignment

Given a set of matched features from images I_k and I_{k+1} whose positions are represented by vectors X_k and X_{k+1}, alignment of I_{k+1} to I_k is to compute the transformation T_k^{k+1} as the solution to the following optimization problem

$$\hat{T}_k^{k+1} = \arg\max_{T_k^{k+1}} D\left(X_k, T_k^{k+1} X_{k+1}\right) \tag{5.11}$$

where $D\left(X_k, T_k^{k+1} X_{k+1}\right)$ is a cost function that characterizes how good the alignment is. For example, a simple form of $D\left(X_k, T_k^{k+1} X_{k+1}\right)$ can be the Euclidean distance function. Features commonly used for image alignment include points, corners, lines, edges, or even closed regions. For nondeformable objects, the following rigid body transformation is commonly used:

$$T_k^{k+1} = \begin{bmatrix} \cos\theta & -\sin\theta & 0 \\ \sin\theta & \cos\theta & 0 \\ d_x & d_y & 1 \end{bmatrix} \tag{5.12}$$

where $\left[d_x, d_y\right]$ and θ denote relative displacement and rotation, respectively. Depending on the computation scheme for optimization, estimation of T_k^{k+1} can be explicit or implicit [63]. Other forms of alignment cost also can be used. An example is the iterative closest point algorithm, now a standard technique for 3D image alignment [6].

After applying the transformation of T_k^{k+1} to I_{k+1}, the transformed image must be resampled using interpolation, a low-pass filtering process that can cause degradation of image signals. Feature-based image alignment is undoubtedly the method of choice when artificial alignment markers can be used. However, its accuracy is limited by the accuracy of feature detection. For live imaging, it also requires sufficiently stable features that remain detectable over multiple frames. This requirement often cannot be satisfied in live microscopy due to factors such as imaging condition changes.

5.3.2.2 Region-Based Image Alignment

Region-based registration methods align I_{k+1} to I_k by solving T_k^{k+1} that maximizes the correlation between selected regions within I_{k+1} to I_k. These methods do not require explicit feature detection, they are much less sensitive to changes in image features, and they have the advantage of being able to achieve subpixel alignment accuracy reliably. Since microscopy images usually have low SNR and lack long-term stable features, region-based methods often are more robust than feature-based methods.

A classical example of region-based registration techniques is to find the transformation T_k^{k+1} that maximizes the normalized cross-correlation defined as follows:

$$C\left(I_k, I_{k+1}\right) = \frac{\iint \left(I_k(x,y) - \mu_k\right)\left(T_k^{k+1} I_{k+1}(x,y) - \mu_{k+1}\right) dxdy}{\sqrt{\iint \left(I_k(x,y) - \mu_k\right)^2 dxdy} \sqrt{\iint \left(T_k^{k+1} I_{k+1}(x,y) - \mu_{k+1}\right)^2 dxdy}} \tag{5.13}$$

where μ denotes average image intensity. Various numerical optimization techniques can be used to find the solution. Fast Fourier transform is commonly used to accelerate the correlation computation.

Another commonly used criterion is mutual-information, which is defined as

$$M\left(I_k, T_k^{k+1} I_{k+1}\right) = E\left(I_k\right) + E\left(T_k^{k+1} I_{k+1}\right) + E\left(I_k, T_k^{k+1} I_{k+1}\right) \tag{5.14}$$

where $E(I) = -\sum_i \sum_j p_I(i,j) \log p_I(i,j)$ denotes the Shannon entropy of an image I. Mutual-information based image registration is now a standard technique in medical image analysis for registering images taken from different modalities. Implementation details are provided in [41].

A potential limitation of region-based image alignment is its high computation cost. However, this limitation often can be alleviated: Hierarchical representations of images using pyramids or wavelets allow image alignment to be performed first at a coarse scale and then gradually refined for higher accuracy [52].

5.3.3   Feature Tracking

Microscopy provides the tool to visualize the localization and interaction of objects of interests within complex biological systems in space and time. To obtain quantitative readouts from this visualization, tracking the motion of the objects using machine vision techniques is essential. A practical description of the entire tracking procedure is given in [34]. Many tracking algorithms were initially developed for military applications [5, 7] and then were adopted for general computer vision tasks [12, 13, 21, 59]. Depending on the geometry of the features being tracked, these techniques can be categorized as either particle tracking or region tracking.

5.3.3.1   Particle Tracking Techniques

A particle is an image feature that can be represented as a geometrical point. Whenever an object has a size smaller than the diffraction limit (~200 nm) of light microscopy and is well separated from the others, it will appear as a particle in microscopy images. Since many components in biological systems satisfy this requirement, particle tracking is widely used in life science studies [10]. Among the many techniques available, three classes of tracking techniques—namely, global nearest neighbor (GNN), joint probabilistic data association filter (JPDAF), and multiple hypothesis testing (MHT)—have been widely applied and proved to perform well for different applications [5, 7, 12]. As the simplest technique, nearest neighbor (NN) tracking associates an arbitrarily given particle i from the set of all particles G_k at frame k to its nearest neighbor particle j from all particles G_{k+1} at the next frame $k+1$ [10]. GNN provides a direct improvement over NN by optimizing this assignment among all possible association choices between G_k and G_{k+1} to resolve conflicts. Although the advantage of low computational complexity of GNN makes it applicable to tracking large numbers of particles, its performance is limited because its association decisions are made based on information from two consecutive frames rather than a longer time interval.

JPDAF and MHT address the limitation of GNN of making association decisions based on two frames by delaying association decisions until more evidence from multiple frames is available. JPDAF performs well in tracking small number of particles under noisy conditions when the particle number remains constant. However, its performance can degrade significantly in tracking dense particle motion when trajectory merging and splitting become frequent [7, 12]. MHT provides a theoretically optimal solution to the tracking problem and in principle can overcome the limitations of JPDAF. However, its implementation is complicated and

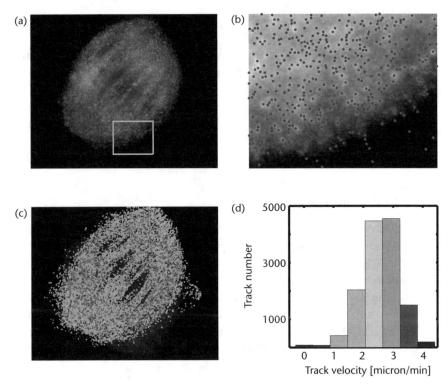

Figure 5.5 Particle tracking application. (a) A fluorescent speckle microscopy image of a *Xenopus* egg extract spindle. (b) Detected particles (a total of 415) within the region marked in (a). (c) Computed particle trajectories are coded based on their magnitude of velocity. Total number of tracks: 13,337. Total number of speckles: 70,396. (d) Histogram of particle velocity and the mapping of velocity range.

requires accurate models of particle motion. Another fundamental limitation of both JPDAF and MHT is their exponential computational complexity, which makes them unusable to tracking large number of particles.

In practice, selection of particle tracking techniques is application specific. An example is the tracking of fluorescent speckles in mitotic spindles[4] [14, 62]. Since the number of speckles in each frame typically ranges between 2,000 and 10,000, JPDAF and MHT are not applicable due to their exponential computational complexity. Furthermore, speckle motion in mitotic spindles is antiparallel: two speckle populations are spatially mixed and move in opposite directions. Tracking this antiparallel motion using simple GNN gives poor results. To solve this problem, we developed a modified GNN algorithm that incorporates local flow vector field so that information of local organization of particle motion is directly used in tracking [9, 62]. This algorithm provides good tracking results, has the same computational complexity as GNN, and is scalable to tracking even larger numbers of particles [62] (Figure 5.5).

---

4. A mitotic spindle is a molecular machinery that separates duplicated chromosomes in cell division.

5.3.3.2   Nondeformable Region Tracking Techniques

When the objects to be tracked cannot be simplified as points or when point features are not available, region tracking techniques are required. For example, when an object has a size greater than the diffraction limit of light microscopy, it appears as a region of connected pixels in images. If the shape of this region is an essential part of the representation of the object, the object should no longer be tracked as a point. As another example, certain biological structures such as the actin network shown in Figure 5.2(b) may not provide point features that are sufficiently stable and distinct for tracking. In this case, region tracking can be used to effectively extract related motion information [4, 25].

Here we review techniques for tracking objects whose shape either is considered to be rigid or undergoes only affine transformation [3]. In the first case, motion of the object is represented by rigid-body transformations (5.12). The second case is a generalization of the first case in which the motion of the object is described by more general affine transformations, which represent not only translation and rotation but also scaling and shear deformation. For both cases, the apparent motion of the object is often referred to as *optical flow* in computer vision [4].

Under the affine transformation framework, a region feature (also referred to as templates) T must be defined before tracking, either manually by the user or automatically by a program. Tracking T from time k to time $k+1$ can then be formulated as the following optimization problem:

$$\hat{W}_k^{k+1} = \arg\min_W C\big(T_k\big(W(x,y;p)\big), T_{k+1}(x,y)\big) \tag{5.15}$$

where $W(x,y; p)$ is the transform between T_k and T_{k+1} and is of the following form [3]:

$$W(x,y;p) = \begin{pmatrix} 1+p_1 & p_3 & p_5 \\ p_2 & 1+p_4 & p_6 \end{pmatrix}\begin{pmatrix} x \\ y \\ 1 \end{pmatrix} \tag{5.16}$$

The cost function $C(\cdot)$ quantifies the similarity between T_k and T_{k+1}. A variety of forms can be used. For example, under the following form

$$C\big(T_k\big(W(x,y;p)\big), T_{k+1}(x,y)\big) = \iint_E \big[T_k\big(W(x,y;p)\big) - T_{k+1}(x,y)\big]^2 \tag{5.17}$$

the gradient of the cost function, which is required by a variety of numerical techniques, can be computed explicitly [3]. For highly dynamic applications, the selection of region template may need to be updated during the tracking process [33].

When the shape of an object is completely rigid, (5.16) simplifies to the form of (5.12). The tracking problem can be solved by various techniques, including the many optical flow algorithms introduced in [4].

5.3.4 Image Segmentation and Deformable Region Tracking

In live microscopy, it is also common for entire specimens to undergo significant shape (morphology) changes. In this case, tracking deformable contours of these specimens is often essential to obtaining biological readouts.

As a first step of the tracking process, the boundary of the specimen must be accurately determined. This process is called image segmentation. It is one of the most fundamental techniques in machine vision and plays a critical role in not only object tracking but also many other high-level image understanding applications such as shape description and object recognition. In the following, image segmentation techniques are reviewed. Then the problem of deformable contour tracking is addressed.

5.3.4.1 Image Segmentation Techniques

Image segmentation techniques can be classified from different perspectives:

- *Contour-based versus region-based segmentation techniques:* Contour-based segmentation techniques start with detection of object boundaries and follow with a process to link boundary segments based on various criteria that characterize the affinity between the segments. Similarly, region-based methods start with selection of seed regions and perform segmentation through seed growth, region merging, and splitting based on affinity criteria [51, 52].
- *Local versus global segmentation techniques:* Early development of contour-based and region-based segmentation techniques relied on heuristic rules for local segmentation decisions. In comparison, techniques such as active contours [26], level sets [38], and graph cuts [48] rely on global optimization. It is now known that local segmentation techniques alone cannot provide satisfactory results since local edge linkage or region growth process often is error-prone due to factors such as noise in boundary and region detection [20, 32, 48]. On the other hand, although the importance of global optimization is established [48], global segmentation techniques are known to have poor local segmentation accuracy due to the difficulty in finding global optima [20]. Therefore, in practice, global optimization techniques are often used to obtain coarse image segmentation. Then local processing techniques are used to obtain higher segmentation accuracy. It should also be pointed out that the requirement for accuracy in boundary determination is application specific. Precise boundary requirements are not necessarily required for object recognition but they are critical for precise quantification of morphological dynamics [30].

5.3.4.2 Deformable Region Tracking

After the boundaries of deformable image objects are determined, the exact meaning of "tracking" must be clearly defined for each specific application. For example, some applications only require object identification. The specimen is considered tracked when its boundary is determined at each time instant for these applications. However, if the precise dynamics of the morphology change must be measured, the precise point-wise correspondence between contours at consecutive time instants must be established. A particular strength of the level-set method is that such correspondence is naturally available in computation of boundary evolution [30, 38]. Although different methods may also be used to define the correspondence, the challenge is in ensuring that these methods truly reflect the often unknown physical and biological processes driving the morphological changes.

5.3.5 Discussion

Due to space limitations, several important topics cannot be covered, including, for example, feature detection techniques under low SNR [34], colocalization analysis techniques [11, 39], super-resolution techniques [43], and flow vector field computation techniques [4, 25]. However, readers interested in these techniques can use the references as a starting point to find related literature.

5.4 A Representative Application

The work reported in [37] provides a representative example of the integration of machine vision with microscopy for life science automation. In this study, the authors developed an automated screening system based on the integration of several techniques: RNAi[5] for gene knockdown, cell array for massive parallel delivery of RNAi, live cell microscopy for dynamic readouts, and machine vision techniques for automated image collection and analysis. This system has been successfully tested on a few human cell lines. The basic process flow includes several steps. First, approximately 400 siRNA transfection mixture spots were directly printed onto the coverslip of each one-well live imaging chamber. After seeding approximately 50 cells on each spot, time-lapse movies were collected on as many as four microarrays in each sampling cycle. All imaging was carried out using a customized automated microscope with cell culture support for live imaging.

Although low resolution (10X) and low sampling rate (every 30 minutes) were used, the volume of image data generated was already high: approximately 100 GB for each cell microarray. The authors developed machine vision techniques to automatically detect and classify chromosomes and to identify cell cycle–related phenotypes as compared to control assays. Quantitative readouts were clustered using machine-learning techniques to identify genes that are likely to function as groups. A total of 49 endogenous genes regulating chromosome segregation or nuclear structure in human cells was studied. Most of the results obtained from screening were consistent with what was obtained using conventional genetic and biochemistry methods.

This work represents a major technical advance because of its use of time-lapse live cell imaging for studying protein dynamics in high-throughput screening. Because of the application of machine vision techniques, this system is scalable to handle an even larger volume of image data under increased sample rates and image resolutions that are required for high-throughput studies at larger scales and finer

---

5. RNAi (RNA interference) is a molecular biology technique that is used to suppress expression of protein-encoding genes with high specificity [19].

resolution. It also demonstrates a few basic guidelines on the integration of machine vision with microscopy. First, integration of different levels of machine vision techniques is often required. In addition to building the automated microscope system, autofocusing, segmentation, classification, clustering, and machine learning techniques all must be integrated for the automated collection and analysis of image data. Second, application of machine vision techniques must be closely integrated with other techniques. Machine vision is just one component of the entire system. Its implementation is dependent on the application of cell culture, cell manipulation, genetics, and biochemistry techniques. As an example, the configuration of the cell microarray and the selection of fluorescence labeling techniques can all strongly influence the implementation of machine vision techniques.

5.5   Future Development and Challenges

Two major developments are expected in microscopy. First, 3D live microscopy will be used much more extensively. Currently, application of 3D live microscopy in life science is limited by factors such as slow sampling, limited imaging depth, and severe photobleaching. Many life science studies are restricted to use thin and flat assays for 2D microscopy. Such restraints will be gradually removed with the development of faster and more sensitive cameras, more stable fluorophores, and novel microscope optics and imaging techniques [22]. Second, significant progress will be made in breaking the resolution limit of light microscopy. Currently, a few promising techniques are being actively pursued [43].

Addressing the image analysis and understanding problems associated with the developments in microscopy will be the primary focus for machine vision. Several additional developments in machine vision are also expected. First, machine vision techniques will become more closely integrated with bioinformatics and biostatistics techniques to serve as a regular part of the quantitative tools for life science studies [46]. Second, significant progress is required and expected in development of machine vision techniques which are robust against imaging condition fluctuations and are self-adaptive in parameter setting for minimal manual tuning. In pursuing these goals, it is important to draw on the experience and lessons from the field of medical image analysis [18]. Finally, application of machine vision techniques will continue to be highly application specific. Real progress in the integration of machine vision with microscopy can only come from in-depth understanding of the targeted applications and effective collaboration between machine vision and life science workers.

Acknowledgments

We thank our colleagues and collaborators for their assistance and support. In particular, we thank Lisa Cameron (University of North Carolina, Chapel Hill), Jedidiah Gaetz (Rockefeller University), and James Lim (The Scripps Research Institute, La Jolla) for providing example images, and Rainer Duden and Irina Majoul (Royal Holloway, University of London) for providing the EB3-GFP construct during the 2006 3D Live Microscopy Course at the University of British Columbia. Ge Yang thanks Dr. Gaudenz Danuser for his encouragement and for sharing his insights into computer vision for microscopy and computational cell biology; he was supported by a Burroughs-Wellcome Fund LJIS interdisciplinary fellowship and NIH grant R01 GM60678 (to Gaudenz Danuser).

References

[1] Alberts, B., et al., *Molecular Biology of the Cell*, 4th ed., New York: Garland Science, 2002.

[2] Axelrod, D., "Total Internal Reflection Fluorescence Microscopy in Cell Biology," *Traffic*, Vol. 2, 2001, pp. 764–774.

[3] Baker, S., and I. Matthews, "Lucas-Kanade 20 Years On: A Unifying Framework," *Int. J. of Computer Vision*, Vol. 56, 2004, pp. 221–255.

[4] Barron, J. L., D. J. Fleet, and S. S. Beauchemin, "Performance of Optical Flow Techniques," *Int. J. Computer Vision*, Vol. 12, 1994, pp. 43–77.

[5] Bar-Shalom, Y., and X. R. Li, *Multitarget-Multisensor Tracking: Principles and Techniques*, Storrs, CT: YBS Publishing, 1995.

[6] Besl, P. J., and N. D. McKay, "A Method for Registration of 3D Shapes," *IEEE Transactions on Pattern Analysis and Machine Intelligence*, Vol. 14, 1992, pp. 239–254.

[7] Blackman, S., and R. Popoli, *Design and Analysis of Modern Tracking Systems*, Norwood, MA: Artech House, 1999.

[8] Born, M., and E. Wolf, *Principles of Optics*, 7th ed., Cambridge, U.K.: Cambridge University Press, 1999.

[9] Cameron, L. A., et al., "Kinesin 5-Independent Poleward Flux of Kinetochore Microtubules in PtK1 Cells," *J. Cell Biology*, Vol. 173, 2006, pp. 173–179.

[10] Cheezum, M. K., W. F. Walker, and W. H. Guilford, "Quantitative Comparison of Algorithms for Tracking Single Fluorescent Particles," *Biophys. J.*, Vol. 81, 2001, pp. 2378–2388.

[11] Chen, X., and R. F. Murphy, "Automated Interpretation of Protein Subcellular Location Patterns," *Int. Rev. Cytology*, Vol. 249, 2006, pp. 193–227.

[12] Cox, I. J., "A Review of Statistical Data Association Techniques for Motion Correspondence," *Int. J. Computer Vision*, Vol. 10, 1993, pp. 53–66.

[13] Cox, I. J., and S. L. Hingorani, "An Efficient Implementation of Reid's Multiple Hypothesis Tracking Algorithm and Its Evaluation for the Purpose of Visual Tracking," *IEEE Transactions on Pattern Analysis and Machine Intelligence*, Vol. 18, 1996, pp. 138–150.

[14] Danuser, G., and C. Waterman-Storer, "Fluorescent Speckle Microscopy—Where It Came from and Where It Is Going," *J. Microscopy*, Vol. 211, 2003, pp. 230–248.

[15] Davies, R., *Machine Vision*, 3rd ed., San Francisco, CA: Morgan Kaufmann Publishers, 2005.

[16] Denk, W., and K. Svoboda, "Photon Upmanship: Why Multiphoton Imaging Is More Than A Gimmick," *Neuron*, Vol. 18, 1997, pp. 351–357.

[17] Diaspro, A., *Confocal and Two-Photon Microscopy*, New York: Wiley-Liss, 2002.

[18] Duncan, J. S., and N. Ayache, "Medical Image Analysis: Progress over Two Decades and the Challenges Ahead," *IEEE Transactions on Pattern Analysis and Machine Intelligence*, Vol. 22, 2000, pp. 85–106.

[19] Echeverri, C. J., and N. Perrimon, "High-Throughput RNAi Screening in Cultured Cells: A User's Guide," *Nature Reviews Genetics*, Vol. 7, 2006, pp. 373–384.

[20] Fowlkes, C., and J. Malik, "How Much Does Globalization Help Segmentation?" *Technical Report No. UCB/CSD-4-1340*, Computer Science Division, University of California Berkeley, 2004.

[21] Genovesio, A., et al., "Multiple Particle Tracking in 3-D+t Microscopy: Method and Application to the Tracking of Endocytosed Quantum Dots," *IEEE Transactions on Image Processing*, Vol. 15, 2006, pp. 1062–1070.

[22] Gerlich, D., and J. Ellenber, "4D Imaging to Assay Complex Dynamics in Live Specimens," *Nature Cell Biology*, Vol. 4 (supplement), 2003, pp. S14–S19.

[23] Giepmans, B. N. G., et al., "The Fluorescent Toolbox for Assessing Protein Location and Function," *Science*, Vol. 312, 2006, pp. 217–224.

[24] Inoue, S., and K. R. Spring, *Video Microscopy*, 2nd ed., New York: Plenum Press, 1997.

[25] Ji, L., and G. Danuser, "Tracking Quasi-Stationary Flow of Weak Fluorescent Signals by Adaptive Multi-Frame Correlation," *J. Microscopy*, Vol. 220, 2005, pp. 150–167.

[26] Kass, M., A. Witkin, and D. Terzopoulos, "Snakes: Active Contour Models," *Int. J. Computer Vision*, Vol. 1, 1987, pp. 321–331.

[27] Krotkov, E., "Focusing," *Int. J. Computer Vision*, Vol. 1, No. 3, 1987, pp. 223–237.

[28] Lippincott-Schwartz, J., N. Altan-Bonnet, and G. H. Patterson, "Photo-Bleaching and Photoactivation: Following Protein Dynamics in Living Cells," *Nature Cell Biology (supplement)*, Vol. 5, 2003, pp. S7–S14.

[29] Lippincott-Schwartz, J., and G. H. Patterson, "Development and Use of Fluorescent Protein Markers in Living Cells," *Science*, Vol. 300, 2003, pp. 87–90.

[30] Machacek, M., and G. Danuser, "Morphodynamic Profiling of Protrusion Phenotypes," *Biophys. J.* Vol. 90, 2006, pp. 1439–1452.

[31] Maintz, J. B. A., and M. A. Viergever, "A Survey of Medical Image Registration," *Medical Image Analysis*, Vol. 2, 1998, pp. 1–36.

[32] Malik, J., et al., "Contour and Texture Analysis for Image Segmentation," *Int. J. Computer Vision*, Vol. 43, 2001, pp. 7–27.

[33] Matthews, I., T. Ishikawa, and S. Baker, "The Template Update Problem," *IEEE Transactions on Pattern Analysis and Machine Intelligence*, Vol. 26, 2004, pp. 810–815.

[34] Meijering, E., I. Smal, and G. Danuser, "Tracking in Molecular Imaging," *IEEE Signal Processing Magazine*, Vol. 23, 2006, pp. 46–53.

[35] Melnick, J. S., et al, "An Efficient Rapid System for Profiling the Cellular Activities of Molecular Libraries," *Proc. Natl. Acad. Sci. USA*, Vol. 103, 2006, pp. 3153–3158, 2006.

[36] Nayar, S. K., and Y. Nakagawa, "Shape from Focus," *IEEE Transactions on Pattern Analysis and Machine Intelligence*, Vol. 16, 1994, pp. 824–831.

[37] Neumann, B., et al., "High-Throughput RNAi Screening by Time-Lapse Imaging of Live Human Cells," *Nature Methods*, Vol. 3, 2006, pp. 385–390.

[38] Osher, S., and N. Paragios, (eds.), *Geometrical Level Set Methods in Imaging, Vision, and Graphics*, New York: Springer, 2003.

[39] Pawley, J., *Handbook of Biological Confocal Microscopy*, 3rd ed., New York: Springer, 2006.

[40] Perlman, Z. E., et al., "Multidimensional Drug Profiling by Automated Microscopy," *Science*, Vol. 306, 2004, pp. 1194–1198.

[41] Pluim, J. P. W., J. B. A. Maintz, and M. A. Viergever, "Mutual Information Based Registration of Medical Images: A Survey," *IEEE Transactions on Medical Imaging*, Vol. 22, 2003, pp. 986–1004.

[42] Pollard, T. D., and W. C. Earnshaw, *Cell Biology*, New York: Elsevier Science, 2002.

[43] Ram, S., E. S. Ward, and R. J. Ober, "Beyond Rayleigh's Criterion: A Resolution Measure with Application to Single-Molecule Microscopy," *Proc. Natl. Acad. Sci. USA*, Vol. 103, 2006, pp. 4457–4462.

[44] Reynolds, G. O., et al., *The New Physical Optics Notebook: Tutorial on Fourier Optics*, Belllingham, WA: SPIE Optical Engineering Press, 1989.

[45] Rieder, C. L., and A. Khodjakov, "Mitosis Through the Microscope: Advances in Seeing Inside Live Dividing Cells," *Science*, Vol. 300, 2003, pp. 91–96.

[46] Sachs, K., et al., "Causal Protein-Signaling Networks Derived from Multiparameter Single-Cell Data," *Science*, Vol. 308, 2005, pp. 523–529.

[47] Shaner, N. C., P. A. Steinbach, and R. Y. Tsien, "A Guide to Choosing Fluorescent Proteins," *Nature Methods*, Vol. 2, 2005, pp. 905–909.

[48] Shi, J., and J. Malik, "Normalized Cuts and Image Segmentation," *IEEE Transactions on Pattern Analysis and Machine Intelligence*, Vol. 22, 2000, pp. 888–905.

[49] Slayter, E. M., and H. S. Slayter, *Light and Electron Microscopy*, Cambridge, U.K.: Cambridge University Press, 1992.

[50] Sluder, G., and D. E. Wolf, *Digital Microscopy: Methods in Cell Biology*, Vol. 72, New York: Elsevier/Academic Press, 2003.

[51] Snyder, W. E., and H. Qi, *Machine Vision*, Cambridge, U.K.: Cambridge University Press, 2004.

[52] Sonka, M., V. Hlavac, and R. Boyle, *Image Processing, Analysis, and Machine Vision*, 2nd ed., Storrs, CT: Thomson-Engineering, 1998.

[53] Stephens, D. J., and V. J. Allan, "Light Microscopy Techniques for Live Cell Imaging," *Science*, Vol. 300, 2003, pp. 82–86.

[54] Subbarao, M., T. S. Choi, and A. Nikzad, "Focusing Techniques," *J. Optical Engineering*, Vol. 32, 1993, pp. 2824–2836.

[55] Subbarao, M., and J. K. Tyan, "Selecting the Optimal Focus Measure for Autofocusing and Depth-From-Focus," *IEEE Transactions on Pattern Analysis and Machine Intelligence*, Vol. 20, 1998, pp. 864–870.

[56] Swedlow, J. R., et al., "Informatics and Quantitative Analysis in Biological Imaging," *Science*, Vol. 300, 2003, pp. 100–102.

[57] Trucco, E., and A. Verri, *Introductory Techniques for 3D Computer Vision*, Upper Saddle River, NJ: Prentice-Hall, 1998.

[58] Tsien, R. Y., "Imagining Imaging's Future," *Nature Rev. Mol. Cell Biology*, 2003, pp. SS16–21.

[59] Veenman, C. J., M. J. T. Reinders, and E. Backer, "Resolving Motion Correspondence for Densely Moving Points," *IEEE Transactions on Pattern Analysis and Machine Intelligence*, Vol. 23, 2001, pp. 54–72.

[60] Yang, G., "Scale-Based Integrated Microscopic Computer Vision Techniques for Micromanipulation and Microassembly," Ph.D. thesis, University of Minnesota Twin Cities, 2004.

[61] Yang, G., J. A. Gaines, and B. J. Nelson, "Optomechatronic Design of Microassembly Systems for Manufacturing of Hybrid Microsystems," *IEEE Transactions on Industrial Electronics*, Vol. 52, 2005, pp. 1013–1023.

[62] Yang, G., A. Matov, and G. Danuser, "Reliable Tracking of Large Scale Dense Antiparallel Particle Motion for Fluorescence Live Cell Imaging," *Proc. Workshop on Computer Vision Methods for Bioinformatics, IEEE Int. Conf. Computer Vision and Pattern Recognition*, 2005.

[63] Zitova, B., and J. Flusser, "Image Registration Methods: A Survey," *Image and Vision Computing*, Vol. 21, 2003, pp. 977–1000.

CHAPTER 6

Control Mechanisms for Life Science Automation

Ruoting Yang, Mingjun Zhang, and T. J. Tarn

This chapter introduces various control mechanisms for life science automation. Discussions on fundamentals and advantages and disadvantages of different control mechanisms are presented along with an illustrative application to blood pressure control. Conclusions and future challenges are presented at the end.

6.1 Introduction

Automation plays an increasingly important role in life sciences. The wide-range applications of life science automation include genomics and proteomics automation, drug delivery system automation, cell and tissue manipulation automation, and medical automation. A key component for the above automation system development is the control mechanism design. For example, controlled drug delivery systems pursue precisely drug targeting and releasing. This obviously requires advanced control mechanisms instead of conventional open-loop control schemes. Cell and tissue manipulation systems demand control mechanisms that are reliable and robust for micro- and nanoscale manipulation. Surgical robots for medical automation require advanced control methods for real-time complex mechanism control.

In this chapter, we will introduce control mechanisms that have been used for life science automation. We will start with an introduction to the basic structure of control systems.

6.1.1 Basic Structure of Control Systems

As shown in Figure 6.1, a typical control system consists of four components: controller, actuator, plant, and sensor. The arrows in the diagram illustrate information flow. The closed loop in the diagram is feedback. Control systems with feedbacks are called closed-loop control systems; systems lacking a feedback element are called open loop. With real-time measurement information, a closed-loop controller can react according to the current desire.

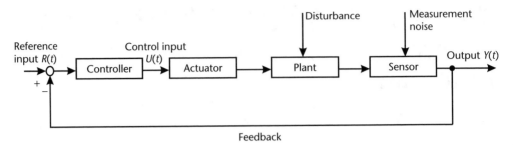

Figure 6.1 Basic structure of a control system.

Blood glucose regulation is a good example of a closed-loop control system for life science automation. To achieve homeostasis, the glucose controller computes the required dose of insulin (control input) needed for normal blood glucose concentration, whenever the internal environment of the body is changing. The information is sent as a control signal to the insulin pump (actuator), which controls insulin injection into the body (plant). The resulting blood glucose level (output) is measured in real time by a glucose monitoring system (sensor). The measured value is sent back to compare with the desired concentration of blood glucose (reference input). The deviation between the output and the reference input is conveyed to the controller for adjusting the future control.

6.1.2 Continuous-Time State-Space Description

For the convenience of analyzing the control system, the plant, actuator, and sensor blocks are often viewed as integral parts and are described mathematically by deterministic ordinary differential/difference equations (ODE) instead of partial differential/difference equations (PDE), which are often obtained by physical models. By collecting all the evolutional parameters as a state vector, the control system can be furthered formulated as a group of first-order ODEs. Thus the system dynamics can be depicted by the evolution of the state. This approach is often called state-space description.

A continuous-time system can be described in the state-space description as follows:

$$\dot{x} = f\bigl(t, x(t), u(t)\bigr), x(t_0) = x_0 \tag{6.1}$$

$$y = g\bigl(t, x(t)\bigr) \tag{6.2}$$

where $x(t) \in R^n$ is the state vector with n components $x_1(t), ..., x_n(t)$ called state variables. The control vector, $u(t) \in R^m$, with m control variables is exerted into the actuator. The output vector, $y(t) \in R^p$, has p output variables, which are measured by sensors. Function vectors f and g have n and p components, respectively.

The solution of the above equations is called a trajectory of $x(t)$ initiated at $x(t_0)$ and end up at some final state $x(t_f)$ $t_f > t_0$. To ensure existence and uniqueness of the trajectory, it is often assumed that all the components of f are continuous functions satisfying the Lipschitz condition.

In (6.1) and (6.2), the function f explicitly contains time t. Therefore, this system is called a time-varying system. In contrast, the system without explicit t is called a constant or time-invariant system. Actually, we can generalize the time-varying system in time-invariant system framework by introducing a new state $x_{n+1} = t$; $\dot{x}_{n+1} = 1$. Thus, a general continuous-time system model can be described in the form:

$$\dot{x} = f(x, u) \tag{6.3}$$

From now on, we will use this model as the general continuous-time system model. This system is a linear system if f is linear with respect to x and u; whereas the system is nonlinear if f is nonlinear. The linear systems have closed analytical solutions due to the principle of superposition. However, for nonlinear systems, it is generally difficult to achieve analytical solutions. One has to use numerical techniques to approximate the solutions. It often happens in practice to apply linearization technique to linearize a nonlinear system.

A common linearization technique is called Taylor expansion about an equilibrium $\bar{x}$. That is,

$$\frac{dx}{dt} = \frac{\partial f(\bar{x}, \bar{u})}{\partial u}(x - \bar{x}) + \frac{\partial f(\bar{x}, \bar{u})}{\partial u}(u - \bar{u}) \tag{6.4}$$

where $\partial f / \partial x = \left[\dfrac{\partial f_1}{\partial x_j} \right]$ and $\partial f / \partial u = \left[\dfrac{\partial f_1}{\partial u_j} \right]$ are the Jacobian matrices with u and x, respectively.

We are redefining the state vector x and control vector u in the equations (6.4) by making $x = x - \bar{x}$ and $u = u - \bar{u}$. Thus a general linear time-invariant (LTI) system can be formulated as follows:

$$\dot{x}(t) = Ax(t) + Bu(t) \tag{6.5}$$

These equations can be easily solved in the following form:

$$x(t) = e^{A(t - t_0)} x_0 + \int_{t_0} e^{A(t - \tau)} B(\tau) u(\tau) d\tau \tag{6.6}$$

where A, B, and C are constant matrices and e^{At} is the transition matrix.

Note that the state vector and control vector are synchronous in all systems stated above. In practice, system state and control input often have certain time delay, which may arise from the transport delay of actuator or sensor. In general, a time-delay system can be described as

$$\dot{x}(t) = Ax(t) + Bu(t - T) \tag{6.7}$$

where A and B are constant matrices and T is the time delay constant of the system.

6.1.3 Discrete-Time State-Space Description

A discrete-time system is a mathematical model described in a time sequence with a constant and fixed interval. The trajectory results in a series of discontinuous points called time series. Gathering the evolutional variables as a state vector, we still can present time-series in the state-space description.

In the last section, we argue that it is difficult to achieve analytical solutions for nonlinear systems. Fortunately, digital computers allow people to utilize numerical techniques instead of calculus to evaluate the solution of a nonlinear ODE (i.e., trajectory). This implies any continuous-time system can be discretized as a discrete-time system.

In fact, the essence of numerical algorithms actually is an approximation of derivative $\dot{x}$ in a small interval of length T, which is called the sampling time. For example, the Euler procedure utilizes $[x((k+1)T) - x(kT)]/T$ to approximate the derivative. Then the one-step future state can be estimated by current state; that is,

$$x((k+1)T) = x(kT) + Tf(x(kT), u(kT)) \tag{6.8}$$

The other algorithms, such as the fourth-order Runge-Kutta method, may look much more complex. However, the derivative algorithms all essentially behave as a one-step estimator based on the current state. Therefore, they can be described in the same frame, as follows:

$$x(k+1) = h(x(k), u(k)); x(0) = x_0 \tag{6.9}$$

where h is a function of the estimator based on the current state and input. For the purpose of notational simplicity, we neglect T in (6.8). It is the discrete analog of the continuous-time version (6.1).

Correspondingly, linear systems also have their discrete-time analog to (6.5). Since the LTI system has the analytical solution (6.6), the one-step trajectory can easily be estimated if we assume constant control efforts within the sampling period. Thus the corresponding discrete-time system becomes

$$x(k+1) = \Phi x(k) + \Gamma u(k) \tag{6.10}$$

where $\Phi = e^{AT}$ and $\Gamma = \int_0^T e^{A(T-r)} B \, d\tau$.

6.1.4 Characteristics of Continuous Linear Time-Invariant System

Based on previous discussions, all lumped-parameter systems can eventually be transformed into linear time-invariant systems. In fact, most modern control mechanisms are based on linear system theory. Moreover, nonlinear system theory actually inherits many essential concepts and principles from linear systems. Therefore, it is necessary for us to comprehend some basic concepts of linear system theory.

Consider the linear time-invariant system:

$$\dot{x}(t) = Ax(t) + Bu(t) \tag{6.11}$$

$$y(t) = Cx(t) \tag{6.12}$$

Controllability.   A state is *controllable* if there exists a control input $u(t)$ such that the trajectory goes to the origin in finite time. If this fact is true for all states, then the system is *completely controllable*.

Note that for continuous-time linear systems, a trajectory moving from x_0 to x_f with control input of $u(t)$ in fact can move backwards by counter action of $-u(t)$. Therefore, completely controllable implies we can manipulate the trajectory to connect any two points in the state space. It is pointless to design a control scheme from an uncontrollable state.

In order for a system to be completely controllable, the controllability matrix,

$$\begin{bmatrix} B & AB & A^2B & \cdots & A^{n-1}B \end{bmatrix}$$

must be full rank.

Observability.   An initial state is *observable* if there exists $T > 0$, such that this state x_0 can be determined given input $u(t)$ and output $y(t)$ over the interval $[0, T]$. If this fact is true for all states, then the system is *completely observable*.

Complete observability means that the initial condition can be observed as long as enough future knowledge of input and output is given. It is also meaningless to design an observer from an unobservable state.

In order for a system to be completely controllable, the observability matrix

$$\begin{bmatrix} C \\ CA \\ CA^2 \\ \vdots \\ CA^{n-1} \end{bmatrix}$$

must be full rank.

Stability.   Consider the continuous-time system,

$$\dot{x} = f(x,u) \tag{6.13}$$

If control input $u = 0$, this system is called an autonomous system. The solution of $f(x) = 0$ is called an equilibrium. The linear system has unique equilibrium in the origin. For a nonlinear system, any equilibrium can be moved to the origin by simple coordinate transformation. Without loss of generality, we assume the origin is equilibrium; that is, $f(0) = 0$.

The equilibrium $x = 0$ is *stable* if, for each $\varepsilon > 0$, there is $\delta = \delta(\varepsilon) > 0$ such that

$$\|x(0)\| < \delta \Rightarrow \|x(t)\| < \varepsilon, \forall t \geq 0$$

The equilibrium $x = 0$ is *asymptotically stable* if it is stable and δ can be chosen such that

$$\|x(0)\| < \delta \Rightarrow \lim_{t \to \infty} x(t) = 0$$

A powerful tool to analyze stability is the well-known Lyapunov's second theory. The idea of Lyapunov theory is to construct an entire energy function $V(x) > 0$ for the system; if the energy gradually vanishes—that is, $\dot{V}(x) < 0$—then the trajectory tends to the equilibrium.

Lyapunov Stability Theory

Suppose that there exists a scalar function $V(x)$ such that $V(0) = 0$ and

1. $V(x) > 0$ if $x \neq 0$
2. $V(x) \to \infty$ as $\|x\| \to \infty$
3. $\dot{V}(x) \leq 0$

Then the equilibrium $x = 0$ is stable. If $\dot{V}(x) > 0$, then the equilibrium is asymptotically stable.

For a continuous LTI system, a simple stability criterion derived from Lyapunov theory is that the state matrix A has only negative real-part eigenvalues.

6.2 Control Mechanisms

Control mechanisms usually mean control algorithms to regulate the behavior of (physical, chemical, or biological) systems. From an engineering point of view, these algorithms must be implemented automatically by analog circuits or digital computers without supervision by human operators.

We need to consider four aspects of a control mechanism:

1. *System dynamics:* What kind of a system it is, linear or nonlinear, time varying or time invariant, with or without time-delay?
2. *Sensor measurement:* What variables of the system can be measured? Real time or time delayed? What type is the noise considered in the measurement?
3. *Constraints and performance:* What are the constraints of the input, state, and output variables? What is the goal of control?
4. *Stability:* Can the control mechanisms guarantee stability of the system? What is the stable range or attractive region?

Next, we will first outline some representative control schemes: proportional-integral-derivative (PID), optimal, model predictive controller, adaptive, intelligent control (fuzzy and neural network), and hybrid.

6.2.1 PID Control

Objective: minimize errors.

Characteristics: freely combined modes (proportional, integral, and derivative).

PID controllers have been extensively applied in the process industries and in robotics since they are inherently robust and easily modified. It can be argued that PID is the most important controller for the common linear single-input-single-output (SISO) system existing in many industrial processes.

The heart of this well-known algorithm is a three-element composition—proportional, integral, and derivative—for compensating for the effect of difference between the actual output and the desired reference input. The elements can be freely combined for different requirements, such as P, PI, PD, PID, and even with other forward feedbacks.

The classical PID controller can be expressed in the form,

$$u(t) = K_p e(t) + K_I \int_{t_1}^{t_2} e(s)ds + K_D \frac{de}{dt} \qquad (6.14)$$

or in Laplace domain,

$$u(s) = \left(K_p + \frac{K_I}{s} + K_D s \right) e(s) \qquad (6.15)$$

where K_p is a proportional gain adjusting the control signal in direct proportion to the difference (error) between the actual output and the reference input; K_I represents integral gain correcting the steady-state error by applying larger compensatory control actions when the error accumulates; and K_D stands for derivative gain balancing rate of error, and improves the transient response. (See Figure 6.2.)

The PID algorithm is the easiest approach to implement in practice, and it can be analyzed in the form of transfer functions. However, PID tuning requires a lot of practice. It is more of an art than a science. Moreover, although they are very popular in various applications, PID controllers cannot guarantee stability for systems with highly nonlinear and long time-delay. In these cases, a nonlinear feedback compensation and prediction algorithm may be better. Researchers have invented hybrid PID controllers combining adaptation (self-tuning regulator) and heuristics (decision rules table) into the classical structure to achieve autotuning and to empirically improve performance.

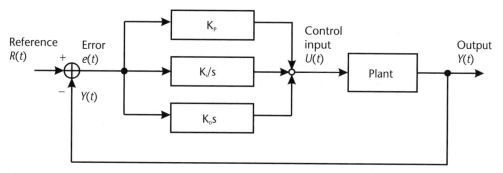

Figure 6.2 PID control loop.

6.2.2 Optimal Control

Objective: minimize performance index, subject to constraints.

Characteristics: trajectory tracking; minimum transition time, minimum energy.

With ubiquitous PID controllers, simply tuning the proportional, integral, and derivative gains is able to regulate most systems. However, we often ask for better performance, although our system has been successfully regulated to some set points. Moreover, in many applications, such as rockets, following a desired trajectory in real-time is considerably more important. The optimal controller is the tool to deal with these problems.

In general, the optimal control is to minimize the deviations from our achievements to the desired manner subject to some constraints. In order to measure the system's performance, it is necessary to have a criterion to specify the deviations. In terminology, this criterion is often called performance index (PI) (it also known as *cost functional* in the literature). There are mainly five types of PIs:

1. *Terminal control*: minimize deviation from final state;
2. *Minimum time*: minimize transition time;
3. *Minimum energy*: minimize control effort;
4. *Regulator*: minimize deviation from desired trajectory;
5. *Pursuit control*: minimize time to achieve a set point.

Example   A well-known linear quadratic (least-squares) index can be defined as follows:

$$J[u] = \int_{t_0}^{f} \left[\|u(t)\|_R^2 + \|x(t) - x_d(t)\|_Q^2 \right] dt + \|x(t_f) - x_d(t_f)\|_S^2 \tag{6.16}$$

where penalty matrices $Q, S \geq 0$ (nonnegative definite), $R > 0$ (positive definite), and $\|x\|_Q^2 = x^T Q x / 2$.

This PI gives us three groups of information:

- Control effort must be minimized.
- Deviation from desired trajectory must be minimized.
- Deviation from final state must be minimized.

In fact, the linear quadratic index is the combination of types 1, 3, and 4. The optimal controller minimizes the linear quadratic index and is called the linear quadratic regulator.

In principle, there are two powerful methods—*maximum principle* (L. S. Pontryagin) and *dynamic programming* (R. Bellman)—to solve optimal control problem. They achieve essentially equivalent solutions although they follow very different ideas. For a specific problem, one method may be significantly easier for solving the problem than another. In this section, we will introduce both methods.

6.2.2.1 Maximum Principle

Let us formulate the optimal control problem as follows.

1. *Control system:* continuous-time system:

$$\dot{x} = f(x,u), \ t \in \left[t_0, t_f\right]; \ x(t_0) = x_0; \ x(t_f) = x_f \tag{6.17}$$

where $x \in R^n$, $u \in R^m$, and $f(x,u)$ is an n-dimensional vector of continuous functions, which does not contain t explicitly. If t is contained explicitly, we can always add a new state x_{n+1} such that $x_{n+1} = t$.

2. *Constraints:* final state conditions:

$$g(x_f) = 0 \tag{6.18}$$

3. *Performance index:*

$$J[u] = k(x_f) + \int_{t_0}^{t_f} l(x(t), u(t)) dt \tag{6.19}$$

where k and l are differentiable scalar functions.

4. *Objective:* Find $u^*(t) = \arg \min J[u]$ subject to constraints.

The heart of maximum principle is to introduce Lagrange multipliers $\lambda(t)$ so that the constrained optimization problem can be reformulated as an unconstrained optimization problem:

$$J[u] = k(x_f) + \int_{t_0}^{t_f} l(x(t), u(t)) + \lambda^T(t)(\dot{x} - f(x,u)) dt \tag{6.20}$$

Applying integration by parts, we have

$$J[u] = k(x_f) + \lambda^T(t_f)x_f - \lambda^T(t_0)x_0 - \int_0^{t_f} H(\lambda(t), x(t), u(t)) dt \tag{6.21}$$

$$H(\lambda, x, u) = -l(x,u) + \sum_{i=1}^{n} \lambda_i f_i(x,u) = -l(x,u) + \lambda^T f(x,u) \tag{6.22}$$

where H is called Hamiltonian, which originates from classical mechanics.

Therefore, minimization of $J[u]$ is equivalent to maximization of $H(\lambda, x, u)$. That is where the name "maximum principle" comes from.

Theorem of Maximum Principle

Let $u^*(t)$ be an admissible control such that starting with initial conditions $x(0)$, the trajectory $x^*(t)$ passes through the point $x(t_f)$ at some time t_f. If $u^*(t)$ is optimal in the sense of minimizing $J[u]$, then there exists a nonzero, continuous vector $\lambda(t)$ satisfying

$$\dot{x}*(t) = \partial H / \partial \lambda \tag{6.23}$$

$$\dot{\lambda}^T(t) = -\partial H / \partial x \tag{6.24}$$

$$\lambda(t_f) = -\partial k(x_f) / \partial x - \mu^T \partial g(x_f) / \partial x \tag{6.25}$$

where μ is a real-value parameter and

$$H(\lambda, x*, u*) = \max_u H(\lambda, x, u) = \text{constant} \geq 0 \tag{6.26}$$

In fact, the maximum principle is a necessary condition to test admissible control candidates. The existence of optimal solutions is not guaranteed by the maximum principle. However, the solution of the maximum principle equations often provides an optimal candidate in explicit form.

Application Example: Linear Quadratic Regulator   Consider the linear system

$$\dot{x}(t) = A(t)x(t) + B(t)u(t) \quad x(t_0) = x_0; \quad x(t_f) = x_f \tag{6.27}$$

with linear quadratic performance index

$$J[u] = \int_0^f \left[\|u(t)\|_R^2 + \|x(t) - x_d(t)\|_Q^2 \right] dt + \|x(t_f) - x_d(t_f)\|_S^2 \tag{6.28}$$

where penalty matrices $Q, S \geq 0$ (nonnegative definite), $R > 0$ (positive definite); and matrix norm $\|x\|_Q^2 = x^T Q x / 2$, $\|u\|_R^2 = u^T R u / 2$.

The Hamiltonian is

$$H(\lambda, x, u) = -\left[\|u(t)\|_R^2 + \|x(t) - x_d(t)\|_Q^2 \right] + \lambda^T(t)(A(t)x(t) + B(t)u(t)) \tag{6.29}$$

$$\dot{x}*(t) = \partial H / \partial \lambda = A(t)x*(t) + B(t)u(t) \tag{6.30}$$

$$\dot{\lambda}^T(t) = -\partial H / \partial x = -\left[-(x*(t) - x_d(t))^T Q + \lambda^T(t)A(t) \right] \tag{6.31}$$

$$\lambda(t_f) = -(x_f - x_d(t_f))^T S \tag{6.32}$$

$$\partial H / \partial u = -u^T(t)R + \lambda^T(t)B(t) = 0 \tag{6.33}$$

Then,

$$u*(t) = R^{-1}B^T(t)\lambda(t), \quad \forall t \in [t_0, t_1] \tag{6.34}$$

Assume $\lambda(t) = P(t)[x_d(t) - x*(t)]$; then the optimal solution is

$$u*(t) = -R^{-1}B^T(t)P(t)[x*(t) - x_d(t)], \quad \forall t \in [t_0, t_f] \tag{6.35}$$

where $P(t) > 0$ is a real-value square matrix.

Combining (6.30) through (6.32), we arrive at a differential equation for $P(t)$,

$$\dot{P}(t) = -P(t)A(t) + A^T(t)P(t) + P(t)B(t)R^{-1}B^T(t)P(t) - Q \qquad (6.36)$$

$$P(t_f) = S \qquad (6.37)$$

This differential equation is called the Riccati equation.

6.2.2.2   Dynamic Programming

Dynamic programming is another powerful technique to solve the optimal control problem. It is based on the following instructive principle: *The principle of optimality: Any tail of an optimal trajectory is still optimal.*

In other words, an optimal trajectory can be achieved backward stage by stage. Define performance index of the trajectory initiating from $x(t_0)$ passing through $x(t_1)$ with control effort $u(t)$ as

$$\Phi\big(x(t_0), x(t_1), u(t)\big) \qquad (6.38)$$

The minimum PI is denoted in the form of

$$I\big[x(t_0), x(t_1)\big] = \min_u \Phi\big(x(t_0), x(t_1), u(t)\big) \qquad (6.39)$$

Thus the principle of optimality can be formulated in mathematical form, as follows:

$$I\big[x(t_0), x(t_f)\big] = I\big[x*(t_1), x(t_f)\big] + \min_u \Phi\big(x(t_0), x*(t_1), u(t)\big) \qquad (6.40)$$

Instead of going into greater detail of theoretical proof, we simply use an illustrative example to present the idea of dynamic programming.

Consider a one-dimensional continuous-time system:

$$\dot{x} = f(x, u), \ t \in \big[t_0, t_f\big]; \ x(t_0) = x_0; \ x(t_f) = x_f \qquad (6.41)$$

where $x \in R, u \in R$, and $f(x, u)$ is a continuous function.

Let us divide the time span $[t_0, t_f]$ into several stages, say five stages. For the purpose of simplifying the explanation, we assume there are only two admissible control values, u_1 and u_2, and three admissible state values, x_1, x_2, and x_3. Moreover, assume the set of state values is complete. Thus, x_0 and x_f are in the admissible set $\{x_1, x_2, \text{and } x_3\}$.

Figure 6.3 is an illustrative roadmap from x_0 to x_f. The brackets (i, c) above the arrow stand for (control value u_i, performance index $\Phi\big(x(t_{start}), x(t_{end}), u_i\big)$.

Start backward from:

Stage 5: $I\big[x_f, x_f\big] = 0.$

Stage 4: Obviously, the minimum PIs for the state values, respectively, are

$$I\big\{[x_1(4), x_2(4), x_3(4)], x_f\big\} = \begin{bmatrix} 4 & 2 & 3 \end{bmatrix} \qquad (6.42)$$

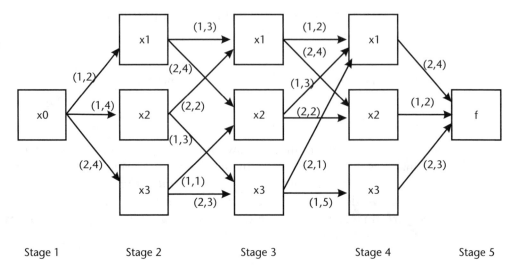

Figure 6.3 Road map of the illustrative example.

Stage 3: There are two paths connected from $x_1(3)$ to x_f:

Path A: $x_1 \rightarrow x_1 \rightarrow x_f$, PI = 2 + 4 = 6.

Path B: $x_1 \rightarrow x_2 \rightarrow x_f$, PI = 4 + 2 = 6.

Therefore the overall minimum PI is 6.

And so on, we can find the minimum PI for both x_2 and x_3.

Optimal path of x_2: $x_2 \rightarrow x_2 \rightarrow x_f$, PI = 4.

Optimal path of x_3: $x_3 \rightarrow x_1 \rightarrow x_f$, PI = 5.

Thus,

$$I\{[x_1(3), x_2(3), x_3(3)], x_f\} = \begin{bmatrix} 6 & 4 & 5 \end{bmatrix} \qquad (6.43)$$

The computation of the other two stages is left to the readers. All the minimum performance indices are summarized in Table 6.1.

From Table 6.1, one can easily generalize that the overall minimum PI is 9; the optimal trajectory is

$$x_0 \xrightarrow{u_2} x_3 \xrightarrow{u_1} x_2 \xrightarrow{u_2} x_2 \xrightarrow{u_1} x_f \qquad (6.44)$$

In this example, we assume the set of state values is complete. However, in reality, the state may go out of bound or yield some values between two discretized states, such as a value $x_3 < x < x_2$. To deal with this problem, some special technique called interpolation is necessary. The reader can refer to [1] for the further information.

Table 6.1 Trace-Back Table

10	8	6	4
12	8	4	2
9	5	5	3

In fact, dynamic programming also can be applied forward according to the principle of optimality; that is,

$$I\left[x(t_0), x(t_f)\right] = I\left[x(t_0), x*(t_1)\right] + \min_u \Phi\left[x*(t_1), x(t_f), u(t)\right] \tag{6.45}$$

Backward and forward dynamic programming approaches essentially have no difference; however, the backward method is more efficient in the case of constrained final state. Moreover, forward method is the only choice if the final time is infinity.

In summary, maximum principle provides the necessary condition while dynamic programming gives sufficient condition. Maximum principle can have explicit analytical solution, which requires solving difficult partial differential equations. Dynamic programming achieves numerical optimal solution. But computational complexity grows exponentially as the dimension of the system grows. Therefore, these two methods may present opposite difficulties in the same optimal control problem, although the results are essentially equivalent.

6.2.3 Model Predictive Control

Objective: minimize quadratic performance index with input-output constraints.

Characteristics: set-point tracking; time-delay compensation; input-output constraints.

Model predictive controller (MPC) is undoubtedly the most influential advanced control scheme [2] in process control. The unique capacity to compensate time-delay and handle input-output constraints helps MPC fast spread many control-related fields. The fundamental difference between MPC and classical optimal problem, such as LQR, is the predictive model. Typically, this model is just the linearized and discretized version of the system model; that is, a discrete-time LTI system:

$$x(k+1) = \Phi x(k) + \Gamma u(k) \tag{6.46}$$

$$y(k) = Cx(k) \tag{6.47}$$

where $\Phi = e^{AT}$ and $\Gamma = \int_0^T e^{A(T-r)} B d\tau, A, B, A, B$ are coefficient matrix after linearization, and T is the sampling period.

Assume all future inputs $\hat{u}(k+i|k)$ are constant, and we can predict any future output based on current state; that is, $\hat{y}(k+i|k)$.

Assign the quadratic performance index in the form,

$$J[u] = \sum_{k=0}^{\infty} \|u(k)\|_R^2 + \|y(k) - y_d(k)\|_Q^2 + \|\Delta u(k)\|_S^2 \tag{6.48}$$

where $Q > 0$, $R > 0$, and $\Delta u(k) = u(k+1) - u(k)$.

The objective is to minimize the above performance index; that is, minimize output error, control energy, and control variation in infinite time, subject to the constraints:

$$u_{\min} \leq u(k) \leq u_{\max}$$

$$y_{\min} \leq y(k) \leq y_{\max}$$

$$\Delta u_{\min} \leq \Delta u(k) \leq \Delta u_{\max}$$

In principle, MPC adapts quadratic programming to solve the optimal control problem.

Because there is no fixed final state, finite-horizon is applied instead of infinite horizon in order to simplify the computational difficulty.

$$J[u] = \sum_{k=0}^{m} \left\| u(k) \right\|_R^2 + \left\| y(k) - y_d(k) \right\|_Q^2 + \left\| \Delta u(k) \right\|_S^2 \tag{6.49}$$

The horizon varies corresponding to transport time-delay in the specific problem. To solve this discrete-time quadratic constrained optimal control problem, we must use numerical optimization techniques. The common means is the well-known quadratic programming, which is one of the most powerful tools in numerical optimization.

6.2.4 Adaptive Control

Objective: dynamically adjust parameters in a changing environment.

Characteristics: identification, model-reference.

The control mechanisms described above require a good understanding and accurate model of systems. The system models are well synthesized based on physical and biological principles and all the system parameters (model coefficients) are known a priori. However, many real systems are too complex to be completely understood. Therefore, instead of searching for precise models, we assume a model structure for the system of interest, and then try to infer the system parameters from a large number of inputs and corresponding system responses. This approach is called identification.

The identification approach is very practical because of the ease in system analysis and control implementation. However, the system parameters often present variability and uncertainty. For example, one can build a good blood pressure control scheme based on the blood pressure variations of one patient. But this controller is individualized and cannot perform reliably in other patients due to physiological variability. In order to apply the same control system in different patients, an adaptation mechanism is essential. We have at least two ideas. The first approach is to identify the system parameters online, and then redesign the controller based on the new model. It is often called self-tuning regulator (STR). In contrast, to identify the parameters online, the second approach proposes a reference model and directly drives the system to behave like the reference model. This approach is called model reference adaptive control (MRAC).

6.2.4.1 Self-Tuning Regulator

The STR (Figure 6.4) in principle is an estimator, which identifies the unknown system parameters online, and a closed-loop controller, such as LQR, which regulates the upgraded model.

In order to devise a simple parameter estimator that can identify parameters from the input and output, linearity between output $y(t)$ and unknown parameters $\theta(t)$ is strongly recommended; that is, $y(t) = \theta^{T}(t)w(t)$.

One straightforward and natural way to achieve fast estimation is to make parameter $\theta(t)$ move in the direction of the steepest descent of a cost function of estimation error, such as the common least squares form,

$$J(\theta) = \frac{1}{2}e^2(\theta) = \frac{1}{2}\left(y(t) - \hat{\theta}^{T}(t)x(t)\right)^{T}\left(y(t) - \hat{\theta}^{T}(t)x(t)\right) \tag{6.50}$$

Thus, we get the following estimator, which is often called gradient algorithm:

$$\frac{d\theta}{dt} = -\gamma\frac{\partial J}{\partial\theta} = -\gamma e\frac{\partial e}{\partial\theta} = \gamma x(t)e \tag{6.51}$$

where γ is a positive parameter representing the adaptation gain.

The gradient algorithm may be the simplest estimator, but it does not necessarily lead to be the best performance. In fact, many new estimators have been proposed for enhancing the performance, including normalized gradient algorithm, recursive least squares (RLS) algorithms, and others [3]. However, formulating a linear relation between output and unknown parameters is tricky. Broadly speaking, one must revert to the Laplace transformation to convert the derivative into

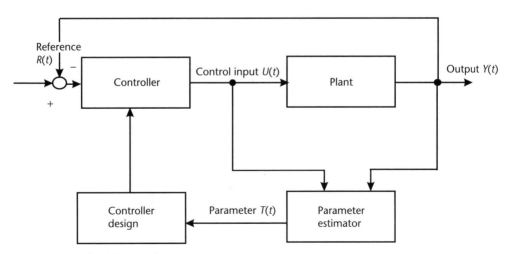

Figure 6.4 Self-tuning regulator.

linear form. The specific procedure can found in Astrom and Wittenmark [3] and Zhang [4].

6.2.4.2 Model Reference Adaptive Control

Rather than having an online estimator, the MRAC (Figure 6.5) compares the output of a real system and a reference model, and directly adjust the controller to drive the system to approximate the desired reference model.

Another key difference is the internal closed-loop controller directly related with the parameter, typically,

$$u(t) = \theta^T(t)y(t) + r(t) \tag{6.52}$$

Gradient algorithms also have good performance to estimate parameters from the output error. The well-known MIT-rule is directly derived from gradient algorithm:

$$\frac{d\theta}{dt} = -\gamma \frac{\partial J}{\partial \theta} = -\gamma e \frac{\partial e}{\partial \theta} \tag{6.53}$$

where $J = e^2(\theta)/2$ and $e(\theta)$ is the output error.

Advanced estimation algorithms use the Lyapunov theory to guarantee the stability. The Lyapunov theory is too complex for an introductory chapter. The reader may refer to Astrom and Wittenmark [3] and Zhang [4] for more information.

6.2.5 Fuzzy-Logic Control

Objective: using human knowledge and experience to control very complex systems.

Characteristics: rule table; inference engine; fuzzification and defuzzification.

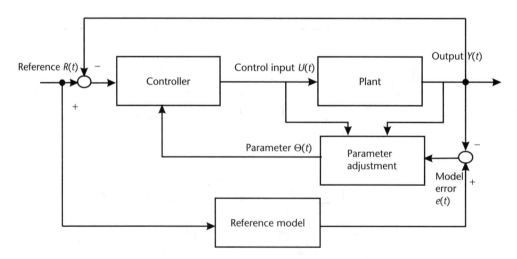

Figure 6.5 Model reference adaptive controller.

Broadly speaking, fuzzy-logic control (FLC) is a technique that uses methods from artificial intelligence to deal with very complex systems whose structure and measurements are difficult to quantify, such as depth of anesthesia. Presently, no one test can define depth of anesthesia; therefore, in practice, many measurements, such as eyelash reflex, pupillary size, swallowing, respiration rate, blood pressure, temperature, heart rate, and electroencephalograph, are taken. At this point in time, we can only rely on conservative clinical practice and knowledge, and then synthesize and analyze all the available clues. Of course, FLC cannot achieve performance as optimal and accurate as the controllers described in previous sections. It can, however, provide possible solutions to control complex systems. Moreover, since FLC is based on a rule base, which can be easily understood and maintained by average customers, it also leads an *intelligent* revolution in home electronics applications.

Conventional FLC mechanisms are *model-free* methods, which in essence are *fuzzy* rule-based regulators with signal processing devices transforming between digital and fuzzy signals. The heart of fuzzy controllers is a fuzzy rule base, which is a collection of empirical rules based on expert knowledge, and an inference engine to mimic the human reasoning process. All the fuzzy rules are formulated by words. For example:

Rule 1: *If* temperature is high, *then* begin cooling.

Rule 2: *If* temperature is normal, *then* no cooling and heating.

Rule 3: *If* temperature is low, *then* begin heating.

If we choose "error" and "change of error" as the condition, FLC is in principle similar to the PD algorithm. (See Figure 6.6.)

Since the computer-controlled systems measure and process information only in digital form (for example, 70 degrees), we need signal processing devices to complete the linguistic/digital (L/D) (and vice versa) transformation. Fuzzification is the process that translates the numerical value to words (i.e., determine the correct

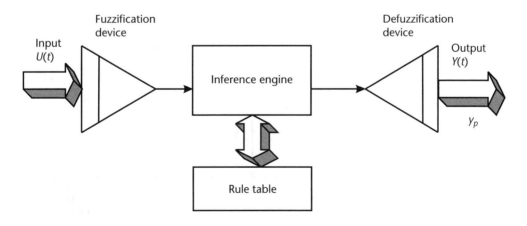

Figure 6.6   Fuzzy controller.

lingustic category where a digital input is placed). Using the temperature example above, we can designate 60 degrees as absolute normal temperature and 100 degrees as high. But 75 degrees falls between normal and high, and seems vague. Therefore, we use a *membership function* (Figure 6.7) to determine the proportion of each category for that number.

Then, from Figure 6.7, 75 degrees can be classified as 50% high and 50% normal. Using rule 1 and rule 2, the output must then come from two fuzzy conclusions, heating or no-heating-no-cooling. Defuzzification is the reverse process, which converts words back to digital values. There are many defuzzification approaches, such as center of gravity and mean of maxima, but they are more or less seeking the average of each conclusion. In our very simple case, we can directly compute the weighted average of each conclusion, such as final output = 50% heating + 50% no-heating-no-cooling.

In the above example, the fuzzy rule-base is quite simple and only SISO is considered. In most life science systems, however, such as depth of anesthesia, a physiologic characteristic is often measured by multiple indices and needs sophisticated empirical rules to regulate. Thus we need an inference engine to combine all the information from different inputs or measurements and provide the specific certainty that each rule can be activated. The inference engine checks the membership values of the inputs in the subcondition of the rule, mathematically translates the connective operation, such as *and, or*, and generates the conclusions of each rule. Recent advancements of inference engines focus on automatic creation of rules by dynamic learning and identification, whereas the conventional inference engine does not alter the rule-base.

Readers can get further detail from some excellent books on fuzzy control, such as [5].

6.2.6 Hybrid Control

Objective: combining multiple controller or models.

Characteristics: switching.

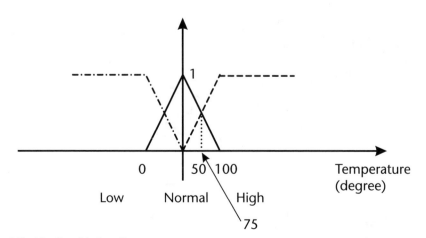

Figure 6.7 Membership function.

The term "hybrid" mainly has two different meanings in literature. First, it means a combination of two existing control methods. For example, a mixed type of adaptive controller and model predictive controller can be called hybrid. Second, "hybrid" means a combination of continuous and discrete system. For instance, suppose a control system contains a group of continuous-time controllers. Each of them is evoked only when some specific condition is satisfied. The activation of controllers is called an event. Thus, in the viewpoint of whole control loop, the continuous-time controller group exhibits a manner of discrete events. The second hybrid control is also known as a "switching" problem.

In this section, we address only the second kind of hybrid system. There are two main classes of control problem addressed in hybrid control area: controller switching and model switching systems.

The principle of controller switching is to use multiple controllers and automatically select the best controller at that moment. The simplest example of the controller switching approach is the sliding mode controller (bang-bang control) for linear minimum-time optimal control. Suppose control $u(t)$ is bounded by -1 and $+1$, and the goal of control is to drive the trajectory to equilibrium (i.e., origin) with minimum-time. Before the origin is reached, the control is just switched between two boundary values, $+1$ and -1, under certain designed conditions.

In contrast, model switching uses a single controller but instead switches between multiple models. The best matching model to real system at that moment can be determined under certain powerful decision making logic. In fact, the stability of a model switching system is not easy to foresee. In general, a switching system consisting of n stable linear systems may turn out to be unstable, while a switching system containing n unstable linear systems can still be stable. It is still an active research area in the literature.

6.3   Applications to Life Science Automation

In general, we suggest the following design procedure for a control mechanism:

1. *Objectives of control:* What is the goal of control? What is the figure of merit?
2. *System modeling:* How to model the system, including plant, sensor, and actuator? What is the input, state, and output variables in the model? Do they have any constraints, such as limited amplitude?
3. *Dynamics analysis:* What are the characteristics of the proposed model? Is it continuous-time, linear and time-invariant? Are the transport time-delays from measurement significant? What type of noise is in measurement?
4. *Controller design:* What are the assumptions under the proposed model? Is the system completely controllable? Are the state variables measurable? If not, check the observability so that the observer can be applied. What is the most appropriate control mechanism considering the system characteristics?
5. *Stability:* Can the control mechanisms guarantee stability under disturbance? What is the stable range?
6. *Experiment and simulation:* Does the designed control mechanism really work? How are the parameters adjusted to satisfy the specifications?

To illustrate how this procedure works, we shall illustrate several selected examples in life science automation.

6.3.1 Blood Pressure Control

The anesthesiologist or critical care physician must monitor and regulate a wide range of physiological states, such as blood pressure, cardiac output (CO), carbon dioxide and oxygen levels, blood acidity, fluid levels, heart contractility, renal function, and more. Among these vital variables, physicians closely examine blood pressure both during cardiac surgery and postsurgery to avoid hypertension. A real-time measurement of blood pressure is mean arterial pressure (MAP), which stands for the average blood pressure in the body. The typical means to reduce MAP is intravenous infusion of vasodilator drugs, such as sodium nitroprusside (SNP). The main side effect of SNP is the reflex tachycardia, which can be ameliorated by applying dopamine (DPM). Notwithstanding, the total amount of SNP infusion must be limited because of drug toxicity.

In general, there are two different modeling techniques: top-down and bottom-up. The top-down approach is to construct a structural model based on physical principles, and then use statistics to test how good the model matches the experimental data. This approach applies typically in mechanical systems. The corresponding model often takes the form of nonlinear high-order differential/difference equations. In pharmacology, two main models that follow the top-down approach are pharmacokinetic (PK) and pharmacodynamic (PD) models. They are often used to study the relationship between drug concentration and drug effects. Pharmacokinetics describes the drug concentrations in body fluids, while pharmacodynamics studies the drug effects on the organism.

In contrast, the bottom-up approach is a black-box method. Assume the model has some kind of structure, such as a simple linear model, and then apply regression techniques to determine the coefficients. This model may be oversimplified and bear less physical meaning than the top-down model; however, it is quite practical in the sense of computational consideration. Particularly, associated with identification techniques, the bottom-up model can adapt to the modeling error and unforeseen disturbances. It seems apparent that the bottom-up approach is more amenable to complex systems than the corresponding top-down approach.

In the blood pressure control problem, the physiological systems remain limit understanding. The pharmacodynamics model results in nonlinear time-varying high-order differential equations, which largely enhance the complexity of control design. Therefore, we have to turn to the bottom-up approach.

Suppose the relationship between variations of blood pressure and drug infusion rate is depicted by a linear time-delay system,

$$\dot{x}(t) = Ax(t) + Bu(t - T_i) \qquad (6.54)$$

where $A = 1/\tau$ and $B = K/\tau$; K(mmHg/min/μg/kg) stands for sensitivity to SNP; T_i(s) represents injection and recirculation transport time delay; and τ(s) stands for time-delay resulting from drug uptake.

The corresponding discrete-time model is as follows:

$$x(k+1) = \Phi x(k) + \Gamma u(k) \tag{6.55}$$

where $\Phi = e^{-T/\tau}$ and $\Gamma = K(1 - e^{-T/r})$. T is the sampling period.

Actually, this example is a simplified Slate's model in 1979 [6], which was identified using pseudorandom binary signals (PRBS). The relationship between variations of blood pressure and drug infusion rate is depicted by a linear time-delay system,

$$\frac{\Delta P(s)}{\Delta I(s)} = \frac{Ke^{-T_i s}\left(1 + \alpha e^{-T_c s}\right)}{1 + \tau s} \tag{6.56}$$

For tutorial purposes, we ignore the models of actuator and sensor; that is, assume both of them in the form,

$$y(t) = x(t) \text{ or } y(k) = x(k)$$

6.3.2   Control Mechanism Design

In this section, we will apply several control schemes based on different assumptions on the same blood pressure model. Both continuous-time and discrete-time control approaches will be discussed in detail.

First of all, we shall point out that the ubiquitous PID control may be feasible for any case. Especially for the set-point regulator problem, the PID controller mostly can achieve the goal by carefully tuning the coefficients. However, PID controllers have to be tuned very conservatively for time-delay or nonlinear systems due to stability consideration. Moreover, PID controllers do not lead to optimal performance.

Assumption 1: There is no transport time delay; uptake time-delay and sensitivity coefficient is constant in all patients.
In response to demands for minimizing control input and deviation from set-point in a linear system, we naturally apply classical LQR with the following performance index:

$$J[u] = \int_0^\infty \|u(t)\|_R^2 + \|x(t) - x_e(t)\|_Q^2 \, dt \tag{6.57}$$

where $Q > 0$, $R > 0$; for instance, $Q = 10$; $R = 1$.

Note that there are no final state constraints; we must set final state as infinity and neglect the penalty term on the final state. In fact, this modified LQR is called steady-state LQR. Furthermore, the matrix $P(t)$ tends to be zero as time t goes to infinity due to the boundary condition $P(t_f) = S$ of the Riccati equation.

The resulting optimal control law and the Riccati equation (algebraic Riccati) are as follows:

$$u(t) = -R^{-1}B^T P(x(t) - x_d(t)) \tag{6.58}$$

$$-PA - A^T P + PBR^{-1}B^T P - Q = 0 \tag{6.59}$$

where $A = -1/\tau$, $B = K/\tau$, and $P > 0$.

It is easy to verify that this system model is completely controllable and observable. Therefore, the algebraic Riccati equation has a unique solution, $P = \tau\left(\sqrt{10K^2 + 1} - 1\right)/K^2$.

If noise from measurement is taken into account, we can use a Kalman filter [1] to filter out the noise.

Discrete-Time Controller Scheme:

Performance index: $J[u] = \sum_{k=0}^{\infty} \|u(k)\|_R^2 + \|x(k) - x_d(k)\|_Q^2$

Optimal control: $u(k) = -L\left(x(k) - x^d(k)\right)$

Where $L = \left(\mathrm{T}^T Q\Gamma + R\right)^{-1}\Gamma^T Q\Phi$

Assumption 2: There is no transport time delay, and sensitivity coefficient is different patient by patient. Uptake time delay is constant.

To deal with the unknown parameter, we must turn to adaptive control (STR or MRAC).

Transform $\dot{x}(t) = Ax(r) + Bu(t)$ into Laplace domain and divide both sides by $s + \delta$ for some $\delta > 0$.

$$\frac{s - A}{s + \delta} x(s) = \frac{B}{s + \delta} u(s) \tag{6.60}$$

It can be rewritten as

$$x(s) = (A + \delta)w_1 + Bw_2 \tag{6.61}$$

where $w_1 = x(s)/(s+\delta)$—that is, $\dot{w}_1(t) = -w_1(t) + x(t)$—and $w_2 = u(s)/(s+\delta)$ —that is, $\dot{w}_2(t) = -w_2(t) + u(t)$.

Let $w(t) = [w_1(t)\, w_2(t)]^T$ and $\theta = [A - \delta\ B]^T$, then $x(t) = w^T(t)\theta$.

By the gradient algorithm,

$$\dot{\hat{\theta}}(t) = \gamma w(t)\left(x(t) - w(t)^T \hat{\theta}(t)\right); \quad \gamma > 0 \tag{6.62}$$

Discrete-Time Controller Scheme: Let $w(k) = [x(k)\, u(k)]^T$, $\theta = [\Phi\ \Gamma]^T$, and $\xi(k) = x(k+1)$; then,

$$\xi(k) = w^T(k)\theta + e(k) \tag{6.63}$$

Gradient algorithm:

$$\hat{\theta}(k) = \hat{\theta}(k-1) + \gamma w(k)\left(\xi(k) - w(k)^T \hat{\theta}(k-1)\right); \quad \gamma > 0 \tag{6.64}$$

Combining identification and steady-state LQR, we finally accomplish an adaptive controller with respect to varying sensitivity coefficient. In iteration, the estimated parameter is conveyed to LQR in order to generate a new control input, and then this input, together with the corresponding new state and current state, are further substituted into the gradient algorithm for the updating parameter.

Assumption 3: There are constant transport time delays, uptake time delays, and sensitivity coefficients in all patients.

The model predictive control is a popular way to deal with the time-delay problem. To compensate the time-delay effect, a predictive block is needed to forecast future outputs. Here, we can directly use the discrete-time model as the predictive model:

$$x(k+1) = \Phi x(k) + \Gamma u(k) \qquad (6.65)$$

where $\Phi = e^{-T/\tau}$ and $\Gamma = K(1 - e^{-T/\tau})$. T is the sampling period.

Consider the quadratic performance:

$$J[u] = \sum_{k=0}^{m} \|u(k)\|_R^2 + \|x(k) - x_d(k)\|_Q^2 + \|\Delta u(k)\|_S^2 \qquad (6.66)$$

subject to

$$0 \leq u(k) \leq 10 \ \mu g / kg / min$$

$$70 \leq x(k) \leq 160 \ mmHg$$

$$-0.2 \leq \Delta u(k) \leq 0.2 \ \mu g / kg / min$$

Assume future inputs are equivalent—that is, $\hat{u}(k+1|k) = $ const—then we can forecast any future output based on current state—that is, $\hat{x}(k+1|k)$. Assume the transport time delay is 45 seconds and the sampling period is 1 minute; thus, the horizon can be set to be $m = 2$.

Assumption 4: Transport time delay varies for different patients. Uptake time delay and sensitivity coefficient are constant.

Because the transport time delay results in an exponential term $e^{-T_i s}$, it is too sophisticated to find an expression $x(t) = w^T(t)\theta$, so that separates the estimate parameter and system dynamics. Therefore, the adaptive control approach is not easy to follow in this case.

One practical approach is to employ the model switching approach in hybrid control. First, we construct a group of submodels with the same structure, but different transport time delays, and then we select the best matching submodel as the predictive model in MPC in iteration under some decision making rule, such as the minimal deviation of output response between submodel and real system,

$$e_i(k) = x_i(k+1) - \hat{x}_i(k+1) \text{ for } i = 1, \ldots, N$$

6.3.3 Modeling and Control of an Insulin Delivery System

Diabetes mellitus is a disease in which the patient has difficulty regulating glucose. Today, 2% of the population—that is about 120 million people worldwide— suffers from diabetes. The diabetes may affect functions of many physiological systems including retinopathy (blindness), circulatory problems (sensory loss, and even amputation), nephropathy (kidney failure), and heart disease. It has been shown that intensive management can significantly lower the risk of developing complications and slow the progress of existing complications. However, the dosage of insulin must be strictly regulated because excess insulin can cause hypoglycemia (shortage of glucose), whereas insufficient insulin may cause hyperglycemia (too much glucose).

A practical objective of glucose control is to build a wearable "artificial pancreas," which achieves automatic painless insulin infusion, accurate dosage management, and reliable real-time monitoring of the blood glucose level. In recent years, microfluid and microfabrication technologies have made the artificial pancreas a promising solution. Advances in micropumps, microneedles, and microsensors have made the device a possible solution. However, building a reliable dynamic model remains a challenge for the artificial pancreas. An ideal strategy is to control the microscale micropump and microneedles based on macroscale sensor information (blood glucose level), which needs to consider the communication delay. Unfortunately, no such dynamic model is available in the open literature. Although several closed-loop control strategies for diabetes care have been studied [7–10] via intravenous or subcutaneous route, they are all based on well-assumed physiological models. Furthermore, most of the control strategies involve specifying appropriate insulin doses that have not taken device dynamics into consideration.

Another challenge encountered is the control strategy. An efficient control strategy must meet three requirements:

1. Fast response to glucose change;
2. Minimize overshoot, which results in hypoglycemia;
3. Adapt to different patients.

Unfortunately, none of the current control strategies proposed in the literature satisfies these three goals. In order to accomplish these challenges, an adaptive backstepping controller with nonlinear observer (ABCNO) will be proposed here. The basic idea is to break the whole system into a series of cascade subsystems. Lyapunov function–based nonlinear control methods are proposed for each subsystem backwards from output to the control input. The goals of designing Lyapunov functions are not only to guarantee the global stability, but also to utilize nonlinearity for control purposes instead of canceling it. As a result, the backstepping approach significantly reduces the control effort while still ensuring stability. In addition, adaptation is embedded in ABCNO to compensate for intra- and inter-patient variations.

6.3.3.1   Microneedle and Micropump Integrated Drug Delivery System for Diabetes Care

Figure 6.8 shows a schematic drawing of the proposed microneedle and micropump integrated system for subcutaneous injection of insulin. The drug delivery system consists of a piezoelectric micropump, multiple silicon microneedles, insulin reservoir, membrane, wireless telemetry, and remote control components. The piezoelectric pump can be triggered through wireless telemetry and remote control components. A feedback control system is then used to regulate the piezoelectric system based on sensor inputs of glucose level information. The insulin is released through the microneedles from the reservoir. The goal is to control insulin for diabetes care in real-time.

Piezoelectric Micropump Dynamics
Since the proposed system is a wearable device and the insulin dosage is limited, the high flow rate is not required. A piezoelectrically driven displacement pump is regarded as a good choice.

To make a piezoelectric micropump, a piezoelectric thin film is usually deposited on silicon nitride structural circular membrane. When an electric field is applied across the piezoelectric thin film, the membrane deflects upward or downward, and the volume of the sealed chamber diagram changes. If the entrance and exit channels have check valves or nozzles, then the piezoelectric micropump creates a one-way net flow.

The unloaded volume change of the pump can be expressed as follows [11]:

$$\Delta V = \frac{3a^4 (5+2\mu)(1-\mu)d_{13}\Delta U}{4h^2 (3+2\mu)} \tag{6.67}$$

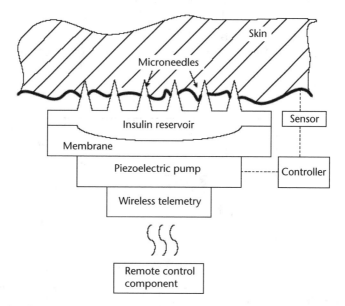

Figure 6.8   A schematic drawing of the microneedle and micropump integrated system for diabetes care insulin controlled delivery.

where U is the voltage applied to the PZT film with piezoelectric coefficient d_{13}. The thickness of the membrane is h and the radius is a. μ is the Poisson ratio.

Note that the volume change is proportional to the voltage change applied to the PZT film. The relationship can be expressed as

$$\Delta V = K\Delta U \tag{6.68}$$

where K is the coefficient.

The purpose of the micropump is to switch the working voltage between zero and a positive value with the difference in voltage by ΔU. Then the flow rate can be expressed as

$$Q = Kf\Delta U \tag{6.69}$$

where f is the working frequency of the micropump.

The above equation is an ideal performance of the piezoelectric diaphragm membrane micropump. When the working frequency f grows, the fluid overpressure in the chamber also increases. This fact will result in a lost volume. The real performance of a micropump can be illustrated by Figure 6.9 [12].

The flow rate Q grows linearly with working frequency f until the maximum flow rate is achieved. After that, the flow rate drops rapidly with f. Therefore, we define the frequency corresponding to Q_{max} as the maximum working frequency f_{max}. Thus, the control frequency is limited between zero and f_{max}.

The silicon nitride membranes are around 100 to 400 μm in diameter and 1 to 2 μm thick. The radius and thickness are assumed to be a = 100 μm and h = 1.2 μm, respectively. The Poisson's ratio μ is equal to 0.3, and the piezoelectric coefficient is $d_{13} = 9.8 \times 10^{-13}$ mN^{-1}. The proposed pump is driven by one AAA battery. Thus, set the working voltage to be 1.5V, then the volume change is $\Delta V = 8.34 \times 10^{-5}$ μl. Assume that the maximum working frequency is 0.1 kHz, then the corresponding maximum flow rate would be 0.00834 μl/s = 30 u/h. In general, the infusion range of current insulin pumps (e.g., Minimed, Animas, and so on) is of 25 to 35 u/h. As a result, we set the insulin infusion rate in the range of 0 to 30 u/h for all the following sections.

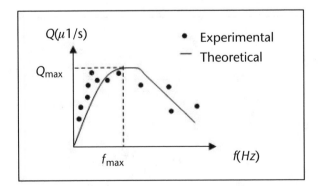

Figure 6.9   The performance of micropump.

Microneedle Dynamics
The human skin has three layers: the stratus corneum layer, the epidermis layer, and the dermis. The tratus corneum layer is usually about 10 to 15 μm thick and has no nerves or living cells. The viable epidermis layer is 50 to 100 μm. It contains a few nerves and no blood vessels. The dermis is rich in nerves and blood vessels. In contrast to the conventional needles, microneedles only penetrate the first two layers and partially into the dermis so that the insulin can be absorbed into bloodstream. Hence, microneedle injection is almost painless and minimizes the damage to a patient's skin. The microneedle array design can reduce the diameter of each microneedle.

The volume flow rate of a microneedle array can be expressed as follows:

$$Q = \frac{\Delta P}{R} \tag{6.70}$$

where ΔP is the pressure drops across the channel and R is the channel resistance.
The resistance of a circular shape microneedle can be further expressed as [13]

$$R = \frac{8\mu L}{\pi r^4} \tag{6.71}$$

where μ is the fluid viscosity, L is the channel length, and r is the channel radius.

The diameter of the microneedles for insulin delivery varies usually from 0.5 to 0.35 mm, which is sufficiently large to allow free flow and cause little resistance. It is reasonable to assume that the microneedle array will not impede the flow rate from the micropump.

Subcutaneous Sensor Dynamics
Subcutaneous sensors tend to alternate the conventional finger stick detectors. The U.S. Food and Drug Administration (FDA) has approved six continuous glucose-monitoring systems. The sensor lifespan varies from 13 hours to 3 months. Most of them are minimally invasive, and they are located in subcutaneous abdomen or arm.

The dynamics of this sensor can be written as [14]

$$\tau \frac{dS}{dt} = -S + \alpha G \tag{6.72}$$

where S stands for sensor value, G represents for the blood glucose, and τ is the time delay constant. The sensitivity of the sensor is characterized by a constant coefficient α. (See Figure 6.10.)

Subcutaneous Insulin Kinetics Model
Current insulin infusion via a subcutaneous route often uses cannula connected to the abdomen. Subcutaneous infusion is far superior to intravenous delivery because it can be much easier and can be safely managed by the patients themselves. However, infusion via the subcutaneous route has been largely affected by time delay of the absorption process. The onset of regular insulin (Humulin R) takes place after

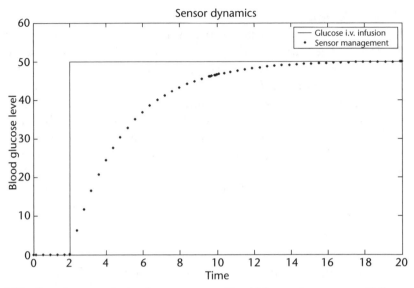

Figure 6.10 Step response of subcutaneous sensor. Time delay $\tau = 3$; sensor sensitivity $\alpha = 1$.

30 to 45 minutes and for a duration of 6 to 8 hours. Patients can suffer either from hyperglycemia or hypoglycemia due to regular insulin infusion. The new short-acting insulin analog, such as Monomeric insulin (Lispro), was proved to be two to three times faster in absorption speed than regular insulin. The onset of action is much quicker and the elimination time is also shorter.

The rapid-acting insulin lispro kinetics with continuous subcutaneous infusion has been studied in [15]. Nine models were compared by the authors. Here, the Shimoda's three-compartmental model [16] (Figure 6.11) is applied as follows:

$$\dot{Q}_1 = kQ_1 + u \tag{6.73}$$

$$\dot{Q}_2 = -(p+o)Q_2 + kQ_1 \tag{6.74}$$

$$\dot{Q}_3 = -k_e Q_3 + pQ_2 \tag{6.75}$$

$$I = Q_3 / V_i \tag{6.76}$$

where Q_1 and Q_2 are insulin mass (mU/kg) in the intraperitoneal insulin pools; Q_3 is the plasma insulin mass (mU/kg); and I is the plasma insulin concentration (mUl). u is the intraperitoneal insulin infusion rate (mU/kg/min). k and p are the transition rate constants (min^{-1}); o and k_e are degradation decay rate. The parameter V_I stands for plasma distribution volume (l/kg).

Subcutaneously infused insulin at the injection site is assumed to diffuse into the secondary intraperitoneal insulin pool by a transition rate k. Then the insulin diffuses to the circulatory insulin pool by rate p. The degradation decay rates in the intraperitoneal pool and circulatory pool are assumed to be o and k_e, respectively.

Using the data of [16], we can identify the model parameters by nonlinear least squares technique as follows: $k_e = 0.015$, $V_i = 0.21$, $p = 0.025$, $o = 0.25$, and $k =$

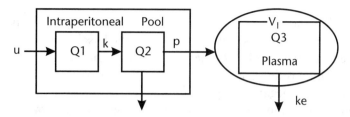

Figure 6.11   Mathematical model for intraperitoneal insulin absorption kinetics.

0.135. The dynamics of plasma insulin level after a 2U dosage intraperitoneal infusion is shown in Figure 6.12.

Glucose Kinetics Model
The glucose kinetics model varies from a simple three-compartment model [17] to a complicated nineteenth-order nonlinear model [18]. Since glucose uptake and excretion actually is a very complicated process, the lower-order model may not predict the glucose-insulin relationship accurately (see Figure 6.13). However, the higher-order models contain discontinuous terms and explicit time dependence, which adds the complexity and difficulty in applying conventional control mechanisms and performing stability analysis. Therefore, a practical alternative would be to choose a lower order model and use online identification technique to identify the coefficients. Here, we use the same dynamic model proposed by Bergman [17].

$$\dot{G} = -XG + P_1\left(G_b - G\right) \tag{6.77}$$

$$\dot{X} = -P_2 X + P_3\left(I - I_b\right) \tag{6.78}$$

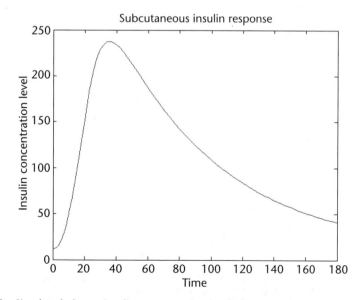

Figure 6.12   Simulated plasma insulin concentration level after Humalog infusion of a 2U dosage. The onset of action is around 15 minutes, and the peak effect occurs in 40 minutes. The total duration will be around 4 hours.

where G is the plasma glucose concentration and G_b is the basal value; I stands for the plasma insulin concentration; and I_b is the basal value. P_1 is a coefficient for glucose effectiveness and P_3/P_2 measures the insulin sensitivity [17]. The glucose effectiveness and insulin sensitivity will affect the glucose disappearance rate K_G, which is defined as the negative slope of ln(glucose). Large disappearance rate ($K_G > 1.5$) is called good insulin tolerance, while small disappearance rate ($K_G < 1.5$) is low tolerance [17].

Complete Dynamic Model
Based on the above discussions, an integrated dynamics model of the insulin delivery system can be summarized as in Table 6.2.

$$
\begin{aligned}
\dot{x}_1 &= -\frac{1}{\tau} x_1 + \frac{P}{\tau} x_2 = \theta_1 x_1 + \theta_2 x_2 \\
\dot{x}_2 &= -x_2 x_3 - P_1 (x_2 - G_b) \\
\dot{x}_3 &= -P_2 x_3 + P_3 (x_4 - i_b) \\
\dot{x}_4 &= p x_5 / V_i - k_e (x_4 - i_b) = \theta_3 x_5 - k_e (x_4 - i_b) \\
\dot{x}_5 &= k x_6 - (p + o) x_5 \\
\dot{x}_6 &= -k x_6 + u
\end{aligned}
\tag{6.79}
$$

where the parameters are summarized in Table 6.2.

The equilibrium of the above dynamic model is $(\rho G_b, G_b, 0, i_b, 0, 0)$. By translating the origin of system (12) in the setpoint $(\rho G_b, G_b, 0, i_b, 0, 0)$, the system equations become

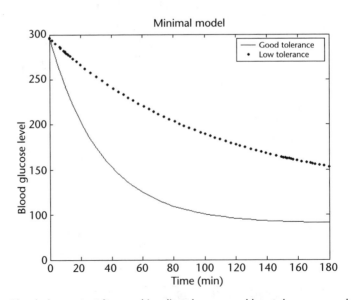

Figure 6.13 Blood glucose test for good insulin tolerance and low tolerance people. The minimal model predicts the glucose kinetics with basal insulin level. The "dot" line illustrates the blood glucose profile of lean, good insulin tolerance group (No. 1, Group I in [17]). The "solid" line describes the obese, low tolerance group (No. 18, Group IV in [17]). The blood glucose level (BGL) of the first group decreases to basal level after 2 hours, while the BGL of the second group declines much slower.

Table 6.2   Physical Variable, Values, and Definition Used in the Simulation

x_1(mg/dL): sensor measurement	k_e, o(min^{-1}): insulin degradation rate
x_2(mg/dL): plasma glucose concentration	V_1(L/kg): plasma distribution volume
x_3(min^{-1}): interstitial insulin	u(mU/kg/min): insulin infusion rate
x_4(mU/L): plasma insulin concentration	G_b(mg/dL): basal plasma glucose concentration
x_5(mU/kg): insulin mass in intermediate site	i_b(mU/L): basal plasma insulin concentration
x_6(mU/kg): insulin mass at the injection site	OF(mg/dL): offset current
P_1(min^{-1}): glucose effectiveness	τ(min): time delay constant of sensor
P_3(min^{-1}/mU/L)/P_2(min^{-1}): insulin sensitivity	ρ: sensor sensitivity
k,p(min^{-1}): insulin transition rate	

$$\dot{x}_1 = \theta_1 x_1 + \theta_2 x_2 \tag{6.80}$$

$$\dot{x}_2 = -\left(x_2 + G_b\right)x_3 - P_1 x_2 \tag{6.81}$$

$$\dot{x}_3 = -P_2 x_3 + P_3 x_4 \tag{6.82}$$

$$\dot{x}_4 = \theta_3 x_5 - k_e x_4 \tag{6.83}$$

$$\dot{x}_5 = kx_6 - \left(p + o\right)x_5 \tag{6.84}$$

$$\dot{x}_6 = -kx_6 + u \tag{6.85}$$

$$y = x_1 \tag{6.86}$$

subject to

$$0 \le u(t) \le u_{\max} \tag{6.87}$$

$$y(t) \ge 60 \text{ mg / dL} \tag{6.88}$$

All the states are bounded.

6.3.3.2   Control System Development

Observing the nonlinear system (6.80) through (6.86), one can describe it in a more general format, as follows:

$$\begin{aligned}
\dot{x}_1 &= f_1(x_1) + g_1(x_1)x_2 \\
\dot{x}_2 &= f_2(x_1, x_2) + g_2(x_1, x_2)x_3 \\
&\cdots \\
\dot{x}_i &= f_i(x_1, x_2, \ldots, x_i) + g_i(x_1, x_2, \ldots, x_i)x_{i+1} \\
\dot{x}_n &= f_n(x_1, x_2, \ldots, x_n) + g_n(x_1, x_2, \ldots, x_n)u \\
y &= x_1
\end{aligned} \tag{6.89}$$

where $x_i \in R$ is the state and $u \in R$ is the control input. The functions f_i and $g_i : R^i \to R$ are smooth in R^i with $f_i(0) = 0$ and $g_i(x_1, \ldots, x_i) \ne 0$, for $i = 1, \ldots, n$.

This system is called a strict feedback systems [19], because the nonlinearities f_i and g_i in the $\dot{x}_i$ equation depend only on $x_1, \ldots, x_i$ and state x_i. Actually, the system exhibits in a cascade connection form as illustrated in Figure 6.14.

In general, there exist two ways to design a nonlinear feedback to regulate the strict feedback systems. The first way is the exact feedback linearization, which uses a nonlinear coordinate transformation to transfer the original system into a linear form. The second approach is the backstepping design, which divides the original system into n subsystems and stabilizes them one by one using Lyapunov functions. By exploiting the extra flexibility existing in scalar systems, backstepping can solve regulation problems under conditions less restrictive than those encountered in exact feedback linearization methods.

The backstepping design can be divided into n steps. Starting from the first component (6.80), one can design a state feedback $x_2 = \alpha_1(x_1)$ to make output $y = x_1$ track a reference signal. Then, stepping back into the second component, one can design another state feedback $x_3 = \alpha_2(x_1, x_2)$ to track α_1. And so on, one designs the state feedbacks recursively towards the final component. Finally, a stabilizing state feedback u is devised to track α_{n-1}.

Now, consider the system (6.80) through (6.86), let us assume the tracking setpoint is r.

Step 1: Starting from the first component (6.80), define the error variable $z_1 = x_1 - r$. Its time derivative is then given by

$$\dot{z}_1 = \dot{x}_1 = \theta_1 x_1 + \theta_2 x_2 \tag{6.90}$$

Select a candidate Lyapunov function

$$V_1 = z_1^2 / 2\gamma_1 > 0 \tag{6.91}$$

where $\gamma_1 > 0$.

Choosing

$$x_2 = \frac{-\theta_1 x_1}{\theta_2} - c_1 z_1 \overset{\Delta}{=} \alpha_1(x_1, z_1)$$

yields

$$\dot{V}_1 = -c_1 \theta_2 z_1^2 / \gamma_1 \leq 0 \tag{6.92}$$

By LaSalle's invariance principle, the only solution that can stay identically in set $S = \dot{V}_1(z_1) = 0$ is the trivial solution $z_1 = 0$. Thus, the origin is asymptotically stable. Here, c_1 is the convergence coefficient of z_1.

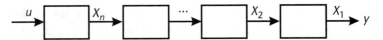

Figure 6.14 Cascade structure of a strict feedback system.

The tracking system becomes

$$\dot{z}_1 = \theta_1 x_1 + \theta_2 \alpha_1 = -c_1 \theta_2 z_1 \tag{6.93}$$

Step 2: Consider the two first components (6.80) and (6.81); we assign x_3 as the virtual control. By selecting x_3, we will make z_1 and new error variable $z_2 = x_2 - \alpha_1(x_1, z_1)$ tends to zero for $t \to +\infty$.

Select a candidate Lyapunov function,

$$V_2 = z_1^2 / (2\gamma_1) + z_2^2 / (2\gamma_2) > 0 \tag{6.94}$$

where $\gamma_1 > 0$, $\gamma_2 > 0$. Its derivative can be calculated as

$$\dot{V}_2 = \dot{V}_1 + z_2 \dot{z}_2 / \gamma_2$$
$$= \frac{\theta_2 z_1 z_2 - c_1 \theta_2 z_1^2}{\gamma_1} + \frac{z_2 \dot{z}_2}{\gamma_2} \tag{6.95}$$

Thus we choose

$$x_3 = (-P_1 x_2 - \dot{\alpha}_1) / (x_2 + G_b) + c_2 z_2 \overset{\Delta}{=} \alpha_2(x_2, z_2) \tag{6.96}$$

such that

$$\dot{V}_2 = -c_1 \theta_2 z_1^2 / \gamma_1 - c_2(x_2 + G_b) z_2^2 / \gamma_2 + \theta_2 z_1 z_2 / \gamma_1 \tag{6.97}$$

where c_2 is the convergence coefficient of z_2. The term $\theta_2 z_1 z_2 / \gamma_1$ will be cancelled later.

Steps 3–5: Repeating the process of Steps 1 and 2, we define error variables $z_i = x_i - \alpha_{i-1}(x_{i-1}, z_{i-1})$ and design virtual controls $x_{i+1} = \alpha_i$ in order to make $z_i(t) \to 0$ as $t \to +\infty$.

Selecting candidate Lyapunov functions as

$$V_i = V_{i-1} + z_i^2 / \gamma_i$$

then the virtual controls can be chosen as follows:

$$x_4 = (P_2 x_3 + \dot{\alpha}_2) / P_3 - c_3 z_3 \overset{\Delta}{=} \alpha_3$$
$$x_5 = (k_e x_4 + \dot{\alpha}_3) / \theta_3 - c_4 z_4 \overset{\Delta}{=} \alpha_4$$
$$x_6 = ((p+o)x_5 + \dot{\alpha}_4) / k - c_5 z_5 \overset{\Delta}{=} \alpha_5$$

Step 6: Define the last error variable $z_6 = x_6 - \alpha_5$, and design the real control u so that all z_i tends to zero as $t \to +\infty$. Choose the Lyapunov function candidate,

$$V_6 = \sum_{i=1}^{6} z_i^2 / (2\gamma_i) \tag{6.98}$$

where $\gamma_i > 0, i = 1, ..., 6$. In this case, we can choose a control law,

$$u = kx_6 + \dot{\alpha}_5 - c_6 z_6 \tag{6.99}$$

and let $\gamma_1 = \theta_2, \gamma_2 = \min(x_2 + G_b), \gamma_3 = P_3, \gamma_4 = \theta_3, \gamma_5 = k$, and $\gamma_6 = 1$, then the time derivative of V_6,

$$\dot{V}_6 = -c_1 z_1^2 - c_2 \frac{x_2 + G_b}{\min(x_2 + G_b)} z_2^2 - \sum_{i=4}^{6} c_i z_i^2 + z_1 z_2 - \frac{(x_2 + G_b) z_2 z_3}{\min(x_2 + G_b)}$$

$$+ z_3 z_4 + z_4 z_5 + k z_5 z_6 \tag{6.100}$$

$$\leq -\sum_{i=1}^{6} c_i z_i^2 + z_1 z_2 - z_2 z_3 + z_3 z_4 + z_4 z_5 + z_5 z_6$$

$$= Z\Gamma Z^T$$

where $Z = [z_1 z_2 ... z_6]$, and

$$\Gamma = \begin{bmatrix} -c_1 & \frac{1}{2} & 0 & 0 & 0 & 0 \\ \frac{1}{2} & -c_2 & -\frac{1}{2} & 0 & 0 & 0 \\ 0 & -\frac{1}{2} & -c_3 & \frac{1}{2} & 0 & 0 \\ 0 & 0 & \frac{1}{2} & -c_4 & \frac{1}{2} & 0 \\ 0 & 0 & 0 & \frac{1}{2} & -c_5 & \frac{1}{2} \\ 0 & 0 & 0 & 0 & \frac{1}{2} & -c_6 \end{bmatrix}$$

Thus, if we choose

$$c_1 > 0, c_2 > \frac{1}{4c_1}, c_3 > \frac{c_1}{4c_1 c_2 - 1} \overset{\Delta}{=} c_3^*,$$

$$c_4 > \frac{1}{4(c_3 - c_3^*)} \overset{\Delta}{=} c_4^*,$$

$$c_5 > \frac{1}{4(c_4 - c_4^*)} \overset{\Delta}{=} c_5^*,$$

and $c_6 > \frac{1}{4(c_5 - c_5^*)}$

then $\dot{V}_6 = Z\Gamma Z^T < 0$, and the origin of system (6.80) through (6.86) is asymptotically stable. With different convergence coefficients, however, the convergence rates and undershoot will be different.

By Steps 1 through 6, we construct a backstepping controller, which has much smaller undershoot and transition time than a linearization approach. A disadvantage, however, is that the approach is very model dependent. It is sensitive to different model parameters. In practice, parameters in minimal model system (6.77) and

(6.78) often exhibit significant difference among patients. Even for the same patient, parameters may vary from day to day. To overcome the intra- and interpatient variability, adaptive controllers are often used.

Suppose P_1, P_2, and k_e are the parameters that we strive to identify in real time; we can embed the adaptive controller into the above backstepping controller.

$$\dot{\hat{P}}_1 = \sigma_1 x_2 z_2$$

$$\dot{\hat{P}}_2 = \sigma_2 x_3 z_3 \tag{6.101}$$

$$\dot{\hat{k}}_1 = \sigma_3 x_4 z_4$$

where $\sigma_i > 0, i = 1, \ldots, 3$ are the convergence rates for the adaptive controller.

Define the following Lyapunov function:

$$V_{ad} = V_6 + \frac{\left(P_1 - \hat{P}_1\right)^2}{2\gamma_2 \sigma_1} + \frac{\left(P_2 - \hat{P}_2\right)^2}{2\gamma_3 \sigma_2} + \frac{\left(k_e - \hat{k}_e\right)^2}{2\gamma_4 \sigma_3} \tag{6.102}$$

Its derivative can be presented as

$$\dot{V}_{ad} = \frac{z_1 \dot{z}_1}{\gamma_1} + \frac{z_6 \dot{z}_6}{\gamma_6} - \frac{\left(P_1 - \hat{P}_1\right) x_2 z_2}{\gamma_2} + \frac{-z_2 \left((x_2 + G_b) x_3 + \hat{P}_1 x_2 + \dot{\alpha}_1\right)}{\gamma_2}$$

$$+ \frac{z_3 \left(P_3 x_4 + \hat{P}_2 x_3 - \dot{\alpha}_2\right) - \left(P_2 - \hat{P}_2\right) x_3 z_3}{\gamma_3} + \frac{z_5 \dot{z}_5}{\gamma_5}$$

$$+ \frac{z_4 \left(\theta_3 x_5 - \hat{k}_e x_4 + \dot{\alpha}_3\right) - \left(k_e - \hat{k}_e\right) x_4 z_4}{\gamma_4} \tag{6.103}$$

$$= \frac{z_1 \dot{z}_1}{\gamma_1} - \frac{z_2 \left((x_2 + G_b) x_3 + P_1 x_2 + \dot{\alpha}_1\right)}{\gamma_2} + \frac{z_5 \dot{z}_5}{\gamma_5} - \frac{z_3 \left(P_3 x_4 + P_2 x_3 - \dot{\alpha}_2\right)}{\gamma_3}$$

$$+ \frac{z_6 \dot{z}_6}{\gamma_6} - \frac{z_4 \left(\theta_3 x_5 + k_e x_4 + \dot{\alpha}_3\right)}{\gamma_4} \tag{6.104}$$

Selecting the same coefficients $\gamma_1 - \gamma_6$, the above equation is equal to (6.100). Thus, the adaptive backstepping controller is also asymptotically stable.

6.3.3.3   Nonlinear State Observer

In the last section, one may notice that the proposed adaptive backstepping controller requires full state information as feedback. However, it is impossible and unnecessary to measure all the states directly. For this reason, we have to construct a nonlinear state observer to estimate the unmeasured states in real time. Asymptotic or exponential error reduction in time is usually desired.

For a nonlinear time-invariant system,

$$\dot{x} = Ax + f(x) + Bu, y = Cx \tag{6.105}$$

where the nonlinearity $f(x)$ satisfies the Lipschitz condition,

$$\|f(x) - f(\hat{x})\| \le \beta \|x - \hat{x}\| \tag{6.106}$$

with a Lipschitz constant β, we still can design a Luenburger-like observer as follows:

$$\dot{\hat{x}} = A\hat{x} + f(\hat{x}) + Bu + L(y - \hat{y}), \quad \hat{y} = C\hat{x} \tag{6.107}$$

This observer is often called the Thau observer or Lipschitz observer due to the Lipschitz nonlinearity encountered in the assumption.

The stability of this observer is guaranteed by Thau's theorem [20]: *If a gain matrix L can be chosen such that*

$$\frac{\lambda_{\min}(Q)}{2\lambda_{\max}(P)} > \beta \tag{6.108}$$

where $(A - LC)^T P + P(A - LC) = -Q$, then the system (6.105) with Thau observer (6.107) yields asymptotically stable.

However, this theorem only serves as stability criteria for known gain matrix **L**. It does not tell how to construct the gain matrix **L**. Raghavan [21] and Rajamani [22] proposed two useful design algorithms for choosing **L** to ensure stability. However, these two algorithms only satisfy when the Lipschitz constant is small. In the considered system (6.80) through (6.86), the nonlinearity $-x_2 x_3$ contains a large variable x_2; therefore, the Lipschitz constant will not be small enough to fit with Raghavan's and Rajamani's algorithms.

Because of the unavailability of the Thau observer, we have to turn to more basic and general ideas for designing nonlinear state observers. The exact feedback linearization that we mentioned before now becomes a feasible choice.

For nonlinear systems in the following format:

$$\begin{aligned} \dot{x} &= f(x) + g(x)u \\ y &= h(x) \end{aligned} \tag{6.109}$$

If the system has relative degree [19], there exists a diffeomorphic transformation,

$$z = T(x) = \begin{bmatrix} h & L_f h & L_f^2 h & L_f^3 h & L_f^4 h & L_f^5 h \end{bmatrix} \tag{6.110}$$

to change the original system (6.109) into a canonical form:

$$\dot{z} = Az + B\phi(z,u), \text{ where } A = \begin{bmatrix} 0 & 1 & 0 & 0 & 0 & 0 \\ \vdots & \vdots & \vdots & \vdots & \vdots & \vdots \\ 0 & 0 & 0 & 0 & 0 & 1 \\ 0 & 0 & 0 & 0 & 0 & 0 \end{bmatrix}, B = \begin{bmatrix} 0 \\ 0 \\ \vdots \\ 1 \end{bmatrix} \tag{6.111}$$

and

$$\phi(z,u) = L_f^6 h + L_f^5 L_g hu$$

The notation $L_f h = \dfrac{\partial h}{\partial x} f(x)$ is called Lie Derivative.

Then one can design a high-gain observer [23] as follows:

$$\dot{\hat{z}} = A\hat{z} + B\phi(\hat{z},u) + H_\varepsilon\left(y - \hat{y}\right) \tag{6.112}$$

where the observer gain matrix H_ε is given by

$$H_\varepsilon = \begin{bmatrix} \alpha_1 / \varepsilon & \alpha_2 / \varepsilon^2 & \cdots & \alpha_n / \varepsilon^n \end{bmatrix}$$

Here $\varepsilon > 0$ and the $\alpha_i s$ are chosen such that the polynomial equation

$$s^n + \alpha_1 s^{n-1} + \ldots + \alpha_{n-1} s + \alpha_n = 0 \tag{6.113}$$

is Hurwitz.

Applying $\hat{x} = T^{-1}(\hat{z},u)$, yields

$$\dot{\hat{z}} = \frac{\partial T}{\partial \hat{x}} \dot{\hat{x}} \tag{6.114}$$

Then,

$$\dot{\hat{x}} = \left[\frac{\partial T(\hat{x},u)}{\partial \hat{x}}\right]^{-1} \left(A\hat{z} + B\phi(\hat{z},u) + H_\varepsilon\left(y - \hat{y}\right)\right)$$

$$= f(x,u) + \left[\frac{\partial T(\hat{x},u)}{\partial \hat{x}}\right]^{-1} H_\varepsilon\left(y - \hat{y}\right) \tag{6.115}$$

Select $\alpha_i = \begin{bmatrix} 0.0461 & 0.5645 & 2.5984 & 5.8800 & 7.0000 & 4.2000 \end{bmatrix}$, and $\varepsilon = 400$, then the observer error is illustrated as shown in Figure 6.15.

6.3.3.4 Simulation of the Control System

To show the effectiveness of the control strategy, simulations have been conducted using MATLAB. Both ABCNO and ABC are simulated in the same figures under the same parameters. In practice, the control input is set to be saturated with 0 and 30. The reference system parameters are listed in Table 6.3. The initial state of system is set to $\begin{bmatrix} 180 & 200 & 0 & i_b & 0 & 0 \end{bmatrix}$, and the observer starts at $\begin{bmatrix} G_b & G_b & 0 & i_b & 0 & 0 \end{bmatrix}$. The control input starts from 10 minutes.

Figure 6.16 shows the simulation results of the proposed control strategies. From the figure, we can see that both approaches have small undershoot, 7.5 mg/dL and 8 mg/dL, respectively. The blood glucose level enters and stays in the acceptable range +90±10 mg/dL at 123 and 130 minutes, respectively. Due to the difference of sensor measurement and real blood glucose level, the first figure shows blood glucose increases in a few minutes. If the initial difference is greater, the increasing effect will be more significant.

From the aspect of insulin infusion rate, ABC starts at a maximum rate, while ABCNO only requires 12 U/h. The ABC approach stops infusing in 30 minutes. But the infusion rate of ABCNO appears a two-phase infusion: gradually to zero at 20 minutes, and then increasing to 0.7 U/h at 35 minutes, before finally vanishing.

Due to the time delay effect of subcutaneous infusion and sensor measurement, the blood glucose level may decrease slower than that via the intravenous infusion. However, the proposed approach still achieves fast decreasing of blood glucose level. There is an inevitable trade-off between fast response and small undershoot during the simulation. To achieve a fast response, a large dosage of insulin must be applied; while large control often leads to large undershoot.

To show the effectiveness of the proposed approaches, linear model predictive control and exact feedback linearization techniques were applied to the same system. The blood glucose level and insulin infusion rate are compared in the following sections.

Comparison with Model Predictive Control
The performance of ABCNO has been compared with that of the MPC [24]. The linear MPC approach was applied based on the linearized model of system (6.80) through (6.86). Assign the quadratic performance index in the form,

$$\min_{\Delta u(k), \Delta u(k+1)} \left\{ \sum_{l=1}^{m} P \Gamma_u \left(\Delta u \left(k + \frac{1}{1} - 1 \right) \right) P^2 + \sum_{l=1}^{p} P \Gamma_y \left(y_{(k+l/k)} - r(k+l) \right) P^2 \right\} \quad (6.116)$$

subject to

$$0 \, U/h \le u \le 30 \, U/h$$

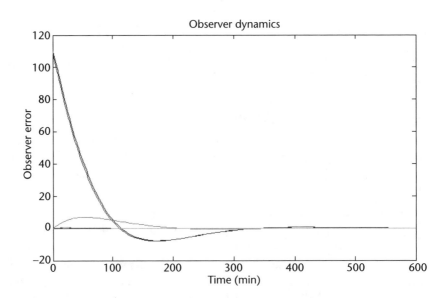

Figure 6.15   Observer estimation errors in high-gain observer.

Table 6.3   Parameters Used in the Simulation

$P_1 = 0.0093$	$P_2 = 0.02$	$k_e = 0.015$	$P_3 = 0.00000778$	$o = 0.25$	$\rho = 1$	$p = 0.025$
$k = 0.135$	$V_i = 0.21$	$r = 90$	$G_b = 90$	$i_b = 0$	$\tau = 3$	

Source: [16, 17].

$$y \geq 60 \text{ mg / dl}$$

where Γ_u and Γ_y are weight matrices.

The comparison of ABCNO and MPC is illustrated in Figure 6.17. From the figures, we can see that the MPC approach has significantly larger undershoot (30 mg/dL), which leads to a long settling time (= 600 minutes). The MPC appears to increase infusion rate gradually and reach a maximum value of 1 U/h at 180 minutes, then vanish slowly at 490 minutes. The total dosage of insulin via the MPC approach is around 4U, while that of ABCNO is 2U. It can be concluded that ABCNO performs significantly better than MPC.

Comparison with the Exact Feedback Linearization Method
For nonlinear systems,

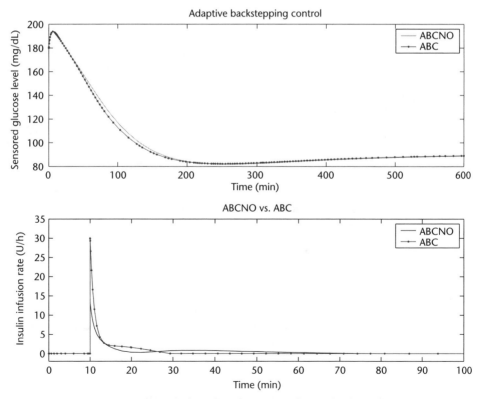

Figure 6.16   Sensor measurement and insulin infusion via adaptive backstepping control with nonlinear observer versus adaptive backstepping control. The solid line presents the dynamics of ABCNO, and the dot-dash line illustrates the ABC approach.

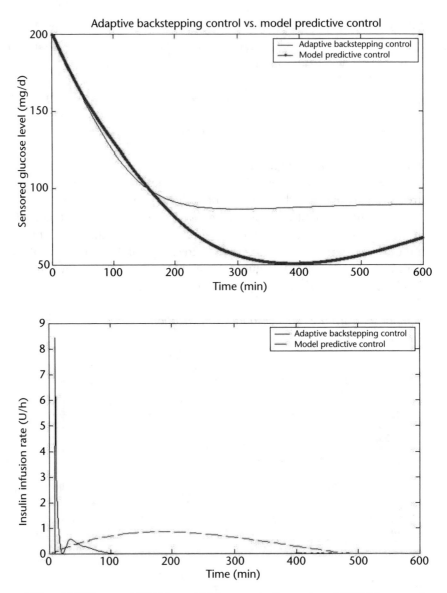

Figure 6.17 ABCNO versus MPC. The solid line presents the proposed ABCNO approach, and the dashed line illustrates the MPC approach.

$$\dot{x} = f(x) + g(x)u$$
$$y = h(x)$$

if the relative degree is n, there exists a diffeomorphic transformation,

$$z = T(x) = \begin{bmatrix} h & L_f h & L_f^2 h & L_f^3 h & L_f^4 h & L_f^5 h \end{bmatrix}$$

to change the original system into a canonical form,

$$\dot{z} = Az + B\phi(z,u), \text{ where } \phi(z,u) = L_f^6 h + L_f^5 L_g h u$$

Obviously, one can design the following controller

$$u = \left(-L_f^6 h + Kz\right) / L_f^5 L_g h \qquad (6.117)$$

to cancel the nonlinearities and assign new poles to the canonical linear system.

The comparison of ABCNO and FLC is represented in Figure 6.18. FLC reduces the blood glucose concentration to 100 mg/dL within 90 minutes, but the undershoot is too large (72 mg/dL), larger even than that of MPC. With respect to insulin infusion, the FLC approach keeps the maximum infusion rate for 30 minutes. That

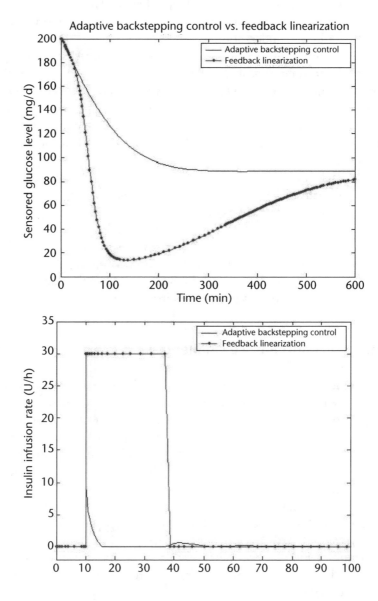

Figure 6.18   ABCNO versus FLC. The solid line presents the proposed ABCNO approach, and the dot-dash line illustrates the FLC approach.

is the main reason for the large undershoots. It can be concluded that the ABCNO approach performs far better than the FLC approach.

6.3.4 Discussion

Subcutaneous infusion systems are nonlinear time invariant systems, which contain a hyperbolic nonlinear term. Three techniques can be chosen from to deal with the nonlinearity: Jacobian linearization, exact feedback linearization, and the Lyapunov-function based backstepping approach. Jacobian linearization is the simplest approach to implement, although it can precisely approximate the original system only in a small neighborhood of equilibrium. Therefore, if the system dynamics start far from the equilibrium, the closed-loop system often experiences slow response and large undershoot.

The feedback linearization technique is accurate since it applies nonlinear change of coordinates to transform a nonlinear system to linear form. The FLC approach compensates for all the nonlinearities and assigns new poles to the canonical linear system. However, for the proposed system, the term $L_f^5 L_g h$ in (6.117) appears to be a very small value, which often results in large inputs. To reduce the magnitude of control efforts, one has to select small poles, which leads to slow convergence.

The backstepping approach designs the control law based on the Lyaponov function directly. The extra flexibility of scalar systems allows the backstepping design to have less restrictive conditions than exact feedback linearization methods. It avoids the dilemma between control magnitude and small poles. Therefore, the backstepping approach can have faster response and smaller undershoot than the first two approaches. In addition, MPC and FLC approaches do not consider patient variability. However, the proposed ABCNO embeds the adaptive controller, which can identify the important parameters in real time.

6.4 Conclusions and Future Challenges

Development of control mechanisms is based upon current knowledge of biological systems. With an increasing understanding of biological systems, researchers will be able to propose new, highly accurate models. The development of real-time sensors, fast response actuators, and even efficient input drugs, combined with improved models, will make the existing control mechanisms in life science automation achieve better performance. However, controlling biological systems is an urgent task that cannot wait until all the ideal technical features mature. We must develop new control mechanisms based on existing technology to tackle our task. Moreover, new applications have exploited new technology with surprising speed. With the development of new technologies in the physical and chemical sciences, many opportunities will arise for them to be utilized in controlling biological systems. It is critical to have a new theoretical framework for control mechanisms to deal with these new challenges.

In general, the ubiquitous PID controllers are often the earliest practical implements. However, most of them do not perform well, since they must sacrifice efficiency to keep the systems stable due to the highly nonlinear characteristics of the biological systems. In order to improve efficiency—achieving short transition times and small overshoots—researchers and engineers have proposed many advanced control mechanisms. The efforts are generally divided into two approaches: the first approach is to linearize the nonlinear models, then apply optimal control; the second approach is to directly presume simple linear models, then use adaptive control. The former eases online computational burdens but requires an accurate model; the latter does just the opposite—that is, eases the modeling complexity but requires real-time parameter adjustment. The choice of these two approaches depends on how well we know the biological system. For extremely complex systems that are not well understood, the fuzzy control method appears more attractive.

The biggest challenge in devising future control mechanism algorithms certainly comes from the newest technology: nanotechnology. It can safely be argued that nanotechnology is the future of a wide spectrum of sciences that will give rise to new applications not attainable by available technologies. Exciting results from exploring micron size systems have encouraged us to scale our devices down into the nanon size (billionth of a meter). Understanding and utilizing this new and sophisticated world is a challenge for physics, chemistry, biology, as well as for engineering. Control engineering, especially, plays an important role in developing new applications, spreading productive tools and urging rapid advancement in nanotechnology. Many new challenging applications in life science have flourished, including nanoelectromechanical systems (NEMS), nanorobotics, and nanomanipulation. These systems may fundamentally alter the conventional means of drug delivery and disease diagnosis, innovate biomaterials and devices, and improve existing tools in medical and industrial applications.

The challenges of nanoscale control mechanisms mainly come from four areas: the complexity coupled nonlinear model, sensitivity to the environment and noises, different length and time scales, and self-assembly.

In summary, control mechanisms play an increasingly important role in life science automation, particularly for micro/nanoscale dynamics systems. On one hand, traditional control mechanisms find exiting applications in life science automation. On the other hand, there are tremendous challenges. Advanced control mechanisms to deal with nanoscale complex dynamics are still an open issue. The issue is about integrating engineering principles with biological sciences.

References

[1]   Lapidus, L., and R. Luus, *Optimal Control of Engineering Processes*, Waltham, MA: Blaisdell Publishing, 1967.

[2]   Qin, S. J., and T. A. Badgwell, "A Survey of Industrial Model Predictive Control Technology," *Control Engineering Practice*, Vol. 11, 2003, pp. 733–764.

[3]   Astrom, K. J., and B. Wittenmark, *Adaptive Control*, Reading, MA: Addison-Wesley, 1989.

[4] Zhang, T., "Valveless Piezoelectric Micropump for Fuel Delivery in Direct Methanol Fuel Cell (DMFC) Devices," Ph.D. Dissertation, Dept. of Mechanical Engineering, University of Pittsburgh, Pittsburgh, PA, 2005.

[5] Passino, K. M., and S. Yurkovich, *Fuzzy Control*, Reading, MA: Addison-Wesley, 1998.

[6] Slate, J. B., et al., "A Model for Design of a Blood Pressure Controller for Hypertensive Patients," *Proc. IEEE Engineering in Medicine and Biology Conference*, Denver, CO, October 1979.

[7] Kienitz, K. H., and T. Yoneyama, "A Robust Controller for Insulin Pumps Based on H-Infinity Theory," *IEEE Transactions on Biomedical Engineering*, Vol. 40, 1993, pp. 1133–1137.

[8] Parker, R. S., F. J. Doyle, and N. A. Peppas, "The Intravenous Route to Blood Glucose Control," *Engineering in Medicine and Biology Magazine*, Vol. 20, 2001, pp. 65–73.

[9] Bellazzi, R., G. Nucci, and C. Cobelli, "The Subcutaneous Route to Insulin Dependent Diabetes Therapy," *Engineering in Medicine and Biology Magazine*, Vol. 20, 2001, pp. 54–64.

[10] Hovorka, R., et al., "Nonlinear Model Predictive Control of Glucose Concentration in Subjects with Type I Diabetes," *Physiological Measurement*, Vol. 25, 2004, pp. 905–920.

[11] Polla, D. L., et al., "Microdevices in Medicine," *Annual Review of Biomedical Engineering*, Vol. 2, 2000, pp. 551–576.

[12] Accoto, D., M. C. Carrozza, and P. Dario, "Modelling of Micropumps Using Unimorph Piezoelectric Actuator and Ball Valves," *J. Micromech. Microeng.*, Vol. 10, 2000, pp. 277–281.

[13] Beebe, D. J., G. A. Mensing, and G. M. Walker, "Physics and Applications of Microfliudics in Biology," *Annual Review of Biomedical Engineering*, Vol. 4, 2002, pp. 261–286.

[14] Steil, G. M., A. E. Panteleon, and K. Rebrin, "Closed-Loop Insulin Delivery: The Path to Physiological Glucose Control," *Advanced Drug Delivery Reviews*, Vol. 56, 2004, pp. 125–144.

[15] Nucci, G., and C. Cobelli, "Models of Subcutaneous Insulin Kinetics: A Critical Review," *Computer Methods and Programs in Biomedicine*, Vol. 62, 2000, pp. 249–257.

[16] Matsuo, Y., et al., "Strict Glycemic Control in Diabetic Dogs with Closed-Loop Intraperitoneal Insulin Infusion Algorithm Designed for an Artificial Endocrine Pancreas," *J. Artif. Organs,* Vol. 6, 2003, pp. 55–63.

[17] Bergman, R. N., L. S. Phillips, and C. Cobelli, "Physiologic Evaluation of Factors Controlling Glucose Tolerance in Man," *Journal of Clinical Investigation*, Vol. 68, 1981, pp. 1456–1467.

[18] Sorensen, J. T., "A Physiologic Model of Glucose Metabolism in Man and Its Use to Design and Assess Improved Insulin Therapies for Diabetes," Ph.D. Dissertation, Dept. of Chemical Engineering, MIT, 1985.

[19] Khalil, H. K., *Nonlinear Systems*, 2nd ed., Upper Saddle River, NJ: Prentice-Hall, 1996.

[20] Thau, F. E., "Observing the State of Nonlinear Dynamic Systems," *Int. J. Contr.*, Vol. 17, 1973.

[21] Raghavan, S., "Observers and Compensators for Nonlinear Systems, with Application to Flexible Joint Robots," Ph.D. dissertation, University of California at Berkeley, 1992.

[22] Rajamani, R., "Observers for Lipschitz Nonlinear Systems," *IEEE Transactions on Automatic Control*, Vol. 43, 1998, pp. 397–401.

[23] Findeisen, R., et al., "Output-Feedback Nonlinear Model Predictive Control Using High-Gain Observers In Original Coordinates," *7th European Control Conference ECC,* Cambridge, UK. Paper ID T35 on CD-ROM, 2003.

[24] Parker, R. S. , F. J. Doyle III, and N. A. Peppas, "A Model-Based Algorithm for Blood Glucose Control in Type I Diabetic Patients," *IEEE Transactions on Biomedical Engineering* Vol. 46, 1999, pp. 148–157.

CHAPTER 7

Robotics for Life Science Automation

Jaydev P. Desai and Anand Pillarisetti[†]

In 1921, Czechoslovakian playwright *Karel Capek* introduced the word *robot* in the play *R.U.R – Rossums Universal Robots*. The word originated from the Czech word *robota*, meaning work.

This chapter focuses on the role of robotics in the life science industry. Automated biomanipulation techniques and their advantages over conventional cell manipulation techniques are discussed. Implementation of robotics in a few of the life science sectors is also presented followed by a brief discussion at the end.

7.1 Introduction

Robotics has had a tremendous influence in life science automation, ranging from biomanipulation to production of DNA and protein microarray. In this chapter we will briefly describe the various techniques for biomanipulation, cell injection, cell culture automation, high throughput processing of biological samples, and production of DNA and protein microarray. With respect to biomanipulation, we will discuss various techniques such as magnetic manipulation, acoustic energy, microelectromechanical systems and mechanical manipulation, to name a few. We will also discuss the various robotic systems for cell injection and compare the success rate between robotic and manual techniques. Similarly, several commercial systems and academic efforts will be discussed in cell culture automation, high throughput processing of biological samples, and production of DNA and protein microarray.

7.2 Cell Manipulation Techniques

Manipulating an individual biological cell is a common process involved in intracytoplasmic sperm injection (ICSI), pronuclei DNA injection, gene therapy, and other biomedical areas. Efforts to micromanipulate cells under the microscope date back to the last half of the twentieth century [1]. Conventional methods of manipulating cells have been prevalent in the past. The principles of microinjection were

†   University of Maryland

developed by Marshall A. Barber. Barber developed the *pipette method* to isolate bacterial cells. Researchers in the field of biology have conducted experiments using conventional microinjection techniques—for example, to understand the role of the nucleus in embryonic differentiation, to suggest that the pronuclei formation from nuclei of a species depends on the activation of egg cytoplasm and developing pronuclei by surgically injecting the egg cytoplasm with a spermatozoa of the same species or different species [2]. Manual manipulation, however, requires long training, and the success rate depends on the experience of the operator. Even for an experienced operator, the injection process results in a low success rate and less reproducibility. Early efforts have been made to automate the injection process. Capillary pressure microinjection (CPM) is one of the supporting technologies for injecting macromolecules into a single living cell. Injection in nuclei or cytoplasm was performed using an ejection system with pressure levels manipulated by a single button, which requires no learning time, and the injection rate obtained can be as high as 70% to 80%. A semiautomatic microinjection system has also been developed to increase the cell survival rate in CPM. The introduction of computer control in manipulating biological cells has improved the efficiency of the process. A computer-controlled microrobotic system with 3 degrees of freedom (DOF) was developed for ICSI in a mouse [3]. The sperm injection was successfully completed without damaging any of the mouse ova. Later on, piezo driven pipette had been used to perform ICSI in a mouse, which demonstrated 80% survival rate of sperm-injected oocytes. Few other researchers have proposed piezo actuators for cell manipulation, which offers highly repeatable motion and increases the chances of the oocytes survival rate. Different control strategies have also been used to develop a visually servoed microrobotic system. For example, Sun et al. [4] developed an autonomous embryo pronuclei DNA injection system by implementing a hybrid visual servoing control scheme. In the sections below, we provide a comprehensive overview of the state of the art in biomanipulation, which includes various techniques such as optic and electric micromanipulation, magnetic micromanipulation, acoustic energy for micromanipulation, microelectromechanical systems (MEMS), and mechanical micromanipulation.

7.2.1 Optical and Electric Micromanipulation

Optical trap is one of the promising methods to manipulate microscopic objects without physical contact. Ashkin [5] was the first to report the acceleration and trapping of micron sized particles by the forces of radiation pressure from visible light. The concept proposed by Ashkin gained importance in the area of cell manipulation. An automated optical manipulator was also developed by Buican [6] for trapping of biological particles. Optical forces are also used for rapid and active control of cell routing on a microfluidic chip. The cell sorter developed by Wang et al. [7] uses two lasers: a near infrared laser for the optical switch and a visible wavelength laser for detection and fluorescence measurement. It was demonstrated that the cell sorter can sort 280,000 cells in 44 minutes.

Apart from optical trapping, another noncontact manipulation technique is electrorotation. Washizu et al. [8] proposed dielectrophoresis (DEP) and electrorotation for generating translational and rotational force on living bacteria.

Dielectrophoresis involves applying a nonuniform electric field to exert an external force on the living bacteria and obtain its force-versus-velocity (F-v) characteristics. Electrorotation involves controlling the phase shift and magnitude of electric fields to exert an external torque on the living bacteria and obtain the torque-versus-speed (T-ω) characteristic. The authors claimed that the characteristics of the biomotor derived will play an important role in understanding its mechanism. A new technique called *optoelectrostatic micromanipulation* (OEMM) developed by Nishioka [9] combines optical pressure and electrical force to manipulate a single cell. Arai et al. [10] proposed a biojig, which uses dielectrophoretic force to manipulate a micro-object. An electric field is applied between two pairs of electrodes and analyzed using the finite element method (FEM). An electroreological (ER) joint was proposed and its application in a safety mechanism and microgripper was demonstrated. The advantage of the ER fluid is fast response time and the ease of miniaturizing the mechanical component. Later, Arai et al. [11] integrated a laser scanning manipulator for local position control of the target (cell) and dielectrophoresis for exclusion of other cells around the target, which proved to be an effective method of selective separation.

7.2.2   Magnetic Micromanipulation

The introduction of magnetic energy to manipulate cells resulted in a high success rate and increased the viability of the cell as reported by Pesce and De Felici [12]. Alenghat et al. [13] developed a magnetic tweezer with the goal of applying focused and quantifiable mechanical stress to individual cells in culture. The tweezer is an electromagnet, and ligand coated magnetic microbeads were attached to the cell membrane. This technique allows one to examine cell mechanics of an individual cell, which can play a major role in quantifying the material properties of the integrin-cytoskeleton linkages. A cell manipulation system using magnetic techniques to analyze individual cells has also been reported by Boukallel et al. [14]. Boukallel et al. developed a force sensing manipulator using permanent magnets and diamagnetic material. The approach required no control loop and the manipulator was highly sensitive to measuring forces in the nN range. Manipulating individual cells in a fluid using magnetic and electric fields was developed by Lee et al. [15]. Experiments were performed on yeast cells labeled with magnetic beads, which were trapped by a microelectromagnet matrix while the unlabelled cells were trapped by microposit matrix generating electric fields. This setup allowed for the possibility of constructing an efficient microfluidic system for sorting cells.

7.2.3   Micromanipulation Using Acoustic Energy

The manipulation of microparticles and biological cells by ultrasonic waves has been investigated by a few researchers. The advantage of this technique is that it involves no mechanical contact and it is noninvasive and efficient. Kozuka et al. [16] used acoustic energy to trap and control the position of micro-objects in two dimensions. Polystyrene particles (100 to 150 μm in diameter) were trapped at the nodes of the standing waves which were formed by three ultrasonic transducers placed at an angle of 120° to each other in the same plane. The position of the particle was controlled by changing the phase of one of the transducers, which caused the particle to be transported along the sound beam axis of phase shifted transducer. A three-dimensional (3D) manipulation technique developed by Kozuka et al. [17] was realized by using four ultrasonic transducers placed at the corners of a regular triangular pyramid. The movement of each particle was captured by two CCD cameras. The nodes of the standing wave play an important role in manipulating micron sized particles, and the position of such nodes can be changed by varying the electronic parameters of the resonator instead of the mechanical parameters [18]. The resonator consists of a fluid filled tube and two piezoelectric transducers. The proposed model of the resonator studied the electronic parameters of piezo devices and proved its importance in affecting the position of the nodes of standing waves. The advantage of this approach is that there is no mechanical movement which would cause unwanted fluid flow. Kim et al. [19] used acoustic forces in an ultrasonic field for concentrating HeLa cells and human mesencymal stem cells (hMSCs).

7.2.4 MEMS and Mechanical Micromanipulation

Microelectromechanical (MEMS) systems technology has emerged as an important tool to manipulate a single cell or an array of cells. The advent of this technology is proving to be successful in developing micro/nanosystems, which allows biomanipulation to be more reproducible and efficient. Microrobot was fabricated from conducting polymer like polypyrrole (PPy) by Jager et al. [20] to manipulate single cells in an aqueous medium. The fabricated structure has PPy in a bilayer configuration with gold acting as structural layer and electrode. Pillarisetti et al. [21] developed a polypyrrole-based hexagonal microgripper (LOTUS microgripper) for holding an individual cell. The LOTUS microgripper is comprised of six polypyrrole-gold (PPy-Au) bilayers. The PPy-Au bilayer is ideal for biomanipulation because it actuates at low voltages (less than 2V) and can operate in aqueous media at room temperature. Successful actuation of the LOTUS microgripper was demonstrated by performing experiments on the egg cell of a zebra fish. Chan et al. [22] took a step ahead and developed a polymer-based microrobotic actuator by estimating the actuation force. The fabricated actuator is trimorph with platinum layer sandwiched between two layers of parylene. The two methods of actuation are: (1) passing current through the resistive heater, and (2) changing the temperature of the medium surrounding the actuator. The successful actuation of the device was demonstrated by grasping follicle of zebra fish (*Danio rerio*) ranging from 500 μm to 1 mm in diameter. Bronson et al. [23] presented the concept and design of MEMS "tensile testing machine" to determine elastic and adhesive properties of cells. Apart from analyzing single cells, MEMS has the advantage of treating an array of cells, thus reducing the time of an operation. Chun et al. [24] proposed a micromachined DNA injection system, which presents the detailed fabrication process for an array of hollow microneedles. A review on microneedles for gene delivery is provided in [25]. Microfluidics plays an important role in transporting cells, which is an integral part of a cell manipulating system [26].

Mechanical micromanipulation referred to as contact manipulation is the common method adopted to inject sperm/genetic material into oocytes, for example in

ICSI and pronuclei DNA injection. Codourey et al. [27] discussed the development of two planar 3-DOF mechanisms using piezo actuators. The first one is a micro-crawling machine, and the second one is Abalone II, which relies on an impact drive. A 3-DOF micromanipulator incorporating split tube flexure was also developed by Goldfarb et al. [28]. The new flexure-based revolute joint exhibited no backlash, and the range of motion was five times more than a conventional flexure resulting in increased reachability workspace. It could also withstand roughly three times more load than a conventional flexure. One of the new techniques developed by Kallio et al. [29] is to use piezohydraulic actuation to construct a 3-DOF parallel micromanipulator. The basic elements of actuator system are piezoelectric actuator, bellows, and hydraulic oil. The displacement of the piezo actuator is amplified by the actuator system. To have better control on manipulating micron sized objects, Gao and Swei developed a 6-DOF micromanipulator [30]. The piezoelectric transducer has high resolution and quick response, and the displacement resolution obtained with the system is 10 nm with a natural frequency of 2 kHz. Parallel mechanisms, such as the 3-DOF finger module developed by Ohya et al. [31], was adopted in designing the actuators due to their high speed, high accuracy, and high stiffness properties. Tanikawa and Arai [32] developed a two-fingered microhand to manipulate a microscopic object, by simulating chopstick manipulation. A precision parallel micromechanism with 6 DOF was developed by Guo et al. [33], which consisted of a macro/micromechanism to position a micro-object.

Even though there has been considerable effort to automate manipulation of biological cells, vision has been the only sensing modality. Recently, there have been efforts aimed at sensing the interaction forces to improve the reliability of biomanipulation tasks [34, 35]. Force sensing in addition to vision would make the manipulation process repeatable and accurate. Few researchers have proposed the concept of *bilateral control*, which involves a master-slave setup. The bilateral control system takes into account the scaling effect in the macro/microworld and maintains a stable, transparent system. Pillarisetti et al. [35] have demonstrated a master-slave teleoperation framework for manipulating egg cells. A force and vision feedback system was also developed by Zhou et al. [36] in which the optical microscope provided vision feedback at the micron resolution, and an optical beam deflection sensor provided force feedback in the nN range. A 3-DOF piezoresistive force sensor was used [37] to measure forces while manipulating the cell of an onion. Ando et al. [38] proposed a telemicromanipulation system with haptic feedback.

Atomic force microscopy (AFM) has also been used to develop a telenanorobotics system. In this system, a 1-DOF haptic device was constructed for tactile sensing. Guthold et al. [39] developed a nanomanipulation system consisting of an AFM and a haptic device such as the PHANToM (manufactured by Sensable Technologies, Inc.) to provide real-time a 2D/3D graphics display along with force feedback. To measure real manipulating forces, MEMS force sensors have been developed which offer the advantage of miniaturization. MEMS force transducers have been developed by Lin et al. [40] by integrating 3D microstructures and signal processing electronics onto a single chip 2 mm^3 in size. A 2-DOF capacitive force sensor developed by Sun et al. [34] is capable of resolving forces up to 490 μN with a resolution of 0.01 μN in the x-direction, and up to 900 μN with a resolution of

0.24 μN in the y direction. The system can be used in aqueous solution where only the probe of the force sensor is immersed in the solution.

7.3 Robotics in the Life Science Industry

7.3.1 Cell Injection

Cell injection involves depositing sperm or genetic material into the cytoplasm/ nucleus of a biological cell. Injecting cells is an important process carried out in ICSI, pronuclei DNA injection, gene therapy, and other biomedical areas. ICSI is used to treat male-factor infertility and involves direct injection of a single immobilized spermatozoon into the cytoplasm of a mature oocyte. Transgenic species are produced by injecting DNA into the pronuclei of an embryo. In gene therapy, a normal gene is inserted into the genome to replace an "abnormal," disease-causing gene.

As discussed in the previous section, in the field of molecular biology, conventional pipette technology is used to carry out cell injection tasks, and the pipette technology was originally developed by Marshall A. Barber. A typical cell injection system is shown in Figure 7.1. Conventional methods of injecting cells require the operator to undergo long training (over a period of 1 year) and the success rates are low (around 10% to 15%) due to poor reproducibility. The fragile nature of a biological cell requires the operator to be efficient; otherwise, he/she may damage the cell. There are also chances of contamination due to direct human involvement. The

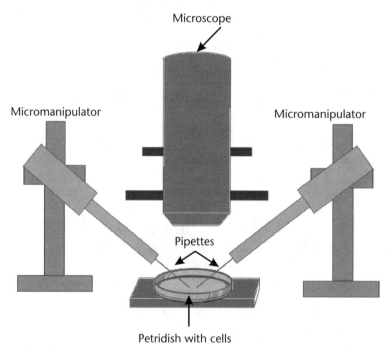

Figure 7.1 Cell injection system.

drawbacks involved in conventional methods have motivated the research community to develop robotic-based tools to perform cell injection tasks.

Tremendous improvements in technology over time has allowed researchers to develop robotic micromanipulation systems. One of the main features of these systems is to have better control on the movement of the pipette used for injecting cells. A piezo-driven micropipette was used to perform ICSI in a mouse [41]. A resolution of 0.5 μm was achieved by piezoelectric actuation. The pipette punctured the cell membrane with minimal distortion of the cell (oocyte). Experiments showed that 80% of sperm-injected oocytes survived and 70% of them developed into blastocysts using the piezo-driven micropipette. By conventional methods, only 16% of the oocytes survived.

Yanagida et al. [42] used piezo micromanipulator to perform ICSI in humans and compared it to conventional ICSI. The results are shown in Table 7.1. The piezo micromanipulation system also allowed a fine resolution of 10 nm for the injecting pipette with repeatable motion. MANiPEN micromanipulator [43] was developed to perform automated capillary pressure microinjection (CPM). The pen-like shape of the manipulator allows simultaneous placement of several manipulators around a microscope. Compared to manual manipulators, computer-controlled micromanipulators are ideal for cell injection tasks performed at the micro/nanoscale.

Advances in the field of computer vision algorithms have facilitated the integration of robotics with visual servoing control schemes. The integration was a boon to the microrobotics community. Pronuclei DNA injection of mouse embryos was performed using an automated microrobot and a hybrid visual servoing control scheme [4]. The injection unit consisted of a high precision 3-DOF microrobot with a step resolution of 40 nm and a travel range of 2.54 cm in x-, y-, and z-axes. The microrobot controlled the movement of injection pipette. The movement of the holding pipette was controlled by a 3D coarse manipulator (manual). An injection success rate of 100% was achieved with the system. An automatic micromanipulation system (AMMS) used a pair of microrobots and employed a real-time visual servo control system to perform automatic gene insertion [44]. The system consisted of two manipulators: 3-DOF parallel mechanism and 3-DOF xyz stage. The parallel mechanism link was driven by piezotranslators and was used to move the injection pipette precisely with an accuracy of 0.1 μm. The xyz stage was driven by a dc motor and was used to move the holding pipette with an accuracy of 1 μm. Microinjection of embryonic stem (ES) cells was carried out efficiently using a semiautomated system [45]. The system was composed of two manual micromanipulators for coarse positioning of injecting and holding pipettes and one computer-controlled piezo injector for fine positioning of the injecting pipette. Vision-based algorithms were developed for the system. A robotic system composed

Table 7.1   Comparison of the Results Obtained from Piezo and Conventional ICSI

Type of ICSI	Survival Rate	Fertilization Rate	Pregnancy Rate
Piezo ICSI	88.1%	79.4%	23.1%
Conventional ICSI	81.4%	66.4%	14.9%

Source: [42].

of a pair of 3D hydraulic micromanipulators [3] was used to insert mouse spermatozoon into mouse oocyte. The hydraulic micromanipulators were driven by pulse
motors. These motors were regulated by pulse signals initiated by a computer program. With the help of image processing, the computer located the tip of the injecting pipette and the oocyte in the microscopic field. A typical schematic of the cell
injection setup is shown in Figure 7.2. A fertilization rate of 68.8% was obtained by
injecting 143 oocytes.

One of the main reasons for the low success rates of cell injection tasks is poor
control of cell injection forces. Biological cells are highly deformable, and excessive
force applied by the operator to penetrate the cell membrane may damage the cell.
Thus, the delicate nature of biological cells pose challenges for the operator. Few
researchers have addressed this problem. Sun et al. [34] used MEMS technology to
develop a 2-DOF capacitive force sensor (resolution in the range of micro-Newton).
Thereafter, the robotics researcher used the force sensor to characterize the mechanical properties of mouse oocytes and embryos [46]. A piezoelectric force sensor was
used to measure the injection force of zebra fish embryos at various developmental
stages [47]. Later on, the group characterized the mechanical properties of zebra fish
embryo chorion. An automated MEMS-based fruit fly injection system utilized a
piezoresistive pressure sensor [48]. The efficiency of the system was more than 96%.

7.3.2 Cell Culture Automation

Cell culture is a laboratory process used to maintain or grow cells in vitro. The process is used to: (1) investigate the normal physiology or biochemistry of cells; (2) test
the effect of various chemical compounds or drugs on specific cell types; (3) generate
artificial tissues (tissue engineering); and (4) synthesize biological entities such as
proteins from large scale cultures. A cell culture laboratory has the following equipment: microbiological safety cabinets, centrifuges, incubators, plasticware, and
consumables. A typical cell culture cycle includes cell growth, harvesting, reseeding,
and analysis.

Conventional cell culture is a manual process and involves long hours of repetitive tasks. It requires a diligent technician to maintain sterility and viability of
cells under favorable environmental conditions. Moreover, the technician has to

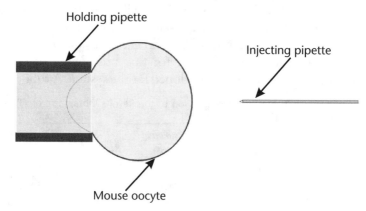

Figure 7.2 Microscopic view of the cell injection setup.

undergo rigorous training to carry out the cell culture cycle. Even for an experienced professional, there are chances of cell contamination due to direct human involvement. Smooth movement is desired while handling cell culture vessels because rough handling can cause cell damage and halt the cell growth process. Considering the disadvantages associated with conventional cell culture techniques, pharmaceutical companies and researchers have turned to automated cell culture systems. Automated cell culture systems use robots to improve consistency and sterility of the cell culture system, which may not be achievable by even the best technician. Normally, a contamination frequency level of 1 in 50 million is achieved by automated systems compared to 1 in 20 with manual handling. Unattended operation over a period of days or weeks is achieved by introducing automation in the cell culture process.

Robots are an integral part of automated cell culture systems. One of the main applications of robots is handling flasks/tubes carrying biological samples. An automated system was proposed for cultivating mammalian cells [49]. The system used a robot (PA-10, Mitsubishi Heavy Industries) to: (1) move sample tube; (2) move Cedex tube (an automated cell counting system, Innovatis Inc.); and (3) move storage tube. The robot was equipped with a camera and a force torque sensor to obtain visual and tactile information, respectively. The dual information was used to guide the robot and compensate for uncertainties in positioning the tubes. RTS Life Science International has created acCellerator, an automated cell culture system [50]. A unique feature of acCellerator is the ability to provide high throughput performance through the use of multiple robots. The use of three staubli robots resulted in parallel processing of flasks for harvesting and plating/passaging. Parallel processing improved the speed, throughput, and viability of the cell culture process. Georgia State University automated titration enzyme immunoassay (t-ELISA) for detecting Herpes B virus antibodies [51]. The system was custom-made by Beckman-Coulter, Inc. and consisted of a Genesis 150 liquid handler (Tecan, Inc.) integrated into Sagian Core system (Beckman Coulter, Inc.). The system was programmed using Gemini (Tecan, Inc.) and Sami (Beckman-Coulter, Inc.) software for carrying out ELISA. Genesis liquid handler performed tube-to-tube dilutions of the standard serum. The automated dilution procedure was superior compared to the manual procedure. The performance of robot automated t-ELISA was satisfactory and was within limits of experimental error. The authors stated that the results were the first published report on a fully automated serological laboratory in an academic facility.

Drug discovery efforts were automated by an integrated software-robotic system. The Tecan Genesis RSP 200 (Tecan, Inc.) was integrated with software package to automate in vitro dose inhibition assays. The workstation has the following key components: (1) positive identification barcode scanner (POSID); (2) carousel; (3) carousel scanner; (4) liquid handling arm (LiHa); (5) robotic manipulator arm (RoMa); (6) Gemini software; (7) Tecan-db database; and (8) system liquid. The function of RoMa was to transport bar-coded incubation plates from the carousel and place them into carriers. The Tecan robotic system can be operated by pushing a single tool bar icon. The assay results were within the experimental error and were similar to the data obtained from manual procedures. The main goal in using the system was to minimize human involvement, which reduced the risk of manual error and contamination involved in traditional methods. A robotic system was used for handling human serum samples. The robot CRS 475 (Burlington, Ontario) was mounted on an inverted track and was enclosed in a biosafety cabinet. The purpose of using the cabinet is to protect the operator from infectious agents such as Hepatitis C present in the biological sample. A swapper device maximized the efficiency of the system. It consisted of Packard Multiprobe II pipetting robot (Packard Instrument) with eight swappable shelves. The primary goal was to carry out simultaneous loading/unloading of one shelf by the robot and pipetting on the other shelf by Packard. The robot was controlled by software, consisting of an executable program, BTA.exe, and an intelligent algorithm. The empty boxes of disposable tips were replaced by the robot without human intervention, which allowed overnight work. The other parts of the robot cell were refrigerator storage and 2D bar code reading. The Twister II (Zymark Corporation), a laboratory robot, was used for handling up to 400 microplates [52]. The microplates were transported to and from plate reader, washer, and pipettors. A visual basic application (VBA) interface was developed for controlling Twister II.

An automated robotic system was used for high throughput fermentation for the cultivation of *Saccharomyces cerevisiae* [53]. The system consisted of a screening robot, a rail mounted robotic arm (Sagian ORCA, Beckman-Coulter, Fullerton, California), and a liquid handler (Biomek 2000; Beckman-Coulter). The robotic arm transported microtiter plates between various laboratory devices. The liquid handler transported the liquid medium into 48 well plates and also inoculated the samples from precultures. The automated system permitted 768 microscale fermentations under high oxygen transfer rates without human involvement. Symex technologies (Santa Clara, California) developed a flexible automated system called the extended core module (XCM) robotic system [54]. The XCM consisted of a three-axis Cartesian robot, pump housing, and deck. The unique architecture of the system allowed the addition of appropriate elements to the robot for liquid handling, material handling, moving plate/vial, mixing heads, and instrumented probing (i.e., temperature, pH). The XCM also allowed the addition of two more robotic arms. The system was controlled by Renaissance Impressionist Software. Thus, XCM has the capacity to automate a variety of tasks depending on the need in the particular life science industry. This feature greatly reduced the time required to reconfigure the workflow for a short-term project. The assay services department at Genentech, Inc., implemented two robotic assay ELISA systems (RAS) [55]. Each system comprised of a CRS robotic arm (CRS Robotics, Inc.), a 96 well dispensing station, a reagent dispensing station, a plate washer, a plate shaker, a microplate reader, and a carousel for holding tips and plates. The robot systems were custom built by Scitec (Wilmington, Delaware). Each RAS was capable of running 10 to 15 plates per assay run. Assays can run overnight without the presence of laboratory personnel. Over the years, there was an increase in running assays on the robot. In 1998, two assays were put on the robot. In 1999, 13 assays were running on the robot, which increased to 17 in 2000. Thus, the two robotic systems increased the throughput of assay and reduced the research associate's bench time by 80%.

Cellmate, an automated cell culture system (manufactured by The Automation Partnership, Ltd.), used a Staubli RX 60 CR robot. The system automated all operations including cell seeding, media change, bottle greasing, and cell sheet rinsing.

The robot arm smoothly moved bottles/flasks to a series of cell culture work-stations. The system was capable of handling cells up to volumes of 1,000 bottles per day. Above all, simple operation was an important feature of Cellmate. A few of the operations—namely, loading/unloading, labeling and scraping flasks—were performed manually using Cellmate. Bernard et al. [56] proposed the integration of three new custom subsystems (FlaskMaster, FlaskLabeler, and FlaskScraper) into the Cellmate robotic system. The FlaskMaster robot provided automated loading and unloading of flasks onto the Cellmate conveyor systems. The FlaskLabeler (a subsystem of the FlaskMaster) was capable of printing, applying, and reading labels marked with barcodes on the flasks. The subsystem was used for flask identification. FlaskScraper allowed automated, unattended scraping and was capable of attaining scraping coverage equal to the manual process. The system's performance was promising. The FlaskMaster increased the total unattended batch throughput from 5 to 168 flasks. FlaskScraper provided a full-time employee (FTE) savings of 1.5 days per week. The Automation Partnership built SelecT, an automated cell culture robot. SelecT was built around a horizontally mounted Staubli RX 60 L robot that performed all the handling, incubating, and sampling movements on up to 160 T-flasks. The robot was able to handle both the flasks and pipettes. The system was capable of processing small batches of up to 40 different cell lines in parallel. This allowed for immediate analysis of overnight assay plates. The system has the capability to prepare up to 300 assay-ready microtiter plates and can operate unattended outside normal working hours. Later on, The Automation Partnership upgraded the SelecT system to CompacT SelecT. The system was able to automatically manipulate up to 175 T-flasks. Soloveva et al. [57] used a robotic team composed of SelecT and the integrated laboratory automation system (LAS) (Thermo Electron Corporation) with Dual FLIPR (fluorescence imaging plate reader). The automated system was custom built for Wyeth screening sciences (Collegeville, Pennsylvania) by Thermo Electron Corporation. Table 7.2 lists some of the parts of the thermo LAS system along with their functions.

Apart from the parts listed in the table, the system has two stericult incubators, a biotek plate washer Elx 40, a Perkin-Elmer flexidrop IV noncontact dispenser, Perkin-Elmer Evolution EP3 pipettor, and FLIPR^{TETRA}. The integrated robotic system maintained cell culture conditions according to the protocol and eliminated variable delays between various steps in the protocol. A high throughput screening of 120 plates per day was achieved with the system. The capillarys (multicapillary zone electrophoresis instrument; Seiba, Inc.) was designed with a three-axis robotic system to handle tubes [58]. The robot picked tubes and arranged them on the racks of the capillarys. It was also able to handle racks of tubes. A qualification plan was

Table 7.2   Parts of the Thermo LAS Robotic System

Parts of Thermo LAS	Quantity	Function
High speed movable belt	1	Moved plate from one part of the system to another (linear plate transport, LPT).
Flip mover robotic arm	8	Moved plate from LPT to peripheral instruments.
Vertical array loader (VAL)	1	Moved compound source plates from the plate carousels to LPT.

Source: [57].

proposed to validate the automated workcell. It was found that the robot did not influence the capillarys throughput but reduced human intervention in handling cell culture operations.

Other companies are also involved in cell culture automation. Juan Robotics manufactures automated cold storage (Molbank), centrifuge (GR4 Auto), and incubator (Autocell) for cell cultivation [59]. The specifications of the Autocell incubator and Molbank storage system are listed in Table 7.3.

Experimental results demonstrated no significant difference between the automated and traditional incubator, but the authors believed that the Autocell 200 could provide meaningful benefits for cell culture operations. The automated incubator was designed to be integrated in a fully automated cell culture system. A robotic system (manufactured by Cytogration) performed cell culture as well as high throughput screening assays for drug candidates [60]. The cell culture system automated cell production, membrane penetration, and/or in vitro screening. The system featured a CRS 465 robotic arm in fixed or rail mounted configuration. The fixed arm configuration was used in smaller feeder systems and was capable of handling up to 168 plates, with one incubator. The rail mounted configuration was used for all systems—namely, coater, seeder, and feeder. The configuration was capable of handling up to 504 plates, with three incubators. An automated system was developed in collaboration between Organon and Scitec laboratory automation. The system consisted of three CRS robots. One of the robots was capable of performing cellular and noncellular assays to produce a maximum throughput of 300 (96 well) or 150 (384 well) microtiter plates in 24 hours. The other two robots were involved in: (1) assay plate preparation; (2) selection of compounds to determine potential drug candidates (cherry picking); and (3) refreshing storage plates by replacing them with newly prepared plates. The storage and distribution system was kept under controlled temperature (4°C to 10°C) and humidity (10% to 15% RH). The system

Table 7.3   Specification of Automated Cell Culture Components

Specification	Molbank (Storage System)	Autocell 44 (Incubator)
Storage capacity (number of microplates)	2,571	198
Temperature control	−20°C to +20°C	30°C to 50°C
Relative humidity	70%	> 95%
Access time	21 to 33 seconds	< 15.3 seconds
Interfaces	RS232, Lonworks network	RS232, V24

Source: [59].

The automated incubator was developed for high throughput screening in cell culture operations. Molbank automates storage of pharmaceutical compounds. It is comprised of a double storage carousel and a gripper to transport plates from the carousels. The stored and retrieved plates were traced by embedded bar code reading. The automated cell culture incubator, Autocell 200, is comprised of a carousel and a robotic arm, which has shovel at its end. It has an external door which protects the internal door. The internal door allows visual inspection of the incubator contents without opening the incubator. This setup reduced contamination of biological samples. A robotic gate permitted retrieval of plates without disturbing the internal climate of the incubator. The shovel at the end of the robotic arm was used to load plates on the stackers. The frequency of plate loading and unloading was measured as 150 per hour and 120 per hour, respectively. The mean time of the robotic gate opening was measured as 7 seconds.

provided high throughput screening and short delays between different operations. The REMP store robot achieved high throughput handling of storage racks. The system was flexible and allowed the handling of vials, microplate format racks, minitube racks, and any of the REMP tube technology formats. Apart from many advantages offered by automated cell culture systems, there are a few disadvantages: potential for mechanical failure, need for technical expertise, and familiarity with the system. However, automation proved to be beneficial for cell culturing.

7.3.3 High Throughput Processing of Biological Samples

Biological samples contain DNA, RNA, proteins, and other biological materials. These samples are generally involved in DNA purification, RNA isolation, and protein purification. Isolation and purification of DNA from blood samples, biopsy samples, cultured cells, and buccal cells is a key step to investigate genetic contribution to human disease. Isolation of RNA from biological samples is a critical process in many fundamental molecular experiments such as nuclease protection assays, RNA mapping, and cDNA library construction. The purification of protein is desired for protein stability measurements. These measurements study the relationship between protein sequence and stability. Traditional laboratory methods adapted to process biological samples offer many challenges. Being manual, these methods lack consistency/reproducibility, involve human error, and are time consuming. Robotic workstations on the other hand perform consistently, 24 hours a day, 7 days a week. Moreover, high throughput is achieved by a robotic system. Janssen Research Foundation (JRF) started using a Zymark robotic system in 1985. Since then, different robotic systems have been used by the organization to increase throughput of screening biological samples. The output of automated screening systems increased six-fold from 260,000 to 1.4 million data points in a 5-year period (1991 to 1996), which proved that the efficiency of the robotic systems increased over the years. JRF validated the Staccato system (Zymark, Inc.) by performing plate replication process and assay formats using 96 or 384 well plates. The Stacatto system consisted of a SciClone, an automated liquid handling workstation capable of pipetting a variety of liquid solutions with a higher degree of accuracy and precision. The incorporation of the Stacatto system increased the throughput of cell and enzymatic assays 5 and 10 times, respectively.

Companies and researchers are heavily involved in automating purification of DNA, RNA, and proteins from biological samples. The AUTOPURE LS robotic system (Gentra Systems, Inc.) automated the DNA purification process from blood samples in the 1- to 10-mL volume range [61]. DNA purified with manual methods and the robotic system resulted in equivalent yields of DNA from human blood samples from a single donor. The system was also used to purify DNA from buccal mouth wash samples collected from 16 different donors. The high yield of DNA from buccal samples suggested that the AUTOPURE LS instrument is capable of purifying DNA from difficult samples. The automation of the DNA purification process provided more time for genotyping compared to the manual method (71% versus 36% of the total time). A multipurpose robot (SX-96GC) with Integrated Magtration Units (IMU) was developed for the purification of the DNA sequencing reactants. The purification process is the most tedious step in DNA sequencing. The robotic protocol used biomagnetic beads and was designed to purify 384 samples within 1 hour. The protocol required least technical skills and manual labor. The main features of the automated system were high throughput and adaptability. A fully automated procedure was developed for the purification of polymerase chain reaction (PCR) products using Wizard Magnesil paramagnetic particles and robotic methods [62]. The Magnesil purification process has been adapted to a number of robotic platforms. The Beckman Biomek FX robotic workstation processed four plates (96 and 384 well formats) in an unattended run of 45 minutes. The procedure was highly reproducible and generated PCR products of high purity with minimal loss of target DNA. Purification of BigDye Terminator DNA sequencing reactions was also performed using Magnesil paramagnetic particles and robotic methods [63]. The Beckman Biomek processed up to four plates in 40 minutes. The Stanford automated sequencing system was capable of processing 10,000 samples per day [64]. The entire system required only three technicians to operate at a throughput of 100 plates per day, and less than 500 ft^2 of space. Meldrum et al. [65] developed an automated biomehatronic fluid handling system for genome analysis, ACAPELLA – 1K. The system was capable of processing 1,000 samples in 8 hours for use in molecular biology. One of the goals was to obtain high-throughput DNA sequencing for the Human Genome Project. Later on, ACAPELLA – 5K was developed, which was capable of processing 5,000 biosamples in 8 hours [66]. The system was tested in the preparation of DNA PCR and DNA sequencing reactions.

Proteomics researchers are depending on automated workstations to purify and prepare protein solutions. Biomek 2000 liquid handling robot (Beckman-Coulter, Inc.) was used to purify his-tagged eglin c proteins [67]. The automated workstation increased throughput as well as precision for protein stability measurements. A throughput of about 20 stability measurements per day was attained. Automation reduced protein measurement variability due to contaminants to around 10% of that obtained by the manual method. Overall, the automated protein purification process was 50 to 100 times better than manual purification process. A protein purification robot was developed by the Genomics Institute of the Novartis Research Foundation. The robot can follow a specific recipe for protein purification and can simultaneously prepare 96 individual samples. The same task would take nearly 30 technicians to complete. Najmabadi et al. [68] proposed *tower-based automation* for automatic execution of various genomics and proteomics protocols. The tower-based configuration has a high throughput-to-footprint ratio, high scalability, and wide protocol flexibility. The system consisted of a central robotic structure with two arms. One arm is used for transporting accessories (tubes, plates) and the other arm is used for handling liquid. The robotic system performed magnetic isolation of tandem affinity purification (TAP) tagged protein complexes. The electron microscopy proteomic organellar preparation (EMPOP) robot [69] was developed to: (1) perform parallel microscopic screening; (2) validate subcellular sample purity; and (3) confirm protein localization for high throughput proteomics. The EMPOP robot consisted of six components: (1) transfer platform; (2) core mechanism; (3) cooling platform; (4) wash station; (5) electromagnetic arm; and (6) dispensing needle manifold. The robot was able to prepare 96 subcellular fraction samples with high quality. The process was totally automated in less than one working day, and reduced the time and labor by approximately a thousand-fold.

Integration of biotechnology protocols with robotic systems is crucial to maximize the efficiency of the purification process. A robotic protocol was developed for high throughput selection of mRNA from total RNA preparations. The presence of ribosomal, transfer, and other RNA species in the sample weakens rare messages in direct and indirect detection assays. The protocol was initially developed on the MultiProbe II HT EX (Perkin Elmer, Inc.) and Biomek 2000 (Beckman-Coulter, Inc.) robotic systems. For the proposed protocol, any robotic workstation would take about a half hour to process 96 samples with an eight-channel pipetting tool. The automated system removed greater than 99% of the rRNA initially present in the biological sample. The MultiProbe II HT was used for high throughput isolation of RNA from cultured cells using the RNA aqueous-96 kit (Ambion, Inc.) [70]. The run time required to complete the procedure was approximately 1 hour, 20 minutes. Results indicated that the system provided total RNA for use in the quantitative analysis of target mRNA. An automated platform was developed for liquid-liquid extraction which isolated natural compounds from mixtures [71]. The robotic system was composed of a Zymark robot (Caliper Life Sciences), a liquid-liquid extraction module, a control system, a programmable logic controller (PLC), and an operator interface. The operator interface was user friendly and had several control screens developed by the supervisory control software (FIX). A robotic system was used to determine optimal reaction conditions for high purity and high yields of an organic chemical. The system also consisted of a Zymark robot, PAL robot (CTC Analytics), and a solid dispensing system (Autodose S.A.). The Zymark robot integrated the instrumentation of the system. The PAL robot and solid dispensing robot were used for handling liquid and solids, respectively. The integrated system was controlled by graphical software and allowed screening of several hundred reaction combinations within a short time. A few other companies also developed liquid handling robots. Zinsser Analytic, Inc., manufactures Lissy 2002, a robotic system for sophisticated and flexible liquid handling. The system is equipped with up to 16 dispensing tips (8 on each arm). Lissy 2002 is an ideal system for high throughput of 96 and 384 plates. Hamilton, Inc., manufactures MICROLAB STAR, a liquid handling robot. Some of the key features of the robot are multiple channels, 0.5- to 1,000-μL volume range, and the use of disposable tips. The use of robotic workstations for processing biological samples certainly offers many advantages but at the expense of a high budget. Cost, space, throughput, and flexibility are the most important parameters for the decision making process in designing or purchasing a robotic workstation.

7.3.4   Production of DNA and Protein Microarrays

Microarrays are grids of tiny spots of DNA or protein on a microscopic slide. In molecular biology, conventional methods generally involve one gene in an experiment and limit the throughput of the process. The microarray technology allows scientists to analyze expression of many genes in a single experiment quickly and efficiently. The main application areas of this technology are gene discovery, disease diagnosis, drug discovery, and toxicological research. The microarrays can be fabricated by fine pointed pins, photolithography, or inkjet technology.

Robotics plays an important role in the production of DNA and protein microarray. In 1995, Schena et al. [72] used a high-speed robot to spot array different complimentary DNAs (cDNAs) on the glass surface. The microarrays were used for quantitative expression measurements of the corresponding genes. In 1998, Joseph DeRisi published a document, "MGuide," on the Web [73]. The document listed all the necessary parts, suppliers, and prices for building a microarray robot. The robot can be employed to print thousands of tiny DNA spots on glass slides. A high precision contact-printing robot was used to fabricate protein microarrays [74]. The robot delivered nanoliter volumes of protein samples onto microscopic glass slides. The proteins were immobilized by covalently attaching them to the slides. A spot diameter of 150 to 200 μm allowed 10,000 samples in half the area of a standard (2.5 cm × 7.5 cm) slide. The protein microarray was used: (1) to study protein-protein interaction; (2) to identify the substrates of protein kinases; and (3) to identify protein targets of small molecules. The probe preparation for DNA microarray was automated by Packard-brand MultiPROBE II Nuclei Acid Purification Workstation (Perkin-Elmer Life Sciences). The preparation involved plasmid DNA purification, PCR setup and cleanup, gel loading, fluorescence DNA quantification, and tracking DNA probes. The robotic workstation provided high throughput production of DNA probes. A pin-based robotic system, termed SmartPin, was developed for DNA and protein microarray fabrication. The system is based on a contactless printing technique. The SmartPin assembly consisted of a fluid reservoir for handling the sample liquid materials, a fluid delivery plunger for dispensing/aspirating liquid, and a fiber probe to determine the distance between the fiber tip and the slide. The functions of the smart pin are spot formation, detection of spots, and production of uniform size microarrays. The diameter of the spot varied from 80 to 200 μm. Microarrays of synthetic heparin oligosaccharides were prepared by an arraying robot. The robot delivered 1 nL of carbohydrate containing solutions to create spots with a diameter of ~200 μm. An array of 460 spots was generated. The construction of heparin microarrays can be used to rapidly screen heparin protein interactions. A high precision robot was used to print small molecules as microarrays for detecting protein-ligand interactions. The robot picked up a small volume of the compound and delivered approximately 1 nL of solution to defined locations on a glass microscopic slide (~150 slides per print area). Microscopic spots formed on the slides were 200 to 250 μm in diameter. The immobilized compound spots were probed with a different tagged protein and the binding events were detected by fluorescence linked assay.

A few companies are also actively involved in the production of microarrays. Agilent Technologies, Inc., proposed automation of cDNA and protein microarray fabrication using inkjet technology [75]. The goal was to mass produce several microscopic slides containing multiple microarrays of different biological features in a single production run. The system was used to make cDNA microarrays from a large number of different gene sequences. The size of the microarrays (ranging from 2,000 to 1 million features) allowed multiple tests to be performed at the same time with minimal sample amounts. ChipWriter Pro (Virtek, Inc.) is a high precision robot designed to collect 100 nl of DNA solution to deposit 0.6 nl per spot. The robot was used to spot 21,376 gene-specific probes onto a single glass slide, "Drosophilla MicroArray." High quality spots were generated with a spot diameter of 100 μm. The robot controlled the spot morphology, the spot diameter, and uniformity. Telechem International, Inc., manufactures SpotBot, a personal microarray robot. The robot can print a maximum of 50,000 spots per chip, with a spot diameter of 100 to 120 μm. Genomic Solutions offered production of DNA and protein microarray in 96 and 384 well formats, which was possible using BioRobotics MicroGrid II and the Gene Machines manufactured by Genomic Solutions. The BioRobotics MicroGrid II has the capacity to hold 16 microplates, while the GeneMachines Omni Grid has the capacity to hold six microplates. At least 1,000 DNA or protein spots can be printed within individual microplate wells. The arrays can be produced using contact printing. The split pin performed the printing by simple surface tension and adhesion, thereby minimizing the contact between the pin and substrate. This enabled printing on a delicate substrate. The plate array technology allows micro-ELISAs, diagnostic arrays, and other protein-protein studies in a higher throughput format. Thus, robots are being used in academics as well as in industries to produce DNA and protein microarrays.

7.4   Discussions

Robotics and automation have influenced a wide variety of areas in the life sciences, ranging from biomanipulation to production of DNA and protein microarrays. There are several promising approaches for biomanipulation. In this chapter, we have covered some of the most common approaches, namely optic and electric micromanipulation, magnetic micromanipulation, acoustic energy for micromanipulation, micro-electromechanical systems, and mechanical micromanipulation. Each of the above techniques has their own advantages and disadvantages. Optic and electric techniques offer the ability to manipulate single cells without contact; however, these techniques do not readily lend themselves to effective manipulation accomplished by MEMS and other mechanical manipulation techniques. Acoustic energy based micromanipulation is perhaps the most noninvasive approach in biomanipulation. However, this approach can be used at best currently for trapping and holding microparticles in a required position. The research into the use of acoustic energy for micromanipulation is currently in its infancy. Effective manipulation of individual cells has been developed through the use of a bilateral tele-operation framework whereby individual cells can be grasped in place and injected with a genetic material by the operator through a vision and force feedback interface. There are several challenges in micromanipulation. One of those challenges lies in accurately fixating the nucleus to deliver a genetic material accurately within the nucleus.

In the past few decades, robotics has had a tremendous impact on life science automation industry. Ongoing efforts in industries and academia provide innovative solutions to carry out life science–based tasks. Automation certainly has improved the efficiency of the cell injection task, biological sample purification, cell culture operation, and DNA/protein microarray fabrication. For example, the success rate of cell injection tasks increased from 10% to 15% to around 100% using robotic systems. Such marvelous achievement motivates companies to automate the entire task at an economical price. Robotic methods perform consistently and ensure safe laboratory operation, thereby reducing labor costs. There are many commercially available robotic systems to carry out life science based tasks, such as: Sutter, Inc.; World Precision Instruments, Inc.; RTS Life Science; The Automation Partnership, Ltd.; Cytogration; Gentra Systems, Inc.; Beckman-Coulter, Inc.; Perkin Elmer, Inc.; and Zinsser Analytic. Despite the competition in the market, there is still research going on to implement and integrate new technology in life science based tasks. For example, integrating haptics with the biomanipulation system improved the success rate of cell injection tasks. Similarly, MEMS technology is used to develop microtools to manipulate biological cells, whose size is in the μm to nm range. Although haptics-based life science systems are yet to be commercialized, it has been proved in academia that haptics plays a critical role in cell manipulation [21]. Robotics can also be integrated with optic and magnetic techniques to manipulate cells without physical contact. On the other hand, mechanical manipulation involves physical contact. In certain applications, automation allows technicians to devote more time to tasks like genotyping while the robot processes the biological samples. In summary, robotic techniques for life science automation is currently in its infancy and there are a wide range of applications and engineering challenges that will be addressed in future research.

References

[1]   Chabry, L., "Contribution a L'embryologie Normal Et Teratologique Des Ascidiens Simples," *Jour. de l'Anat. et. de. Physiol*, Vol. 25, 1887, pp. 167.

[2]   Uehera, T., and R. Yanagimachi, "Microsurgical Injection of Spermatozoa into Hamster Eggs with Subsequent Transformation of Sperm Nuclei into Male Pronuclei," *Biology of Reproduction*, Vol. 15, 1976, pp. 467–470.

[3]   Kobayashi, K., et al., "Subzonal Insemination of a Single Mouse Spermatozoon with a Personal Computer-Controlled Micromanipulation System," *Molecular Reproduction and Development*, Vol. 33, No. 1, 1992, pp. 81–88.

[4]   Sun, Y., and B. J. Nelson, "Microrobotic Cell Injection," *IEEE Int. Conference on Robotics and Automation*, Seoul, Korea, May 2001, pp. 620–625.

[5]   Ashkin, A., "Acceleration and Trapping of Particles by Radiation Pressure," *Physical Review Letters*, Vol. 4, 1970, pp. 156–159.

[6]   Buican, T. N., et al., "Automated Single-Cell Manipulation and Sorting by Light Trapping," *Applied Optics*, Vol. 26, No. 24, 1987, pp. 5311–5316.

[7]   Wang, M. M., et al., "Microfluidic Sorting of Mammalian Cells by Optical Force Switching," *Nature Biotechnology*, Vol. 23, No. 1, 2005, pp. 83–87.

[8]   Washizu, M., et al., "Dielectrophoretic Measurement of Bacterial Motor Characteristics," *IEEE Transactions on Industry Applications*, Vol. 29, No. 2, 1993, pp. 286–294.

[9]   Nishioka, M., et al., "Evaluation of Cell Characteristics by Step-Wise Orientational Rotation Using Optoelectrostatic Micromanipulation," *IEEE Transactions on Industry Applications*, Vol. 33, No. 5, 1997, pp. 1381–1388.

[10]   Arai, F., et al., "Bio-Micro-Manipulation (New Direction for Operation Improvement)," *IEEE Int. Conference on Intelligent Robots and Systems*, Grenoble, France, September 7–11, 1997, pp. 1300–1305.

[11]   Arai, F., et al., "Teleoperated Laser Manipulator with Dielectrophoretic Assistance for Selective Separation of a Microbe," *IEEE Int. Conference on Intelligent Robots and Systems*, Kyonjyu, Korea, October 1999, pp. 1872–1877.

[12] Pesce, M., and M. De Felici, "Purification of Mouse Primordial Germ Cells by Mini-Macs Magnetic Separation System," *Developmental Biology*, Vol. 170, 1995, pp. 722–725.

[13] Alenghat, F. J., et al., "Analysis of Cell Mechanics in Single Vinculin-Deficient Cells Using a Magnetic Tweezer," *Biochemical and Biophysical Research Communications*, Vol. 277, 2000, pp. 93–99.

[14] Boukallel, M., E. Piat, and J. Abadie, "Micromanipulation Tasks Using Passive Levitated Force Sensing Manipulator," *IEEE Int. Conference on Intelligent Robots and Systems*, Las Vegas, NV, October 27–31, 2003, pp. 529–534.

[15] Lee, H., T. P. Hunt, and R. M. Westervelt, "Magnetic and Electric Manipulation of a Single Cell in Fluid," *Materials Research Society Symposium Proc*, 2004, pp. 2.3.1–2.3.8.

[16] Kozuka, T., et al., "Two-Dimensional Acoustic Micromanipulation Using Three Ultrasonic Transducers," *Int. Symposium on Micromechatronics and Human Science*, Nagoya, Japan, 1998, pp. 201–204.

[17] Kozuka, T., et al., "Three-Dimensional Acoustic Micromanipulation Using Four Ultrasonic Transducers," *Int. Symposium on Micromechatronics and Human Science*, 2000, pp. 201–206.

[18] Haake, A., and J. Dual, "Micro-Manipulation of Small Particles by Node Position Control of an Ultrasonic Standing Wave," *Ultrasonics*, Vol. 40, 2002, pp. 317–322.

[19] Kim, D. H., et al., "High Throughput Cell Manipulation Using Ultrasound Fields," *Proc. 26th Annual Int. Conference of the IEEE EMBS*, San Fransisco, CA, September 2004, pp. 2571–2574.

[20] Jager, E. W. H., O. Inganas, and I. Lundstrom, "Microrobots for Micrometer-Size Objects in Aqueous Media: Potential Tools for Single-Cell Manipulation," *Science*, Vol. 288, 2000, pp. 2335–2338.

[21] Pillarisetti, A., et al., "Evaluating the Role of Force Feedback for Biomanipulation Tasks," *14th Symposium on Haptic Interfaces for Virtual Environment and Teleoperator Systems*, Alexandria, VA, March 25–26, 2006, pp. 11–18.

[22] Chan, H., and W. J. Li, "A Thermally Actuated Polymer Micro Robotic Gripper for Manipulation of Biological Cells," *IEEE Int. Conference on Robotics and Automation*, Taipei, Taiwan, September 14–19, 2003, pp. 288–293.

[23] Bronson, J. R., G. J. Wiens, and R. Tran-Son-Tay, "A Feasibility Study on MEMS Test-Structures for Analysis of Biological Cells and Tissue," *Proc. 2004 Florida Conference on Recent Advances in Robotics*, Orlando, FL, May 6–7, 2004, pp. 1–6.

[24] Chun, K., et al., "Fabrication of Array of Hollow Microcapillaries Used for Injection of Genetic Material into Animal/Plant Cells," *Japanese Journal of Applied Physics, Part 2: Letters*, Vol. 38, No. 3A, 1999, pp. 279–281.

[25] McAllister, D. V., M. G. Allen, and M. R. Prausnitz, "Microfabricated Microneedles for Gene and Drug Delivery," *Annual Review of Biomedical Engineering*, Vol. 2, 2000, pp. 289–313.

[26] Beebe, D., et al., "Microfluidic Technology for Assisted Reproduction," *Theriogenology*, Vol. 57, 2002, pp. 125–135.

[27] Codourey, A., et al., "A Robot System for Automated Handling in Micro-World," *IEEE Int. Conference on Intelligent Robots and Systems*, Pittsburgh, PA, August 1995, pp. 185–190.

[28] Goldfarb, M., and J. E. Speich, "Design of a Minimum Surface-Effect Three Degree-of-Freedom Micro-Manipulator," *IEEE Int. Conference on Robotics and Automation*, Albuquerque, NM, April 1997, pp. 1466–1471.

[29] Kallio, P., et al., "A 3 DOF Piezohydraulic Parallel Micromanipulator," *IEEE Int. Conference on Robotics and Automation*, Leuven, Belgium, May 1998, pp. 1823–1828.

[30] Gao, P., and S.-M. Swei, "A Six-Degree-of-Freedom Micro-Manipulator Based on Piezoelectric Translator," *Nanotechnology*, Vol. 10, 1999, pp. 447–452.

[31]   Ohya, Y., et al., "Development of 3-DOF Finger Module for Micro Manipulation," *IEEE Int. Conference on Intelligent Robots and Systems*, Kyonjyu, Korea, October 1999, pp. 894–899.

[32]   Tanikawa, T., and T. Arai, "Development of a Micro-Manipulation System Having a Two-Fingered Micro-Hand," *IEEE Transactions on Robotics and Automation*, Vol. 15, No. 1, 1999, pp. 152–162.

[33]   Guo, S., H. Zhang, and S. Hata, "Complex Control of a Human Scale Tele-Operating System for Cell Biology," *4th World Congress on Intelligent Control and Automation*, Shanghai, China, 2002, pp. 71–76.

[34]   Sun, Y., et al., "A Bulk Microfabricated Multi-Axis Capacitive Cellular Force Sensor Using Transverse Comb Drives," *Journal of Micromechanics and Microengineering*, Vol. 12, No. 6, 2002, pp. 832–840.

[35]   Pillarisetti, A., et al., "Force Feedback Interface for Cell Injection," *World Haptics Conference*, Pisa, Italy, 2005, pp. 391–400.

[36]   Zhou, Y., B. J. Nelson, and B. Vikramaditya, "Fusing Force and Vision Feedback for Micromanipulation," *IEEE Int. Conference on Robotics and Automation*, 1998, pp. 1220–1225.

[37]   Arai, F., et al., "3-D Viewpoint Selection and Bilateral Control for Bio-Micromanipulation," *IEEE Int. Conference on Robotics and Automation*, San Francisco, CA, April 2000, pp. 947–952.

[38]   Ando, N., P. Korondi, and H. Hashimoto, "Development of Micromanipulator and Haptic Interface for Networks Micromanipulation," *IEEE/ASME Transactions on Mechatronics*, Vol. 6, No. 4, 2001, pp. 417–427.

[39]   Guthold, M., et al., "Controlled Manipulation of Molecular Samples with the Nano-manipulator," *IEEE/ASME Transactions on Mechatronics*, Vol. 5, No. 2, 2000, pp. 189–197.

[40]   Lin, G., et al., "Miniature Heart Cell Force Transducer System Implemented in MEMS Technology," *IEEE Transactions on Biomedical Engineering*, Vol. 48, No. 9, 2001, pp. 996–1006.

[41]   Kimura, Y., and R. Yanagimachi, "Intracytoplasmic Sperm Injection in the Mouse," *Biology of Reproduction*, Vol. 52, No. 4, 1995, pp. 709–720.

[42]   Yanagida, K., et al., "The Usefulness of a Piezo-Micromanipulator in Intracytoplasmic Sperm Injection in Humans," *Human Reproduction*, Vol. 14, No. 2, 1998, pp. 448–453.

[43]   Lukkari, M., and P. Kallio, "Multi-Purpose Impedance-Based Measurement System to Automate Microinjection of Adherent Cells," *Int. Symposium on Computational Intelligence in Robotics and Automation*, Espoo, Finland, June 27–30, 2005, pp. 701–706.

[44]   Li, X., G. Zong, and S. Bi, "Development of Global Vision System for Biological Automatic Micro-Manipulation System," *IEEE International Conference on Robotics & Automation*, Seoul, Korea, May 2001, pp. 127–132.

[45]   Mattos, L., E. Grant, and R. Thresher, "Semi-Automated Blastocyst Microinjection," *IEEE Int. Conference on Robotics and Automation*, Orlando, FL, May 2006, pp. 1780–1785.

[46]   Sun, Y., et al., "Mechanical Property Characterization of Mouse Zona Pellucida," *IEEE Transactions on Nanobioscience*, Vol. 2, No. 4, 2003, pp. 279–286.

[47]   Kim, D. H., et al., "Cellular Force Measurement for Force Reflected Biomanipulation," *IEEE Int. Conference on Robotics and Automation*, April 2004, pp. 2412–2417.

[48]   Zappe, S., et al., "Automated Mems Based Fruit Fly Embryo Injection System for Genome-Wide High-Throughput Rnai Screens," *8th Int. Conference on Miniaturized Systems for Chemistry and Life Sciences*, Malmo, Sweden, September 26–30, 2004.

[49]   Lutkemeyer, D., et al., "First Steps in Robot Automation of Sampling and Sample Management During Cultivation of Mammalian Cells in Pilot Scale," *Biotechnology Progress*, Vol. 16, 2000, pp. 822–828.

[50] Timoney, C. F., and R. A. Felder, "A Collaborative Approach to Creating an Automated Cell Culture System," *The Journal of the Association for Laboratory Automation (JALA)*, Vol. 7, No. 1, 2002, pp. 55–58.

[51] Katz, D., et al., "Automation of Serological Diagnosis for Herpes B Virus Infections Using Robot-Assisted Integrated Workstations," *The Journal of the Association for Laboratory Automation (JALA)*, Vol. 7, No. 6, 2002, pp. 108–113.

[52] Pechter, D. S., "Controlling a Twister II with VBA and the Zymark Zyrobot ICP," *The Journal of the Association for Laboratory Automation (JALA)*, Vol. 7, No. 6, 2002, pp. 92–95.

[53] Zimmermann, H. F., and J. Rieth, "A Fully Automated Robotic System for High Throughput Fermentation," *The Journal of the Association for Laboratory Automation (JALA)*, Vol. 11, No. 3, 2006, pp. 134–137.

[54] Chandler, W., et al., "The Extended Core Module Robot: A Flexible, Modular Platform for High Throughput Development of Workflows," *The Journal of the Association for Laboratory Automation (JALA)*, Vol. 10, No. 6, 2005, pp. 418–422.

[55] Cheu, M., and L. Myers, "Increasing Productivity Through a Combination of Automation and Robotics: A Case Study of Assay Services," *The Journal of the Association for Laboratory Automation (JALA)*, Vol. 6, No. 3, 2001, pp. 48–51.

[56] Bernard, C. J., et al., "Adjunct Automation to the Cellmate Cell Culture Robot," *The Journal of the Association for Laboratory Automation (JALA)*, 2004, pp. 209–217.

[57] Soloveva, V., J. LaRocque, and E. McKillip, "When Robots Are Good: Fully Automated Thermo Las Robotic Assay System with Dual Flipr^tetra and Tap Select Robotic Cell Culture System," *The Journal of the Association for Laboratory Automation (JALA)*, Vol. 11, No. 3, 2006, pp. 145–156.

[58] Neel, T. L., et al., "Development and Qualification of an Automated Workcell That Includes a Multicapillary Zone Electrophoresis Instrument," *The Journal of the Association for Laboratory Automation (JALA)*, Vol. 10, No. 1, 2005, pp. 54–58.

[59] Triaud, F., et al., "Evaluation of an Automated Cell Culture Incubator: The Autocell 200," *The Journal of the Association for Laboratory Automation (JALA)*, Vol. 8, No. 6, 2003, pp. 87–95.

[60] Kempner, M. E., and R. A. Felder, "A Review of Cell Culture Automation," *The Journal of the Association for Laboratory Automation (JALA)*, Vol. 7, No. 2, 2002, pp. 56–62.

[61] O'Brien, D. P., et al., "Automated Purification of DNA from Large Samples: A Study of Effectiveness and Labor Efficiency," *The Journal of the Association for Laboratory Automation (JALA)*, Vol. 7, No. 4, 2002, pp. 38–42.

[62] Otto, P., B. Larson, and S. Krueger, "Automated High-Throughput Purification of PCR Products Using Wizard® Magnesil™ Paramagnetic Particles," *The Journal of the Association for Laboratory Automation (JALA)*, Vol. 7, No. 6, 2002, pp. 120–124.

[63] Otto, P., B. Larson, and S. Krueger, "Automated High Throughput Purification of Bigdye™ Terminator Fluorescent DNA Sequencing Reactions Using Wizard® Magnesil™ Paramagnetic Particles," *The Journal of the Association for Laboratory Automation (JALA)*, Vol. 7, No. 5, 2002, pp. 75–79.

[64] Marziali, A., et al., "An Automated Sample Preparation System for Large-Scale DNA Sequencing," *Genome Research*, Vol. 9, 1999, pp. 457–462.

[65] Meldrum, D. R., et al., "Acapella-1 K: A Biomechatronic Fluid Handling System for Genome Analysis That Processes 1000 Samples in 8 Hours," *IEEE/ASME Transactions on Mechatronics*, Vol. 5, No. 2, 2000, pp. 212–220.

[66] Meldrum, D. R., et al., "Sample Preparation in Glass Capillaries for High Throughput Biochemical Analyses," *Int. Conference on Automation Science and Engineering*, Edmonton, Canada, 2005.

[67] Yi, F., et al., "High-Throughput, Precise Protein Stability Measurements Facilitated by Robotics," *The Journal of the Association for Laboratory Automation (JALA)*, 2005, pp. 98–101.

[68] Najmabadi, P., A. A. Goldenberg, and A. Emili, "A Scalable Robotic Based Laboratory Automation System for Medium Sized Biotechnology Laboratories," *Int. Conference on Automation Science and Engineering*, Edmonton, Canada, 2005, pp. 166–171.

[69] Waterbury, R. G., et al., "Design and Automated Control of the Electron Microscopy Proteomic Organellar Preparation Robot," *The Journal of the Association for Laboratory Automation (JALA)*, Vol. 10, No. 4, 2005, pp. 246–253.

[70] Harris, N., et al., "Automated Rna Isolation Using the Ambion Rnaqueous™-96 Kit with the Packard Multiprobe Ii Ht Liquid Handling System," *The Journal of the Association for Laboratory Automation (JALA)*, Vol. 7, No. 2, 2002, pp. 75–77.

[71] Bostyn, S., et al., "An Automated Liquid-Liquid Extraction System Supervised by an Industrial Programmable Logic Controller," *The Journal of the Association for Laboratory Automation (JALA)*, Vol. 6, No. 1, 2001, pp. 53–57.

[72] Schena, M., et al., "Quantitative Monitoring of Gene Expression Patterns with a Complementary DNA Microarray," *Science*, Vol. 270, 1995, pp. 467–470.

[73] Marshall, E., "Companies Battle over Technology That's Free on the Web," *Science*, Vol. 286, No. 5439, 1999, p. 446.

[74] MacBeath, G., and S. L. Schreiber, "Printing Proteins as Microarrays for High-Throughput Function Determination," *Science*, Vol. 289, No, 2000, pp. 1760–1763.

[75] Fisher, W., and M. Zhang, "An Automated Biological Fluid Dispensing System for Microarray Fabrication Using Inkjet Technology," *IEEE Int. Conference on Robotics and Automation*, Orlando, FL, May 2006, pp. 1786–1793.

PART III

Device Design, Simulation, and Fabrication for Automation

Sensors and Actuators for Life Science Automation

Eniko T. Enikov and Geon S. Seo[†]

This chapter introduces currently used and emerging sensing and actuation modalities for life science automation. Classification of sensors and actuators (transducers) and their main characteristics are presented. The discussion is then focused on technologies amenable to miniaturization via the application of microfabrication techniques, MEMS, and microfluidics. Both the principles of operation and applications are discussed. Illustrative examples include a pressure-volume catheter sensor, potentiometric pH and CO_2 sensors, an amperometric O_2 sensor, and electromagnetic sensors for 3D ultrasound. The examples of actuators include electrostatic chemical printers, thermal microactuators for electrochemical cells, and electroactive polymer actuators used in catheter steering.

8.1 Introduction

Sensors and actuators are collectively referred to as transducers (i.e., devices that convert one type of energy into another). Sensors are transducers that produce electrical energy (signals) as a result of some type of stimulation, while actuators produce mechanical energy in response to some type of energy input. The history of sensors dates back to 1860, when Wilhelm von Siemens built a temperature sensor using the temperature dependence of the conductivity of a Cu wire [1]. Since then, and especially after the invention of solid-state amplification based on transistors, the development of sensors has expanded tremendously. Approximately 100 years after von Siemens' thermometer, the development of a planar integrated circuit technology in 1956 allowed researchers at Westinghouse Research Labs to demonstrate the first electrostatic actuator compatible with integrated circuit (IC) technology [2]. By then, Arnold Beckman had developed the modern glass electrode (1932) [3], and Kolthoff and Sanders (1937) had demonstrated the use of solid-state fluoride-selective electrodes based on silver halides [4]. With the maturation of Si processing technology, additional sensors emerged, most notably the Si pressure sensor, first demonstrated by Kurtz and Goodman in the period of 1961–1970 [5]. During the same period, Clark and Lyons developed a dissolved oxygen sensor

---

[†]   University of Arizona

commonly known as the Clark-type oxygen electrode [6], and King demonstrated a piezoelectric quartz crystal sensor [7]. The period of 1970–1980 resulted in demonstrations of the first micromachined accelerometer [8], the ink-jet printer nozzle [9], and various other solid-state sensors. In 1980, Howe demonstrated polycrystalline-silicon fingers forming capacitive structures on the surface of a Si substrate [10]. Several years later, the first surface micromachined resonant cantilever gas sensor was demonstrated by the same group [11]. Thus, the fields of sensor and actuator development merged into what is known today as microelectromechanical systems (MEMS). Many of these devices are used to detect mechanical quantities such as pressure, acceleration, and force, while others utilize electrochemical processes to detect chemical and biological analytes. The latter, now called bio-MEMS, are not necessarily based solely on Si processing techniques.

Microsensors and actuators are an integral part of most automation problems, since closed-loop control of an instrument requires action (actuation) based on measurements of the output of the system (sensing). This chapter will present several commonly used or emerging transducers for life science automation applications. Due to the large number of such devices, this review will not be exhaustive. Instead, it is focused on those that are amenable to miniaturization.

8.2 Sensors

Based on the measured signal, sensors can be classified into several main groups: mechanical, chemical and biochemical, electrical, thermal, and radiation sensors. Sensors can be further subdivided into self-modulating (i.e., producing output voltage solely dependent on the quantity of the measured analyte) and modulating, which requires additional energy input, such as current, that is then modulated by the measured analyte. Examples of self-modulating sensors include thermocouples and pH electrodes, while piezoresistive pressure sensors and the Clark-type oxygen electrodes are modulating sensors.

Actuators are classified according to the principle and type of energy conversion. Most common types are electrostatic (uses Coulomb forces), electromagnetic (based on Lorentz force), piezoelectric (based on electric field-induced strain), thermal (based on thermal expansion), and chemical or electrochemical (utilizing combustion or gas generation to produce volume expansion and work).

This chapter is focused on several examples of sensors and actuators that in our view have become well established, or have a great potential to become important transducers in the field of life science automation.

8.2.1 Pressure and Volume Sensors for Medical Applications

Pressure sensors are now well-established tools in life science automation. Pressure sensors are used in artificial lung-heart machines to monitor and adjust the blood perfusion parameters, in telemedicine and digital blood pressure sensors to monitor the patient's vital signs [12], and in cardiology for recording the pressure-volume characteristics of a pumping heart [13]. Most pressure sensors utilize a deformable membrane with integrated strain- or displacement-sensing elements. The classical Si pressure sensor was first developed by Kurtz and Goodman in the early 1970s [5]. It is based on anisotropic etching of Si, which results in a truncated pyramid forming the cavity of the sensor [see Figure 8.1(a, b)]. The pressure is measured by (typically) four piezoresistive elements, placed at the location of highest stress and forming a Wheatstone bridge [see Figure 8.1(c)].

For a square diaphragm with size $a \times a$ and thickness h, the change of resistance, ΔR, as a function of the applied pressure, p, can be calculated through

$$\frac{\Delta R}{R_0} = -\frac{1}{2}\pi_{44}\sigma, \qquad \sigma = 0.0513 pa^2 \frac{12}{h^2} \tag{8.1}$$

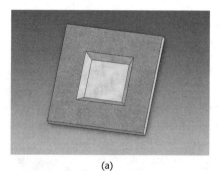

(a)

(b)

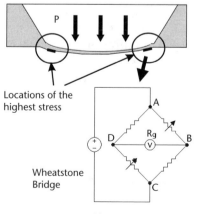

(c)

Figure 8.1 (a–c) Typical Si piezoresistive pressure sensors.

where π_{44} is the piezoresistive coefficient. Typical values of π_{44} for lightly doped Si are -13.6 Pa^{-1} and 138.1 Pa^{-1} for n-type and p-type Si, respectively [14]. Due to its high temperature sensitivity, the piezoresistors are almost always connected in a Wheatstone bridge configuration to reduce the effect of temperature variation. In this case, the output voltage of the bridge is a function of the input voltage through

$$V_{out} = V_{in} \frac{\Delta R}{2R_0} = -0.0513 V_{in} \pi_{44} p a^2 \frac{3}{b^2} \tag{8.2}$$

It is instructive to notice that the output voltage in (8.2) is not a function of the value of the resistors. The resolution of the sensor, however, is determined by its noise level, which is a function of the value of the resistors. The lowest discrimination value is typically twice the root mean square (RMS) of the thermal noise [15]:

$$V_{out}^{min} = 2v_{rms} = 2\sqrt{4kTR\Delta f} \tag{8.3}$$

where $k = 1.38 \times 10^{-23}$ W-s/K is the Boltzmann's constant, T is the absolute temperature, and Δf is the frequency bandwidth of the amplifier used to amplify the signal. Therefore, using (8.2) and (8.3), one can estimate the ultimate pressure resolution for a given sensor design.

Fabrication of Si pressure sensors is now a well-established process. The individual steps needed to produce the sensor from Figure 8.1(b) are shown in Figure 8.2. The process starts with the formation of p-type resistors by ion implantation of Boron into the n-type Si substrate. Aluminum metallization is used to make ohmic contacts to the resistors and form the bonding pads of the chip. Subsequently, the chip is placed in a protective fixture to protect the front-side resistors while etching the backside and forming the diaphragm. Upon completion of the etch step, the sensors are separated by sawing the wafer and are then packaged in stand-alone packages or inside of hollow catheters [16].

Steady increased miniaturization of the pressure sensors has led to recent attempts to continuously monitor the intraocular pressure of the eye [17]. A somewhat more established use is the simultaneous pressure and volume measurements in the heart of a mouse. An example of such catheter is shown in Figure 8.3.

The measurement of arterial and heart pressure-volume changes is an invasive method for assessing the myocardial function. The measurement is based on impedance measurements using multiple (at least four) electrodes placed along the length of the catheter. The original method was developed by Baan et al. [18] and is based on conductance measurements. The conductance of a cylindrical volume of blood with length, L, and cross-section, A, is given by

$$g_B = \frac{i}{v} = \frac{A}{\rho L} \tag{8.4}$$

where ρ is the resistivity of the blood inside the cylinder, and i and v are the applied current and potential, respectively. The volume, therefore, can be found through

$$Vol = g_B \rho L^2 \tag{8.5}$$

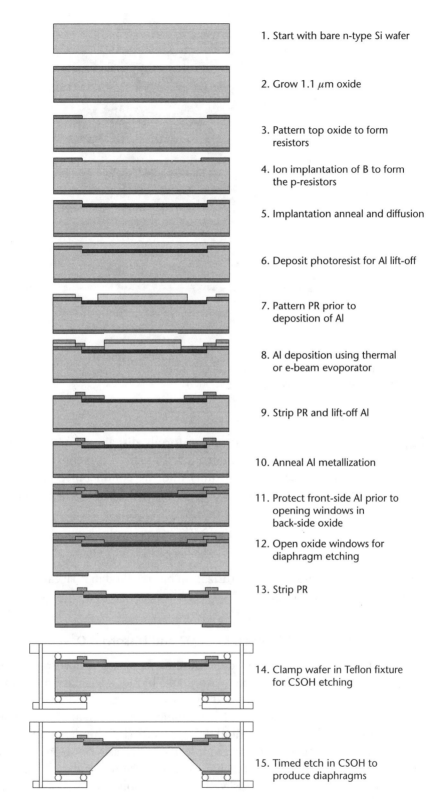

1. Start with bare n-type Si wafer

2. Grow 1.1 μm oxide

3. Pattern top oxide to form resistors

4. Ion implantation of B to form the p-resistors

5. Implantation anneal and diffusion

6. Deposit photoresist for Al lift-off

7. Pattern PR prior to deposition of Al

8. Al deposition using thermal or e-beam evoporator

9. Strip PR and lift-off Al

10. Anneal Al metallization

11. Protect front-side Al prior to opening windows in back-side oxide

12. Open oxide windows for diaphragm etching

13. Strip PR

14. Clamp wafer in Teflon fixture for CSOH etching

15. Timed etch in CSOH to produce diaphragms

Figure 8.2 Fabrication process for Si pressure sensors.

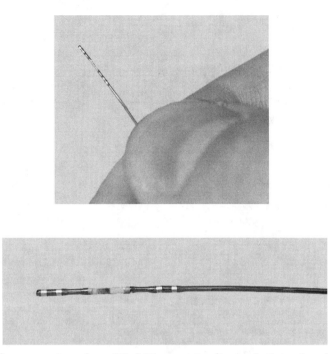

Figure 8.3   Millar pressure transducer (1F, 0.33 mm outer diameter). Outer ring electrodes are used for current injection, while the inner electrode pair is used for potential sensing. (Photograph used by permission from Millar Instruments, Inc.)

In practice, the electric current is not uniformly distributed and the electrodes are not planar. Therefore, Baan et al. introduced a calibration constant, α, which takes into account the current nonuniformity and the conductance of the surrounding myocardial tissue. With the calibration factor, the volume is determined by [18]

$$Vol = \frac{g_B \rho L^2}{\alpha} \tag{8.6}$$

Since the 1980s, continued miniaturization has resulted in a 1F catheter with similar functionality, shown in Figure 8.3.

8.2.2   Electrochemical Sensors: pH, CO_2, and Dissolved Oxygen

Many automated systems require measurements of gas concentrations and acidity (pH). Examples include CO_2 incubators and blood oxygenation sensors in heart-lung machines. Elevated acidity and hypoxia are common characteristics of many tumors. Interstitial hypertension, for example, is a universal characteristic of solid tumors [19, 20]. Elevated interstitial fluid pressure (IFP) hinders transvascular transport of therapeutic agents with large molecules, such as monoclonal antibodies [21]. Poor perfusion is also indicated by hypoxia and elevated acidity [22, 23]. Therefore, pH and oxygen concentration, as well as IFP of a solid tumor, could be indicative of increased drug resistance [24], increased mutations [25], chromosomal instability [26], or increased invasion and metastasis [27]. Thus, implantable microsensors capable of continuous monitoring of pH, dissolved oxygen, and pressure can be of great utility in better understanding the tumor growth and in assessing the effectiveness of a cancer treatment.

Most electrochemical sensors (pH, ion-selective sensors) fall into two categories: potentiometric (voltage sensing) and amperometric (current sensing). In the former case, a voltage drop across an electrochemical electrode is measured, while in the latter case a current under given potential is measured.

Potentiometric sensors are based on the Nernst equation, which gives the electrode potential with respect to a reference electrode potential, V^{ref},

$$V^{electrode} = V^{ref} + \frac{RT}{nF} \ln \frac{\prod_i a_{O_i}}{\prod_j a_{R_j}} \qquad (8.7)$$

where $a_{O_i} = \gamma_i [O_i]$ and $a_{R_j} = \gamma_j [O_j]$ are the activities of the oxidized (O_i) and reduced (R_j) species, respectively, participating in the electrode reaction [28],

$$\sum_i \gamma_i [O_i] + ne^- \rightarrow \sum_j \gamma_j [R_j] \qquad (8.8)$$

The pH sensors can be constructed by using the oxidation of an array of metals (Pt, Ir, W), as discussed by Hendrikse et al. [29]. In these cases, the hydrated metal oxide, MeO(OH), participates in the reaction,

$$MeO(OH) \rightarrow MeO_2 + H^+ + e^- \qquad (8.9)$$

in which the electrode potential is a function of the proton activity,

$$V^{electrode} = V^{ref} + \frac{RT}{F} \ln \frac{a_{MeO(OH)}}{a_{MeO_2}} + \frac{RT}{F} \ln a_{H^+} \qquad (8.10)$$

The challenge associated with this type of sensor is to eliminate the contribution of the terms in (8.10) that are not directly related to the activity of the measured analyte. In addition to pH, potentiometric sensors can by used to detect other analytes, most notably carbon dioxide. One commonly used technique for measuring CO_2 is based on the use of an aqueous solution in contact with the unknown sample. Since CO_2 reacts with water to produce carbonic acid, the pH of the aqueous solution can be used to measure the amount of CO_2. The electrode based on this principle was developed by Severinghaus and Bradley [30], and it is shown in Figure 8.4. A thin silicone membrane is used to separate the test solution (whole blood) from the inner electrolyte consisting of a sodium bicarbonate-KCl solution. The change in pH is measured by a pH-sensitive glass membrane in close proximity to the silicone membrane. The total cell electrode potential is given by [31]

$$V^{electrode} = V^{ref} + s \ln p_{CO_2} - s \ln HCO_3^- + s \ln Cl^- \qquad (8.11)$$

where p_{CO_2} is the carbon dioxide partial pressure and s is the sensitivity of the electrode.

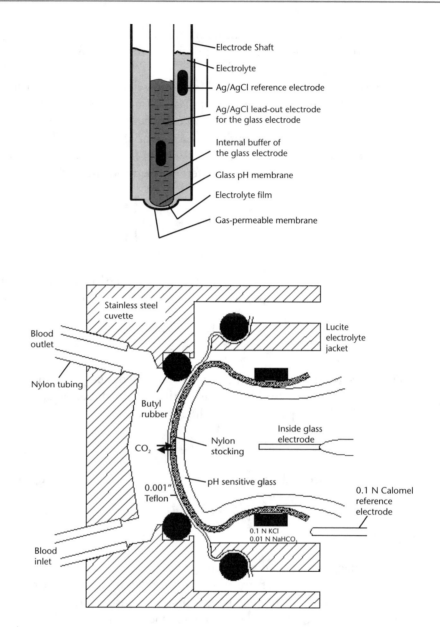

Figure 8.4 Principal design of CO_2 electrode.

Amperometric sensors utilize a fast electrochemical reaction that depletes the electrode of the analyte of interest. The electric current in this case is modulated by the diffusion of the analyte towards the electrode, which is a function of the concentration outside of the depletion region. Clark's dissolved oxygen sensor is the most well-known sensor of this type. The limiting current of amperometric sensors, i, is modulated by the bulk concentration of species, C, according to the following theoretical model [32]:

$$i = nFAC\left(\sqrt{\frac{D}{\pi t}} + \frac{D}{r}\right) \tag{8.12}$$

where n is the number of electrons transferred in the reaction, A is the area of the electrode, r is the radius of the electrode, and D is the diffusion coefficient of the species of interest. The time-dependent (transient) term accounts for the initial growth of a depletion layer and is know as the Cottrell term. Equation (8.12) is valid for hemispherical electrodes. In the case of a circular planar electrode, the steady-state term has been shown to be [28]

$$i = 4nFACD \tag{8.13}$$

Clark's dissolved oxygen electrode consists of a platinum electrode immersed in the test medium and biased negatively with respect to a reference electrode. The cathode current is produced by a two-step oxygen reduction process described by the following two reactions [33]:

$$O_2 + 2H_2O + 2e^- \rightarrow H_2O_2 + 2OH^-$$
$$H_2O_2 + 2e^- \rightarrow 2OH^- \tag{8.14}$$

At low potentials, the electrode current is governed by the exchange current density and the overpotential of the reactions above. At higher potentials, the oxygen concentration (activity) and the surface become negligible and the current limiting step is the diffusion of oxygen in the solution. Under these conditions, the current becomes independent of the voltage and is given by (8.12). Instead of directly exposing the Pt electrode to the test solution, Clark's electrode contains an oxygen-permeable polyethylene membrane, separating the electrode from the test solution (see Figure 8.5). Using this arrangement provided greater stability of the electrode and an ability to sterilize it, and most importantly, it allowed measurements in nonconductive media (gas mixtures). Clark-style electrodes have undergone significant development and miniaturization over the last 40 years, resulting in commercial micromachined versions produced by Fujitsu, Inc., of Japan [34].

8.2.3 Impedimetric Sensors

Impedimetric sensors measure the change of electrical impedance (or its frequency spectrum) as a result of the growth of a biofilm, adsorption of proteins on the surface of an electrode, an increase of cell concentration in a cell suspension, or a change in volume of the test chamber. When a harmonically varying potential is applied across a sample, it establishes a current with the same frequency but a different phase. Due to the harmonic nature of the excitation, both signals can be conveniently represented by the complex quantities,

$$v(\omega) = V_o e^{j\omega t} \tag{8.15}$$

$$i(\omega) = I_o(\omega)e^{j(\omega t + \phi)} \tag{8.16}$$

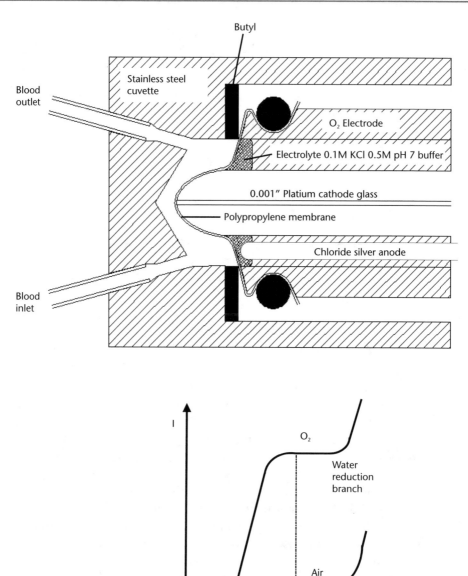

Figure 8.5 Clark's oxygen electrode.

The electrical impedance is defined as the ratio (transfer function) of the two signals,

$$Z(\omega) = \frac{v(\omega)}{i(\omega)} = \frac{V_\phi}{I_\phi(\omega)} e^{-j(\phi(\omega))} = z' + z''j, \text{ where } j^2 = -1 \qquad (8.17)$$

Both the amplitude ratio, $V_o / I_o = |Z(\omega)|$, and the phase shift, $\phi = \angle Z(\omega)$, are functions of the test frequency and the properties of the substrate being tested. For example, if the test structure can be represented by a resistor, R, in series with a capacitor, C, the equivalent impedance is

$$Z(\omega) = R - j\frac{1}{\omega C} \tag{8.18}$$

If the resistance and/or the capacitance is a function of the analyte of interest, the complex impedance spectrum can be used to deduce its amount. There are many examples of impedimetric sensors. These include heart-rate monitors, blood flow and volume sensors, and bioimpedance sensors for measuring the concentration of proteins or whole cells (biomass). Impedance measurements have also been used in the detection of breast, cervical, and prostate cancers, based on the fact that the electrical properties of cancerous lesions are not the same as those of normal tissue.

Two or more electrodes are used to determine the impedance of a sample. At low frequencies, the electrodes' impedance contributes significantly to the overall impedance of the sample. In many cases, this proves to be the dominant contribution, resulting in a screening of what is happening in the bulk of the sample. Because of this, many impedimetric sensors operate in a surface-detection mode (i.e., measure the changes in the electrode impedance due to adsorption and desorption of an analyte or cell attachment). Interpretation of impedance spectra and its correlation to the quantity of interest could be a difficult task in the presence of multiple electrochemical reactions or a mixture of biological cells and proteins. A classical model of electrode-electrolyte impedance is due to Wartburg (1899) [35], and, later, to Cole and Cole [36]. In their model, the electrode is modeled by two elements: (1) a capacitor accounting for the charge accumulated in the double layer, C_{dl}, in parallel with a resistor, R_{ct}, and a constant-phase element, Z_w, cumulatively representing the faradic resistance of the electrode [see Figure 8.6(a)]. The double-layer capacitance is a pure capacitor, while the constant-phase element arises from the mass transfer resistance associated with transport of ions to and away from the interface. For one-step, one-electron redox reactions $O^+ + e^- \leftrightarrow R$, the Warburg impedance Z_w is given by

$$Zw = \frac{\sigma}{\omega^{1/2}}(1-j) = \frac{\sqrt{2}\sigma}{(j\omega)^{1/2}} \tag{8.19}$$

with

$$\sigma = \frac{RT}{F^2 A\sqrt{2}}\left(\frac{1}{\sqrt{D_o}C_o} + \frac{1}{\sqrt{D_R}C_R}\right) \tag{8.20}$$

where D_O and D_R are the diffusion constant of the oxidized and reduced species, respectively, and C_O^* and C_R^* are their equilibrium concentrations [37]. It is apparent, therefore, that this impedance is due to mass transfer (diffusion related) and is always present at the interface. The second term of the faradic impedance is the charge transfer resistance, R_{ct}, which can be related to the equilibrium ion exchange current, i_0,

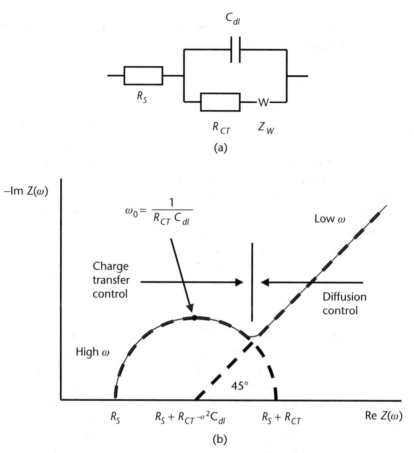

Figure 8.6 (a) Randels equivalent circuit for electrode-solution impedance, and (b) combined Nyquist plot of the total impedance.

$$R_{ct} = \frac{RT}{Fi_0} \tag{8.21}$$

If the reaction is very fast, the equilibrium exchange current is high and the charge transfer resistance is low. Under these conditions, the mass transfer represented by Z_W is the current limiting step. Alternatively, if the reactions are slow in comparison to the diffusion, the charge transfer resistance will be the dominant contribution to the faradic impedance. It is also worth nothing that the Warburg impedance has a constant phase of $-45°$, hence the name "constant-phase element."

The analysis of the combined impedance plot can be understood by examining the low- and high-frequency extremes. At low frequencies, the Warburg impedance will be large and dominant. Under these conditions, the double-layer capacitance will not influence the current and the overall impedance will be

$$Z_{TOT}^{low\,fr} = R_S + R_{ct} + Z_W \tag{8.22}$$

which is marked by the dashed line in Figure 8.7(b). At high frequencies, the Wartburg impedance is negligible in comparison to R_{ct}. Under these conditions, the overall impedance is

$$Z_{TOT}^{high\,fr} = R_S + \frac{R_{ct}}{1 + j\omega C_{dl}R_{ct}} \tag{8.23}$$

which is marked by the dashed circle in Figure 8.7(b).

The combined impedance plot of an electrode-solution interface is the a combination of the two frequency limits [continuous line in Figure 8.7(b)]. Using such impedance plots, it is possible to extract data about the conductivity of a solution, the diffusion constants, and the exchange currents. In the case of protein solutions, cell suspensions, and tissues, impedance spectra deviate from this ideal case. Cole and Cole [36] proposed a more generalized constant-impedance element for modeling the impedance of biological tissues and cell suspensions, which is given by

$$Z_{TOT}^{high\,fr} = R_S + \frac{R_0 - R_S}{1 + (j\omega T)^\alpha}, \quad 0 < \alpha < 1 \tag{8.24}$$

The Nyquist plot of the element above (also know as a Cole-Cole plot) is also a circular arc, but with a center below the real axis. Since most biological solutions are ionic, the electrodes will also contribute to the overall impedance. Therefore, the combined plot will have multiple relaxations resulting in overlapping arcs on the impedance plot (see Figure 8.7). Despite these difficulties, impedance-based sensors are still the subject of active research and development, including DNA sensors [38, 39] and pathogenic viruses and bacterial microorganisms [40–43]. Many of these rely on antibody-antigen recognition [44] and are therefore based on electrode (surface) impedance measurements. For a more extensive review, the reader is referred to [45] and the references therein.

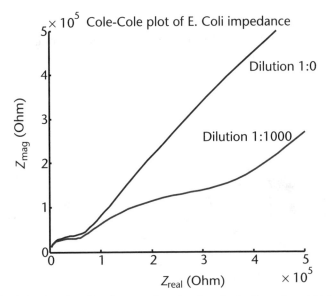

Figure 8.7 Cole-Cole plot of an *E. coli* suspension showing single dispersion at high concentration (as grown) and double dispersion of the diluted (1:1,000) sample.

8.2.4 DNA/Antigen-Antibody Sensors (Cantilever Sensors)

Microfabricated cantilevers similar to the ones used in atomic force microscopy (AFM) can be used as biochemical sensors [46–48]. Since microcantilevers are the simplest microstructure, they could lead to a new class of inexpensive, miniature sensors. Through fabrication of multielement sensor arrays and immobilization of different detection probes, the specificity of the sensor can be increased. The low stiffness of the microfabricated cantilevers offers high sensitivity and fast response. When molecular adsorption is confined to one side of a functionally coated surface of a microcantilever, it undergoes bending due to adsorption-induced surface stress, as shown in Figure 8.8. Fritz et al. [46] have demonstrated the detection of DNA hybridization and receptor-ligand binding where DNA sequences can be distinguished. Godin et al. [49] measured the surface-stress-caused bending during the formation of alkanethiol self-assembled monolayers (SAMs) on gold from the vapor phase of alkanethiols. Also, similar studies have recently been reported on the adsorption of low-density lipoprotein [50], antigen-antibody binding [51], and on the detection of odor molecules (artificial nose) [52]. Cantilever bending is traditionally monitored by measuring the microcantilever deflection based on the optical lever technique. Optical detection requires alignment and calibration for different testing species and is limited to transparent liquids. Therefore, microcantilevers with an embedded piezoresistive deflection readout have been proposed recently. Unfortunately, the sensor performance of these devices is adversely influenced by the self-heating of piezoresistor, and the associated thermal stress seriously contaminates the surface stress measurement [53].

As described earlier, the difference in the surface stress yields a moment M given by

$$M = \frac{\Delta\sigma Wt}{2} \tag{8.25}$$

where $\Delta\sigma$ is the difference in the surface stresses between the top and the bottom side of the cantilever. For a uniform moment M exerted along the length of a rectangular cantilever, a deflection angle, θ_M, at a position $x(x = aL, 0 \leq a \leq 1)$ along the cantilever length is given by [54, 55]

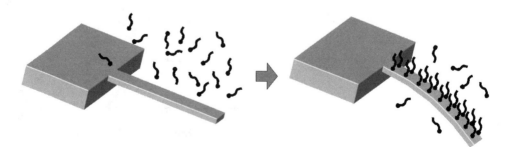

Figure 8.8 Microcantilever with self-assembling alkenethiol molecules.

$$\theta_M = \frac{Mx}{EI} = \frac{12ML}{EWt^3} a \tag{8.26}$$

where L is length, W is width, t is thickness, and E is Young's modulus.

By substituting (8.25) into (8.26), the surface stress can be expressed in terms of θ_M as

$$\Delta\sigma = \frac{Et^2}{6La} \theta_M \tag{8.27}$$

The deflection angle can be expressed the radius of the cantilever through

$$\theta_M = \frac{aL}{R} \tag{8.28}$$

When (8.28) is substituted into (8.27), one obtains the Stoney formula [56],

$$\frac{1}{R} = \frac{6}{Et^2} \Delta\sigma \tag{8.29}$$

The Stoney formula is now widely used in surface stress measurements and in the design of cantilever sensors. Commercially available SiN_x sensors (thickness: 0.6 μm; length: 150 μm; width: 18 μm; spring constant: about 0.5 N/m) with a 20-nm gold receptor layer achieve a surface stress resolution for SAMs of 10^{-7} N/m, which corresponds to a change of zeptomolar (10^{-21} mol) quantities [57]. Further reduction of sensor thickness will provide the detection limits of smaller surface stress variations. The sensitivity of the sensors is determined by (8.29), evaluating against the magnitude of random vibrations due to thermal excitation.

The microcantilevers are fabricated using conventional microfabrication techniques. Figure 8.9(a) shows the simple process for producing gold-coated silicon nitride cantilevers for alkanethiol SAMs. First, a silicon nitride layer is deposited by low-pressure chemical vapor deposition (LPCVD) on silicon wafer. A thin gold layer is then deposited using an e-beam evaporator, and the surface is patterned by

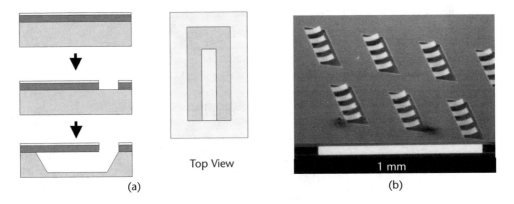

(a)   Top View   1 mm   (b)

Figure 8.9 (a) Schematic of fabrication process; and (b) SEM photograph of microcantilevers.

photolithography. The cantilever itself is formed by wet etching with alkaline etchant, such as CsOH or KOH. The large lateral etch rate of convex corners results in undercutting and produces a freestanding cantilever, shown in Figure 8.9(b).

8.2.5 Electromagnetic Sensors

Electromagnetism arises from electric current moving through a conducting material. Electromagnetic fields arise and disappear rapidly (thus permitting devices with very fast operation speeds) and propagate through the human body and other obstructions with very little attenuation [58]. Also, electromagnetic devices are highly efficient in converting electrical energy into mechanical work [59], and they provide stronger forces over a longer distance compared with electrostatic driving mechanisms [60, 61]. Therefore, the electromagnetic devices are used in actuators and sensors, such as tactile displays, hearing instruments, microvalves, drug delivering pumps, catheter flow sensors, position detectors, and so on [62–66]. Conventional electromagnetic actuators are well established. The reader is referred to [67] for technical details. Here, we describe a relatively new use of an electromagnetic field for generating 3D ultrasound images. The concept is based on augmenting ultrasound (2D) images with positional information of the detector probe, which allows computer reconstruction of the image in three dimensions. The key elements in position measurements are three or more beacon coils producing a low-frequency electromangetic field that is used to determine the location and orientation of the mobile sensor (detector). The sensor unit (to be worn on the finger) contains three miniature coils, one for each of the three orthogonal directions (see Figure 8.10). These are used to measure the magnitude and direction of the induced magnetic field and subsequently calculate the position and orientation of the sensor [68].

The developed magnetic field around each beacon coil is described by [69]

$$\overline{B} = \frac{\mu_0 NIa}{4\pi r^3}\left(2\cos(\theta)\hat{r} + \sin(\theta)\hat{\theta}\right) \tag{8.30}$$

where μ_0 is permeability of the free space, N is the number of turns of the wire, I is the current, θ is the elevation angle, r is the distance from the center of the loop, and a is the area of the loop. For the case where all beacon coils are coplanar with an axis pointing in the z-direction, the three components of the magnetic field created by the ith beacon are related to the position of the sensor by the following equations [68]:

$$B_{ix} = 3k(x - x_i)(z - z_i)/r_i^5 \tag{8.31}$$

$$B_{iy} = 3k(y - y_i)(z - z_i)/r_i^5 \tag{8.32}$$

$$B_{iz} = k\left(2(z - z_i)^2 - (x - x_i)^2 - (y - y_i)^2\right)/r_i^5 \tag{8.33}$$

where $k = \mu_0 NIa / 4\pi$; i is the identify number of each beacon (1, 2, and 3); x, y, and z are sensor positions; x_i, y_i, and z_i are positions of beacon i; and r_i is the distance between the sensor and beacon $i\left(r_i = (x - x_i)^2 + (y - y_i)^2 + (z - z_i)^2\right)^{1/2}$. These estimates are written with respect to the beacon-fixed coordinate frame $(\hat{x}, \hat{y}, \hat{z})$, but

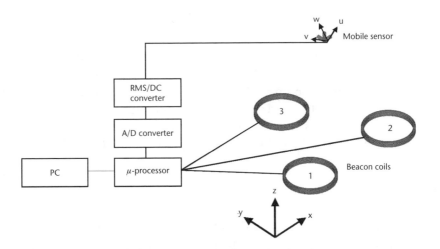

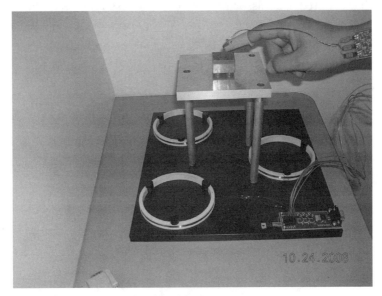

Figure 8.10 A distributed magnetic local positioning system.

actual measurements are taken in the local sensor coordinate frame $(\hat{u}, \hat{v}, \hat{w})$. A direction matrix containing three Euler angles is used to relate these two frames as follows:

$$\begin{bmatrix} B_{iu} \\ B_{iv} \\ B_{iw} \end{bmatrix} = \begin{bmatrix} A_{11} & A_{12} & A_{13} \\ A_{21} & A_{22} & A_{23} \\ A_{31} & A_{32} & A_{33} \end{bmatrix} \begin{bmatrix} B_{ix} \\ B_{iy} \\ B_{iz} \end{bmatrix} \tag{8.34}$$

where B_{iu}, B_{iv}, and B_{iw} are actual measurements from a mobile sensor. Since $\left| A_{ij} \right| = 1$, the magnitudes of the two vectors have to be equal, and three magnitude equations can be obtained from (8.31) through (8.33):

$$|B_i| = k \left(3(z - z_i)^2 + r_i^2 \right)^{1/2} / r_i^4, \quad i = 1,2,3 \qquad (8.35)$$

Equation (8.35) represents three equations for the three unknowns (x, y, z), and can be solved using the Gauss-Newton method to obtain the position of the sensor. Figure 8.11 shows experimental results from scanning a 10 mm × 10 mm × 10 mm cube and three parallel lines. The demonstrated resolution was better than 0.2 mm along all three axes. This position-detecting system can be used for medical examinations, for example, in a 3D ultrasound system (see Figure 8.12) [70, 71].

8.3 Actuators

Micromechanical actuators have become the hallmark of MEMS since their inception more than 30 years ago. The search for an efficient and useful actuation modality has resulted in an assortment of devices based on electrostatic [72], piezoelectric

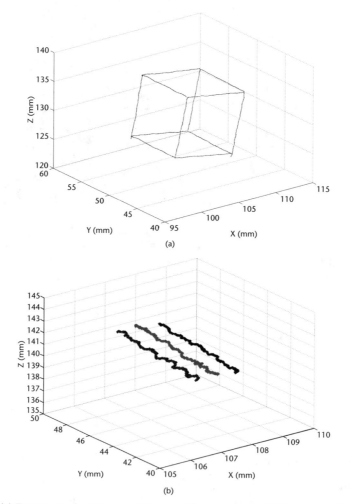

Figure 8.11 (a) Test scans of a 10 mm × 10 mm × 10 mm cube and (b) three parallel lines spaced at 0.8 mm from one another.

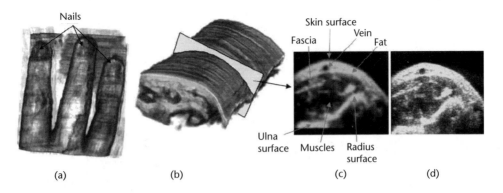

Figure 8.12 Results of 3D ultrasound imaging for the subjects in vivo: (a) the volume of three fingers; (b) the volume of part of a forearm; (c) a typical slice obtained from the volume with the location and orientation indicated by the plane in (a); and (d) the corresponding original B-scan image collected at approximately the same location of the slice showed in (c). (*From:* [71]. © 2005 Elsevier. Reprinted with permission from Elsevier.)

[73], electromagnetic [74], electrothermal [75, 76], thermopneumatic [77], electrochemical [78], electro- and magnetostrictive [79], shape memory [80], and mass transport effects [81] Actuator miniaturization is especially important in life sciences, as it permits the development of minimally invasive surgical tools and microrobots capable of autonomously navigating through the human body [82]. In protein analysis and drug discovery, electromagnetic, electrostatic, and piezoelectric actuators have been used extensively to deliver small droplets of reagents such as antigens or enzymes. Because different enzymes cleave the polypeptides at different locations, by identifying the fragments, one can deduce the amino-acid sequence in the original polypeptide. Clearly, this type of combinatorial analysis requires numerous cleaving reactions with a variety of enzymes, which can only be accomplished by automated fluid dispensing equipment. Microactuators and micronozzles are increasingly becoming the technology of choice for this application since they allow droplets of a few picoliters to be deposited directly onto the substrate containing the unknown proteins.

8.3.1 Micronozzles for Reagent Printing

Proteins are polymers of amino acids joined by a covalent amide linkage called a peptide bond. Polypeptides can contain many amino acids in a chain, which is also folded onto itself into secondary, tertiary, or quaternary structures. Analysis of the amino acid structures is the key to understanding the properties and function of a given protein. Prior to identification, the proteins have to be separated and purified. The most common technique for protein separation is based on their charge and molecular weight [see Figure 8.13(a)]. Each amino acid comprising the polypeptide chain can undergo various degrees of ionization based on the pH of the suspending medium. Therefore, a polypeptide chain possesses an overall positive, negative, or neutral charge, according to the pH of the suspending medium. Using electrophoresis in gels with a built-in gradient of the pH forces the proteins to migrate to locations where the pH of the gel is equal to their isoelectric point (pI). Under these conditions, the proteins have no net charge and are not affected by further

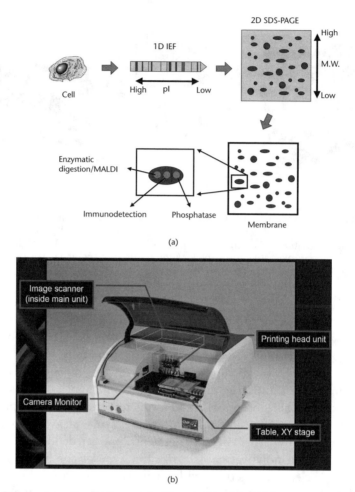

Figure 8.13   Protein separation and reagent microprinting: (a) protein separation in 2D; and (b) Shimadzu Chemical Printer ChIP1000. (Photo courtesy of Shimadzu, Inc., Japan.)

application of the external electric field. This constitutes the separation in the first dimension (1D), also known as isoelectric focusing (IEF) [see Figure 8.13(a)]. In the second step, the proteins are separated according to their molecular weight. To accomplish this, the proteins are transferred to a sodium dodecyl sulfate containing polyacrylamide gel for a second electrophoresis step (SDS-PAGE). The SDS ions bind to the proteins in proportion to the number of amino acids (one SDS ion for every two amino acid residues), thus impacting an overall negative charge proportional to the molecular weight of the polypeptide. Using electrophoretic separation, the proteins are sorted according to their mobility (size). The separated proteins can be visualized by adding a dye (Coomassie blue), which binds to the proteins, but not to the gel. Upon separation, the proteins are transferred onto an inert membrane and each protein is identified using one or more identification techniques. Among these, the more common are time of flight mass spectroscopy (TOF-MS), which typically uses laser pulses to ionize the proteins and eject them from a support matrix (matrix assisted laser desorption ionization, MALDI). Alternative techniques involve addition of fluorescently labeled antibodies in which antibody recognition is detected optically (immunodetection) or using enzyme-specific fragmentation of the peptides, followed by mass spectroscopy. Using microdeposition techniques, it is possible to perform several of these tests on the same blot of separated proteins, thus increasing the speed and throughput of the test. A commercial piezoelectric printing system from Shimadzu, Inc. (Japan), produces spot sizes with diameters of 200 μm and reagent volumes of 200 pL [Figure 8.13(b)]. Figure 8.14 shows a commonly used MEMS nozzle for the generation of microdroplets. Electrostatic attraction or piezoelectric actuation can be used to deflect the back-plate of the ejector. A combination of inertial and capillary forces results in the formation of a droplet, which is ejected from the small orifice. During the back-stroke, the capillary tension prevents entry of air into the chamber. Instead, it is refilled from the reservoir, as indicated by the arrow.

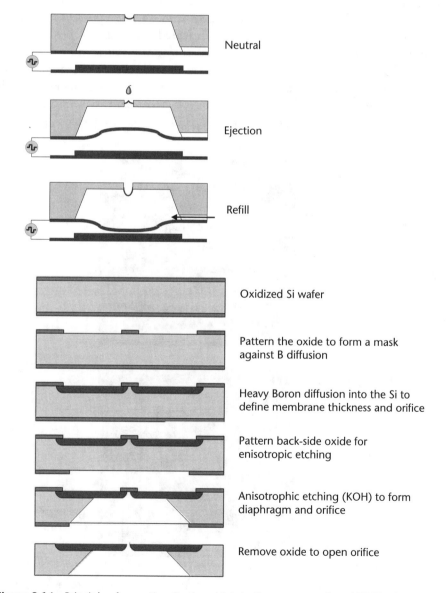

Figure 8.14 Principle of operation (top) and fabrication sequence for a MEMS microprinter.

8.3.2 Thermal Microactuators

Electrothermal microactuators are simple to design and fabricate and provide significant advantages, such as large displacements, large forces, and low operating voltages. These characteristics can be particularly useful in the development of miniature instruments. For example, Chu et al. [83] demonstrated a precision integrated positioner for scanning microscopy. The device they developed utilizes V-beam thermal actuators and a capacitive feedback and was fabricated via deep reactive ion etching using silicon-on-insulator (SOI) wafers. A similar device fabricated using single-mask UV photolithography and an inexpensive nickel electroplating step is shown in Figure 8.15.

This device integrates a folded-beam thermal actuator and a movable mechanical frame coupled to a capacitive comb sensor. This sensor allows closed-loop position control and eliminates the displacement uncertainty, which makes it suitable for electrochemical cells with variable geometry.

Thermal microactuators can be fabricated via electrodeposition in positive photoresist mold (AZ4903.) Figure 8.16 shows the fabrication sequence. A seed layer of 5-nm titanium/100-nm copper was deposited on a thermally oxidized silicon wafer via electron-beam evaporation. Photoresist with a thickness in the range of 25 to 30 μm was spun and patterned in order to create an electroplating mold for the devices. Approximately 20 μm of nickel was then electroplated using a

(a)

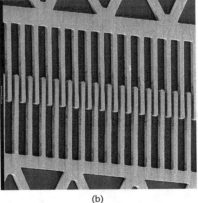

(b)

Figure 8.15 (a) Folded-beam actuator driving interdigitated comb electrodes and (b) a magnified view of the interdigitated comb electrodes.

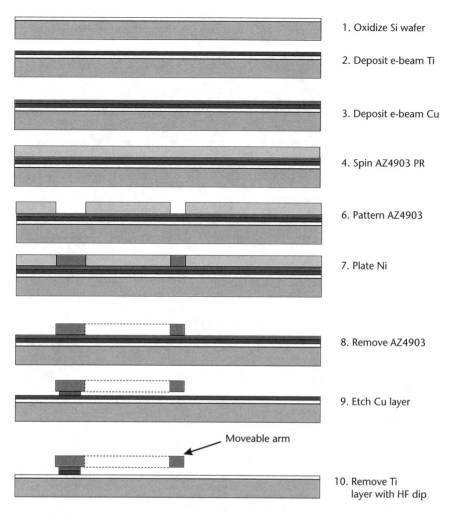

1. Oxidize Si wafer

2. Deposit e-beam Ti

3. Deposit e-beam Cu

4. Spin AZ4903 PR

6. Pattern AZ4903

7. Plate Ni

8. Remove AZ4903

9. Etch Cu layer

Moveable arm

10. Remove Ti layer with HF dip

Figure 8.16 Fabrication sequence.

commercially available bath (Microfab Ni 100, Enthone-OMI, Inc.). The photo-resist was removed and the actuators were released via wet etching in an aqueous solution of ammonium persulfate, which preferentially attacks the copper seed layer. During this etch process, the narrower structures are freed up from the substrate, while the wider segments of the pattern remain anchored to the substrate, as illustrated in Figure 8.16 (steps 9 and 10).

There are two common types of thermal actuators: folded beam and V-shaped. In both devices, a lateral bending moment is created, causing the structure to move laterally. The geometry of a typical folded beam thermal microactuator is shown in Figure 8.17(a). This actuator consists of two arms with different cross sections. When current is passed through the two arms, the higher current density occurs in the smaller cross-section beam and thus generates more heat per unit volume. The displacement is a result of the temperature difference induced in the two arms. Similar to the folded-beam device, V-shaped actuators also use thermal expansion of a

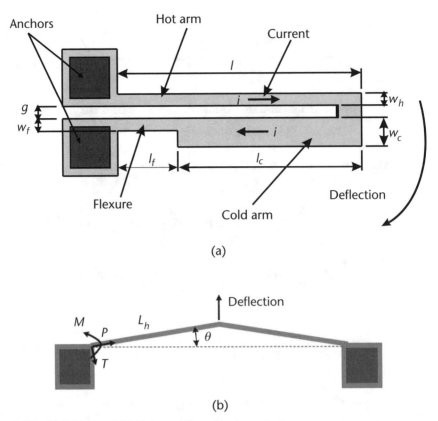

Figure 8.17   (a) Folded and (b) V-shaped thermal microactuators.

beam, anchored between two fixed points [see Figure 8.17(b)]. The mechanical analysis of each of these actuator types is presented next.

The deflection analysis of these actuators requires mechanical and electro-thermal analysis. Analytical solutions developed by Enikov et al. can be found in [84–86].

Thermal microactuators could have a significant role in the development of miniature detectors for biological cells. Whole-cell biosensors are emerging as a new technological tool for anticancer drug screening [87], toxicity monitoring [88], pathogen detection [89], glucose monitoring [90], and inflammatory agent detection [91]. Most of these devices rely on electrical impedance measurements of a suspension of biological cells. At low frequencies, however, electrode polarization effects interfere with the measurement of the electrical impedance of the cell suspension. To overcome this problem, electrochemical cells with variable electrode separations can be used. Making sure that the electrode polarization is the same for all configurations (i.e., maintaining constant current conditions at the electrodes), the solution impedance can be obtained by subtracting the impedance data from two consecutive measurements conducted at electrode separations d_1 and d_2. Figure 8.18 illustrates this concept based on parallel plate electrodes driven by two bistable V-shaped thermal actuators. The bistable operation allows the actuator to be turned off during measurements and eliminates the need for electrical isolation. Furthermore, the distances of a bistable actuator are fixed, which eliminates the need for

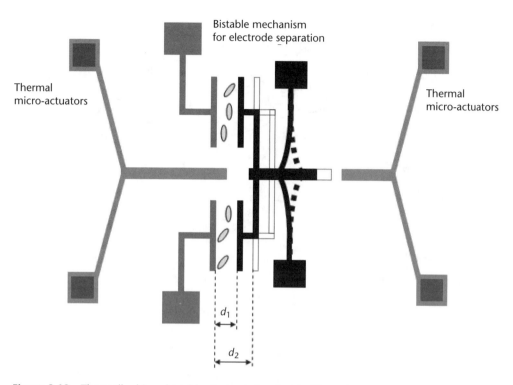

Figure 8.18 Thermally driven bistable electrode for electrical impedance spectroscopy of bacterial suspensions.

separate position detection. The difference between the two impedance values is given by the electrode separations, d_1 and d_2, and effective area, A,

$$z_{sol}(\omega) = z_1(\omega) - z_2(\omega) = \frac{d_1 - d_2}{(c + j\omega\varepsilon)A} \qquad (8.36)$$

where ω is the circular frequency. Calculating the solution conductivity or its dielectric constant is therefore reduced to solving (8.36) for c and ε, respectively. For example, for the dielectric constant of the solution, one has

$$\varepsilon_{rel}(\omega) = \frac{d_1 - d_2}{\varepsilon_0 \omega A} \operatorname{Im}\left(\frac{1}{z_1(\omega) - z_2(\omega)}\right) \qquad (8.37)$$

The application of this technique to the measurement of the concentration of *E. coli* bacteria has been demonstrated using macroscopic electrodes by Prodan et al. [92]. The Advanced Microsystems Laboratory at the University of Arizona is currently developing a MEMS version of an *E. coli* detector operating on this principle.

8.3.3 Electroactive Polymer Actuators

The most attractive features of ionic electroactive polymer (EAP) materials are their mechanical characteristics, which are similar to those of biological muscles in terms

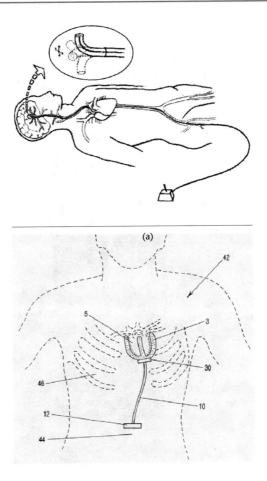

Figure 8.19 Medical applications of EAP actuators. (a) Active streering catheter. (*From:* [96]. © 2000 IEEE. Reprinted with permission.) (b) Heart compression device concept. (*From:* [97]. © 2005 IOP. Reprinted with permission.)

of compatibility with water, light weight, flexibility, softness, and large displacements (see Figure 8.19). The specific characteristics of an ionic EAP are as follows [93]:

- Driving voltage is low (1V to 6V).
- Response time (down to 10 ms).
- It works in water and in wet conditions.
- It is flexible and silent.
- It is durable and chemically stable (it is able to bend over 1×10^6 times).
- Miniaturization and weight saving are possible.

The similarity between the characteristics of natural muscles and ionic EAP actuators has led some to explore using the latter as artificial muscles [94]. Ionic polymer actuators have been used in various practical applications, such as heart and circulation assist devices, artificial smooth muscle actuators, peristaltic pumps, active bandages, and active streering catheters [95–98].

Ionic EAPs include ionic polymer/metal composites (IPMC) and conducting polymers. An IPMC actuator consists of a bulk region (the polymer) bound by a thin conductive metal surface. The most commonly used polymer is Nafion (Du Pont), which is a fluorocarbon polymer containing three main parts: fluorocarbon backbone; side groups containing oxygen, carbon, and fluorine; and sulfonic acid ion clusters. In the presence of water, the polar sulfonic sites dissociate and produce fixed anionic groups and mobile H^+ ions [see Figure 8.20(a)]. These protons are able to migrate under an external electric field. Upon application of such field, the cations in the bulk of the polymer migrate towards the cathode, carrying along water molecules. The large volume of water molecules causes local matrix expansion and deformation of the composite, as shown in Figure 8.20(b).

The molar fluxes have a component opposite to the pressure gradient, as well as the usual electrodiffusion law, generalized with a strain-induced diffusion term. The

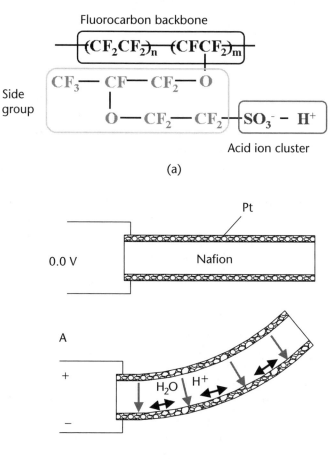

(a)

(b)

Figure 8.20 (a) Structure of Nafion; and (b) IPMC actuator.

molar fluxes in the membrane are predicted from the Onsager theory as follows [99]:

$$j_b^p = -\frac{\sigma}{F}\nabla\phi - n_d^{pw}\left[D^w\nabla c^w - \tilde{K}c^w\nabla L_{kk}^{poly}\right] \tag{8.38}$$

$$j_b^w = -n_d^{wp}\frac{\sigma}{F}\nabla\phi - D^w\nabla c^w - \tilde{K}c^w\nabla L_{kk}^{poly} \tag{8.39}$$

where F is Faraday's constant, σ is the proton conductivity coefficient of the membrane, ϕ is electric potential, n_d is drag coefficient, D^w is the diffusion coefficient of water in the membrane, c is concentration, $\tilde{K}$ is the modified bulk modulus, and L_{kk}^{poly} is the elastic deformation of the polymer network. The upper indexes p and w denote protons and water molecules, respectively, and the lower index b denotes bulk. The molar fluxes at both electrodes are predicted from a chemical reaction using the overpotential theory as follows [99]:

$$j_a^p = \frac{i_a}{F} = \frac{k_+}{F}\left(c_a^w\right)^{1/2}\exp\left(\frac{\alpha F}{R\theta}(\phi_{ext}-\phi_a)\right), \quad j_a^w = \frac{j_a^w}{2} \tag{8.40}$$

$$j_c^p = \frac{i_c}{F} = \frac{k_-}{F}c_c^p\exp\left(-\frac{(1-\alpha)F}{R\theta}\phi_c\right), \quad j_c^w = 0 \tag{8.41}$$

where i is the current density, k_+ and k_- are the anodic and cathodic reaction coefficients, and α is the charge transfer barrier or symmetry coefficient. Using conservation of current and mass, all fluxes and concentrations of protons and water can be calculated. From integration of the resulting water concentration gradient, the final deformation shape is obtained using a small-strain, large-deformation solution [99].

Conducting polymers are another class of ionic polymers that undergo a reversible volume change, driven by redox processes accompanied by ion movement. The most commonly used polymer of this group is polypyrrole (PPy). When ion and/or solvent enter the polymer, it expands; and when they exit, it contracts. Ions can enter the polymer in either the oxidized or the reduced states:

$$P^+(A^-) + e^- \leftrightarrow P^0 + A^- \tag{8.42}$$

$$P^+(A^-) + C^+ + e^- \leftrightarrow P^0(AC) \tag{8.43}$$

where P^+ represents the oxidized (doped) state of the polymer and P^0 the reduced (undoped, neutral) state. The process in the first reaction is typical in polymers prepared with a small, mobile anion, and the polymer expands in the oxidized state due to the uptake of the anion A^-. In the polymers prepared with a large, immobile anion (the second reaction), the polymer expands in a reduced state due to the uptake of the mobile counter ion C^+ [100]. The actuators based on these polymers consist of a conducting polymer with the support of an electrode layer, forming a bilayer. If the volume of polymer is changed while the volume of the other layer is unchanged, the actuator will bend towards one side.

IPMC actuators can be produced by an electrochemical platinization method based on the ion-exchange properties of the polymer [101]. The method consists of two main steps: step 1, ion exchange of the protons H^+ with metal cations (Pt^{2+}); and step 2, chemical reduction of the metal ions onto the membrane surface with $NaBH_4$ (see Figure 8.21). In Figure 8.21, an SEM microphotograph of such a composite, clearly shows the white Pt particles at the boundary, where the electrode surfaces are approximately 0.8-μm-thick Pt deposits. A more efficient actuator can be produced by repeating these two steps several times, which results in a dendritic growth of the Pt electrodes into the polymer, producing a larger surface area [102].

For fabrication of the conducting polymer actuators, the standard surface micromachining technology is used. Thin Cr and Au are deposited on Si wafer surface by evaporation. The polymer is electrochemically grown on the gold surface from a 0.1M pyrrole, 0.1M NaDBS (dodecyl benzene sulfonic acid, sodium salt) solution at a voltage of 0.5V versus Ag/AgCl with a gold counter electrode [103]. Then, the actuators are formed by lithograph and etching.

Figure 8.22 shows the peak moment and efficiency of a fully hydrated actuator [104]. While the peak moment is proportional to the external applied voltage, the efficiency reaches the steady value after 5V. The predicted efficiency is much smaller in comparison with the reported 40% characteristic for natural muscles [105].

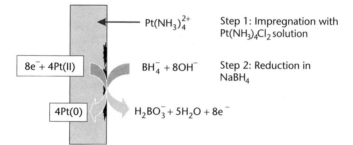

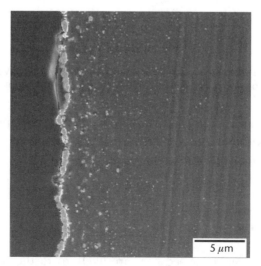

Figure 8.21 Takenaka method for platinization.

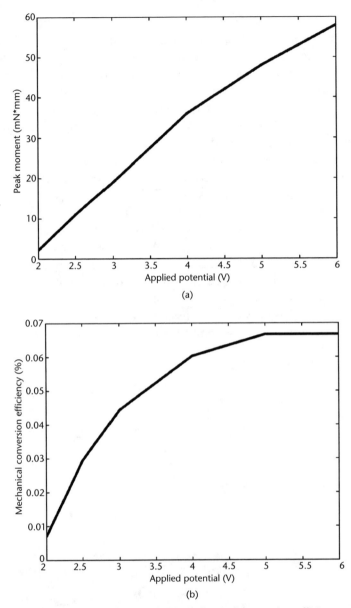

Figure 8.22 (a) Predicted peak moment and (b) mechanical conversion efficiency of IPMC actuator. (*From:* [104]. © 2006 ACS. Reprinted with permission.)

Therefore, multiple efforts have been undertaken to improve their efficiency: incorporation of counterions with larger hydration shells such as Li^+ and Na^+, increasing the surface areas of the anode and the cathode, and reducing surface resistance by multiple plating steps [106–108].

8.4 Future Trends

The microdevices described here are likely to further benefit from the continued miniaturization into the submicrometer domain. The main technological obstacle to be overcome stems from the poor energy coupling efficiency of classical actuation modalities at the nanometer scale. Therefore, the new devices are likely to utilize short wavelength well into the deep UV part of the spectrum, or they would be based on an agglomeration of individual agents operating collectively into a larger system. An example of such concept has already been demonstrated in the area of controlled release of drugs where nanometer magnetic particles are incorporated into a spherical shell carrying the active ingredient [109]. Upon stimulation with an external magnetic field, the particles in the shell break away, providing a diffusion path for the release of the drug. Similar concepts could be envisioned for triggering electrochemical actuators performing some useful function in the body. A description of a magnetically steered microrobot that could perform such function can be found in [110].

Acknowledgments

The authors acknowledge the financial support of the National Science Foundation under Grant Nos. DMI-0134585, DUE-0633312, and CBET-0603198. Further acknowledged is the support of the University of Arizona under grants from the Bio5 Institute and from the University of Arizona Foundation.

References

[1]   Gopel, W., J. Hesse, and J. N. Zemel, "Sensors: A Comprehensive Review," *Sensors*, Vol. 2, Weinheim: VCH Verlagsgesellschaft, 1991, pp. 1–16.

[2]   Nathanson, H., and R. A. Wickstrom, "A Resonant Gate Transistor with High-Q Bandpass Properties," *Applied Physics Letters*, Vol. 7, No. 4, August 15, 1965, pp. 84–86.

[3]   Oggenfuss, P., et al., "Neutral-Carrier-Based Ion-Selective Electrodes," *Anal. Chim. Acta*, Vol. 180, 1986, pp. 299–311.

[4]   Frant, M. S., "Historical Perspective: History of the Early Commercialization of Ion-Selective Electrodes," *Analyst*, Vol. 119, 1994, pp. 22–93.

[5]   Kurtz, A. D., and S. J. Goodman, "Apparatus and Method for Interconnecting Leads in a High Temperature Pressure Transducer," U.S. Patent No. 3,800,264, 1974.

[6]   Clark, Jr., L. C., and C. Lyons, "Electrode System for Continuous Monitoring in Cardiovascular Surgery," *Annals of the New York Academy of Sciences*, Vol. 148, 1962, pp. 133–153.

[7]   King, W. H., "Piezoelectric Sorption Detector," *Anal. Chem.*, 1964, Vol. 36, pp. 1735–1739.

[8]   Roylance, L. M., "Miniature Integrated Circuit Accelerometer for Biomedical Applications," Ph.D. dissertation, Stanford Univ., Stanford, CA, 1978.

[9]   Petersen, K. E., "Fabriaction of an Integrated Planar Silicon Ink-Jet Structure," *IEEE Transactions on Electron Devices*, Vol. ED-26, 1979, pp. 1918–1920.

[10]   Howe, R. T., "Resonant Microsensors," *Technical Digest Transducers '87, 4th Int. Conf. on Solid-State Sensors and Actuators*, Tokyo, Japan, 1980, pp. 834–849.

[11]   Howe, R., and U. Muller, "Resonant Microbridge Vapor Sensor," *IEEE Trans. on Electron Devices*, ED-33, 1986, pp. 499–506.

[12]   Millar Instruments, "Mikro-Tip® Pressure Transducer Catheters," http://www.millarinstruments.com/products/cardio/cardio_sngldual.html.

[13] Millar Instruments, "Multi-Segment Pressure-Volume Mikro-Tip® Catheters," http:// www.millarinstruments.com/products/cardio/cardio_pv.html.

[14] Bao, M.-H., "Micromechanical Transducers: Pressure Sensors, Accelerometers, and Gyro-scopes," in *Handbook of Sensors and Actuators, Vol. 8,* S. Middeelhoek (ed.), New York: Elsevier, 2000.

[15] Motchenbacher, C. D., and J. A. Connelly, *Low-Noise Electronic System Design,* New York: John Wiley & Sons, 1993.

[16] Wu, X., et al., "A Miniature Piezoresistive Catheter Pressure Sensor," *Sensors and Actuators A,* Vol. 35, 1993, pp. 197–201.

[17] Moorhead, L. C., et al., "Dynamic Intraocular Pressure Measurements During Vitrectomy," *Archive of Ophthalmology,* Vol. 123, No. 11, 2005, pp. 1514–1523.

[18] Baan, J., et al., "Continuous Measurement of Left Ventricular Volume in Animals and Humans by Conductance Catheter," *Circulation,* Vol. 70, 1984, p. 812.

[19] Jain, R. K., "Barriers to Drug Delivery in Solid Tumors," *Sci. Am.,* Vol. 271, 1994, pp. 58–65.

[20] Jain, R. K., "The Next Frontier of Molecular Medicine: Delivery of Therapeutics," *Nat. Med.,* Vol. 4, 1998, pp. 655–657.

[21] Netti, P. A., et al., "Enhancement of Fluid Filtration Across Tumor Vessels: Implication for Delivery of Macromolecules," *Proc. Natl. Acad. Sci. USA,* Vol. 96, 1999, pp. 3137–3142.

[22] Dewhirst, M. W., et al., "Microvascular Studies on the Origins of Perfusion—Limited Hypoxia," *Br. J. Cancer Suppl.,* Vol. 27, 1996, pp. 247–251.

[23] Gillies, R. J., et al., "Causes and Effects of Heterogeneous Perfusion in Tumors," *Neoplasia,* Vol. 1, 1999, pp. 197–207.

[24] Raghunand, N., et al., "Enhancement of Chemotherapy by Manipulation of Tumour pH," *Br. J. Cancer,* Vol. 80, 1999, pp. 1005–1011.

[25] Yuan, J., and P. M. Glazer, "Mutagenesis Induced by the Tumor Microenvironment," *Mutat. Res.,* Vol. 400, 1998, pp. 439–446.

[26] Yuan, J., et al., "Diminished DNA Repair and Elevated Mutagenesis in Mammalian Cells Exposed to Hypoxia and Low pH," *Cancer Res* Vol. 60, 2000, pp. 4372–4376.

[27] Martinez-Zaguilan, R., et al., "Acidic pH Enhances the Invasive Behavior of Human Mela-noma Cells," *Clin. Ex. Metastasis,* Vol. 14, 1996, pp. 176–186.

[28] Kovacs, G. T. A., *Micromachined Transducers Sourcebook,* New York: McGraw-Hill, 1998.

[29] Hendrikse, J., W. Olthuis, and P. Bergveld, "A Drift Free Nernstian Iridium Oxide pH Sen-sor," *Proceedings of Transducers '97, the 1997 International Conference on Solid-State Sensors and Actuators,* Vol. 2, Chicago, IL, June 16–19, 1997, pp. 1367–1370.

[30] Severinghaus, J. W., and A. F. Bradley, "Electrodes for Blood PO_2 and PCO_2 Determina-tion," *J. Appl. Physics.,* Vol. 13, 1958, pp. 515–520.

[31] Buehler, H. W., and R. Bucher, "Applications of Electrochemical Sensors," *Sensors in Bioprocess Control,* J. V. Twork and A. M. Yacynych, (eds.), New York: Marcel Dekker, 1990.

[32] Janata, J., "Chemical Sensitivity of Fluid-Effect Transistors," *Sensors and Actuators,* Vol. 12, No. 2, August/September 1987, pp. 121–128.

[33] Davies, P. W., "The Oxygen Cathode," in *Physical Techniques in Biological Research,* Vol. IV, W. L. Nastuk, (ed.), New York: Academic, 1962, Chapter 3.

[34] Karube, I., and H. Suzuki, "Miniaturized Oxygen Electrode and Miniaturized Biosensor and Production Process Thereof," U.S. Patent No. 497517, 1990.

[35] Warburg, E., "Uber das Verhalten sogenannter unpolarisierbarer Electroden Gegen Wechselstrom," *Ann. Phys.,* Vol. 67, 1899, pp. 493–499.

[36] Cole, K. S., and R. H. Cole, "Dispersion and Absorption in Dielectrics. I. Alternating Cur-rent Characteristics," *J. Chem. Phys.,* Vol. 9, 1941, pp. 341–351.

[37] Bard, A. J., and L. R. Faulkner, *Electrochemical Methods: Fundamentals and Applications*, 2nd ed., New York: John Wiley & Sons, 2000.

[38] Seo, K., S. A. Jelinsky, and E. L. Loechler, "Factors That Influence the Mutagenic Patterns of DNA Adducts from Chemical Carcinogens," *Mutat. Res-Rev. Mutat.*, Vol. 463, 2000, p. 215.

[39] Toniolo, P., and E. Taioli, "Development of Biomarkers of Human Exposure to Carcinogens: The Example of DNA-Protein Cross-Links," *Toxicol. Lett.*, Vol. 77, 1995, p. 231.

[40] Zhang, M., et al., "Rapid Detection of Hepatitis B Virus Mutations Using Real-Time PCR and Melting Curve Analysis," *Hepatology*, Vol. 36, 2002, pp. 723–728.

[41] Wang, J., G. Rivas, and X. Cai, "Screen-Printed Electrochemical Hybridization Biosensor for the Detection of DNA Sequences from the Escherichia Coli Pathogen," *Electroanalysis*, Vol. 9, 1997, p. 395.

[42] Patolsky, F., A. Lichtenstein, and I. Willner, "Detection of Single-Base DNA Mutations by Enzyme-Amplified Electronic Transduction," *Nature Biotechnology*, Vol. 19, 2001, pp. 253–257.

[43] Bardea, A., et al., "Sensing and Amplification of Oligonecleotide-DNA Interactions by Means of Impedance Spectroscopy: A Route to a Tay-Sachs Sensor," *Chemical Communications*, Vol. 21, 1999, pp. 21–22.

[44] Tang, D., et al., "A Novel Immunosensor Based on Immobilization of Hepatitis B Surface Antibody on Platinum Electrode Modified Colloidal Gold and Polyvinyl Butryal as Matrices Via Electrochemical Impedance Spectroscopy," *Bioelectrochemistry*, Vol. 65, 2004, pp. 15–22.

[45] Isaac, O., et al., "Review. Impedance Spectroscopy: A Powerful Tool for Rapid Biomolecular Screening and Cell Culture Monitoring," *Electroanalysis*, Vol. 17, No. 23, 2005, pp. 2101–2113.

[46] Fritz, J., et al., "Translating Biomolecular Recognition into Nanomechanics," *Science*, Vol. 288, 2000, pp. 316–318.

[47] Shea, S. J., et al., "Atomic Force Microscopy Stress Sensors for Studies in Liquids," *J. Vac. Sci. Technol. B*, Vol. 14, No. 2, 1996, pp. 1383–1385.

[48] Raiteri, R., et al., "A Novel Transducer For Direct Sensing of Bioaffinity Interactions Based on Microfabricated Cantilevers," *J. Phys. Chem.*, Vol. 99, 1995, pp. 15728–15732.

[49] Godin, M., et al., "Surface Stress, Kinetics, and Structure of Alkanethiol Self-Assembled Monolayers," *Langmuir*, Vol. 20, 2004, pp. 7090–7096.

[50] Moulin, A. M., S. J. O'Shea, and M. E. Welland, "Microcantilever-Based Biosensors," *Ultramicroscopy*, Vol. 82, 2000, pp. 23–31.

[51] Raiteri, R., et al., "Micromechanical Cantilever-Based Biosensors," *Sens. Actuators. B*, Vol. 79, 2001, pp. 115–126.

[52] Baller, M. K., et al., "A Cantilever Array-Based Artificial Nose," *Ultramicroscopy*, Vol. 82, 2000, pp. 1–9.

[53] Yin, T. I., and S. M. Yang, "Double-Microcantilever Design for Surface Stress Measurement in Biochemical Sensors," *Proceedings of the 2005 IEEE/ASME, International Conference on Advanced Intelligent Mechatronics*, Monterey, CA, July 24–28, 2005.

[54] Timoshenko, S. P., and J. M. Gere, *Mechanics of Materials*, New York: Van Nostrand Reinhold, , 1972.

[55] Sarid, D., *Scanning Force Microscopy*, Oxford, U.K.: Oxford University Press, 1991.

[56] Stoney, G. G., "A Chemical Sensor Based on a Micromechanical Cantilever Array for the Identification of Gases," *Proc. R. Soc. London, Ser. A*, Vol. 82, 1909, p. 172.

[57] Berger, R., et al., "Surface Stress in the Self-Assembly of Alkanethiols on Gold Probed by a Force Microscopy Technique," *Appl. Phys. A*, Vol. 66, 1998, pp. S55–S59.

[58] Prigge, E. A., and J. P. How, "Signal Architecture for a Distributed Magnetic Local Positioning System," *IEEE Sensors Journal*, Vol. 4, No. 6, 2004, pp. 864–873.

[59] Gilbertson, R. G., and J. D. Busch, "A Survey of Micro-Actuator Technologies for Future Spacecraft Missions," *The Journal of The British Interplanetary Society*, Vol. 49, 1996, pp. 129–138.

[60] Busch-Vishniac, I. J., "The Case for Magnetically Driven Microactuators," *Sensors and Actuators A*, Vol. 33, 1992, pp. 207–220.

[61] Jeong, M. C., and L. Busch-Vishniac, "A Submicron Accuracy Magnetic Levitation Micromachine with Endpoint Detection," *Sensors and Actuators*, Vol. A29, 1991, pp. 225–234.

[62] Benali-Khoudja, M., et al., "Electromagnetically Driven High-Density Tactile Interface Based on a Multi-Layer Approach," *DSV IS*, June 4–6, 2003, pp. 147–152.

[63] Chowdhury, S., et al., "A Modular MEMS Electromagnetic Actuator for Use in a Hearing Instrument," *Proc. 43rd IEEE Midwest Symp. on Circuits and Systems*, Lansing, MI, August 8–11, 2000, pp. 240–243.

[64] Bohm, S., et al., "A Micromachined Silicon Valve Driven by a Miniature Bi-Stable Electromagnetic Actuator," *Sensors and Actuators* Vol. 80, 2000, pp. 77–83.

[65] Rinderknecht, D., A. I. Hickerson and M. Gharib, "A Valveless Micro Impedance Pump Driven by Electromagnetic Actuation," *Journal of Micromechanics and Microengineering* Vol. 15, 2005, pp. 861–866.

[66] Kolin, A., "A New Principle for Electromagnetic Catheter Flow Meters," *Proc. Natl. Acad. Sci. USA*, Vol. 63, No. 2, June 1969, pp. 357–363.

[67] Nasar, S. A., and I. Boldea, *Linear Electric Motors: Theory, Design, and Practical Applications*, Upper Saddle River, NJ: Prentice-Hall, 1987.

[68] Prigge, E. A., and J. P. How, "Signal Architecture for a Distributed Magnetic Local Positioning System," *IEEE Sensors Journal*, Vol. 4, No. 6, 2004, pp. 864–873.

[69] Inan, U. S., and A. S. Inan, *Electromagnetic Fundamentals and Applications*, Reading, MA: Addison-Wesley, 1997.

[70] Boctor, E., et al., "PC-Based System for Calibration, Reconstruction, Processing and Visualization of 3D Ultrasound Data Based on a Magnetic-Field Position and Orientation Sensing System," *ICCS 2001: International Conference in Computational Science*, Vol. 2074, San Francisco, CA, 2001, pp. II.13–II.22.

[71] Huang, Q. H., et al., "Development of a Portable 3D Ultrasound Imaging System for Musculoskeletal Tissues," *Ultrasonics*, Vol. 43, No. 3, 2005, pp. 153–163.

[72] Baltzer, M., T. Kraus, and E. Obermeier, "A Linear Stepping Actuator in Surface Micromachining Technology for Low Voltages and Large Displacements," *Transducers '97: Ninth International Conference on Solid-State Sensors and Actuators*, 1997, pp. 781–784.

[73] Damjanovic, D., and R. Newnham, "Electrostrictive and Piezoelectric Materials for Actuator Applications," *J. Intelligent Material Systems and Structures*, Vol. 3, No. 2, 1992, p. 190.

[74] Tilmans, H. A. C., et al., "A Fully Packaged Electromagnetic Microrelay," *Proceedings of IEEE Micro Electro Mechanical Systems*, Orlando, FL, 1999, pp. 25–30.

[75] Butler, J., V. Bright, and W. Cowan, "Average Power Control and Positioning of Polysilicon Thermal Actuators," *Sensors and Actuators*, Vol. 72, No. 2, 1999, pp. 88–97.

[76] Enikov, E., and K. Lazarov, "PCB-Integrated Metallic Thermal Microactuators," *Sensors and Actuators A: Physical*, Vol. 105, No. 1, 2003, pp. 76–82.

[77] Jeong, O., and S. Yang, "Fabrication of a Thermopneumatic Microactuator with a Corrugated p+ Silicon Diaphragm," *Sensors and Actuators A: Physical*, Vol. 80, No. 1, 2000, pp. 62–67.

[78] Neagu, C., et al., "An Electrochemical Active Valve," *Electrochimica Acta*, Vol. 42, No. 20, 1997, pp. 3367–3373.

[79] Quandt, E., and A. Ludwig, "Magnetostrictive Actuation in Microsystems," *Sensors and Actuators A: Physical*, Vol. 81, Nos. 1–3, 2000, pp. 275–280.

[80] Ray, C., et al., "A Silicon-Based Shape Memory Alloy Microvalve," *Smart Materials Fabrication and Materials for Micro-Electro-Mechanical Systems*, Vol. 276, San Francisco, CA, 1992, pp. 161–166.

[81] Enikov, E. T., and G. S. Seo, "Large Deformation Model of Ion-Exchange Actuators Using Electrochemical Potentials," *Electroactive Polymer Actuators and Devices (EAPAD), SPIE 9th Annual International Symposium on Smart Structures and Materials*, Vol. 4695, San Diego, CA, 2002, pp. 199–209.

[82] Yesin, K. B., K. Vollmers, and B. J. Nelson, "Modeling and Control of Untethered Biomicrorobots in a Fluidic Environment Using Electromagnetic Fields," *The International Journal of Robotics Research*, Vol. 25, Nos. 5–6, 2006, pp. 527–536.

[83] Chu, L. L., and Y. B. Gianchandani, "A Micromachined 2D Positioner with Electro-Thermal Actuation and Sub-Nanometer Capacitive Sensing," *J. of Micromechanics and Microengineering*, Vol. 13, 2003, pp. 279–285.

[84] E. T. Enikov, S. S. Kedar, and K. V. Lazarov, "Analytical Model for Analysis and Design of V-Shaped Thermal Microactuators," *Journal of Microelectromechanical Systems*, Vol. 14(4), 2005, pp. 788–798.

[85] Lazarov, K. V., and E. T. Enikov, "Design of Electro-Thermal Micro-Actuators: Mechanics and Electronics Position Detection," in *Microsystems Mechanical Design*, F. DeBona and E. T. Enikov, (eds.), New York: Springer, 2006.

[86] Enikov, E. T., "Electrodeposited Micro-Actuators: A Simple Tool for Impedance-Based Sensing," *ECS Transactions*, Vol. 3, No. 10, 2006, pp. 339–350.

[87] Torisawa, Y., et al., "Scanning Electrochemical Microscopy-Based Drug Sensitivity Test for a Cell Culture Integrated in Silicon Microstructures," *Anal. Chem.*, Vol. 75, 2003, pp. 2154–2158.

[88] Choi, S. H., and M. B. Gu, "A Portable Toxicity Biosensor Using Freeze-Dried Recombinant Bioluminescent Bacteria," *Biosensors & Bioelectronics*, Vol. 17, 2002, pp. 433–440.

[89] Nagamine, K., et al., "Application of Microbial Chip for Amperometric Detection of Metabolic Alteration in Bacteria," *Sensors and Actuators, B*, Vol. 108, 2005, pp. 676–682.

[90] Ito, Y., et al., "Escherichia Coli and Its Application in a Mediated Amperometric Glucose Sensor," *Biosensors and Bioelectronics*, Vol. 17, 2002, pp. 993–998.

[91] Abdelghani, A., et al., "Cell-Based Biosensors for Inflammatory Agents Detection," *Materials Science and Engineering C*, Vol. 22, 2002, pp. 67–72.

[92] Prodan, C., et al., "Low-Frequency, Low-Field Dielectric Spectroscopy of Living Cell Suspensions," *J. App. Phys.*, Vol. 95, No. 7, 2004, pp. 3754–3756.

[93] Yamakita, M., et al., "Development of an Artificial Muscle Linear Actuator Using Ionic Polymer-Metal Composites," *Advanced Robotics*, Vol. 18, No. 4, 2004, pp. 383–399.

[94] Bar-Cohen, Y., *Electroactive Polymer [EAP] Actuators as Artificial Muscles*, Bellingham, WA: SPIE Press, 2001.

[95] Shahinpoor, M., et al., "Ionic Polymer-Metal Composites (IPMC-S) as Biomimmetic Sensors, Actuators and Artificial Muscles—A Review," *Smart Mater. Struct.*, Vol. 7, 1998, pp. R15–R30.

[96] Onishi, K., et al., "Biomimetic Micro Actuators Based on Polymer Electrolyte/Gold Composite Driven by Low Voltage," *MEMS 2000: The Thirteenth Annual International Conference*, January 23–27, 2000, pp. 386–390.

[97] Shahinpoor, M., and K. J. Kim, "Ionic Polymer–Metal Composites: IV. Industrial and Medical Applications," *Smart Mater. Struct.*, Vol. 14, 2005, pp. 197–214.

[98] Bar-Cohen, Y., "Transition of EAP Material from Novelty to Practical Applications: Are We There Yet?" *Proceedings of EAPAD, SPIE's 8th Annual International Symposium on Smart Structures and Materials*, Newport, CA, March 5–8, 2001, Paper No. 4329–02.

[99] Enikov, E. T., and G. S. Seo, "Analysis of Water and Proton Fluxes in Ion-Exchange Polymer-Metal Composite (IPMC) Actuators Subjected to Large External Potentials," *Sensors and Actuators A*, 122, 2005, pp. 264–272.

[100] Smela, E., "Conjugated Polymer Actuators for Biomedical Applications," *Adv. Mater.*, Vol. 15, No. 6, 2003, pp. 481–494.

[101] Liu, R., W.-H. Her, and P. S. Fedkiw, "In Situ Electrode Formation on a Nafion Membrane by Chemical Platinization," *Journal of the Electrochemical Society,* Vol. 139, No. 1, 1990, pp. 15–23.

[102] Oguro, K., et al., "Polymer Electrolyte Actuator with Gold Electrodes," *Proceedings of SPIE,* Vol. 3669, 1999, pp. 64–71.

[103] Jager, E. W. H., et al., "Polypyrrole Microactuators," *Synthetic Metals,* Vol. 102, 1999, pp. 1309–1310.

[104] Enikov, E. T., and G. S. Seo, "Numerical Analysis of Muscle-Like Ionic Polymer Actuators," *Biotechnol. Prog.,* Vol. 22, 2006, pp. 96–105.

[105] Full, R., and K. Meijer, "Metrics of Natural Muscle," in *Electroactive Polymers (EAP) as Artificial Muscles: Reality Potential and Challenges*, Y. Bar-Cohen, (ed.), Bellingham, WA: SPIE Press, 2001, pp. 67–83.

[106] Abe, Y., et al., "Effect on Bending Behavior of Countercation Species in Perfluorinated Sulfonate Membrane-Platinum Composite," *Polymers for Advanced Technologies*, Vol. 9, No. 8, 1998, pp. 520–526.

[107] Onishi, K., et al., "Morphology of Electrodes and Bending Response of the Polymer Electrolyte Actuator," *Electrochimica Acta*, Vol. 46, 2000, pp. 737–743.

[108] Shahinpoor, M., and K. Kim, "The Effect of Surface-Electrode Resistance on the Performance of Ionic Polymer-Metal Composite (IPMC) Artificial Muscles," *Smart Materials and Structures*, Vol. 9, No. 4, 2000, pp. 543–551.

[109] Lu, Z., et al., "Magnetic Switch of Permeability for Polyelectrolyte Microcapsules Embedded with Co@Au Nanoparticles," *Langmuir,* Vol. 21, 2005, pp. 2042–2050.

[110] Yesin, K., et al., "Design and Control of In-Vivo Magnetic Microrobots," *Medical Image Computing and Computer-Assisted Intervention,* MICCAI2005, Pt. 1/Lecture Notes in Computer Science, Vol. 3749, 2005, pp. 819–826.

Dynamics Modeling and Analysis of a Swimming Microrobot for Drug Delivery

Huaming Li, Jindong Tan, and Mingjun Zhang

Dynamics modeling and analysis of a tiny swimming robot are presented in this chapter [1–27]. The microrobot is designed for controlled drug delivery as well as various life science automation applications. It is at the micrometer scale and is suitable for a swimming environment under low Reynolds number (Re). This chapter focuses on analyzing the dynamics of the microrobot, since this is crucial for further developing advanced controller and detailed implementation of the system. Resistive force theory is used to model and analyze the dynamics of the microrobot under two different situations when the elastic tail is with two different statuses. Finally, the propulsion performance of the proposed microrobot is compared with other rigid-body microrobots.

9.1 Introduction

Inventing new drugs and precisely delivering drugs to target tissues or organs for treatment are equally important for disease treatment. Drug delivery has drawn many researchers' attentions in recent years in the medical and engineering communities. Drug delivery is one of the fastest growing health sectors. Sales of the drugs that incorporate drug delivery systems are increasing at an annual rate of 15%. By 2003, the U.S. drug delivery market alone will be worth $24 billion. Controlled drug delivery aims to deliver the drug to the desired site directly, without affecting other tissues or organs, which helps to keep side effects to a minimum [17].

Approaches for drug delivery include traditional approaches as well as many new methods that use advanced materials and/or new physical or chemical effects. Controlled drug delivery is a definite trend for drug delivery. The goal of controlled drug delivery systems is to directly deliver drugs to tumor sites, while controlling the delivery speed and the drug's release, and thus reducing harm to other tissues. Controlled drug delivery with good targeting, precise amount, and exact speed of release are critical for drug effects. This requires considerations of not only side effects of the drug, but also the stability of the drug during the delivery process. Two primary features of controlled drug delivery systems are:

- *Controlled drug release, including delayed release or extended release:* In an ideal case, drugs are gradually released from a depot so that the drug concentration is maintained at an effective level over a long period.

- *Localized drug targeting:* The objective is to achieve a desired pharmacological response at a selected site without undesirable interactions at other sites. This is especially important in cancer chemotherapy and enzyme replacement treatment.

9.1.1 Routes of Administration for Drug Delivery

Traditionally, after a drug is liberated from its dosage form, it must pass through several barriers before it arrives at the site of action. All these barriers consist of membranes. From absorption to elimination, every phase of a drug's change is associated, directly or indirectly, with its passage through cell and basement membranes. These membranes are interfaces and, in a biological sense, form barriers between morphological and functional entities. The main force responsible for transmembrane movement of most active substances is their concentration gradient across the membrane. The main motion of dissolved molecules produces a net movement of molecules from a region of high concentration to one of low concentration. This process of transmembrane movement of molecules is called *diffusion*. The diffusion rate depends essentially on the size of the concentration gradient across the membrane, on the nature of the membrane surface through which the dissolved molecules must pass, and on the thickness of the membrane. The following are some possible routes of administration for drug delivery:

- *Oral delivery:* Oral delivery is one of the most common drug delivery approaches and also the easiest one. The drawback of oral drug delivery is its relatively poor bioavailability. Some better delivery could be obtained by lessening degradation in the stomach and small intestine and by increasing drug absorption in the upper gastrointestinal track.

- *Injection delivery:* Injection drug delivery has a number of drawbacks. Injected vaccine is not effective at eliciting mucosal immunity, and many purified, synthetic, or inactivated antigens are poorly immunogenic.

- *Nasal delivery:* Nasal delivery offers a faster rate of absorption and provides the rapid onset of action that is very important for pain relief, nausea, sleep induction, and other diseases of the central nervous system.

- *Parenteral delivery:* Parenteral delivery is a primary method for the administration of peptide/protein-based pharmaceutical products.

- *Transdermal delivery:* Skin can be used for systematic drug delivery via a transdermal patch, which is called transdermal delivery. The main barrier for transdermal delivery is the stratum corneum. A light gas gun can be used to accelerate drugs in microparticle form to a velocity at which they will enter the skin painless and needle-free. However, stability of drug particles may be affected by high-speed accelerations, which may cause a loss of the drug's effects.

- *Ocular delivery:* An example of ocular drug delivery is ocular tumor treatments. Ocular tumor treatments face problems of getting the drug through the blood vessel wall and into the tumor, and getting the drug to penetrate deep into the tumor.
- *Implantable delivery:* Implantable drug delivery is used for long-term, continuous drug administration. The purpose is to deliver drugs directly into the bloodstream at a controlled rate of transmission. Currently available implantable systems include Norplant and various pumps, such as insulin pumps.

Some other possible routes of drug administration include rectal, pulmonary, vaginal, and intrauterine. Most of them are still in development stage and lack successful commercial applications.

There exist potential problems with all the major routes of drug administration: inhalation, injection, and transdermal delivery, including the most well-known route, oral drug delivery, which accounted for about 50% of the market as of 2003. Some of their disadvantages include:

- *Oral delivery:* Drug degradation in the stomach;
- *Nasal delivery:* Gas and powder only;
- *Trasdermal delivery:* Stratum corneum barrier;
- *Injection delivery:* Hard to operate;
- *Ocular delivery:* Blood vessel wall.

Other limitations include drug effective area, applicable drug types, drug releasing speed, and poor bio-availability.

9.1.2 Controlled Drug Delivery

Controlled drug delivery is the phasing of drug administration to the needs of a particular condition so that an optimal amount of drug is used to cure or control the condition in a minimum amount of time. The area of controlled drug delivery is increasingly being based on molecular biology. Controlled drug delivery systems can be classified as diffusion controlled systems (polymer/membrane/lipsomes controlled drug release), chemically controlled systems, solvent activated systems, and self and externally controlled systems using magnetically or ultrasonically triggered systems.

Biology-driven drug development will continue to create large molecule pharmaceuticals, such as recombinant proteins, monoclonal antibodies, antisense drugs, and gene medicines. As a result, drug release becomes complex. One way of achieving complex drug release is pulsatile release from polymeric materials in response to specific stimuli, such as electric or magnetic fields, exposure to ultrasound, light or enzymes, and changes in pH or temperature. Santini et al. [26] reported a solid-state silicon microchip that can provide controlled release of single or multiple chemical substances on demand. The release mechanism is based on the electrochemical dissolution of thin anode membranes covering microreservoirs filled with chemicals in solid, liquid, or gel form. One nice thing about the chip is that controlled release from the microchip involves no moving parts. Release from a particular reservoir is initiated by applying an electric potential between the anode membrane covering that reservoir and a cathode. The device was fabricated by a sequential process using silicon wafers and microelectronic processing techniques including ultraviolet photolithography, chemical vapor deposition, electron beam evaporation, and reactive ion etching.

Examples of critical needs for controlled drug delivery are cancer, AIDS, and brain disease treatments. Controlled drug delivery can help parenteral, oral, and transdermal route drug delivery. For the conventional methods of drug administration, usually, drug delivery at a constant controlled rate is preferable. However, a better method may be in response to the physiological needs of the body. This kind of system has not been implemented yet for drug delivery, even though it is quite common for many engineering control systems.

Pharmacokinetics and pharmacodynamics are important areas to look at. Pharmacokinetics deals with changes in the concentration of substances in the organism. It is concerned primarily with the kinetics of absorption, distribution, metabolism, poisons, and metabolites produced by the body itself. Pharmacodynamics is concerned with the reactions of the biosystems to the drug. In contrast to pharmacokinetics, which is concerned with what happens to a drug in the organism after it has been released from the supply form, pharmacodynamics deals with the question of what happens to the organism under the influence of the drug.

Recent developments of microelectromechanical systems (MEMS) and nanoelectromechanical systems (NEMS) make it possible to fabricate miniature robots that can be implanted into human bodies to accomplish certain complex tasks, including controlled drug delivery and minimal invasive surgery. In [18], three-dimensional helical nanostructures including nanospirals and nanocoils are fabricated using hybrid nanorobotic approaches. In [8], a biomimetic swimming microrobot for destroying kidney stones in real human bodies is proposed. Because of their miniature size and biocompatible properties, microrobots are able to reach and function within regions that are unsuitable for traditional medical devices, which make them a good choice for controlled drug delivery carriers.

9.1.3 Swimming Microrobots

In order to safely inject a microrobot into a human body, the diameter of the microrobot needs to be 0.8 mm or smaller [24]. In this case, the Reynolds number (i.e., the ratio of inertial force to viscous force) is very low (less than one) [25]. The Reynolds number (Re) defined by the propulsive velocity, U, and the typical dimension of the organism L is UL/μ, where μ is the kinematic viscosity of the liquid environment. The Re of a flow is a dimensionless parameter that measures the relative significance of inertial forces to viscous forces. When Re is low, the viscous force involved in propulsion is dominant and the inertial force often becomes negligible.

In a low Reynolds number environment, traditional propulsive methods, such as fish-like actuators, scallops, and rigid oars, which depend on reciprocal motions, cannot work efficiently or cannot function at all due to their underlying inertial forces propulsion mechanism. In [6], the effect of Reynolds number on propulsive performance of a fish-like wiggling micromachine is numerically analyzed. When Re is between 10 and 100, the decrement of Re reduces the propulsion efficiency and increases the power consumption. When Re is less than one, the micromachine loses thrust force completely due to the increased friction and pressure around the micromachine.

Inspired by the biology, several swimming microrobots were proposed recently. In [4], the authors present a fish-like microrobot which can swim with 3 degrees of freedom (DOF). It is 45 mm long, 10 mm wide, and 4 mm thick. A 3-mm swimming robot which is driven by Ferromagnetic polymer (FMP) actuators under magnetic field is developed in [5]. The size of the swimming microrobot is 3 mm × 2 mm × 0.4 mm. Since both of them use the propulsion mechanism relying on inertial force, they are not suitable for working under low Reynolds environments. In [6], numerical analysis for the effect of Re on propulsive performance of a submerged wiggling micromachine is given. Studies show that the decrease of Re deteriorates the propulsive performance when Re > 10. When Re = 1, the thrust force from wiggling motion is completely lost.

In order to accommodate to the special low Re environments, spiral-like microrobots that imitate bacteria have been proposed in several recent papers. A micromachine that can operate under a wide range of Re conditions is proposed by Ishiyama et al. in [7]. The micromachine contains a rigid cylindrical body and spiral blade. It is able to swim in a rotational magnetic field under different Re conditions from 430 to 6×10^{-7}. It demonstrates that the spiral-type design is a promising approach with good adaptability to different Reynolds number.

In [8], a surgical microrobot that swims inside the human ureter to perform kidney stone destruction is proposed. The microrobot uses multiwall carbon nanotubes which are driven into a helical shape to provide propulsion. Carbon nanotubes are grown vertically on a substrate which will provide the driving torque. While it gives the fluid mechanics modeling, the analysis is limited to the situation that the conformation of the helical shape is done.

Behkam and Sitti present a swimming microrobot inspired by *E. Coli* in [9, 10]. An analytical model is developed to predict the thrust force, required torque, velocity, and efficiency. The research shows that all the parameters are only related to geometrical characteristics. A scaled-up prototype for verifying the predicted values is also fabricated. The measured propulsive force of the prototype is in agreement with theoretical predictions.

In the aforementioned research, the swimming microrobots that can properly function in low Re environment have a rigid cylindrical body with spiral blade, a rigid spiral-like body propelling a payload such as a sphere-like head, or elastic tails which are grown from a rotating substrate and shaped into helix to generate propulsion.

The microrobot discussed in this chapter is composed of a helix-type head and an elastic tail. The head of the swimming robot is driven by an external rotating magnetic field, which enables it to be operated wirelessly. The spiral-type head accommodates communication and control units and serves as the base for the elastic tail. When a rotating magnetic field is applied, the head rotates synchronously with the field, generating and propagating driving torque to the straight elastic tail. When the driving torque reaches a threshold, dramatic deformation takes place on the elastic tail. The tail then transforms into a helix and generates propulsive thrust. The entire tail also serves as a drug reservoir.

9.1.4 Propulsion Under Low Re Number Environment

Bacteria live in low Reynolds environments in nature. In order to gain the mobility for better living conditions or food (chemotaxis), they often evolve to develop certain kinds of handedness which enables them to generate the propulsion needed for the mobility. Many microorganisms use flexible rotating flagella to provide propulsion, such as the bacterium *Escherichia Coli*, *Bacillus Megaterium* and *Trichomonas* [2]. Their flagellum is known as the smallest rotary propeller that can function effectively under low Reynolds environment. A bacterium usually consists of a basal body with a rotary motor, a universal joint allowing for the transmission of rotary motion, and the flagellum which convert the rotary motion into translational thrust.

Inspired by the spirochaeta and flagella of many bacteria in nature, a swimming microrobot suitable for swimming in low Re environments is proposed. The unique feature of this microrobot is the integration of a spiral-type head and an elastic flagellum-like tail. The spiral-type head of the microrobot can carry control and communication units. It also serves as the base and provides driving torque for the attached elastic tail. Film magnet could be deposited on the spiral-type wire and magnetized. When put in a rotating magnetic field, the head rotates synchronously with the external field. The rotation arouses different liquid pressure distribution around the microrobot and generates driving torque and propulsive thrust. The propulsive force is produced because the resistive force in the direction normal to the wire is larger than that in the tangential direction.

The elastic tail functions like the flagellum of bacteria. It transforms from a thin straight cylinder into helical geometry when enough driving torque is provided through the head and the dramatic deformation (bifurcation) is triggered. After the deformation, helical waves diffuse from the base of the head to the open end of the elastic tail. Hydrodynamic friction converts the rotational motion into thrust to the opposite direction of the propagating helical wave along the helix axis.

9.1.5 Features of the Proposed Swimming Microrobot

The swimming robot is designed to be in the micrometer scale so that it can swim in a human's body freely. The special geometric features make it suitable for swimming under low Re environment. The schematic views of the microrobot (before and after the bifurcation) are shown in Figure 9.1.

Compared with the rigid body microrobots, a microrobot with an elastic tail that can generate the same thrust has several advantages. First, the microrobot can swim more easily in the constrained areas of human bodies, due to the flexible dimensions. Second, the elastic tail design makes it possible for several tails to form a bundle and give larger thrust, which is very hard for rigid body microrobots. In nature, bacteria often rely on more than one flagellum to generate propulsion since one single flagellum may not provide enough thrust. Finally, the tail itself can be utilized as the drug reservoir in controlled drug delivery. The tail containing the drug can be released directly to the targeted area for best medical performance without

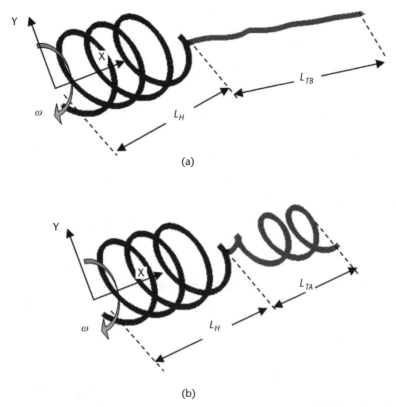

Figure 9.1   Schematic drawing of the spiral-type head and the elastic tail of the swimming microrobot: (a) before the bifurcation and (b) after the bifurcation.

affecting other tissue or organs. The rigid head part can then be recycled to accomplish other missions. During the drug delivery mission, the elastic tails makes the microrobot operate more efficiently because it provides helpful thrust as a propeller, rather than resistive force as a pure payload.

9.1.6   Implementation Issues

Due to the features of medical applications, wireless operation of the microrobots is always desired. For example, RF energy is used in [8]. In this design, an elastic tail is attached to the spiral-type head, whose wire is magnetized and enables the head to rotate synchronously with an external rotating magnetic field. In this way, the head can provide both the driving torque needed for the elastic tail and the translational thrust needed for propelling the microrobot. Furthermore, by changing the frequency and amplitude of the rotational magnetic field, the swimming velocity of the microrobot can be controlled. The swimming direction can also be controlled by changing the direction of the rotating magnetic field.

Fabrication of nanocoils and coil-shaped carbon nanotubes are proposed in [22, 23]. In [18], 3D helical nanostructures including nanocoils and nanospirals are fabricated using hybrid nanofabrication approaches based on nanorobotic manipulation. Processes that can effectively handle, structure, and assemble these nanostructures into NEMS are developed. Experiments also show that the radial stiffness of the helical structure can be controlled. Those controllable helical nanostructures could provide ideal building blocks for the microrobot proposed in this chapter.

9.2 Nomenclature

For the convenience of the following discussion, typical variables used in the rest of this chapter are listed:

L_H Length of the head part;

L_{TB} Length of the tail part before the bifurcation (contour length);

L_{TA} Length of the tail part after the bifurcation;

A_H Helical amplitude of the head part;

A_T Helical amplitude of the tail part after the bifurcation;

μ Viscosity;

p_H Pitch of the helical head;

p_T Pitch of the tail after the bifurcation;

a Cross-sectional radius of the elastic tail;

b Cross-sectional radius of the wire of the helical head;

d Radius of the payload in the head part;

V Translational velocity of the swimming microrobot;

ω Angular velocity of the swimming microrobot;

D_L Translational resistance coefficient of a sphere;

D_R Rotational resistance coefficient of a sphere;

E_L Translational resistance coefficient of the elastic tail (before the bifurcation);

E_R Rotational resistance coefficient of the elastic tail (before the bifurcation);

C_N RFT coefficient along the direction normal to the body center line;

C_S RFT coefficient along the direction tangential to the body center line.

9.3 Dynamics Modeling and Analysis

9.3.1 The Governing Equations

Under low Reynolds number environment, the flow of a fluid is called Stokes flow, where the viscous forces are much larger than the inertial forces [16]. The fluid in which the microrobot swims is assumed to be incompressible. The Navier-Stokes equations are the fundamental partial differentials equations that describe the flow of incompressible fluids. The equations describing Stokes flow are the Navier-Stokes equations with the inertial and body force terms equal to zero:

$$\mu\nabla^2 v = \nabla p, \nabla \cdot v = 0 \tag{9.1}$$

where p is the hydrostatic pressure, μ is the constant viscosity of the fluid, and v is the velocity vector.

Since any time-dependent terms which might account for any imbalance in the Navier-Stokes equation are negligible, the dynamics can be considered to degenerate into a blend of kinematics and statics. The fluid acts on the swimming microrobot with a system of forces that must effectively be in static equilibrium at all times [15]. Following the governing law, two sets of balance equations can be set up: one is force balance, the other is torque balance.

9.3.2 Tail Bifurcation

The dynamics modeling and analysis are divided into two parts depending on the states of the elastic tail: before bifurcation and after bifurcation. Bifurcation is known as remarkable deformation of rotating filaments. The studies of propulsive performance and viscous dynamics of rotating elastic filaments in [11, 12] reveal that when applying increasing driving torque to straight elastic polymers clamped on one end and open the other, the polymer filament exhibits a strongly discontinuous shape deformation at a finite torque value (N_C) and a finite rotation frequency (ω_C).

For a driving torque smaller than N_C or a rotation frequency smaller than ω_C, the filament almost remains straight and the generated propulsion is negligible. For a driving torque larger than N_C or a rotation frequency larger than ω_C, the propulsive thrust or flagella velocity shows significant improvement and increases proportionally with the driving torque or rotation frequency.

9.3.3 Before the Bifurcation

Before the deformation, the elastic tail remains almost straight. The propulsion generated by the elastic tail is negligible [11]. The tail can be regarded as a pure payload.

In order to rotate the entire system, the driving torque needs to overcome the viscous friction brought by the spiral-type head and the elastic tail. The resistive torque T_R is composed of two parts:

$$T_R = T_H + T_T \tag{9.2}$$

where T_H and T_T are resistive torques from the head and tail. The resistive torque from the head can be decomposed into two parts:

$$T_H = T_P + T_S \tag{9.3}$$

T_P and T_S are the resistive torques from the payload (the control or communication unit) and the spiral-type head, respectively.

From (9.3), we have the moment balance of the microrobot:

$$T_D + T_T + T_P + T_S = 0 \tag{9.4}$$

The driving torque T_D, acting on the head by the external magnetic field, can be represented by

$$T_D = mH \sin \beta \qquad (9.5)$$

where m is the magnetic moment of the magnetized head, H is the amplitude of the magnetic field, and β is the angle between m and H.

We model the payload in the spiral-type head as a sphere with a radius d. According to Stokes' law, the resistive torque introduced by the sphere is

$$T_P = D_R \omega \qquad (9.6)$$

where D_R is the rotational resistance coefficient of a sphere. According to Stokes' law,

$$D_R = 8\pi\mu d^3 \qquad (9.7)$$

Since the radius of the elastic tail is chosen to be much smaller than its length ($a \gg L_{Tb}$), it can be modeled as an elongated rod. Therefore, the axial rotational resistance torque of the tail can be represented as

$$T_T = E_R L_{Tb} \omega \qquad (9.8)$$

where E_R is the rotational resistance coefficient of an elongated rod,

$$E_R = 4\pi\mu a^2 \qquad (9.9)$$

Assume F_Y is the per-unit length viscous force along the y-direction brought by the spiral-type head. T_S can be represented by

$$T_S = A_H F_Y L_H \qquad (9.10)$$

Therefore, the moment balance equation (9.4) can be represented as

$$T_D + E_R L_{Tb} \omega + D_R \omega + A_H F_Y L_H = 0 \qquad (9.11)$$

Similarly, the force balance equation is set up:

$$F_T + F_P = F_S = 0 \qquad (9.12)$$

where F_T, F_P, F_S is the translational resistive force brought by the circular cylinder tail, the payload carried by the head, and the spiral-type head itself, respectively.

Before the bifurcation, the elastic tail is treated as an elongated rod. F_T can be obtained from

$$F_T = E_L V \qquad (9.13)$$

where E_L is the translational resistance coefficient of the elastic tail. According to [15],

$$E_L = \frac{2\pi\mu L_{TB}}{\ln(L_{TB}/2a) + \ln 2 - 1/2} \tag{9.14}$$

F_P is the drag force from the payload of the head; therefore, we have

$$F_p = D_L V \tag{9.15}$$

where D_L is the translational resistance coefficient of a sphere. According to Stokes' law,

$$D_L = 6\pi\mu d \tag{9.16}$$

Assume F_X is the per-unit length viscous force along the x-direction brought by the spiral-type head, then F_S can be given as

$$F_S = F_X L_H \tag{9.17}$$

Therefore, the translational force balance equation is

$$E_L V + D_L V + F_X L_H = 0 \tag{9.18}$$

Resistive force theory (RFT) [2] is adopted in this chapter to give an approximate solution for F_X and F_Y, provided that the radius of the microrobot body is extremely small compared with other relevant lengths, including its bending radius of curvature.

In RFT, the velocity relative to the fluid at infinity of each small segment on the spiral-type head is decomposed into normal and tangential components. Similarly, the force on each small segment is decomposed into components involving two force coefficients, C_N and C_S, which are tangential force and normal force coefficient, respectively. C_N and C_S are only related to the geometrical parameters of the analyzed object. Then the force on any segment of the spiral-type head can be determined by the motion of each small segment and C_N and C_S.

C_N and C_S can be calculated as follows [14]:

$$C_S = \frac{2\pi\mu}{\ln\left(\frac{2l^*}{b}\right) - \frac{1}{2}} = \frac{2\pi\mu}{\ln\left(\frac{2p_H}{b}\right) - 2.90}, \quad C_N = \frac{4\pi\mu}{\ln\left(\frac{2l^*}{b}\right) + \frac{1}{2}} = \frac{4\pi\mu}{\ln\left(\frac{2p_H}{b}\right) - 1.90} \tag{9.19}$$

where l^* is the so-called effective length, which equals $0.09\, p_H$.

The local velocity of each segment is decomposed into velocities along tangential and normal directions (see Figure 9.2). According to RFT, the relation between velocities and forces along tangential and normal direction is set up:

$$\begin{aligned} F_S &= C_S\,(\omega A \sin\theta - V\cos\theta) \\ F_N &= C_N\,(\omega A \cos\theta + V\sin\theta) \end{aligned} \tag{9.20}$$

where θ is defined by

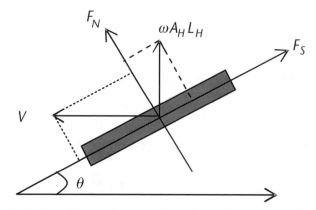

Figure 9.2 Local velocity and forces decomposition of a small segment on the spiral-type head.

$$\tan \theta = \frac{2\pi A_H}{p_H} \tag{9.21}$$

The forces along the x- and y-directions can be represented by F_S and F_N:

$$\begin{aligned}
F_X &= F_S \cos \theta - F_N \sin \theta \\
F_Y &= F_S \sin \theta + F_N \cos \theta
\end{aligned} \tag{9.22}$$

Using (9.18) through (9.22), the translational velocity of the microrobot can be solved:

$$V = \frac{\omega A_H L_H \sin \theta \cos \theta (C_N - C_S)}{D_L + E_L - L_H \left(C_N \sin^2 \theta + C_S \cos^2 \theta\right)} \tag{9.23}$$

The force along the x-direction is

$$F_X = -\frac{E_L + D_L}{L_H} V = \frac{\omega A_H \sin \theta \cos \theta (C_S - C_N)(D_L + E_L)}{D_L + E_L - L_H \left(C_N \sin^2 \theta + C_S \cos^2 \theta\right)} \tag{9.24}$$

Then, the angular velocity of the microrobot can be solved as

$$\omega = -\frac{T_D}{\dfrac{A_H^2 L_H^2 \sin^2 \theta \cos^2 \theta (C_N - C_S)^2}{\left(D_L + E_L - L_H \left(C_N \sin^2 \theta + C_S \cos^2 \theta\right)\right)}} \tag{9.25}$$

where $S = E_R L_{Tb} + D_R + A_H^2 L_H \left(C_S \sin^2 \theta + C_N \cos^2 \theta\right).$

9.3.4   After the Bifurcation

The elastic tail as a circular cylinder is characterized by its geometrical and material properties. There are two geometrical quantities: the radius of the tail a, the contour length L_{TB}, and two material quantities: the bending modulus B, and twist modulus T. Although twist modulus affects the internal force of the elastic tail, it is not related to the bifurcation directly. For simplification, the twist behavior of the elastic tail is not taken into account in this chapter.

Before the bifurcation, the angular velocity ω of the microrobot is proportional to the driving torque (9.25). When ω is low, the elastic tail twists along its centerline while remaining straight. When the angular velocity reaches a certain value (i.e., bifurcation frequency), dramatic deformation appears and the straight elastic tail shapes into a helix. The bifurcation frequency, ω_C, of the elastic tail is related to its bending modulus B, contour length L_{TB}, and rotational resistance coefficient E_R [12]:

$$\omega_C = \frac{B}{E_R L_{TB}^2} \tag{9.26}$$

in which ω_C is independent of the twist modulus T, proportional to the bending modulus B, and inversely proportional to its contour length and rotational resistance E_R. This can be explained intuitively as follows: When B is larger, the tail is stiffer and it becomes harder to bend. When E_R and L_{TB} are larger, the fluid introduces larger friction to the tail and makes it easier to bend.

After the bifurcation, the torque and force balance equations (9.4) and (9.12) still hold. The difference is that the elastic tail is no longer a payload but transforms to helical geometry and becomes a propulsive unit. The balance equations are represented by

$$T_D + A_T F_Y^* L_{TA} + D_R \omega + A_H F_Y L_H = 0 \tag{9.27}$$

$$F_X^* L_{TA} + D_L V + F_X L_H = 0 \tag{9.28}$$

where F_X^* and F_Y^* are the per-unit length viscous force along the x and y direction brought by the elastic tail. V and ω take the form

$$V = P\omega \tag{9.29}$$

where

$$P = \left(M + M^*\right) / \left(D_L - N - N^*\right)$$

$$M = A_H L_H \sin\theta \cos\theta (C_N - C_S)$$

$$N = L_H \left(C_N \sin^2\theta + C_S \cos^2\theta\right)$$

$$M^* = A_T L_{TA} \sin\theta^* \cos\theta^* \left(C_N^* - C_S^*\right)$$

$$N^* = L_{TA} \left(C_N^* \sin^2\theta^* + C_S^* \cos^2\theta^*\right)$$

$$C_S^* = \frac{2\pi\mu}{\ln\left(\dfrac{2p_T}{a}\right) - 2.90}$$

$$C_N^* = \frac{4\pi\mu}{\ln\left(\dfrac{2p_T}{a}\right) - 1.90}$$

in which p_T is the pitch of the helical tail, A_T is the amplitude, and L_{TA} is the length. F_Y and F_Y^* in (9.27) can be represented as $Q\omega$ and $Q^*\omega$, respectively:

$$Q = A_H\left(C_S \sin^2\theta + C_N \cos^2\theta\right) + P\sin\theta\cos\theta(C_N - C_S)$$

$$Q^* = A_T\left(C_S^* \sin^2\theta^* + C_N^* \cos^2\theta^*\right) + P\sin\theta^*\cos\theta^*\left(C_N^* - C_S^*\right)$$

By substituting F_Y and F_Y^* into (9.27), the angular velocity ω can be solved as the following:

$$\omega = -\frac{T_D}{A_T L_{TA} Q^* + A_H L_H Q + D_R} \tag{9.30}$$

9.4   Performance Analysis

Simulations have been conducted using MATLAB. The parameter values used in the simulation are as follows: L_H is 700 microns, L_{TB} is 800 microns, A_H is 60 microns, μ is 0.004 Ns/m^2, p_H is 250 microns, a is 20 nm, b is 20 nm, and d is 60 microns. After the bifurcation, the geometrical parameters of the elastic tail are assumed as follows: A_T is 40 microns, and p_T is 200 microns.

In Figures 9.3 and 9.4, the relationships between linear, angular velocity and driving torque are given. The dot and cross lines represent the situation before and

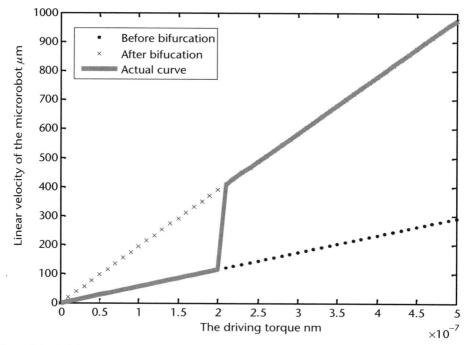

Figure 9.3   Driving torque versus linear velocity of the microrobot before (.) and after (x) bifurcation.

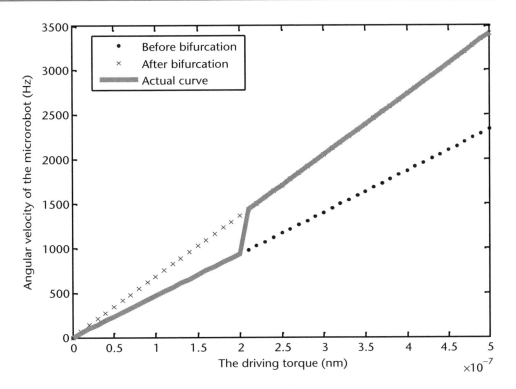

Figure 9.4 Driving torque versus angular velocity of the microrobot before (.) and after (x) bifurcation.

after bifurcation, respectively. The microrobot actually follows the thick gray line. There is a transition at the driving torque T_C. T_C can be obtained from (9.25) and (9.26). It is apparent that the microrobot has better performance after the bifurcation because the elastic tail transforms from a pure payload to a propulsion unit. Both the linear and angular velocity increases proportionally with the driving torque.

Figure 9.5 shows the effect of changing viscosity. When the elastic tail of the microrobots is in the same state of bifurcation status and geometrical parameters, a lower viscosity value results in a higher velocity. This can be explained because a lower viscosity value introduces less friction to the microrobot and so improves its velocity and efficiency. The figure also shows that the driving torque needed for triggering the bifurcation is different for different environments. According to (9.26), the critical torque and angular velocity needed to trigger the bifurcation decreases when the viscosity increases because when the fluid becomes more viscous, larger friction acts on the elastic tail and makes it easier to bend (bifurcate). So the microrobot in the highest viscosity environment shows the lowest velocity until its bifurcation, which makes its velocity exceed others that have not bifurcated yet. With the driving torque increasing, bifurcations happen in all of them and the microrobot in the lowest viscosity leads again.

Besides providing thrust, another important task of the elastic tail is to serve as the drug container. In a rigid-body spiral-type micromachine, the drug needs to be carried as a payload, which introduces extra friction and energy cost. Suppose the rigid-body microrobot and the one with an elastic tail carry the same volume drug

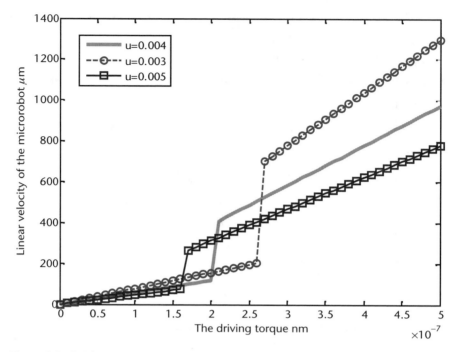

Figure 9.5 Driving torque versus linear velocity of the microrobot with different environmental visocity.

and the drug payload in the rigid-body microrobot is modeled as a sphere, the relation between the driving torque and linear velocity of these two robots is shown in Figure 9.6. We can see that before the bifurcation, the sphere drug payload robot performs better than the robot with an elastic drug tail. This is because a sphere

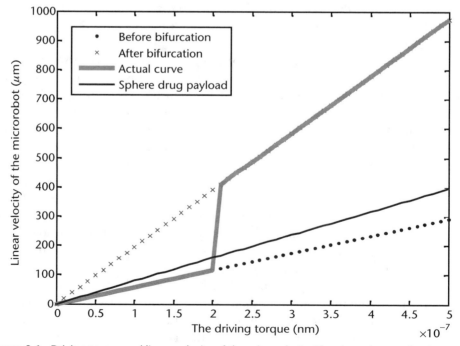

Figure 9.6 Driving torque and linear velocity of the microrobot with sphere drug payload.

introduces less friction than a circular cylinder when they are at the same volume. However, after the bifurcation, since the elastic tail transforms to a helix and generates thrust while the sphere remains its shape, the microrobot with an elastic tail get a much better performance.

Figure 9.7 shows that increasing the length of the elastic tail and keeping the driving torque unchanged will reduce the angular and linear velocity of the microrobot. This can be explained intuitively in that carrying more payloads will consume more energy and slow down the velocity of the system.

We define the efficiency as

$$\eta = \frac{FV}{T_d \omega} \tag{9.31}$$

which is a ratio of the useful work to the total energy input by the external rotating fields [8].

The efficiency ratio of the robot with an elastic tail to the robot with a sphere drug payload is shown in Figure 9.8. When the tail length increases, the driving torque unchanged, although the velocity of the elastic tail microrobot decreases as shown in Figure 9.7, its efficiency ratio to the sphere payload microrobot increases. This indicates that the microrobot with a spiral-type head and an elastic tail exhibits advantages over a rigid body spiral-type microrobot with sphere-type payload when more drugs are going to be carried. The reason is understandable. In order to carry more drugs, both the elastic tail length and the diameter of the sphere drug payload need to be increased. Because of its geometric structure, the elastic tail can

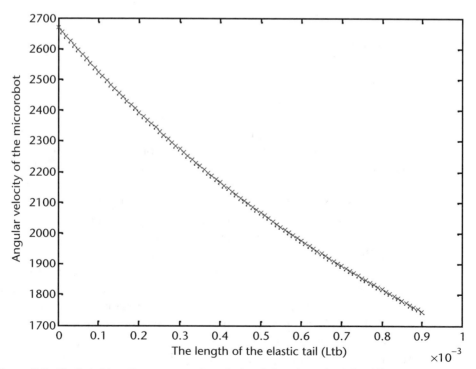

Figure 9.7   Elastic tail length versus angular velocity of the microrobot after bifurcation.

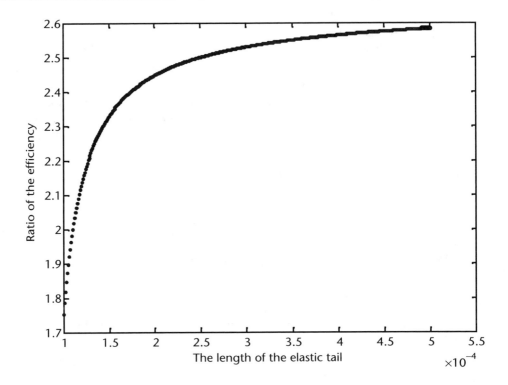

Figure 9.8 Efficiency ratio of the elastic tail robot and the sphere payload robot

generate propulsive thrust, while the sphere drug payload only introduces larger friction to its carrier, which further reduces the efficiency.

9.5   Discussions and Conclusions

There are underlying differences between the balance equations discussed in this chapter and the equations in [9, 10], in which a microrobot with a capsule-like body propelled by a flagellum is presented. In the work of [9, 10], a step motor located between the body and the flagellar tail is used to rotate the tail and generate linear thrust. When the motor rotates, the propagation of the helical wave along the flagellar tail induces a net torque on the tail about the y-axis. This causes the capsule-like body and the flagellar tail to rotate in *opposite* directions to balance the entire system. A balanced torque on the capsule-like body is generated. Hence the total moment on the system remains zero. The capsule-like body has to overcome friction and consume extra energy in order to rotate and keep the system moment balanced. The body angular velocity value is assumed to be smaller than the tail.

In this chapter, the discussed spiral-type head is designed to be driven by external rotating magnetic fields. Under a proper frequency, the entire microrobot rotates synchronously with the rotating field. The angular velocity of the head (including the payload) and the tail is always in the *same* direction and at the same value. The entire system has zero total moment because the driving torque is balanced by the resistive torque from the head and the tail. It is also possible to make use of the external magnetic field to locate the swimming microrobot by analyzing the potential magnetic field change.

In order to make the microrobot work properly, the elastic tail must be stiff enough to keep its helical shape while elastic enough to allow the bifurcation. Micro/nanotubes and some polymer are promising candidates. Since the elastic tail is used as a drug reservoir and will be released in the human body, its building material also needs to be biocompatible and biodegradable. In [21], the biocompatibility and biofouling of several microfabrication materials, such as metallic gold, silicon nitride/dioxide, silicon, and SU-8 photoresist, for a MEMS drug delivery device were evaluated in terms of in vivo inflammatory and wound healing response. The results showed that most of the materials mentioned above are biocompatible. This makes it promising to use traditional MEMS processing techniques to fabricate the biocompatible microrobot, though better biodegradable materials are desired.

A dynamic model for the swimming microrobot composed of a spiral-type head and an elastic tail has been proposed in this chapter. The above works address issues of the dynamics of the microrobot in a stable state, before and after the bifurcation, using resistive force theory. The design takes the low Reynolds number of the environment into consideration and avoids traditional propulsion mechanisms that depend on inertial forces. Driven by external rotating magnetic fields, the microrobot can be operated wirelessly, which is highly desirable in medical applications. When the angular velocity reaches a certain value, the elastic tail shapes into a helix and generates propulsive force.

Compared with a rigid body spiral-type microrobot, the microrobot with an elastic tail is able to move more freely in a constrained area due to the flexible dimensions. The elastic tails also make it possible to form a flagella bundle to produce larger thrust than a single tail. Moreover, as a part of the propeller instead of a pure payload, the elastic tail helps to generate propulsion so that higher energy efficiency is achieved. Although designed for controlled drug delivery applications, it may also be used for a wide range of medical applications including blood vessel enlargement and cleaning, vital sign monitoring, and minimal invasive surgical operations.

The simulation results show that when the driving torque increases to a certain value, a jump can be observed in both linear and angular velocity of the microrobot due to the bifurcation on the elastic tail. Furthermore, the microrobot with an elastic tail shows better performance in terms of speed and energy efficiency than a rigid body spiral-type microrobot carrying a sphere drug payload due to the tail's propulsive function.

References

[1]   Purcell, E. M., "Life at Low Reynolds Number," *American Journal of Physics*, Vol. 45, 1976, pp. 3–11.

[2]   Brennen, C., and H. Winet, "Fluid Mechanics of Propulsion by Cilia and Flagella," *Annual Review of Fluid Mechanics*, Vol. 9, 1977, pp. 339–398.

[3]   Purcell, E. M., "The Efficiency of Propulsion by a Rotating Flagellum," *PNAS*, Vol. 94, 1997, pp. 11307–11311.

[4]   Guo, S., T. Fukuda, and K. Asaka, "A New Type of Fish-Like Underwater Microrobot," *IEEE/ASME Transactions on Mechatronics*, Vol. 8, Issue 1, 2003, pp. 136–141.

[5]   Zhang, Y., et al., "Dynamic Analysis and Experiment of a 3mm Swimming Microrobot," *Proceedings of 2004 IEEE/RSJ International Conference on Intelligent Robots and Systems*, Vol. 2, 2004, pp. 1746–1750.

[6]   Uchiyama, T., and K. Kikuyama, "Numerical Simulation for the Propulsive Performance of a Submerged Wiggling Micromachine," *Journal of Micromech and Microeng.*, Issue 14, 2004, pp. 1537–1543.

[7]   Ishiyama, K., et al., "Swimming of Magnetic Micro-Machines Under a Very Wide Range of Reynolds Number Conditions," *IEEE Transactions on Magnetics*, Vol. 37, Issue 4, Part 1, July 2001, pp. 2868–2870.

[8]   Edd, J., et al., "Biomimetic Propulsion for a Swimming Surgical Micro-Robot," *Proceedings of 2003 IEEE/RSJ International Conference on Intelligent Robots and Systems*, Vol. 3, 2003, pp. 2583–2588.

[9]   Behkam, B., and M. Sitti, "E. Coli Inspired Propulsion for Swimming Microrobots," *Proceedings of the IMECE*, November 2004.

[10]   Behkam, B., and M. Sitti, "Modeling and Testing of a Biomimetic Flagellar Propulsion Method for Microscale Biomedical Swimming Robots," *Proceedings of the 2005 IEEE/ASME International Conference on Advanced Intelligent Mechatronics*, Monterey, CA, 2005, pp. 37–42.

[11]   Manghi, M., X. Schlagberger, and R. R. Netz, "Propulsion with a Rotating Elastic Nano-Rod," *Physical Review Letters*, Vol. 96, Issue 6, 2006.

[12]   Wolgemuth, C. W., T. R. Powers, and R. E. Goldstein, "Twirling and Whirling: Viscous Dynamics of Rotating Elastic Filaments," *Physical Review Letters*, 2000, pp. 1623–1626.

[13]   Gray, J., and G. Hancock, "The Propulsion of Sea-Urchin Spermatoza," *Journal of Experimental Biology*, Vol. 32, 1955, pp. 802–814.

[14]   Nakano, M., S. Tsutsumi, and H. Fukunaga, "Magnetic Properties of Nd-Fe-B Thick-Film Magnets Prepared by Ablation Technique," *IEEE Transactions on Magnetics*, Vol. 38, 2002, pp. 2913–2915.

[15]   Lighthill, M. J., *Mathematical Biofluiddynamics*, Philadelphia, PA: SIAM, 1975.

[16]   Happel, J., and H. Brenner, *Low Reynolds Number Hydrodynamics with Special Applications to Particulate Media*, Upper Saddle River, NJ: Prentice-Hall, 1965.

[17]   Zhang, M., T.-J. Tarn, and N. Xi, "Micro-/Nano-Devices for Controlled Drug Delivery," *Proceedings of the IEEE International Conference on Robotics and Automation*, 2004, pp. 2068–2073.

[18]   Dong, L., et al., "Hybrid Nanorobotic Approaches for Fabricating NEMS from 3D Helical Nanostructures," *Proceedings of the IEEE International Conference on Robotics and Automation*, 2006, pp. 1396–1401.

[19]   Li, H., J. Tan, and M. Zhang, "Dynamics Modeling and Analysis of a Swimming Microrobot for Controlled Drug Delivery," *Proceedings of the IEEE International Conference on Robotics and Automation*, 2006, pp. 1768–1773.

[20]   Laiirent, G., and E. Piat, "Efficiency of Swimming Microrobots Using Ionic Polymer Metal Composite Actuators," *Proceedings of the 2001 IEEE International Conference on Robotics & Automation*, 2001, pp. 3914–3919.

[21]   Voskerician, G., et al., "Biocompatibility and Biofouling of MEMS Drug Delivery Devices," *Biomaterials*, Vol. 24, No. 11, 2003, pp. 1959–1967.

[22]   X. B. Zhang, et al., "The Texture of Catalytically Grown Coil-Shaped Carbon Nanotubes," *Europhysics Letter*, Vol. 27, 1994, pp. 141–146.

[23]   Golod, S. V., et al., "Fabrication of Conducting GeSi/Si Micro- and Nanotubes and Helical Microcoils," *Semiconductor Science and Technology*. Vol. 16, 2001, pp. 181–185.

[24]   Ishiyama, K., et al., "Spiral-Type Micro-Machine for Medical Applications," *Proceedings of 2000 International Symposium on Micromechatronics and Human Science*, 2000, pp. 65–69.

[25]   Cortez, R., et al., "Simulation of Swimming organisms: Coupling Internal Mechanics with External Fluid Dynamics," *Computing in Science & Engineering*, Vol. 6, Issue 3, 2004, pp. 38–45.

[26]   Santini, Jr., J. T., M. J. Cima, and R. Langer, "A Controlled-Release Microchip," *Nature*, Vol. 397, No. 28, January 1999, pp. 335–338.

[27]   Kendall, M. A. F., P. J. Wrighton Smith, and B. J. Bellhouse, "Transdermal Ballistic Delivery of Micro-Particles: Investigation into Skin Penetration," *Proceedings of the 22th Annual EMBS International Conference*, Chicago, IL, July 23–28, 2000, pp. 1621–1624.

CHAPTER 10

DNA and Protein Microarray Fabrication Automation

Timothy Chang[†] and Mingjun Zhang[‡]

This chapter discusses various aspects of automation for the DNA and protein microarrays fabrications [1–29]. Discussions on fundamentals of the microarray technology, spotting methods, elements of the automation, and application case studies are included.

10.1 Introduction

DNA and protein microarrays are matrices of microscopic spots deposited onto a flat substrate surface for the purpose of genetic or proteomic expression profiling analysis. The microarrays provide researchers with molecular signatures for specific cells, tissues, and disease states that can be used for disease identification, prediction, prevention, drug discovery, and diagnostics [24]. Microarray technology has accelerated the process of understanding how genes and proteins work at the genomic and proteomic levels.

Research activities on microarrays have grown steadily since the publication of the seminal paper by Schena et al. in 1995 [29], as evident in the number of microarray publications shown in Figure 10.1.

It is evident that microarray has matured into a major multidisciplinary discipline that includes biology, engineering, and material science.

Prior to generating the microarray, known DNA segments are produced and amplified by, for example, polymerase chain reaction (PCR). The deposited DNA segments are usually immobilized to the surface by hydrogen bonds supplied by a coating agent such as poly-L-lysine. These segments are generally referred to as probes, and the sample materials to be tested are referred to as targets. The extent of pairing up between DNA probes and targets depends on the complementary nature of the two sequences. By labeling the target with a fluorofore, the hybridization pattern can be readily detected using a confocal laser scanner. The current state of technology includes two types of microarray approaches: single channel and two channel microarrays. For single channel microarrays, the probes are oligonucleotide DNAs (oDNAs) chemically synthesized in situ. The length of each strand of

† Department of Electrical & Computer Engineering, NJIT, Newark, New Jersey
‡ Life Sciences and Chemical Analysis Division, Agilent Technologies, Santa Clara, California

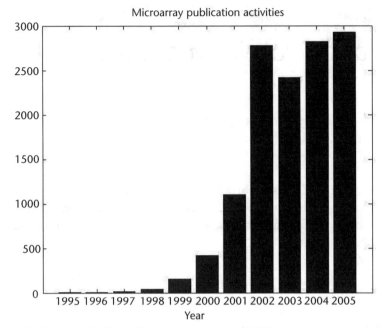

Figure 10.1 Publication activities of DNA microarrays since 1995.

oDNA microarray is normally less than 100 bases. They are synthesized to match parts of the sequence of known messenger RNAs (mRNAs). An example of such microarray is the Affymetrix GeneChip. The single channel microarray has the advantage of being able to determine the absolute value of the gene expression. However, if two expression conditions are needed, two separate single channel microarrays must be used and therefore may result in a higher cost. The two channel microarrays, on the other hand, utilize probes of either complimentary DNA (cDNA) or oDNA. Probes of the two channel microarray can be significantly longer, extending up to thousands of bases. The two channel microarrays allow two sets of sample materials (such as a control and a test object) to be compared. However, two different sets of fluorofores must be used to generate different colors for image analysis. The fluorescent image reveals positions of the probes where hybridization was successful, and identifies targets present in the original samples. The color of the scanned image indicates relative expression of the particular gene between the compared states. Usually, red, green, yellow, and black represent, respectively, *under expression, over expression, similar expression*, and *no expression* in the test samples compared to the control samples. This process makes it possible to simultaneously analyze a mixture of many different genes in one array. However, background noise may complicate interpretation and various statistical analysis methods are usually used in postprocessing to improve reliability and data quality.

In addition, DNA microarrays can also be used for genotyping or detecting subtle sequence variations. This idea been considered as a potential method for disease diagnosis, prognosis, and drug development. For the above applications, DNA microarrays are used to interrogate sample sequence information through complimentary recognition of DNA strands (hybridization) in a highly parallel fashion.

While the DNA microarrays can effectively measure the levels of mRNA expressed in a cell, they cannot directly measure the amounts or functions of the proteins that those messengers produce. This requirement leads to the use of protein microarrays, which can identify interactions between proteins as well as targets of biological molecules. Applications and potential applications of protein micro-arrays include: studying disease susceptibility, performing diagnosis, monitoring progression, and discovering potential points of therapeutic interference faster and more accurately [11]. At present, antibody is the most common used protein microarray.

Protein microarrays essentially extend the similar concepts and principles of DNA microarrays, except proteins have fragile three-dimensional structures and must maintain their native structures through the fabrication process. Even though it is still early in identifying all of the biological applications for protein micro-arrays, they have shown potential promise in genomic and proteomic research.

With the extensive applications of automation and high-speed computation techniques, genomic research has been developed into an unprecedented era. The amount of available gene sequences has increased dramatically while the cost for one finished base has dropped from $5 to about 10 cents in 10 years. It presents a big challenge to the downstream processes to improve both their productivity and quality. The features of microarray such as the capacity of huge number of genes being analyzed, the substantial reduction in sample size requirements, and the use of fluorescence detection schemes for high signal-to-noise ratios has made the way for its wide variety of applications in expression profiling of the human disease states, identification of transcriptional profiles for disease diagnosis, comparative genome hybridization to detect gene duplications and deletions in cancer, and identification of specific biomarkers in crude protein samples.

As for the fabrication of DNA microarrays, there are currently two major approaches: commercial predesigned chips and lab-made custom chips. Affymetrix has been a leader in supplying predesigned chips using a photochemical fabrication process. The spatial specificity of this manufacturing process can generate a high inherent degree of reproducibility. However, this large-scale oriented character makes it costly for many small/medium-scaled laboratories to conduct innovative research. On the contrary, the conventional printing technique is more flexible and less expensive. There are two major schemes of printing technique represented prev-alently by impact and nonimpact principles, respectively. The impact printing tech-nique utilizes a hollow pin to draw up a small amount of sample solution and "stamp" a series of small dots on a glass slide. In practice, it is difficult to overcome the limitations on minimum spot size and reliability. With this approach, a standard microscope slide can accommodate more than 40,000 spots. On the other hand, the 120,000 types of the entire human transcriptome imply that the current spot size of approximately 100 microns must be decreased accordingly. The waste of valuable material is another major drawback of the contact printing that significantly increases the cost of chip production. Besides, the time cost of fabrication with 48 pins and 384 well microplates (19,200 spots) is approximately 5 to 10 hours, which must be significantly reduced. This has been improved by using advanced automated systems [13]. Another inherent problem is the clogging of the impact pins by dust or viscous material, such as those containing glycerol which are useful for printing proteins. Efforts have been made to improve pin performance using MEMS silicon, ceramic materials, and microfabrication [6, 20].

A critical issue for massive microarray fabrication is automation of the fabrication process, which faces many challenges. First, the operation is in microscale. A high precision motion system is clearly needed. Second, the process involves large varieties of biological samples that need to be precisely delivered to the specific locations. This is hard without a robotic system. Third, there are many factors affecting the fabrication process. Dufva [5] summarized 13 factors affecting the quality of the microarray fabrication, including robotics, fabrication method, pin type, humidity, temperature, probe concentration, spotting buffer, immobilization chemistry, blocking technique, stringency during hybridization/washing, hybridization conditions, probe sequence, and target preparation. This is a great opportunity for advanced automation. We will start with an introduction of the microarray printing technologies.

10.2 Microarray Printing Technologies

A number of printing techniques have been developed over the past several years. They can be generally classified as contact and noncontact printing.

10.2.1 Contact Printing

One of the first approaches used for microarray fabrication is contact printing [27]. The method is still used for most small/medium-scale in-house microarray fabrications. A key concern for the technique is the spotting pin design. Figure 10.2 shows examples of selected spotting pins used for microarray contact printing.

Most pins are split pins that have various gap shapes: constant gap distance and graduated gap distance. The diameters of the tips run from 60 to 200 microns. Small pin diameters directly translate into smaller spots. However, small pins are expensive to fabricate and are more fragile. Solid pins are primarily used in the handling of high viscosity liquids such as proteins.

Figure 10.3 is a schematic drawing of the microarray spotting process using both solid and split impact pins.

First, the pins are wetted with DNA solutions in wells of the microtiter plate. Due to the surface tension of the liquid and capillary action (for split pins), microliter or nanoliter fluid adheres to the top of the pins. When the pin touches a substrate surface, the liquid will be transferred onto the substrate surface. Depending on materials, shapes, distances of the pin tip to the substrate surface, and diameters of the pins, nanoliter or even subnanoliter liquid may be printed onto a substrate surface. After spotting a loaded pin and before wetting the pin for the next round of printing, which usually deals with different samples, the pin must be washed and cleaned to avoid possible cross-contamination. This process is repeated until the expected number of spots is printed on the substrate surface. Most pins are free loaded onto the holder so that the maximum print acceleration cannot exceed gravity. Surface tension exerted by pin geometry is a driving force to form a spot [18]. Substrate chemistry contributes to this spotting process significantly. It has been

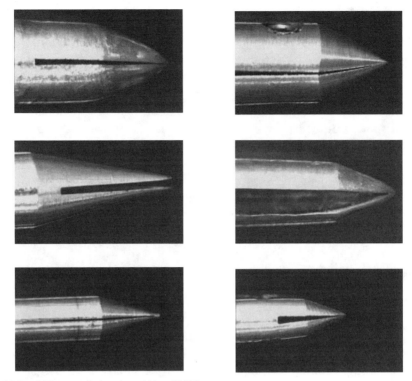

Figure 10.2 Different pin layouts. (*After:* [25].)

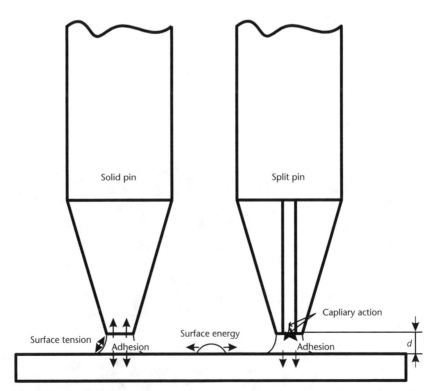

Figure 10.3 Pin spotting process.

shown that the spot size increased by nearly 50% and spot geometry transitioned from square to round as the array progressed from hydrophobic to hydrophilic for a wide range of pins. Decreasing the contact angle or parameter of the contact angle also reduces the surface tension force. The increase in spot size is also proportional to log(viscosity) of the printing solution [15]. The contact pin spotting technique is often used to make cDNA or protein microarrays in low quantities. The concerns for spotted microarrays are spot uniformity and positional accuracy, which affect microarray hybridization and data extraction results. An example of defective spots is shown in Figure 10.4 where smearing and uneven material distributions are visible.

Operations of the contact pins require prerun calibration. Due to its open loop nature, the contact pin technology is difficult to automate precisely and can quickly reach a point of diminishing return.

10.2.2 Self-Sensing Pins

From a control automation point of view, both impact and inkjet printing methods are open loop and require heavy manual interference for each run where missing spot and variable spot morphology (both of which reduce the reliability of micro-array data) are possible. This consideration motivates the design of a noncontact, self-sensing printing technique (referred to as SmartPin [2]) to achieve the specific objectives of smaller size, uniform spot morphology, and compatibility with a wide range of liquids. Besides, this is also able to monitor the spotting in situ and in real-time so that the process can be controlled and regulated.

Figure 10.5 shows the configuration of the SmartPin. The SmartPin assembly consists of two major parts: a capillary tube with tapered tip and an optical sensor with cylindrical probe. The probe is a mixture of illuminating and sensing optic fibers bundled and wrapped by a cylindrical clad. Part of the emitted light from the probe passes through the microreservoir of liquid, is reflected by the surface of the slide, and is carried on by the sensing optic fibers to the electronic unit for signal

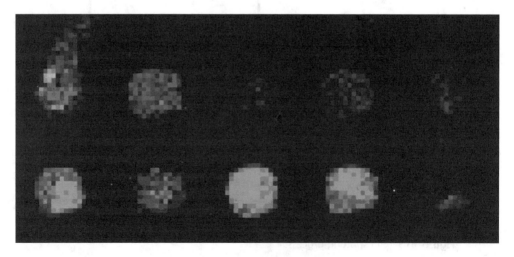

Figure 10.4   Nonuniform impact printing spot formation. (*After:* [2].)

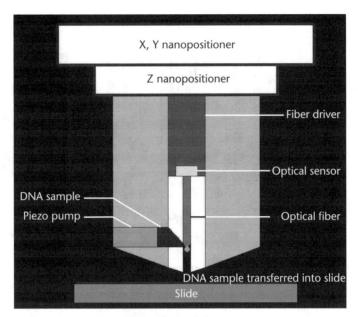

Figure 10.5   SmartPin layout and prototype. (*After:* [2].)

generation. The intensity of the generated signal has patterned relationship with the gap between the bottom end of the tube tip and the upper surface of the microscope slide during the process of spotting as shown in Figure 10.6. A linear drive mounted on the z-axis of the robot implements the relative movement of the probe to the tube. Before each run for spotting a dot, this drive is commanded to push a tiny distance to form a liquid bulge at the bottom end of the tube such that during the whole process, the tip of the tube never touches the surface of the slide. Instead, the liquid bulge at the tip formed by pushing the probe by microdisplacement touches the slide. This feature makes it possible to control the spot size and morphology since the volume of the liquid bulge and the contact depth of the liquid bulge with the surface of the slide is fully controllable.

The uniqueness of the SmartPin design is the patterned intensity curve during the process of spotting. The curve is so called because it has five distinguishing phases associated with the physical steps (shown in Figure 10.6), and the characteristics of these segmented phases have good stability and repeatability through different experimental parameters, such as viscosity and transparency of the material, geometry of the tapered tip of the pin, and the ambient illumination. However, a set of such patterned curves based on the raw data from different experiments have to be unified for the control system design.

Figure 10.7 illustrates a set of curves with raw data from experiments on different liquids. Phase I corresponds to the situation when the pin is filled with liquid and at certain distance from slide surface. It represents the combined effects of variable parameters such as transparency of liquid and geometry of pin. Phase II shows that as the pin approaches the surface, the intensity starts to increase until the end of the liquid bulge touches the slide surface (i.e., beginning of the spot formation). The amount of intensity drop, due to the disappearance of total internal reflection, is shown in Phase III. Phase III represents the area of contact section of the liquid bulge

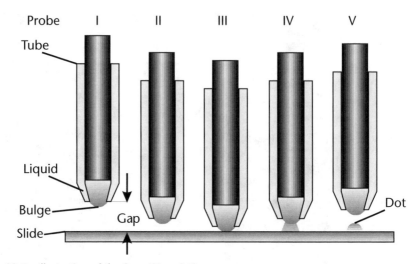

Figure 10.6 Illustration of the SmartPin printing process.

with the slide surface. The length of Phase IV is the length of elongation of the liquid column and implies the volume of the formed spot. Phase V is the pinch off of the liquid column. The location where Phase V occurs is used as a reference point for the next spot.

It is of interest to compare the SmartPin's spots to those obtained using standard impact pins. A morphology comparison of spots produced by GeneMachine OmniGrid 100 with SMP3 pin and SmartPin is shown in Figure 10.8. The spotting material is usually made of 100% glycerol and 150 μM of Cy3 random 9-mer in a 100:1 ratio. Microarrays of 30 spots are printed on poly-l-lysine coated slides by both the SmartPin system and the OmniGrid 100 system. Microarray images were acquired by GenePix 4000B scanner (Axon Instruments) and data was extracted by GenePix Pro 5.1 software (Axon Instruments). Under the same circumstances, mean of spots diameter from SmartPin system is 114.3 microns, which is almost only half of mean of spots diameter (215.8 microns) from OmniGrid 100 system, the capability of producing significantly smaller spots endows the SmartPin system a major advantage in improving the yield and throughput of the current microarray technology. The standard deviation of spots diameter from SmartPin is also less than the one from OmniGrid 100, which means SmartPin produced more uniform and consistent spots, thus it could significantly reduce the biases of the microarray data. The mean of intensity from SmartPin is slightly bigger than the mean of intensity from OmniGrid 100, while the standard deviation of intensity from SmartPin is less than half of the standard deviation of intensity from OmniGrid 100; this strongly supports that SmartPin produced more homogeneous spots and has potentially better reproducibility. New pins with 50-micron diameters are being fabricated to generate smaller spots. A new platform with higher speed and z-axis resolution is also being acquired to improve the cycle time and pin-off control.

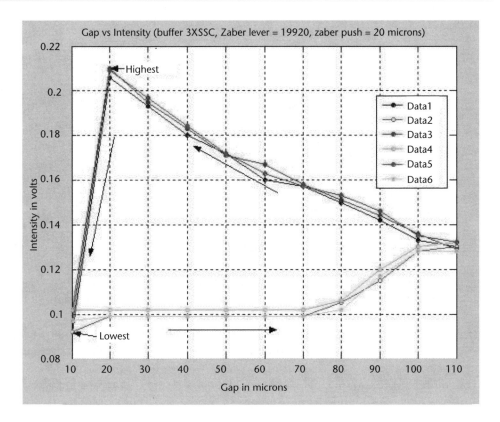

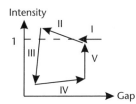

Figure 10.7   A set of patterned curves corresponding to the printing process.

	OmniGrid100 (SMP3)	SmartPin
Mean of dia (μm)	215.8	114.3
SD of dia	+/−8.3	+/−6.6
Mean of intensity	1,693.1	1,747.3
SD of intensity	+/−784.5	+/−317

Figure 10.8   Comparison of the spots formed by SMP3 impact pin and by the SmartPin (SD: standard deviation; Dia: diameter; μm: micron). (*After:* [2].)

10.2.3 Semicontact Printing

Semicontact photolithography technology is generally used only for in situ oDNA microarray fabrication. The process can be described as follows. A wavelength of 365-nm ultraviolet light is shone through a mask to illuminate a subset of regions on a substrate, which is coated with a photosensitive capping chemical. The light releases the capping chemical, exposing parts of the substrate. A solution containing the same type of single nucleotide that is attached to a photosensitive chemical is then washed over the surface. The single base nucleotides then react to the unprotected sites, adding their own capping layer. The process is repeated multiple times to build up sequences of oDNA. The drawback of this technology is that the cost to make the physical mask is expensive and time-consuming. In addition, the semicontact photolithography technology is not capable of making protein or tissue microarrays. An example illustrating the Affymetrix photochemical process is shown in Figure 10.9 where typically a 25-base sequence is synthesized with about 1.3 million unique features per array. Arrays produced by this method tend to be consistent and reliable. However, relatively high cost and inflexibility tend to limit the use of such method to larger facilities.

One critical issue with using photolithography for achieving automated, massive oligonucleotide DNA synthesis is the ability to control each individual reaction at a specific address on a substrate surface. In a typical process, the addition of a monomer containing a 5'-O-photo-labile protection group is preceded by the cleavage of a photo-labile protection group on the 5'-O position of the growing chain. Thus, a reaction that occurs at a specific site is modulated by light irradiation and photo-mask patterning. Physical devices/components used to realize the main function of the photolithography technology based platform usually consist of a reagent manifold, reactor, light source, and masks, in addition to conventional motion, robot, vision and computer control system. The masks can be either traditional physical masks, or digital photolithographic masks. The idea of digital

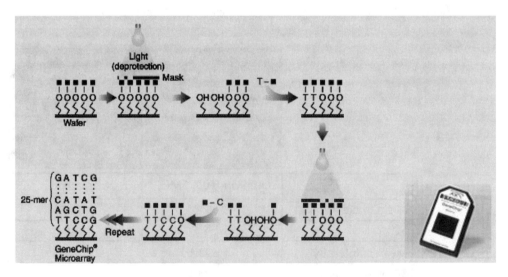

Figure 10.9 Affymetrix photochemical synthesis process and the GeneChip probe array. (Images courtesy of Affymetrix.)

photolithographic masks is to use computers to control reflections of the lights through thousands of micromirrors. By doing this, traditional physical masks can be replaced and the synthesis sequence may be easily programmed. A schematic drawing of the digital photolithographic directed oligonucleotide DNA microarray synthesis system is shown in Figure 10.10, where the optical modular consists of a light source, condenser lens, and filter.

The mirror usually consists of thousands of micromirrors. Reflection light of the micromirrors can be programmed and controlled by a digital computer, which is also used for reagent manifold control and substrate handing. Photo-generated acids take place at the reactor with the help of reflected light from the digital micromirror. By controlling the reagent manifold, the DNA in situ chemical synthesis process can be controlled at the reactor. Protein patterning using this approach has been reported [28].

10.2.4 Inkjet Technology for Fluid Dispensing

Inkjet technology has a long history of fluid dispensing. A well-known example is to print images on paper. Early types of inkjets were primarily continuous-printing that generated a steady stream of ink, and broke the stream apart electronically into droplets during its travel from aperture to target. Continuous type flow printing, however, makes it difficult to control image resolution. The need to better control the droplets resulted in the invention of drop-on-demand inkjet printing, which ejects single drops of ink only when activated. Based on the way that the drops are created, commercially available drop-on-demand inkjets are classified as either thermal inkjet or piezoelectric inkjet. Acoustic wave and electrostatic methods are currently being investigated for ink ejection, but have yet to be commercialized.

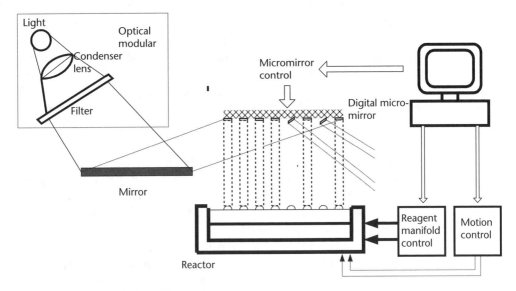

Figure 10.10 Digital photolithography process.

Thermal inkjets use silicon based heating elements to generate vapor bubbles for creating the actuating force and ejecting tiny uniform droplets through nozzles. In a basic configuration, a thermal inkjet consists of an ink chamber having a heater element with a nozzle nearby. With a current pulse of less than a few microseconds through the heater, the ink becomes superheated to the critical temperature for bubble nucleation. When the nucleation occurs, a water vapor bubble instantaneously expands to force the ink out of the nozzle. Once all the heat is used, the ink droplet breaks off and excels toward the paper. The ink then refills back into the chamber and the process is ready to begin again.

The trend of thermal inkjet in the industry is in jetting smaller droplets for improving image quality, faster drop frequency, and a higher number of nozzles [27]. These trends force further miniaturization of the inkjet design [9]. Thermal inkjet has been used for cost-effectively dispensing droplet of smaller than 10-picoliter liquid with speeds greater than 10 kHz and nozzle spatial resolution over 600 dpi [18]. Piezoelectric inkjet uses electric fields to cause shape change of piezoelectric materials and generate pressure to eject tiny uniform droplets through nozzles. Varying the piezoelectric material movement by controlling the electrical field allows the print head to supply various amounts of ink for different conditions. For the piezoelectric drop-on-demand inkjet, deformation of the piezoceramic material causes the ink volume change in the pressure chamber to generate a pressure wave that propagates toward the nozzle. This acoustic pressure wave overcomes the viscous pressure loss in a small nozzle and the surface tension force from ink meniscus so that an ink drop can begin to form at the nozzle. When the drop is formed, the pressure must be sufficient to expel the droplet toward a recording media. In general, the deformation of a piezoelectric driver is in the submicron scale. To have large enough ink volume displacement for drop formulation, the physical size of a piezoelectric driver is often much larger than the ink orifice. Therefore, miniaturization of piezoelectric inkjet has been challenging. In general, printing resolution and firing frequency differ between thermal and piezoelectric inkjet printers, but they are not significant. On the other hand, the piezoelectric inkjets do not generate much heat. This may be an advantage for dispensing temperature-sensitive materials. One of the critical components in a print head design for both thermo-inkjet and piezoelectric inkjet is nozzle design. Nozzle geometries such as diameter and thickness directly affect drop volume, velocity, and trajectory angle. Variations in the manufacturing process of a nozzle plate can significantly reduce the resulting print quality. The two most widely used methods for making the orifice plates are electroformed nickel and laser ablated polyimide material. Other known methods are electro-discharged machining, micropunching, and micropressing. The geometries for the inkjet print head consist of cylindrical orifice, convergent orifice, tapered orifice, tapered with cylindrical orifice, triangle orifice, and square orifice [9]. Usually, inkjet drop volumes range from a few picoliters to 300 picoliters or a few nanoliters. This allows for variable resolution and speed settings. The size of the droplets on the surface of the substrate can be as small as 10 microns in diameter [21] for both thermal and piezoelectric inkjet heads. The number of nozzles per head ranges from 1 to 128 or more. Nozzles can all be fired at one time or controlled individually.

Spatial resolution can be higher than 600 dpi. Firing voltage, droplet velocity, printing resolution, printing distance, drop volume, and nozzle pitch can be customized in wide ranges for various applications. Life of an inkjet head can be up to billions of drops per nozzle with no failure or degradation in performance. The basic components of an inkjet head include a nozzle array, a tubing connection, and a driving apparatus. Usually, a nozzle exit is aligned at one end of the channel, and a capillary tube intake connected to a reservoir is fitted at the other. Each component is built separately through different fabrication processes and assembled together using bonding technology [4]. Inkjet offers feasibility to print a large variety of fluids, including water, oil, pigment, chemical reagents, conductive fluids, and organic polymer electronics, such as light-emitting diodes and field effect transistors. In [12], inkjet technology was used to fabricate microoptic lens arrays. In [15], precise drop-on-demand inkjet has been used to print ultra-fine silver nanopaste with a width of a few microns. In [21], inkjet has been used to produce magnetic layers and structures consisting of nanosized magnetic particles with printed patterns of minimal structure dimensions in the range of 50 to 100 microns. Inkjet technology for biological and biomedical applications is promising. Various biological fluids can be dispensed at microscale or even nanoscale using inkjet technology.

Figure 10.11 shows a picture of two droplets ejected from a drop-on-demand thermal inkjet print head. Figure 10.12 shows a picture of the thermal inkjet [26]. In the figure, each print head consists of two columns of reservoirs on either side of the silicon die and these are used to load biological solutions. Each column consists of five reservoirs, each of which contains multiple nozzles. Before printing, various biological fluids are first loaded into different reservoirs. The print head is then moved to different substrate locations and prints the biological fluids onto the substrate surface. The microarray fabrication process mainly involves dispensing biological particles at precise locations. Material content and printing position as well as the sequence of printing are tightly controlled.

10.2.5   Bead-Based Microarray

The bead-based microarray (BeadArray) is a low-cost microarray offered by Illumina, Inc. The technology uses array probes attached to 3-micron beads and arrays the spot randomly from the microwells onto the slide. The Illumina Sentrix Array Matrix currently contains 96 individual arrays each with 50,000 subarrays. A chemically etched optical fiber provides the optical excitation to each bead, as shown in Figure 10.13. The fiber optic array bundles, each with 50,000 fibers, are fabricated into the standard microtiter plate format. With 96 fiber bundles, about 150,000 individual assay data points are obtained in each Array Matrix experiment. Beadchip, a lower throughput version of the bead array, is also available for experimental applications (Figure 10.14).

Advantages of the BeadArray include high sensitivity, high reproducibility, and low sample input requirements. It provides an alternative experimental and production outlet to the standard photolithography approaches. The technology can be regarded as an extension of semicontact printing; but with low cost, compared with conventional photolithography technique.

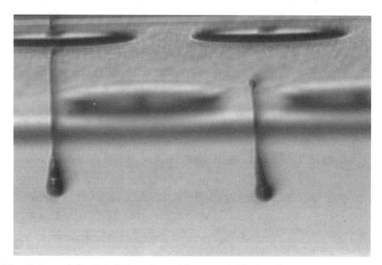

Figure 10.11   Picture of droplets being ejected from an inkjet head.

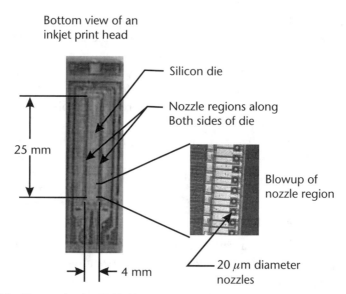

Bottom view of an
inkjet print head

Silicon die

Nozzle regions along
Both sides of die

25 mm

Blowup of
nozzle region

4 mm

20 μm diameter
nozzles

Figure 10.12   Picture of a thermal inkjet.

Excitation beam

Fluorescence
emission

Figure 10.13   Excitation and data acquisition of each bead in the Sentrix Array Matrix. (Photo property of Illumina, Inc.)

Figure 10.14 Illumina Sentrix Array Matrix (left) and Beadchip (right). (Photo property of Illumina, Inc.)

10.3 Microarray Fabrication Automation

Generation of high density, high throughput microarray requires a high degree of automation. At present, most microarray methods are discrete steps with limited *walk-away automation* features. Furthermore, although the present methods are adequate for limited laboratory use where the sample size is generally low and processing time/cost are high, it is critical that the fabrication and scanning technologies be further improved to achieve a broad spectrum impact to the entire medical community and in particular, to expedite our national effort in genetic research.

A microarray automation system comprises of a number of elements: platform, sensors, actuators, and controller, as shown in Figure 10.15.

Platform is generally a robotic workcell consisting of at least three axis of motion capability. Optional theta axis (for rotation about the vertical z-axis) and microstage for fine positioning are also possible. The platform can be considered as a high speed human operator arm responsible for reaching the microwells where the target materials are aspired, moving the pins/inkjets to proper positions for liquid dispensing, and other house-keeping activities such as cleaning.

Presently, most platforms are the so-called Cartesian or linear type where three or more independent perpendicular axes are combined. An example block diagram of x- and y-axes is shown in Figure 10.16.

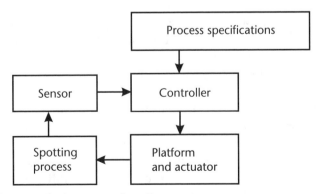

Figure 10.15 Elements of microarray automation.

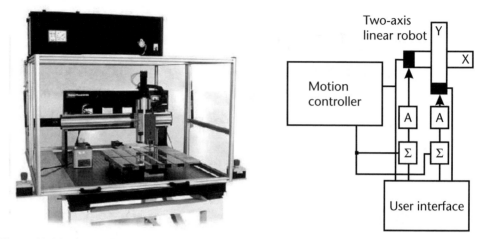

Figure 10.16 A typical two axes view of the components of a contact printing microarrayer.

In the microarrayer shown in Figure 10.16, the x-axis drives the slide tray, while the y-axis positions the z-axis robot to the desired location in the corresponding x-y plane. The advantages of such a workcell are its high speed and relatively low cost. However, it is also less dexterous than, for example, a SCARA robot, which resembles a human arm more closely. The drive mechanism of the workcell is typically a servomotor or a stepper motor. Servomotors are more accurate and suitable for feedback control while stepper motors are inherently an open loop device. To translate the motion from rotary into linear, a ball-screw transmission is generally used. Depending on the pitch of the ball-screw assembly, typically 5- to 25-micron positioning accuracy can be obtained. Higher accuracy is difficult to obtain due to the material properties and grinding requirements of the ball-screw. The position sensors can be as simple as a home switch for a stepper motor based system. For servomotors, the sensors are typically rotary encoders mounted at the motor shaft. The fact that the sensors do not directly measure the load (pin/inkjet) position is another limitation in performance. For positioning accuracy beyond 1 micron, two options are possible: a linear motor or a microstage. Linear motor consists of a linear magnetic track and the "rotor" travels linearly over the track. A linear encoder with diffraction grating is typically used as the position sensor for the linear motor. Such arrangements greatly increase the cost of production. A microstage, on the other hand, is a fine positioning device with typically a range of several microns and submicron accuracy. The most popular microstages are made with piezoelectric drivers. Friction and hysteresis compensation are usually carried out in the microstage controller. In general, bandwidth of a microstage varies from tens to hundreds of hertz.

Aside from the kinematic sensors built into the platform, a number of additional sensors are used to improve the detection of process status. Cameras are commonly used to monitor process procedures, spot formation, and position of the critical components. Being similar to the human operator's eyes, cameras are relatively expensive to install and to process. Due to diffraction limitations, a camera typically works well to about a wavelength of light. The massive amount of sensing elements requires a very high bandwidth real-time signal processor. However, with the advances made in microprocessor design as well as cellular sensing networks, real-time, high speed, high resolution image processing will be an affordable reality.

Other sensors commonly used in the automation process include vibration\acceleration sensors, pressure\tactile sensors, and various position sensors. Aside from being able to report process status, the use of additional sensors tends to improve the robustness of the workcell operations. Selection and location of sensors are therefore critical to the success of the process. Refer to Chapter 8 for a detailed introduction of the subject.

A typical microarray platform control configuration is shown in Figure 10.17. The motion control processor(s) implements the control system that brings the end effecter (pin or print head) to the proper position by sensing the encoder readouts.

In order to automate the platform, a number of issues need to be considered including performance and operating conditions and constraints. A properly designed controller is crucial to the automation.

Performance is a mixed description of speed, accuracy, and robustness. To spot the large amount of microarrays, the platform must be capable of very high speed in a stop-and-go manner. Any mechanical vibration and mismatch can greatly compromise this speed. Furthermore, the spotting accuracy is impacted by the ever-increasing density and decreasing spot size. Other factors such as variations in operating conditions, external vibrations, thermal effects, and so on must also be attenuated. In the present design cycle, most of this burden falls on the controller (Figure 10.13) whose properties are described in Chapter 6. This approach imposes additional constraints on system design and performance. For example, limits on

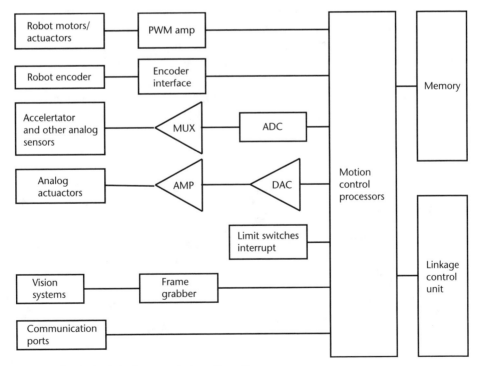

Figure 10.17 Typical platform control configuration.

mechanical platform and sensor performances tend to require overcompensation by the controller. It is desirable to design future generations of microarrayers with a specification driven approach.

10.4 Examples of Automated Microarray Systems

10.4.1 Genomic Solutions OmniGrid 100

OmniGrid 100, shown in Figure 10.18, is a 100-slide contact printing microarrayer. An optional plate loader, the Server Arm3, can be added for automated handling of up to 72 input plates. This microarrayer supports both 96 and 384 well support plates. Feature size is pin dependent and currently runs from 85 to 500 μm with more than 100,000 features per slide. With a 48-pin print head, a 384 well plate can be deposited on 100 slides in less than 7.5 minutes, or equivalently 10,000 spot/100 slides in about 3.5 hours. More compact microarrayers are also available for lower throughput applications.

10.4.2 POSaM Inkjet Microarrayer

The Piezoelectric Oligonucleotide Synthesizer and Microarrayer (POSaM) system [8] shown in Figure 10.19 is an open source system capable of producing multiple microarrays each with 9,800 reporter sequences. It combines synthesis, hybridization, and scan operations within a working day at low cost.

Similar to other microarrayers, the POSaM is based on a three-axis platform with servomotors. The print head is converted from an Epson color printer and features 192 piezoelectric pumps. A laser beam/photodiode assembly detects droplets formation. Nozzles that fail to generate the droplets are taken off-line by the control software named Lombardi. Other components of POSaM include: wash nozzles, reagent storage/delivery, and nitrogen flush. In [8], the authors reported a spot size of 156 +/−16 microns. It is noted the POSaM is constructed with off-the-shelf components and the design is freely available to academic laboratories. The print head is low cost and disposable.

Figure 10.18   Omnigrid 100 and Telechem 946 print head. (Images properties of Genomic solutions.)

Figure 10.19 (a) The POSaM platform; (b) top view showing the array holder; (c) front view; and (d) lower-front view of the inkjet print heads. (*After:* [8].)

10.4.3 ACAPELLA Automated Fluid Handling System

ACAPELLA [13] is a capillary-based fluid handling system, automated to aspire, dispense, mix, transport, and thermal process biological samples. The 1K version, shown in Figure 10.20, is reported to process 1,000 samples in less than 8 hours, whereas the 5K version is expected to handle 500 samples [13].

As shown in Figure 10.20, the ACAPELLA system consists of a dispenser with a 3,800 capillary capacity, an input station with a robotic gripper that dips the capillary into a microwell for piezopump-driven aspiration, a transport comb, piezoelectric reagent dispenser, mixer, and off-load station. Reaction processes such as preparation of cocktail, mixing, thermal cycling, and purification, take place within the capillaries, with the first two steps automated in the 1K system and all four steps fully automated in the 5K system. Operation of the system is monitored by a camera. It is noteworthy that the system software is modular. Each component interacts on an Internet-based client server format. Test results reported in [12] include PCR, reagent stability, restriction enzyme digest, and sequencing reaction.

10.4.4 Agilent's Biological Fluid Dispensing System

A picture of the actual microarray fabrication system is shown in Figure 10.21 [26]. On average, it takes about 30 seconds to print a single slide microarray with 16,000 cDNA features. The dimensions of the overall system are about 8 feet in length, 6 feet in height, and 5 feet in width [26].

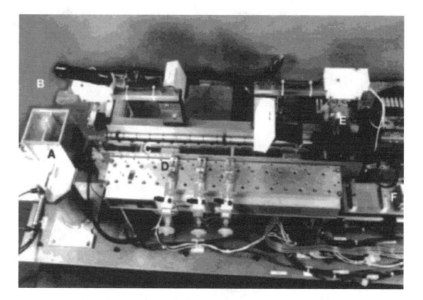

Figure 10.20   ACAPELLA-1K automated fluid handling system. (*After:* [13].)

Figure 10.21   Microarray fabrication system using inkjet technology. (*After:* [26].)

The microarray fabrication system can be regarded as a manufacturing work-cell. To move substrates around the fabrication cell, a robot is used for precise and fast material handing. To deposit biological samples at precise locations, a high precision motion system is used for moving the print heads and substrates. In the meantime, high magnification area array cameras and a line scan camera are used to guarantee reliable system operation and real-time inspection of the printing quality. Fluid and vacuum lines run almost everywhere in the system for print head washing and substrate holding. In addition to the basic operations, multiple electrical sensors and locks are used to make sure that the work cell is under temperature, moisture, and safety control. Interested readers may refer to [26] for details.

10.5 Future Trends of Microarray Fabrication

During the past decade, microarray fabrication has created an industry as well as a blossoming research community. The enabling factors for such success are low cost and high quality microarrays. It is clear at this point that the industry has been very effective in making progress towards very high density arrays. One of the evidences is that the commercialized microarray is continuously gaining market shares. In the meantime, the cost of the commercialized microarray has significantly decreased with increased quality. Though quality of the lab-made microarray has also been increased, most labs face growing cost pressure for running microarray fabrication facilities. The trend is clear that the commercialized microarray will dominate over lab-fabricated microarrays, althrough the lab-made microarrays will still exist for a while.

This is partially due to the nature of the fabrication techniques. Most of lab-made microarrays are fabricated using pin-printing technique, whose quality is hard to control. Though alternative techniques and highly automated systems may help to improve the fabrication quality, the cost does not show advantage over commercialized microarrays. An advantage of the lab-made microarray is its flexibility—that is, researchers can easily design an array format, and put whatever they want to fabricate. However, microarray fabrication industries have started to offer more flexibility for custom-designed microarrays.

The dominant fabrication technique in industry is still the photolithography technique, which occupies the largest market share. The photolithography technique also faces a growing cost challenge for its flexibility, due to the high cost to make physical masks. More recently, a digital mask based photolithography technique has been proposed. The technique adds good flexibility to the conventional photolithography technique. Unfortunately, there seem to be some obstacles for automating the system. The second most-popular technique is inkjet printing, which has a market share around 15% to 18%. The rest is divided among pin-printing and other methods.

In the authors' opinion, a new technique may be accepted for the commercialized microarray fabrication only if the technique can make even lower cost microarray with same or better quality. Unfortunately, no emerging technique shows such a potential. It is challenging as there is still expanding space for current techniques to reduce cost and improve quality as well as providing better flexibility. The authors believe that the status quo will not change until such a technique emerges or the microarray itself is replaced by better technology. This creates opportunities for fabrication and automation professionals to continue research efforts into advanced microarray fabrication techniques.

10.6 Conclusions

The fabrication of microarrays has become an industry of its own. It is difficult to conclude from the literature which fabrication method is absolutely superior. However, it appears that photolithography and inkjet printing dominate the market currently. It is clear that new technology is continuously emerging and likely that the cost of per feature will continue to decrease. Nevertheless, the majority components of automated component remain similar, except difference in the main function components. It is expected that microarray technology will make a major impact on genomic and proteomic research in the next decade.

Acknowledgments

This work is supported, in part, by a National Science Foundation Grant 0243302, "High Resolution, High Density Microarrayer for Genetic Research."

References

[1]   Allain, L., D. Stratis-Cullum, and T. Vo-Dinh, "Investigation of Microfabrication of Biological Sample Arrays Using Piezoelectric and Bubble-Jet Printing Technologies," *Analytica Chimica Acta,* Vol. 518, Issue 1–2, 2004, pp. 77–85.

[2]   Chang, T. N., et al., "Automated Liquid Dispensing Pin for DNA Microarray Applications," *IEEE Transactions on Automation Science and Engineering,* Vol. 3, No. 2, February 2006, pp. 187–191.

[3]   Chen, J., and K. D. Wise, "A High-Resolution Silicon Monolithic Nozzle Array for Inkjet Printing," *IEEE Transactions on Electron Devices,* Vol. 44, No. 9, September 1997, pp. 1401–1409.

[4]   DeRisi, J., V. Iyer, and P. O. Brown, *The MGuide: A Complete Guide to Building Your Own Microarrayer,* Stanford, CA: Stanford University, 1998.

[5]   Dufva, M., "Fabrication of High Quality Microarrays," *Biomolecular Engineering,* Vol. 22, No. 5–6, December 2005, pp. 173–184.

[6]   George, R., J. P. Woolley, and P. T. Spellman, "Ceramic Capillaries for Use in Microarray Fabrication," *Genome Research,* Vol. 11, Issue 10, 2001, pp. 1780–1783.

[7]   Hsieh, H. B., et al., "Ultra-High-Throughput Microarray Generation And Liquid Dispensing Using Multiple Disposable Piezoelectric Ejectors," *Journal of Biomolecular Screening,* Vol. 9, Issue 2, March 2004, pp. 85–94.

[8]   Lausted, C., et al., "POSaM: A Fast, Flexible, Open-Source, Inkjet Oligonucleotide Synthesizer and Microarrayer," *Genome Biology,* Vol. 5, No. R58, 2004.

[9]   Le, H. P., "Progress and Trends in Inkjet Printing Technology," *J. Imaging Science and Technlogy,* Vol. 42, No. 1, January 1998, pp. 1–12.

[10] LeProust, E. M., "Parallel Synthesis of Oligonucleotide Microarrays Using Photogenerated Acids and Digital Photolithography," Ph.D. dissertation, Department of Chemistry, University of Houston, May 2001.

[11] Martinsky, T., "Protein Microarray Manufacturing," *PharmaGenomics,* March/April 2004, pp. 42–46.

[12] MacFarlane, D. L., et al., "Microjet Fabrication of Microlens Arrays," *IEEE Photonics Technology Letters,* Vol. 6, No. 9, September 1994, pp. 1112–1114.

[13] Meldrum, D. R., et al., "ACAPELLA-1K, A Capillary-Based Submicroliter Automated Fluid Handling System for Genome Analysis," *Genome Research*, Vol. 10, Issue 1, January 2000, pp. 95–104.

[14] Moore, S. K., "Making Chips to Probe Genes," *IEEE Spectrum*, March 2001, pp. 54–60.

[15] Murata, K., "Super-Fine Inkjet Printing for Nanotechnology," *Proceedings of the International Conference on MEMS, NANO and Smart Systems*, 2003.

[16] Smith, J. T., and W. M. Reichert, "The Optimization of Quill-Pin Printed Protein and DNA Microarrays," *Proceedings of the 2nd Joint EMBS/BMES Conference*, Houston, TX, October 2002, pp. 1630–1631.

[17] Stafford, P., "An Expression Platform Comparison," *Cambridge Healthtech Institute's Third Annual Microarray Data Analysis*, Baltimore, MD, September 21–23, 2003.

[18] Tseng, F. G., C. J. Kim, and C. M. Ho, "A High-Resolution High-Frequency Monolithic Top-Shooting Microinjector Free of Satellite Drops—Part I: Concepts, Design, and Model," *J. of Microelectromechanical Systems*, Vol. 11, No. 5, October 2002, pp. 427–436.

[19] Tsai, J., and C. J. Kim, "A Silicon Micromachined Pin for Contact Droplet Printing," *Proceedings of the 2nd Joint EMBS/BMES Conference*, Houston, TX, October 2002, pp. 1632–1633.

[20] Tsai, J., et al., "Silicon Microarray Pin with Selective Hydrophobic Coating," *Proceedings of IMECE, American Society of Mechanical Engineers, Micro-Electro Mechanical Systems Division*, 2004, pp. 443–446.

[21] Voit, W., et al., "Application of Inkjet Technology for the Deposition of Magnetic Nanoparticles to Form Microscale Structures," *IEEE Proc.-Sci. Meas. Technol*, Vol. 150, No. 5, September 2003, pp. 252–256.

[22] Wu, R. Z., S. N. Bailey, and D. M. Sabatini, "Cell-Biological Applications of Transfected-Cell Microarrays," *Trends in Cell*, Vol. 12, No. 10, 2002, pp. 485–488.

[23] Zhang, M., et al., "A Function Model Based Approach to Develop Micro-Titer Tray Manufacturing Automation System for DNA Microarray Fabrication," *Journal of Advanced Manufacturing*, Vol. 4, No. 1, 2005, pp. 83–101.

[24] Zhang, M., et al., "An Industrial Solution to Automatic Robot Calibration and Workpiece Pose Estimation for Semiconductor and Gene-Chip Microarray Fabrication," *Industrial Robot*, Vol. 33, No. 2, 2006, pp. 88–96.

[25] Zhang, M., et al., "DNA Microarray Manufacturing Factory Automation," *Proceedings of the 2003 IEEE International Conference on Robotics and Automation*, Taipei, Taiwan, September 2003, pp. 2895–2900.

[26] Fisher, W., and M. Zhang, "A Bio-Chip Microarray Fabrication System Using Inkjet Technology," *IEEE Transactions on Automation Science and Engineering*, 2007.

[27] Barbulovic-Nad, I., et al. "Bio-Microarray Fabrication Techniques," *Critical Reviews in Biotechnology*, December 2006.

[28] Lee, K. N., et al., "Protein Patterning by Virtual Mask Photolithography Using a Micromirror Array," *J. Micromech. Microengineering*, Vol. 13, 2003, pp. 18–25.

[29] Schena, M., et al., "Quantitative Monitoring of Gene Expression Patterns with a Complementary DNA Microarray," *Science*, Vol. 270, 1995, pp. 467–470.

Appendix 10.A List of Commercial Microarray Technology Companies

Affymetrix (http://www.affymetrix.com)

Agilent Technologies (http://www.chem.agilent.com)

Applied Biosystems (https://www2.appliedbiosystems.com)

"ArrayIt" TeleChem International Inc. (http://www.arrayit.com/)

BioMicro Systems, Inc. (http://www.biomicro.com/)

CombiMatrix (http://www.combimatrix.com)

Eppendorf International (http://www.eppendorf.com/)

febit biotech gmbh (http://www.febit.de)

GE Healthcare (http://www4.amershambiosciences.com/)

Genetix (http://www.genetix.com)

Genomic Solutions (http://www.genomicsolutions.com)

Illumina, Inc. (http://www.illumina.com/)

LabNEXT Inc. (http://www.labnext.com)

Microarrays, Inc. (http://www.microarrays.com/)

Micronit Microfluidics (http://www.micronit.com/)

NimbleGen (http://www.nimblegen.com/)

PerkinElmer (http://las.perkinelmer.com/)

Ocimum Biosolutions (http://www.ocimumbio.com)

Roche Diagnostics (http://www.roche-diagnostics.com)

SCHOTT Nexterion (http://www.us.schott.com/nexterion/)

PART IV
System Integration

Automation of Nucleic Acid Extraction and Real-Time PCR: A New Era for Molecular Diagnostics

Patrick Merel[†]

11.1 Introduction

There are many new technologies from which to choose for performing molecular diagnostics in the routine laboratory. Economics, however, are driving the selection of technologies that will ultimately be used in the laboratory.

The time when molecular assays required days or weeks of labor is over. Fortunately, this is occurring at a time when molecular diagnostics are growing in popularity. Because of the higher costs associated with molecular diagnostics as compared to immunoassay, increases in the use of molecular testing cannot be excessive in laboratories.

One of the early technological improvements in the cost-effectiveness of molecular testing has been the use of batch sample processing and the maximal use of the microtiterplate format (a multiwell dish of approximate dimensions of 3 x 5 inches). However technically demanding, steps still remain in the process of molecular procedures. For example, the tedious task of extracting nucleic acids (NA) from whole blood, sputum, saliva, buccal smears, and other clinical specimens is being ameliorated through the use of automated devices designed to perform this procedure automatically.

Like immunochemistry, which has moved to random access robotics a few years ago, molecular testing has just started its revolution. The diagnostic market evolution is certainly driving this revolution, but also the technological jump, brought about by polymerase chain reaction (PCR), has been a major catalyst. Recently, real-time PCR, which can monitor the accumulation of product as the procedure progresses, has been the most rapidly evolving method in the history of PCR. The reasons are simple: real-time PCR simplifies the whole process and presents the advantage of being more amenable to automation.

To take full advantage of this new level of automation in a molecular diagnostic laboratory, two different pathways may be considered.

† University Hospital of Bordeaux, France, CHU Pellegrin, Virology and Immunology Department

First, one can consider modular automation. By looking at this rapidly growing market, there is now the option to choose from reagents providers whose technologies are amenable to automation, and to use generic automated platforms, generally dedicated to high throughput molecular assays. Of course, this degree of automation is quite sophisticated and it requires some additional development and systems integration when one brings it into a functioning laboratory. These sophisticated automated systems are designed for particular needs and thus are not really suited for multipurpose assays.

Another option, more recently introduced on the market, is the industrial solution for molecular diagnostics. As we shall see, the diagnostic industry began 2 years ago to provide a full line of products to automate molecular testing from A to Z, from reagents to robotics. This is clearly a new era for molecular diagnostics, which has changed visions and strategies for the development and the full integration of these tests in a modern clinical laboratory.

Thus, the combination of robotic platforms for NA extraction and real-time PCR instrumentation is clearly driving the evolution of molecular diagnostics.

Most of the molecular tests performed in a molecular diagnostic laboratory today begin with the extraction of nucleic acids (DNA or RNA), which can be called *molecular preanalytics*, or the steps before the analytic step. Molecular preanalytics is immediately followed by an amplification technique, PCR being the most popular. But transcription mediated amplification (TMA), nucleic acid sequence based amplification (NASBA), strand displacement amplification (SDA), and the Invader technology are also target or signal amplification assays which can be employed.

11.2   Nucleic Acids Extraction Automation

NA extraction has always been time consuming and technically demanding. Early NA extraction procedures required the relatively slow and laborious use of a centrifuge; currently, however, this procedure is mainly performed by filtration or magnetic based protocols. These more recent methods have employed various levels of automation. As soon as robotic platforms started to use gripper tools to handle accessories like vacuum manifold, protocols for NA extraction automation based on filtration hit the laboratories in the mid-1990s. Then magnetic beads reagents became popular and emerged on automated platforms because of the ease of use and rapidity of this process.

In this context, the industry which provides robotic platforms has developed different automated instruments to accommodate predominantly the genomics market but also the molecular diagnostic field as well. Thus modern robotic platforms can be divided into two types: multifunctional robotic systems that may be programmed to perform NA extraction and other procedures, and dedicated NA extraction platforms.

11.2.1   Generic Robotic Platforms Which Can Perform Automatic NA Extraction

Robotic platforms which run NA extraction methods are available from the major robotics vendors. See Table 11.1. Basically, these platforms require the use of selected accessories (i.e., vacuum manifold, magnetic stand, gripper tool) which allow them to run NA extraction protocols by filtration or magnetic beads mainly. Of course, the automation platform has to be set up with labware, accessories, and samples, and then be launched by ready-to-use application software, or after significant programming. Manual steps and between-sample variability are greatly reduced once the automation system is established.

It is usual to find platforms that will perform batched based NA extraction by filtration on 96 samples in 2 hours. Other extractors can process this number of specimens even faster, if a robotic pipetting head containing 96 tips is used. Efficiency gains can also be realized using magnetic beads based procedures.

It is worth noting that some vendors are comarketing preferred NA reagents for their robotic platforms. For example, Beckman Coulter, Inc., is providing solutions with the Biomek series and Promega's or Agencourt's reagents; Protedyne is having an association with Qiagen for its BioCube series; and Tecan is recommending various combinations with Macherey-Nagel, Agowa, and others.

Among these robotic companies, Protedyne and Velocity11 have brought the concept of BioCube and BioCell to their instruments. These platforms are enclosed into a hermetic cabinet including a robotic arm with multiple axes of motion. This approach brings an industrial level to NA extraction/purification, combining safety and very high throughput.

Making a final selection of instrument and reagents can be daunting. For the instrument choice, the following elements should be considered:

- Sample batch size;
- Sample volume;
- Sample type;
- Platform capacity;
- Syringes or tips;
- Number of arm/gripper;
- Available accessories;
- Speed;
- Cost of platform;
- Cost of accessories;
- Cost of maintenance;
- Applications available;
- Program files available;
- User friendly software;
- IVD-D or CE marked;
- Support team;
- Application team.

Table 11.1 Generic Robotic Workstations with Available Molecular Testing Applications

Company	Instrument	IVD-D or CE-IVD Compliant; Comments	NA Extraction Applications Available	PCR Applications Available	Web Address
Beckman Coulter, Inc.	Biomek 2000	No	Yes	Yes	http://www.beckmancoulter.com/
	Biomek 3000				
	Biomek NX				
	Biomek FX				
Caliper LifeSciences	SciClone ALH3000	No	Yes	Yes	http://www.caliperls.com/
	SciClone inL10				
	SciClone i1000				
Cybio	CyBi	No	Yes	Yes	http://www.cybio-ag.com/
Eppendorf	epMotion 5070	Yes; MC with integrated PCR	Yes	Yes	http://www.eppendorf.com/
	epMotion 5075 LH, VAC, MC				
Hamilton	STAR, STARlet	Yes; NAT pooling system	Yes	Yes	http://www.hamiltoncomp.com/
Perkin-Elmer	Multiprobe II	No	Yes	Yes	http://las.perkinelmer.com/
Protedyne	BioCube	No; very high throughput	Yes	Yes	http://www.protedyne.com/
Tecan	Freedom Evo	Yes	Yes	Yes	http://www.tecan.com/
ThermoCRS	CRS DNA	No; very high throughput	Yes	Yes	http://www.thermo.com/
Velocity11	Biocell	No; very high throughput	Purification only	Yes	http://www.velocity11.com/
Xiril	Xiril 75, 100, 150	No	Yes	Yes	http://xiril.com/

NAT: nucleic acid testing.

The following companies have developed and subsequently published application notes that describe the use of their reagents and procedures on the platforms listed in Table 11.1: Beckman Coulter-Agencourt, Agowa, Applera-Ambion, Bilatec, Clonetech, Invitrogen-DRI, Invitrogen-Dynal, Macherey-Nagel, Promega, Qiagen, Qiagen-Gentra, Hitachi-Rnature, Sigma-Aldrich, and Whatman.

If a small amount of DNA/RNA is needed, magnetic beads based reagents will be the best way to create rapid and high throughput procedures on automation instruments. Filtration based kits will allow a larger amount of DNA/RNA to be collected.

Finally, the use of generic robotic workstations often brings the concern of separation for extraction and amplification operations. The use of the same platform for NA extraction, amplification, and postamplification has been proposed. However, in the molecular diagnostic routine laboratory, this is a situation that is avoided and not recommended by good laboratory practice. It is thus more common to find assigned platforms for NA extraction and preamplification setup in one room, while platforms dedicated to postamplification steps are located in a separate and different room.

11.2.2   Dedicated Robotic Platforms for Automated NA Extraction

Applied Biosystems (Foster City, California) was one of the very first companies to introduce in 1986 the 340A, a DNA extractor based on the phenol/chloroform procedure. At that time it was hard to predict that this system would inspire the creation of more than 40 platforms in less than 10 years.

Qiagen has been an early adopter of fully dedicated instrumentation for NA extraction with the BioRobot series 3000, followed by the 9604. Qiagen has remained a strong player in this market releasing before everyone else the first IVD-D labeled instrument, the BioRobot MDX.

Today Qiagen has to face numerous competitors, including Roche Diagnostics, which was the first company to introduce magnetic beads based platforms and which has contributed to the spread of these automated instruments for NA extraction.

In this category of platforms, choice of which instrument to use becomes complicated, considering the all various technologies used, from centrifugation to filtration or magnetic beads. Table 11.2 lists the major companies that provide dedicated instruments for NA extraction. It is interesting to see that among the 42 instruments listed, more than 50% of them are using magnetic beads and 25% are using filtration based procedures.

Magnetic beads have started to be a common choice for low to middle throughput in NA extraction, because of ease of use and rapidity. Perhaps maybe the major reason has been the cost of these instruments, which has rapidly decreased with the introduction of magnetic beads based procedures. It has allowed a simpler instrument design, with a size (in term of sample size batch capacity) that is more adaptable to the needs of the majority of molecular diagnostic laboratories.

Precision System Science, Co. (PSS) (Chiba, Japan) has been a leader with their Magtration technology, allowing the concept of various instruments that are now found in the product lineup of various vendors under different designs. Obviously, variation in design will allow various numbers of samples to be processed simultaneously, and various volumes of sample to work with. Careful evaluation of the quality and specification of the magnetic beads will lead to less assay variability. All magnetic beads are not equal and various origins of beads may allow various yields of extracted DNA.

For example, a large proportion of magnetic reagents make use of the Boom patent (Biomérieux, Marcy l'Etoile, France), which is known as a very efficient procedure for NA extraction. Nevertheless, from vendor to vendor, magnetic particles can be different in size, shape, and efficiency. Most of the platforms today are locked into a specific reagent vendor, so it is crucial to evaluate the combination of automation platform as well as reagent.

Table 11.2   Dedicated Platforms for Automated NA Extraction Available as of January 2007

Company	Instrument	Extraction Technology; Additional Features	Sample Batch	Throughput	Sample Volume	Kit Available	Web Address
Applied BioSystems	ABI Prism 6100	Filtration	96	96/30 min	Up to 1m cells	Total RNA, Genomic DNA	http://www. appliedbiosystems. com/
	ABI Prism 6700	Filtration; Archiving, PCR setup	96	96/75 min	Up to 1m cells	Total RNA, Genomic DNA	
Autogen	AutoGenflex 3000	Centrifugation	48	48/4–6 hrs	Up to 5 mL	Genomic DNA, RNA, plasmid, cosmid	http://www. autogen.com/
	AutoGenPrep 2000		48	48/4–6 hrs	1 mL	DNA, tissue, plasmid, cosmid, yeast, plant	
	AutoGenPrep 245		24	24/3.5 hrs	0.7 mL	RNA, DNA, tissue, plasmid	
	AutoGenPrep 965		4×96	384/4–6 hrs	Up to 250 μL or 1 mL	Genomic DNA, plasmid, cosmid	
Bioer	Gene-Pure NPA-32	Magnetic/rod	32	NS	50 to 800 μl	Genomic DNA, and RNA	http://www. bioer.com.cn/
Biomérieux	Extractor	Boom method	10	120/8 hrs	DNA Genomic 100 μL, up to 2 mL for sera	DNA, RNA	http://www. biomerieux.com/
	EasyMag	Boom method/ Magnetic	24	60 min	Up to 1 mL		
Chemagen	Chemagic	Magnetic/rod	12, 96	200/8 hrs (12/40 min), 4000/day (96/20 min)	1 μL to 10 mL	Genomic DNA	http://www. chemagen.de/
Corbett Life Science	X-tractor Gene	Filtration	8–96	8/45 min, 96/95 min	180–200 μL	Genomic DNA, viral DNA and RNA	http://www. corbettlifescience. com/
FujiFilm	QuickGene- 610	80 μm Porous membrane filtration	6	12 min	2 mL	Genomic DNA	http://www. fujifilm.com/

Table 11.2   (continued)

Company	Instrument	Extraction Technology; Additional Features	Sample Batch	Throughput	Sample Volume	Kit Available	Web Address
	QuickGene-810		8	6 min	200 μl	Genomic DNA, RNA, tissue, plasmid,	6,0.583
Genomic	ExtraGene	Centrifugation	48	240/day	0.5 to 10 mL	Genomic DNA, RNA, Proteins	http://www.genomics-tools.com/
Gen Systems	GeneExtract	Filtration	6	Semi-automated	> 1L water	DNA	http://www.genesystems.fr/
Promega	Maxwell 16	Magnetic rod; Reagents prefilled cartridges	16	30 min	50–400 μL	Genomic DNA	http://www.promega.com/
Precision System Science	Magtration System 6GC	Magnetic/ magtration; Reagents prefilled cartridges	6	35 min	100–200 μl	Genomic DNA, RNA, plasmid	http://www.pssbio.com/

NS: not specified

There is a clear intellectual property challenge in this area, and this is one of the reason why other competitive options in magnetic beads can be found, such as the SPRI technology from Beckman Coulter-Agencourt or the ChargeSwitch technology from Invitrogen-DRI (Carlsbad, California) among the most popular.

Maybe the major disadvantage of magnetic beads procedures is the sample volume that can be used during the process. For the last several years, sample volume has been limited to less than 500 μl, but volumes up to 10 ml of blood can now be processed by instruments like the Chemagic from Chemagen (Baesweiler, Germany), or up to 7 ml by the 8LX platform from PSS.

Also, small instruments (PSS design based) like the Roche Diagnostics, MagnaPure Compact, or the Qiagen EZ-1 have allowed up to 1 ml of blood or plasma to be processed with their magnetic bead based reagents. Moreover, these companies have introduced a new concept in the automation of NA: the prefiled reagent rack, which allows rapid setup of the platform. This has promoted a move toward extraction process standardization, which will be important for growth in the molecular diagnostic field.

In addition to the rapidly growing market of magnetic bead based platforms are the centrifugation based instruments. There is still a need for NA extraction from large volumes (for analysis requiring southern blot experiments or for DNA repository and banking purposes). Innovations are quickly emerging into the market from small companies like Genomic (Archamps, France), with their Mega-Extractor. The industry still shows an interest for large-scale extraction, like the recent acquisition of the Gentra Autopure LS instrument (and company) by Qiagen.

The Mega-Extractor from Genomic is a centrifugation based instrument, but of a new kind, as it allows a very high throughput (240 samples per day) for sample volumes up to 10 ml. Also, the design of the instrument enables the use of home brew procedures for large volume sample extraction.

It is important to note that many molecular diagnostic tests have been validated according manual procedures which make use of filtration based technologies for NA extraction. Thus, filtration based platforms still represent 25% of the instruments for NA extraction. In order to capture as much market share as possible, some of them now allow the use of filtration procedures or magnetic beads based protocols.

An excellent example of this category of instrument is the introduction of the Vidiera NSP by Beckman Coulter in 2005. This unique instrument allows NA extraction by filtration or magnetic beads, quantitation of NA by spectrophotometry (a UV spectrophotometer is integrated into the platform), PCR (or any kind of reaction) setup in various kinds of labware, as well as sample archiving. Beckman is marketing this device to the clinical laboratory market since it makes use of exactly the same kind of racks found in Beckman Coulter big clinical chemistry instruments (i.e., Dxi and others) and allows introduction of primary tubes into this NA extraction platform.

A unique feature in all these instruments is their eventual location in clinical laboratories, since they come with a hardware and software natural link to clinical chemistry automated platforms.

11.3   Real-Time PCR Automation

The rapid adoption of real-time PCR in molecular diagnostic laboratories has resulted from assay procedures that are simple, fast, sensitive, and amenable to automation.

The major argument for the conversion of laboratories from performing end point assays, like PCR, is because real-time methods tend to be homogeneous and obviate the need for postamplification processing. These facts have considerably increased the interest for molecular testing, in addition to the reduction of equipment pieces required to perform a complete test. If one combines an automated platform for NA extraction with a real-time PCR instrument, you essentially have a full molecular diagnostic laboratory in terms of instrumentation.

A rough estimate should presumably show us that today 70% of molecular tests could be performed this way, while 5% still require Southern blotting experiments, and 25% require DNA sequencing procedures.

The industry has rapidly followed this interest of laboratories for real-time procedures and currently provides more than 30 different instruments. Table 11.3 lists these available instruments.

Real-time NA amplification instruments are generally built around two technology platforms: heated pulsed air, or thermally regulated Peltier blocks.

Until recently, regular cyclers only used the 96 well plate standard format and allowed high throughput real-time PCR. A major difference with the fast cyclers is the capability of very short PCR cycle time, bringing a whole PCR program to close to 30 to 45 minutes, compared to 2.5 to 3 hours for the regular instruments.

With the availability of the Roche Diagnostics LightCycler 480 in 2005, a new era has started: the era of fast cyclers with the capacity of 96 or 384 samples. It is also noteworthy that robot friendly version of these instruments represent only 15% of the market. Some instruments failed to keep up with the times; for example, the Applied Biosystems Prism 7900 had the same capability but the use of a Peltier block requires relatively lengthy PCR cycles.

For complete automation to be realized, one must integrate real-time PCR based robot friendly instruments with a NA extraction platform. There is, however, one element missing today in such a combination: the plate sealing step, which is neither a simple nor economical step.

It is indeed possible to automate PCR setup in microtiterplate, dispense oil to cover the PCR reaction volume, and bring the plate to the thermocycler with a gripper tool. But the most convenient solution is to use an automatic plate sealer. These instruments are still overpriced for routine molecular diagnostic laboratories and thus are more suitable for very high throughput genomics laboratories.

Table 11.3   Real-Time PCR Instruments Available on the Research and Diagnostic Markets as of January 2007

Company	Instrument	Capacity; Labware	Type of Thermocycler	Robotic Friendly (Yes, No, Not Specified)	Web Address
Applied Biosystems	StepOne	48 well plate	Peltier based, fast cycler	N	http://www. appliedbiosystems. com/
	Prism 7000, 7300	96 well plate	Peltier based	N	
	Prism 7500	96 well plate	Peltier based, fast cycler option	N	
	Prism 7900HT	96 well plate, or 384 well plate or array	Peltier based	Y	
Bioer Technology	Line-Gene	33 tubes	Sandwich thermal-electronic Peltier	N	http://www.bioer. com.cn/en/
	Line-Gene2	66 tubes			
	Line-GeneK	48 tubes			
Biogene	Insyte	96 well plate	Fast cycler	NS	http://www.biogene. com/
Bioneer	Exicycler	96 well plate	Peltier based	NS	http://www.bioneer. com/
Bio-rad	iCycler iQ, MyiQ	96 well plate	Peltier based	N	http://www.bio-rad. com/
	MiniOpticon, Opticon2, Chromo4	96 well plate	Peltier based	N	

Table 11.3   (continued)

Company	Instrument	Capacity; Labware	Type of Thermocycler	Robotic Friendly (Yes, No, Not Specified)	Web Address
Cepheid	SmartCycler	16 independent units; proprietary tubes	fast cycler	N	http://www.cepheid.com/
	GeneXpert	4 independent units; proprietary tubes	fast cycler	Y; Integration of sample preparation, amplification and detection	
Corbett Research	Rotor-Gene 3000	36 tubes; 72 proprietary tubes	fast cycler	N	http://www.corbettlifescience.com/
	Rotor-Gene 6000	36 tubes; 72, 100 proprietary tubes			
Eppendorf	Mastercyler ep realplex	96 well plate	Peltier based	N	http://www.eppendorfna.com/
Evogen	Evocycler	12 tubes-proprietary	fast cycler	N	http://evogen.com/
GeneSystems	GeneDisc Cycler	6 samples-6 tests; CD format	fast cycler	N	http://www.genesystems.fr/
AlphaHelix	QuanTyper	48 tubes	SupeConvection, fast cycler	N	http://alphahelix.com/
Osmetech	Opti Gene	3 independent positions x4 samples, plastic capillaries	fast cycler	N	http://www.osmetech.com/
Roche Diagnostics	LightCyler V1, V2	32 glass capillaries;	fast cycler	N	http://www.roche-applied-scince.com/
	LightCyler 480	96 or 384 well plate		Y	
	Cobas Taqman 48	2×24 proprietary tubes	Peltier based	N	
	Cobas Taqman 96	4×24 proprietary tubes		Y	
Stratagene	Mx4000, Mx3000, Mx3005P	96 well plate	Peltier based	N	https://www.stratagene.com/

A different approach in this category of instrument is represented by the GeneXpert from Cepheid (Sunnyvale, California). This instrument, which allows the simultaneous processing of four samples, integrates all steps from NA extraction to detection by fast real-time PCR. This unique instrument personifies what should probably be considered as a point of care instrument in the near future.

For the laboratory with a higher throughput, the savoir faire of a big company like Roche Diagnostics has brought to the market another level of automation with their series of the Cobas Taqman real-time PCR instruments, among which the Cobas Taqman 96 is the fully automated PCR instrument. This instrument may be linked, robotically, with the Cobas Ampliprep, their fully automated NA extraction platform. These instruments are discussed in the next section on automated molecular diagnostic instruments.

In order to complete this section on instrumentation, we consider small companies, which are bringing new kinds of formats for sample processing. How they will impact automation in molecular diagnostics is not known, but they may offer new alternatives in the very near future. For example, the GeneDisc Cycler from GeneSystems (Bruz, France) is using a CD format with 36 reservoirs prefilled with PCR or real-time PCR reagents for up to six samples. Also, Biogene, Kimbolton (United Kingdom) is bringing an innovation in the technology of 96 independently controlled wells in a Peltier block, by providing the integrated heating technology, which allows each single well of the block to run an independent PCR protocol. This is a very interesting option for laboratories with a small throughput, but numerous markers need to be tested per sample. Similar technology is available in the form of the Opti Gene from Osmetech (Pasadena, California), a fast real-time cycler, using plastic capillaries in groups of four in three independent cassettes, which can run different protocols. With companies like GeneSystems, Biogene, and Osmetech, we have technologies that may lay the foundation for real-time PCR instrumentation design in the very near future.

11.4 Molecular Diagnostic Labeled Automated Platforms

Turnkey reagents and instrumentation solutions are appearing in the market (see Table 11.4). Diagnostic companies used to provide only reagents. The needs for extraction technologies were so great due to the high costs of labor, the industry responded with small extraction systems. The availability of reagents with adapted and automated instruments is clearly a new strategy in the diagnostics market that will now be beneficial to the end user.

Organon-Teknika (Boxtel, the Netherlands), now part of Biomérieux, was the very first company to introduce a diagnostic labeled NA extraction platform, the Extractor, for the Nuclisens product line, which makes use of NASBA for amplification of targets. Now with the EasyMag or the Extractor, combined with the EasyQ for real-time NASBA, Biomérieux offers a semicomplete solution which automates both steps of molecular testing. Many automated platforms still require minor but significant manual interventions and thus miss the full integrated platform advantages.

Gen-Probe (San Diego, California) has also been an early adopter of semi-automated molecular diagnostics platforms. The DTS 800 and DTS 1600, two Tecan-based instruments, allow the process of 800 to 1,600 samples on an 8-hour shift for Chlamydia trachomatis (CT) and Neisseria gonorrhoeae (GC), a very large throughput. Gen-Probe solidified their place in the history of molecular diagnostics by being the first in the industry to release a single automated instrument for molecular testing, the Tigris DTS. The Tigris is one of only two instruments in this industry that contains an automated platform which performs NA extraction, setup amplification by TMA, and processes detection steps, all in a single product. It is worth mentioning that sample tubes do not need to be uncapped when loaded on the instrument. Of course, it performs qualitative testing, but with the capacity to process 100 samples per hour, after an initial processing time of 3.5 hours, the Tigris can handle more than 800 samples per working day, a relatively large capacity by molecular diagnostics standards.

A different approach taken by Roche Diagnostics has been to provide two different instruments to perform the two steps of NA extraction and real-time PCR, which can then be assembled by a docking station to provide a fully integrated platform. The Cobas Ampliprep performs NA extraction and setup PCR on dedicated

Table 11.4 Major Molecular Diagnostics Companies with Available Automated Platforms

Company	Instrument for NA Extraction	Instrument for Amplification/ Detection	Menu	Complete Robotic Integration	Web Address
Abbott Molecular	M2000sp M1000	M2000rt	HIV, HCV, CT, GC	No	http://www. abbottmolecular.com/
Siemens	Centrifuge	Versant 440	HIV, HBV, HCV	No	http://diagnostics. siemens.com/
BD Diagnostics	Viper	ProbeTec ET	CT, GC	No	http://www.bd.com
Biomérieux	Easy-Mag	Easy-Q	HIV; Enterovirus	No	http://www.biomerieux. com/
	Extractor				
Cepheid	GeneXpert	Anthrax	Yes		http://www.cepheid.com/
Gen-Probe	Tigris DTS	CT, GC; HIV/ HCV/HBV; WNV	Yes		http://www.gen-probe. com/
	DTS800, DTS1600	SB100, Leader HC+	CT, GC, GAS, GBS	No	
Qiagen Diagnostics	BioRobot MDX	Artus 3000	16 IVDD markers (viruses, bacteria, parasites)	No	http://www.artus-biotech2.com
Roche Diagnostics	Manual	Cobas Amplicor	HIV, HCV, HBV, CMV, CT, GC, MTB	No	http://www.roche-diagnostics.com/
	Cobas Ampliprep	Cobas Taqman 48	HIV, HCV, HBV	No	
		Cobas Taqman 96 + Docking Station	HIV, HCV, HBV, WNV	Yes	

racks, the K-carriers of which hold 24 PCR tubes. Then, making the link between the Cobas Ampliprep and the Cobas Taqman 96, the docking station transfers the K-carriers to the Cobas Taqman 96, which can hold four racks on four on-board different real-time thermocyclers. This new Cobas product line has introduced a few new innovations to molecular diagnostics. First, Roche Diagnostics has designed a simple tube cap, which can be held by robotic arms, whether it is a sample tube or a PCR tube. This allows the Cobas Ampliprep to keep closed every tube on board, and to open the sample tubes or PCR tubes only when it is needed during the process. Everyone who is concerned with potential tube-to-tube contamination on robotic platforms will appreciate this feature. A feature common to the Cobas Ampliprep and Gen-Probe's Tigris DTS is the employment of continuous flow of samples on a molecular diagnostic instrument. It allows the user to feed any one of the four sample racks containing anywhere from 1 to 24 specimens. The Cobas Ampliprep opens and recaps the PCR tubes—a unique feature in today's market—which also avoids the use of a plate sealer.

Abbott Molecular has chosen to assemble various instruments: the NA extraction platform from Tecan (Männedorf, Switzerland), and a real-time PCR instrument from Celera Diagnostics (Alameda, California). The combination of the M2000sp and M200rt is thus making use of generic robotic workstations but still requires hands-on time to seal PCR plates and transfer plates to the real-time PCR instrument. However, these are the kinds of robust solutions that can be used with Abbott's increasing molecular diagnostics business offerings. One clear benefit of adopting Abbott's platform is the ability to employ in-house reagents for "home grown" assays.

Siemens Diagnostics (previously Bayer Diagnostics) has entered the high sample number market with their semiautomated instruments and running the branched DNA (bDNA) assay. To defend and promote the robustness of bDNA procedures against the real-time PCR procedures, Siemens Diagnostics recently introduced an automated platform, the Versant 440, to automate most of the bDNA procedure. Of course, sample handling still requires hands-on time during the full bDNA procedure, but it is a significant enhancement to this very competitive technology. Also, Siemens Diagnostics has recently announced their partnership with Hamilton (Reno, Nevada) and Stratagene (La Jolla, California) for the development of an automated solution that will be part of an upcoming offer in real-time PCR assays.

BD Diagnostics (Sparks, Maryland) has a solution for real-time SDA which makes use of two instruments to provide semiautomated procedures. The BD Viper instrument takes care of sample preparation, processing, microwell transfers, and incubation steps, while the BD ProbeTec ET System performs the amplification and detection steps. Worth noting is BD's approach to laboratory automation with the Viper; that is, the use of an industrial class of robotics known as Selective Compliance Assembly Robot Arm (SCARA), a classification which indicates that the robot is multijointed (four-axis), similar to the human arm. A new generation of the Viper instrument will be soon available and will take charge of both steps of the process for the complete automation of CT and GC assays.

Finally, companies like Qiagen, which has been selling selected solutions (i.e., NA extraction platforms), may now be considered in this category with the recent acquisition of Artus Biotech (Hamburg, Germany). Renamed Qiagen Diagnostics, this company now offers instruments for NA extraction, real-time PCR (via an agreement with Corbett Research), and additional molecular diagnostic reagents. Qiagen Diagnostics has recently become one of the leaders in the molecular diagnostics market with more than 25 different markers for molecular testing of infectious diseases and 3 in pharmacogenetics, 16 of which are considered in vitro diagnostic device (IVDD) compliant.

11.5   Conclusions

Molecular diagnostics is the fastest growing segment in the in vitro diagnostics industry. TSG partners have recently estimated this segment to reach $2 billion in sales and to grow a 15% compound annual growth rate (market insights for molecular diagnostics can be found at http://tsg-partners.com/).

Many factors are influencing the adoption of molecular testing by laboratories, such as knowledge of disease related markers, FDA and European regulations, reimbursement, standardization, instruments, and reagents availability. Furthermore, the automation of molecular procedures will facilitate adoption of these relatively operator-sensitive and complex procedures. Real-time PCR, with the help of automated platforms for NA extraction and for PCR, has truly boosted laboratories' interest for molecular testing. Reducing technical requirements, hands-on time, and cost, while increasing throughput are the typical arguments that are now part of the challenge for the full automation of molecular procedures.

Steps have been reduced to NA extraction and PCR/detection, and most have been automated, but not every single test requires real-time PCR. Some still need electrophoresis and others require sequencing. Fortunately, the development of capillary electrophoresis has mainly solved the technical challenges associated with automating the electrophoresis molecular separation process.

Moving forward, rapid adoption of molecular diagnostics will require full integration of NA extraction and the amplification/detection steps. Early visionaries like Cepheid, with their GeneXpert instrument, have shown that technology will allow even low throughput laboratories to take advantage of the benefits of automation. Similarly, Iquum (Marlborough, Massachusetts) is developing the Liat Analyzer, a rapid sample-to-result automated instrument for single molecular assay by real-time PCR. Cepheid and Iquum are engineering the next move of the molecular industry, which will undoubtedly be a point-of-care device. In the meantime, other companies will try to minimize the number of steps required by molecular assays. In this context, nanotechnologies hold great promise for process miniaturization and consolidation. Presumably, the use of nanoparticules, as shown by Nanosphere (Northbrook, Illinois) with their BioBarcode technology, will simplify the process by obviating the extraction procedure altogether by employing a silver-enhanced ultra-sensitive detection method.

Thus, following the progress of integration of NA extraction with real-time amplification and/or DNA sequencing, and the progress of nanotechnologies will keep the field of molecular diagnostics one of the most exciting in modern molecular medicine.

CHAPTER 12

Bio-Instrumentation Automation

Norbert Stoll,[†] Stefanie Hagemann,[‡] and Kerstin Thurow[‡]

This chapter provides an overview of automation systems for biological screening applications, which includes the detection of simple absorption and fluorescence signals as well as automated solutions for the determination of complex parameters. Conclusions and future challenges are presented at the end.

12.1   Introduction

12.1.1   Current Trends in Drug Development

Drug development is expensive in terms of costs and time involved in the entire process. Given the high pressure on pharmaceutical companies to maintain strong pipelines of novel therapeutic agents and the high cost of failure in drug development, it is crucial that researchers have tools that can provide high-quality information.

Typical workhorses in current drug development are methods of high-throughput screening and high-content screening.

High-throughput screening (HTS) refers to the integration of technologies to assay thousands of compounds rapidly in search of biological activity in a disease target for drug discovery. HTS has become the workhorse of pharmaceutical and biotechnology companies' drug discovery efforts, with expanding responsibilities and increasing pressure to screen more targets with better compound libraries to find high-quality leads. Increasing the throughput was designed to decrease the overall drug development time by both rapidly identifying potential drug targets and more rapidly identifying promising drug candidates (primary screening). In the meantime, HTS has been widely implemented throughout pharmaceutical R&D [1].

High-throughput screening assays are typically performed within a single well of a microplate; each well represents a single data point. The definition of high-throughput screening is generally accepted to mean performing more than 80,000 wells per month. The term *moderate-throughput screening* (MTS) is used for lower throughput rates. Throughputs above 80,000 wells per month are considered to be ultrahigh-throughput screening (u-HTS).

†   Institute for Automation, University of Rostock
‡   Center for Life Science Automation, Rostock

The high cost of drug failure contributes significantly to the cost of developing new drugs. Significant fallout of compounds from the drug development pipeline occurs due to toxicity reasons and a lack of information early on concerning a compound's effect on the entire biological system. Although significant advances have been made in drug discovery in recent years, a large number of bottlenecks still exist throughout the process (e.g., in target identification, qualification and validation, lead discovery, selection, optimization and prioritization, as well as in preclinical and clinical testing). *High-content screening* assays (cell-based assays) offer the potential to address many of these bottlenecks.

High-content screening systems in contrast to HTS assays (biochemical assays) provide researchers with massive amounts of biological information. They show how a compound is likely to interact in a biological system, not just about how it interacts with a potential drug target. As a result, one can also obtain information about other interactions that may occur within the cell—interactions that may potentially impact efficacy and/or safety of the compounds being evaluated [2].

As compounds that are identified early in the development process as being toxic, for example, will save countless dollars in preclinical and clinical trials [3], the need for high-content assay technologies arose.

12.1.2 High-Throughput Screening Market and Trends

The instrumentation and reagent HTS market in 2002 was approximately $1.1 billion, with equipment accounting for $380 million and consumables accounting for $729 million. Instrumentation sales in this market are dominated by liquid-handling robots that fill, move, and store 96 and 384-well microplates. With the push for higher throughputs, a trend toward lower per-well volume, higher density microplates is emerging. Additionally, detection equipment needed to read the signals generated by the screen assays make up approximately 25% of the equipment market [3]. The growth in each segment is shown in Figure 12.1.

Certain trends that are gaining momentum in HTS will have a more long-term impact on the spending in this area [3].

- *Smaller-well volume screens:* This requires initial capital outlay for equipment to dispense lower volumes and handle higher density microplates or micro-arrays in microplates. Cost per well for the assay is expected to stay the same.
- *Consolidation of resources:* In an effort to conserve overall spending and as a result of consolidation in some large pharmaceutical companies, HTS screens will move toward a more centralized core facility. This will result in an overall reduction in the amount of human resources allocated. It is anticipated that the equipment growth trend in HTS will be maintained at a 10% growth rate.

12.1.3 High-Content Screening Market and Trends

Pharmaceutical companies have made heavy investments in compound chemistry and in standard high-throughput screening approaches. Consequently, increasing numbers of leads are being generated that have to be run through lower-throughput secondary screens for aspects such as specificity and mechanism of action. This

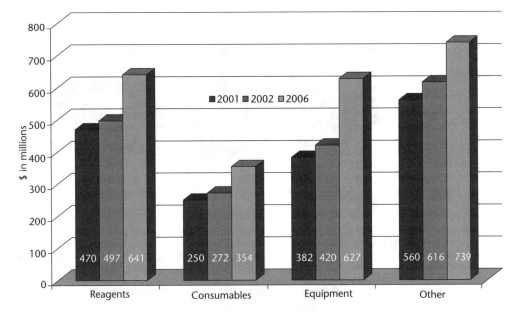

Figure 12.1 HTS market size and growth. (*From:* [1]. © 2002 HighTech Business Decisions. Reprinted with permission.)

disparity in throughput between primary and secondary screening has created serious bottlenecks in drug discovery and development. Consequently, a major challenge is to increase the throughput of secondary screening to be more in line with that of primary screening, while still generating the quality of data needed [2]. High-content screening (HCS) can be directly compared to high-content throughput screening in terms of adoption and growth rates in biopharmaceutical and biotechnology companies [3].

The HCS market is subdivided into hardware systems, reagents, application software, and information technology infrastructure. The hardware market in 2003 was at $55 million (see Figure 12.2 for hardware market share). The total number of systems installed worldwide was estimated at 120. The total reagent revenue for HCS was an additional $100 million. The total market is projected to grow at an overall cumulative annual growth rate (CAGR) of 45% from 2003 to 2007. Main drivers of this growth are the implementation of HCS into secondary screening, primary research, and toxicology in pharmaceutical and drug development companies [3].

12.1.4 Comparison Between HCS and HTS

The advantages of HCS compared to HTS are that the biological information yielded by the HCS approach is many times as complex and detailed. Automated microscopy allows a direct view of the metabolism in intact, three-dimensional cells, in contrast to the more conventional cell-based assays configured for HTS, where only indirect evidence of metabolism in the cell is determined by cell lysate analysis or biochemical assay where the target is examined in complete isolation from the cell complex. Since this method also analyses individual cell events, it has

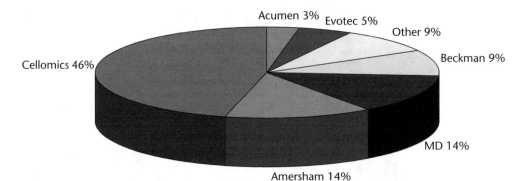
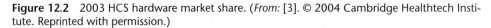

Figure 12.2 2003 HCS hardware market share. (*From:* [3]. © 2004 Cambridge Healthtech Institute. Reprinted with permission.)

the advantage of uncovering events that could be obscured by other signals resulting from homogenizing the cell. In addition, automated microscopy enables the analysis of several parameters (cell number, shape, size, fluorescence distribution, density, granula, and so on) during the course of the experiment. The raw data produced consist of images or histograms. The parameters relevant to the screening are only later extracted by further image and data processing. This yields information on the complex correlation of biochemical and morphological parameters.

In contrast to a high-throughput screening, which consists of parallel screening of many substances on a single target, the HCS process presents the advantage of enabling parallel screening of many metabolic events. The main feature of HCS processes is the simultaneous quantitative recording of several parameters on cellular and subcellular level. This requires the use of costly reagents such as antibodies and fluorescent dyes in several time-consuming process steps. This reduces throughput and increases costs in comparison with HTS. While analyzing an HTS assay plate in a fluorescence, absorption, or luminescence reader—depending on the application—only takes seconds, HCS usually takes a few minutes per microtiter plate. The increased time is a result of a variable number of individual microscopic images per well. The emphasis is on throughput in HTS, which means that one-step assays are used wherever possible with a homogeneous evaluation process and low costs per data point.

HCS is therefore an orthogonal method to HTS; the future will likely see a combination of both technologies.

12.2 Detection Systems for High-Throughput Screening

12.2.1 Typical HTS Assays

High-throughput screening has traditionally involved *biochemical assays* that measure how compounds bind to targeted molecules or how compounds inhibit enzyme activities. These assays can be performed as part of automated, high-throughput procedures using 96-well or 384-well microwell plates for ultra HTS (UHTS) such as 1,536-well plates or even higher density. While these high-throughput biochemical assays rapidly produce large quantities of information about positive reactions on one system, these systems provide little information about interactions of the pathway of interest with other components of a biological system [2].

The other 40% not accounted for by Table 12.1 is accounted for by cell-based assays that are also used in primary high-throughput screening.

Successful assay design, development, and validation are essential for the success of HTS operation [4]. The assay design affects the extent to which automation can be used, the cost of the screen, the sensitivity, and the ability to find hits. Improvements and innovations in assay development include the development of technologies that reduce the number of steps involved, have increased biological relevance, or provide more information per single point of detection [1].

12.2.2 Detection Systems for High-Throughput Screening

Detection systems for high-throughput screening include radiometric and fluorescence-based methods, absorbance/colorimetry, luminescence, and others (see Table 12.2). The choice of detection method impacts the speed, cost, efficiency, and accuracy of the entire operation and is therefore an integral part of the process.

Those detection methods that are expected to increase in use the most are fluorescence polarization (FP), fluorescence intensity (FI), chemoluminescence, and fluorescence resonance energy transfer (FRET). Almost all of the fluorescence-based and luminescence-based detection methods are expected to increase in use.

Detection systems continue to evolve. HTS users are interested in new innovations such as automated patch clamp technology, single-molecule detection, advanced imaging systems, and label-free technologies [1]. Drivers of detection mode changes include:

- Avoiding radioactivity and using safer methods with less disposal costs;
- Increasing sensitivity (achieve better signal-to-noise ratios);
- Increasing speed and throughput;
- Facilitating miniaturization and smaller volumes.

Table 12.1 Assay Types as a Percentage of All HTS Assays

Biochemical Assay	Average Percentage of HTS Assay Types Used	
	2001	2003
Direct enzyme activity	30.48	27.29
Immunoassay (e.g., ELISA)	6.76	6.94
Receptor binding	13.47	9.97
Other protein-protein binding	2.36	2.96
Second messenger (e.g., calcium, cAMP)	2.46	3.52
Transcription (endogenous mRNA expression)	0.5	0.56
Transcription (reporter gene)	1.48	1.33
Others	4.38	4.82
Total	61.89%	57.39%

Source: [1].

Table 12.2 Average Percentage of Detection Modes Used in Primary Screening

Assay Detection Technologies	% of All Assays 2001	% of All Assays 2003
Absorbance/Colorimetric	10.8	4.4
Fluorescence correlation spectroscopy (FCS)	0.6	1.7
Fluorescence intensity (FI)	13.4	16.4
FLIPR	10.7	8.7
Fluorescence polarization (FP)	5.9	11.9
Fluorescence resonance energy transfer (FRET)	4.6	6.6
Time resolved fluorescence (TRF)	5.2	5.4
TR-FRET (HTRF/LANCE)	5.8	6.7
BRET	0.7	1.5
Other fluorescence*	2.7	4.2
Label free	0.3	0.5
Luminescence: Chemi-	5.8	8.5
Bio-	3.3	3.6
Electrochemi-	0.2	0.6
Radiometric: SPA	14.4	10.5
Filter	8.3	3.4
FlashPlate	2.8	1.8
Other radiometric	1.7	0.8
Other**	2.7	3.0

*Includes AlphaScreen, Fluorescence LifeTime Measurement, FMAT, environmentally safe methods.

12.2.2.1 Absorption Readers

Absorption readers use atomic absorption spectroscopy, which covers wavelengths ranging from 200 to 900 nm. Depending on the equipment used, the measuring times will be around 5 seconds for a 96-well plate or 20 seconds for a 384-well plate. Typical systems and their suppliers include Biotrak (GE Healthcare), Flash Scan S12 (Analytik Jena), Ultramark System (Bio-Rad), Victor (Perkin Elmer), and Sunrise (Tecan). Some systems such as Flash Scan 530 (Analytik Jena), Ultramark (Bio-Rad), and GENios plus (Tecan) are also able to measure 1,536-well plates.

12.2.2.2 Fluorescence Readers

Fluorescence readers are based on the principle of fluorescence spectroscopy. The wavelengths that can be applied range between 300 and 800 nm; measuring times per plate are much the same as for absorption readers.

12.2.2.3 Other Reader Systems

Luminescence and nephelometry are two other measuring processes that are especially used in biochemical quantification of reporter gene assays using luciferase, or for particle size determination in solutions.

Table 12.3 shows a selection of reader systems suitable for a variety of applications.

The application areas for optical reader methods especially encompass clinical chemistry and diagnostics, molecular and cell biology, and foodstuff analysis. Quality control, active agent research, substance identification, and, of course, high-throughput screening are the preferred areas of application in pharmacy. Most of these applications cover Ca-flux assays, enzyme activity, immunoassays, ATP quantification and cell toxicity, proliferation and viability.

12.3 Detection Systems for High-Content Screening Measurements

12.3.1 Principles of High-Content Screening

HCS is a collective term for modern microscopy technologies for cell analysis. The possibility of automating a microscope with autofocus, region of interest (ROI) recognition—which might apply to the cell nucleus or cytoplasm—in combination with liquid handling and/or live-cell chamber systems has led to the development of a wide range of cell-based assays specifically for HCS, some of which are now commercially available. Assay development has been based on applications originally developed and carried out on fluorescence microscopes. High-content analysis is used in both primary screening and secondary screening, and in combination. According to one particular definition [3], high-content technologies are those that: "a) analyse translocation fluorescence-marked proteins caused by assay-related activity and status changes; b) enable subcellular analyses of signal cascades; c) are able to measure several parameters in the course of one experiment; and d) simultaneously record electrophysiological single-cell responses in a large cell population."

HCR represents the state of the art in modern microscopy; all of the equipment involved is based on highly modern, automated fluorescence microscopes that use software to record several measurement parameters automatically and simultaneously (number of cells, shape, size, fluorescence distribution, density, granula, and so on), and also for evaluating the data with the same software implemented for data processing. Manual fluorescence microscopy basically offers the same methodological options as HCR, but is not practicable for examining large numbers of samples due to its time-consuming nature and the likelihood of error in evaluating image data by the operator.

The development of confocal microscopy has additionally led to the possibility of three-dimensional analysis of cells and other objects, allowing specialized biological examination. For example, protein trafficking within the cell may be observed from three-dimensional reconstructions of living cells assembled from rapid acquisition of confocal image "slices."

HCS is used for a variety of applications. The exact biological effect of substances on cells and their subsystems is analyzed in secondary screening. This

Table 12.3   Selection of Automatable Reader Systems

Supplier	Product	Measuring Technology	Plate Size
Biotrak II Visible Plate Reader	GE Healthcare Bio-Science	Absorption	96
FLASH Scan S12	Analytik Jena	Absorption	96, 384
FLASH Scan 530	Analytik Jena	Absorption and fluorescence	96, 384, 1536
Expert Plus	Asys Hitech	Absorption	96
Mithras LB940	Berthold Technologies	Luminescence, absorption, fluorescence	6–1,536 (also Petri dishes and Terasaki plates)
Twinkle LB970	Berthold Technologies	Fluorescence	96, 384 (also Petri dishes and Terasaki plates)
Ultramark Microplate System (Bio-Rad Laboratories)	Bio-Rad Laboratories	Absorption	6–1,536 (also Petri dishes and Terasaki plates)
Synergy HAT	Bio-Tek Instruments	Absorption, fluorescence, luminescence (bio- and chemiluminescence)	6–384 (also Terasaki plates and PCR tubes)
FLUOstar OPTIMA	BMG LABTECH	Luminescence, absorption, fluorescence intensity	6–1,536
NEPHEL Ostar Galaxy	BMG LABTECH	Nephelometry	96, 384
CyBi-Lumax flash HAT	CyBio	Luminescence	96, 384, 1,536
Opera	Evotec Technologies	Confocal fluorescence imaging	96, 384, 1,536
Clarina II	Evotec Technologies	Molecular fluorescence, based on confocal single-molecule spectroscopy	24, 96, 384, 1,536, 2,080
Victor	Perkin Elmer	Absorption, fluorescence, TRF/HTRF, fluorescence polarization, luminescence	1–1,536 wells, Petri dishes, Terasaki plates, slides
Viewlux	Perkin Elmer	Absorption, fluorescence, fluorescence polarization, TRF, luminescence, radioactivity	1–1,536 wells
CellLux	Perkin Elmer	Fluorescence	96, 384, 1,536 wells
Sunrise	Tecan Deutschland	Absorption	96 wells
GENios plus (Tecan Deutschland)	Tecan Deutschland	Absorption, fluorescence, glow luminescence	Up to 1,536 wells
Varioskan	THERMO Electron	Absorption, fluorescence, TRF	6–1,536 wells
SPECTRAmax Gemini XS	Molecular Devices	fluorescence, luminescence, chemiluminescence	6–384 wells

includes both tests on differentiation processes in stem cells as well as biogenic active agents for potential use in medication. These areas also include processes for examining certain undesirable side-effects in the cell metabolism such as induction of apoptosis or cytotoxicity.

The second important application area is specific characterization of the interaction of cell components based on proximity to one another, such as in proteins, where confocal microscopy provides an excellent scientifically recognized method. Since confocal microscopy is a significantly more time-consuming method than conventional two-dimensional microscopy, this method is unsuitable for use in screening large numbers of samples, but still offers import and resources for fundamental research work in a research laboratory.

Most of these experiments are carried out on live cells in culture. On the one hand, experiments are required to largely eliminate the need for further culturing (end-point analysis). On the other, one should be able to observe the cells continuously or intermittently, which means they must remain in HCR over a longer period of time, or be carried backwards and forwards from the incubator to the HCR equipment a number of times. Cultured cells are extremely vulnerable to changes in culture conditions; temperature and pH value of the culture medium play an especially important role in the health and survival of the cells [5]. For this reason, some HCR systems have subsystems that enable the storage and analysis of the cells under suitable conditions; these are referred to as live-cell chambers.

12.3.2 Typical HCS Assays

High-content analysis covers general proliferation, cell cycle, apoptosis, and cytotoxicity assay. Examples of where such assays might be used maybe in screening substance libraries for particular cytostatics in cancer research, or in secondary screening in a library for cytotoxicity of hit compounds. Apart from that, high-content screening also caters to more complex analysis methods such as receptor binding and internalization studies (GPCR-based assays or G-protein-coupled seven-transmembrane receptor-based assays) [6], which are particularly important in pharmaceutical research [7]. GPCRs are especially used in the pharmaceutical industry in drug screening methods, as many important intracellular functions can be controlled by GPCR (hormone synthesis and secretion, glycogen and bone metabolism, thrombocyte aggregation, and so on) while their dysfunction may be the cause of a variety of diseases such as asthma, hypertension [8], thyroid cancer, and obesity. Other assay methods that were not available without high-content screening include examinations on intracellular protein translocation or phosphorylation status in individual signal components within intact cells. Besides phosphorylation of mediator and target proteins, translocations between cell compartments also play an important role in signal processes. In addition to cytoplasm-to-nucleus translocation, other translocation forms include membrane-to-cytoplasm and cell compartment-to-cytoplasm translocation (endosome, endoplasmic reticulum (ER), mitochondria). Translocation and phosphorylation assays are often referred to when examining signal cascades in diseases and development processes such as in stem-cell research [9–11].

In addition to the above, further developments in analysis software have enabled examinations on angiogenesis and neurogenesis [12, 13]. Identification of molecules that, for example, are able to induce neurogenesis is immensely important in research on neurodegenerative diseases.

Another application area is the analysis of a cell reaction after gene knockout within the cell using chemically synthesized small interfering RNAs (RNAi or siRNA) [14, 15].

12.3.3   Detection Systems for High-Content Screening

High-content screening assays are performed in large numbers—and instrumentation to automate the procedures is essential. Cellomics was an early leader in the development of high-content screening systems for automated analysis of multiple interacting or independent components within and between living cells, including instruments, image analysis software and, reagents [16].

Table 12.4 gives an overview on currently available HCS reader systems (selection).

12.4   Automation Systems for Mass Spectrometric Measurement

12.4.1   Introduction

Technological innovations in ionization techniques allowed for multiple biological entities—including formerly untraceable high molecular weight molecules—to be analyzed by mass spectrometry (MS). The diversity of suitable biological applications, including molecular weight measurements, chemical synthesis, protein sequencing, SNP genotyping, and protein structure determination, has resulted in MS becoming one of the few quintessential analytical research techniques available today in all of biology [17, 18].

Protein identification and protein characterization remain the top applications in mass spectrometry. There is a broad second tier comprising diverse applications including peptide sequencing, the identification of post-translation modifications, the characterization of multiprotein complexes, small-molecule analysis, the analysis of protein digests, biomarker discovery/validation, and quantitative proteomics.

Chemists working in life sciences have different needs for mass spectrometry as compared to biologists. The differences are seen in the types of applications that they most often perform, their preferred instrumentation cofigurations, and throughput levels [19].

12.4.2   Preferred Ionization Techniques

The underlying principle of MS involves the interplay of magnetic and electric force fields on charged particles (see Section 2.4.4 in Chapter 2). Chemists and biologists prefer different ionization techniques. Table 12.5 gives an overview of the influence of MS application on ionization technique.

Table 12.4   Selection of Available HCS Reader Systems

Company/System	Features
BD Biosciences Pathway	Microscope with options for 4×, 10×, 20×, 40×, or 60× magnification lenses; assays possible with transmitted light, fluorescence, a confocal microscopy; Nipkow-spinning disc with 16 measurable wavelengths (340 nm–IR); acquisition of 3D data possible with confocal optics; image recording using a 12-bit CCD camera [Hamamatsu ORCA ER]; 100-nm positioning accuracy of the lens in the *x*- or *y*-axis, 50 nm in the *z*-axis; 96- or 384-well plates, object carriers, and object carriers with culture chambers may be used; chambers with controllable temperature and air humidity for long-term cell observation; single-head pipette mechanism with disposable pipette tips within the chamber; HCR can be automated or integrated as a subunit into an existing system; various software modules for data analysis.
Cellomics ArrayScan VTI	Four lenses with magnification strengths of 5×, 10×, 20×, 40× may be integrated at the same time; analysis of microtiter plates with 6 to 1,536 wells complete with object carriers; 500-nm precision in the *x*- and *y*-axes; integrated 12-bit CCD camera (see *BD Biosciences*); confocal unit available for the equipment; environment control module for storage and analysis of cell cultures planned; liquid handling system planned; system may be automated.
Evotec Technologies Opera	Similar to Pathway, but without transmitted light microscopy (planned); four lasers and three CCD cameras; environmental control, dispenser, and/or UV lamp may be integrated; dispenser: steel needle with simple dispenser function, pipetting for various substances or mixing steps not possible; hardware and software may be interfaced to automated high-throughput equipment; 500-nm precision in the *x*- and *y*-axes, 100 nm in the *z*-axis.
GE Healthcare— Amersham Biosciences InCell Analyzer 1000	Analysis of plates with 6 to 384 wells complete with object carriers; lenses up to 40× magnification may be integrated; image recording with a 12-bit CCD camera; 500-nm positioning accuracy in the *x*- and *y*-axes; confocal unit planned; partial environment control system for maintaining stable conditions for cells analyzed (only temperature, gas regulation planned); single-head pipetting system and washing station integrated for addition of substances to the cells during analysis; equipment may be automated with any of the major robotic systems.
Molecular Devices Image Xpress Micro Ultra	Automated control of four lenses up to 100× magnification strength; 100-nm accuracy in movement of lens in the *x*-, *y*-, or *z*-axis; confocal optics available from 2006; environment control subsystem for analyzing live cells together with a pipetting station planned; may be integrated in robotic systems.

12.4.3   Throughput Levels

When looking at throughput levels as quantified by number of MS injections per month, most scientists working at lower throughput levels (10 or fewer injections per month) tend to perform more protein identification and characterization, peptide sequencing, and multiprotein complex and conformational change characterizations. Conversely, scientists working at higher throughput levels (100 injections or more per month) tend to perform more small-molecule analysis, biomarker discovery/validation, post-translational modification detection, quantitative proteomics, and metabolomics.

Biologists and chemists differ in how they process their experiments. Biologists typically work at low to medium-throughput levels while chemists expend their

Table 12.5 Preferred Ionization Source by Application

Preferred Ionization Source	Application
ESI or MALDI	Biomarker discovery/validation
	Multiprotein complex characterization
	Peptide sequencing
	Post-translational modification identification
	Protein digest analysis
	Quantitative protein expression
ESI only	Chemical synthesis verification
	Conformational change characterization
	Metabolomics
	Small molecule analysis
CI or EI	Chemical synthesis verification
	Conformational change characterization
	Metabolomics
	Small molecule analysis
APCI	Chemical synthesis verification
	Conformational change characterization
	Metabolomics
	Small molecule analysis

efforts at medium to high-throughput levels. Despite these differences, however, one must be aware that all MS users will, in the future, process smaller sample sizes, require faster and multidimensional runs, and need fully automated and computerized controls for sample setup through data interpretation [18].

12.4.4 Mass Spectrometry Instrumentation

There are numerous mass spectrometric instruments available on the market which can be used for bioanalysis in the drug development process. Table 12.6 summarizes a variety of systems available.

Of the three top mass spec suppliers, Applied Biosystems/MDS SCIEX has defended their position with their popular LC/MS and MALDI/TOF instruments. Quadropole-TOF (QTOF)/MS and MS/MS instruments form Applied Biosystems are also among the most commonly used. Agilent takes the foremost position with their GC/MS, GC-TOF/MS, and inductive coupled plasma (ICP)/MS. Thermo Finnigan is a leader with their instruments in multiple stage MS (MS^n).

12.4.5 Automation of Mass Spectrometry for Bioanalytics

Mass spectrometric measurements, especially in combination with preanalytical chromatographic separations, involve sequential and time-consuming methods. Automation is required for increasing the throughput of mass spectrometric systems. Automation can increase the throughput of a laboratory by parallel processing, process integration, extended hours of operation, and reduced cycle times [17].

Many examples exist for the integration of robotic mechanisms and analytical instruments. Automation has been successfully used in sample preparation

Table 12.6 Available Mass Spectrometry Instrumentation (Selection)

Company	System	Method
Applied Biosystems	QSTAR Elite Hybrid LC/MS/MS	Quadrupole-TOF
	Q TRAP LC/MS/MS	Hybrid Triple Quadrupole Linear Ion-Trap
	API 5000 LC/MS/MS	Triple Quadrupole
	Voyager-DETM STR Biospectrometry	MALDI-TOF
	4800 MALDI TOF/TOF	MALDI TOF/TOF
Agilent Technologies	6100 Series	Quadrupole (LC/MS)
	6210 Series	TOF (LC/MS)
	6300 Series	Ion Trap (LC/MS)
	6410 Series	Triple Quadrupole
	6510 Series	Quadrupole-TOF
	5975B Series MSD	Quadrupole (GC/MS)
	7500a, 7500ec, 7500cs	Quadrupole (ICP-MS)
Thermo Finnigan	TSQ Quantum Hybrid	Triple Quadrupole (LC/MS/MS)
	LTQ FT Ultra Hybrid	Ion Trap / FTICR (LC/MS)
	LTQ XL Linear IonTrap	Ion Trap (LC/MS)
	DSQ Single Quadrupole	Quadrupole (GC/MS)
	PolarisQ Ion Trap	Ion Trap (GC/MSn)
Waters	Quattro Premier XE	Tandem Quadrupole (LC/MS/MS)
	Quattro micro API	Tandem Quadrupole (LC/MS/MS)
	Q-Tof Premier	Quadrupole-TOF
	Q-Tof micro	Quadrupole-TOF
	MALDI Q-Tof Premier	MALDI Quadrupole-TOF
	MALDI micro MX	MALDI-TOF
	GCT Premier	TOF (GC/MS)
Bruker Daltonics	Microflex LT	MALDI-TOF
	Ultraflex II and ultraflex II TOF/TOF	MALDI-TOF and TOF/TOF
	HCTultra PTM Discovery System	ESI-Ion Trap
	APEX IV	FTICR
	APEX-Qe	Hybrid: FTICR, MALDI-TOF

procedures required prior to analysis. Automated solutions are available for classical sample preparation as well as for MALDI-based analysis.

The last decade has seen significant progress by MS vendors for the automation of data acquisition processes. This includes not only an automated acquisition of the data, but also autotuning and calibration scripts to tune and calibrate the systems against a series of reference peaks resulting from tuning standards. The creation of optimized multiple-reaction monitoring tables for quantitative analysis is essential in the drug discovery process [20]. Most commercial mass spectrometry software packages enable data-dependent scanning procedures. Drug discovery experiments generate large amounts of data that have to be quickly interpreted and filtered.

A necessity at the front of MS-based instrument systems are autosamplers that automatically provide samples to the MS systems used for analysis. The main requirements for automated samplers are robustness (unattended and error-free operation, low maintenance), speed (in the range of seconds), lack of memory effects (carryover less then 0.2%), plate capacity (96, 384, or higher), integration with MS software, and flexibility. Commercial autosamplers may also involve integrated sample preparation and can be integrated with column-switching technologies [21]. Newer developments include the design and application of chip-based electrospray systems with applications in quantitative bioanalysis [22], metabolite profiling, and proteomics [23].

Raw data from MS measurements typically undergo a series of processing steps to provide the required information. The required processing steps largely depend on the application. Data analysis can include the automated generation of quantitative data based on peak integration and calibration regression or the determination of elemental compositions from high resolution MS data. Determination of elemental composition (EC) from high-accuracy measured mass is a widely accepted method for gaining information on organic molecules determined in mass spectrometry [24, 25].

12.5   Current Developments in Parallel Chromatography

12.5.1   Parallel Chromatography

The chromatographic principle dictates that chromatographic separation of a complex mixture consisting of many components takes longer with increasing number of components contained or needing analysis. Apart from that, components with a wide range of boiling points or varying substance groups very often need to be separated in the process samples; this requires a temperature program or the use of separation columns with varying polarity. This leads to the use of complex circuitry and applications with long analysis times on the one hand, and a heavy maintenance requirement on the other.

Parallelization may result in shorter analysis periods. This involves separating complex applications into several simple, parallel, simultaneous subapplications. The results from the subapplications are then compiled into an overall result.

However, this requires that certain conditions be met by the equipment for several chromatographic separation processes to be carried out in parallel and simultaneously, which in turn requires that different valves can be controlled at the same time.

In addition, a multichannel detection system that can detect several analytical processes at the same time is needed. Since parallelization requires that several separation systems are installed in the oven space at the same time, the hardware should be designed in as compact a way as possible in order to save space. This may be achieved by miniaturization or integration of several functions into one component. Ten-port valves that combine dosing and back-flushing are a typical example of where several functions are integrated. The multichannel detection principle not only allows the detection of several parallel separation processes, but also the use of additional detectors in distribution circuits as well as inline detectives for optimizing and controlling circuit functions in column circuits [26].

Distribution functions may be necessary in parallel separation processes for complex analytical situations such as the separation of trace components eluted straight after the matrix. Multichannel detection allows applications with distribution functions where very high concentrations along with the trace concentrations may be detected at the same time; this would not be possible with classical single-channel detection.

Since parallel chromatography by design allows several simply constructed and easily matched separation processes for solving complex tasks, a certain level of standardization may be reached. This makes it possible to define modular standard separation processes consisting of dosing, backflushing, and corresponding separation columns for certain frequently repeated applications. Modules can easily be duplicated.

12.5.2 Parallel Capillary Electrophoresis

The worldwide human genome project encouraged work on parallel capillary electrophoresis (CE) all over the world. The emphasis of the studies published was placed on capillary electrophoresis [27–43].

12.5.2.1 Components in a Parallel CE System

Parallel CE systems require different components. Since the parallel use of a capillary array involves the use of an additional dimension, realization requires different approaches, some completely new.

Main Voltage Supply Unit.   The voltage level necessary for analysis is mainly limited by the heat resistance that occurs in the capillary, therefore also the conductance of the buffer and external cooling systems. Since increasing the voltage and field strength leads to increased electrophoretic ion migration speeds, thus increasing analysis speed, the voltage should be as high as possible. The upper limit to be aimed for should be at least 30 kV. The current per capillary must not exceed a maximum value of 100 μA. The current strength per capillary must be monitored and recorded. Apart from shutting the system down if the current exceeds the critical value, this allows further information on the quality of the analysis. For example, bad measurements caused by gas bubbles in the capillary may be recognized while checking that the maximum current value is not exceeded.

Detectors with Corresponding Optical Equipment.   A CE detection system sets heavy requirements on equipment used. To realize a parallel CE system, a detection principle must be selected to allow parallel or serial evaluation of the capillary array. The sampling rate must suffice for the peaks occurring during the course of analysis to be reliably recorded. Depending on the typical migration times and peak widths, a sampling rate of at least 10 Hz should be possible. Optical detection methods are usually used; for example, fluorescence and absorption detection.

The injection and detection ends of the capillaries must be pressure-tight and electrically terminated. At the same time, the capillary ends must be protected from mechanical damage. The most favorable solution seems to be to feed the quartz capillary through a short stainless-steel capillary that doubles as an electrical terminal and as mechanical protection. The capillaries may be terminated via a common mounting plate; however, this will only allow monitoring for the total current through the capillaries. If individual current levels need to be monitored, the capillaries will need to be individually terminated.

The heat loss may be dissipated by an active ventilation system or active thermostat control.

Control and Evaluation Unit.   A computer equipped with additional hardware is used for controlling the individual CE modules and collecting the raw data. The interface to the optical sensor may take the form of a multi-IO card (such as PCI-MIO family by National Instruments). This card may be used to create the clock signals (synchronization, lighting control, and pixel reading) as required by the sensor, while performing A/D sensor signal conversion.

Depending on the form that the high-voltage supply takes, the control system may use the serial interface (such as RS 232 or RS 485) or a multi-IO card with corresponding A/D-D/A conversion.

A separate unit should be used for monitoring individual capillary currents (such as μController base), and the unit should be connected to the computer via a completely isolated interface.

12.5.2.2 Equipment Implementations

The fluorescence process is usually used for detection. Besides the commercially available equipment, there have been some interesting solutions developed by research institutes. Although fluorescence detection is highly sensitive, this method presents a slew of disadvantages; for example, the number of spectral lines that can be realized is limited by economic considerations. Most of the substances to be analyzed do not possess natural fluorescence characteristics, requiring the suitable derivation solutions. On the other hand, many substances have vaporization properties in the UV or visible-light range, thus enabling direct analysis without chemical modification. In contrast with fluorescence detection, the spectral range may be determined by a broadband light source using optical filters.

The detection system presents a fundamental problem, as light from the source is refracted in the capillary in the transmitted-light system, inevitably causing interference with signals from neighboring capillaries. Apart from that, all of the light components reduce the sensitivity if they are not aimed straight at the core of the detector. Masking systems around the detector window are impracticable when large numbers of capillaries are involved; in addition to the mechanical problems in manufacturing and handling, they may also lead to significant power losses.

Gong and Yeung [44] have developed a very elegant solution. The detector window is exposed to a light source that is as even and linear as possible and transmitted onto a detector series via an optical system in such a way that each capillary covers an area of around 10 detector cells. The differences in light intensity between the intermediate space, capillary wall, and capillary membrane allow the identification of the core ranges to be analyzed.

Various companies supply commercially available systems. Table 12.7 shows a summary of the most important systems with their characteristic parameters.

12.6   Other Methods

12.6.1   Lab-on-a-Chip Systems

12.6.1.1   Principle

There is high demand for complete automated systems to enable savings in reagent use as well as staffing and time requirements for many analytical and diagnostic issues. Lab-on-a-chip systems are one promising solution, and are also suitable for portable measurement systems in many cases due to their compact size. The main feature of lab-on-a-chip systems is that all of the process steps from sample preparation to the final detection procedure are automated in a single disposable system.

Lab-on-a-chip systems offer the complete functionality of a macroscopic lab on a polymer substrate about the size of a chip card. This technology has seen much development over the past years, and it enables fully automated and integrated analyses with very small amounts of original substances and reagents in the picoliter to milliliter range on one single chip. Liquids are migrated on the chip surface using capillary action and the surface tension properties of the chip surface. Application-specific biological, chemical, and physical processes take place in reaction chambers (cavities) that are filled according to predefined requirements through micro-channels [45–48].

12.6.1.2   Applications

Miniaturization of bioanalytical processes is immensely important in the rapidly expanding field of life sciences. For example, miniaturization allows the realization of cost-effective, portable, stand-alone analysis systems. Lab-on-a-chip processes are used in a variety of applications including the following:

- Chemosensitivity testing: Applications include the testing of active agents. Since the specific effect of medication depends on morphological and functional cell properties, sensitivity (e.g., in the selection of chemotherapeutic drugs) is an important issue.
- Pharmaceutical screening;
- Water analysis in rivers, lakes, and so on;
- Biosensor technology;
- Biological systems research.

Table 12.7   Parallel Capillary Electrophoresis Systems

Company	Model	Number of Capillaries	Detection
Applied Biosystems	ABI PRISM 3100-Avant Genetic Analyzer and ABI PRISM 3100 Genetic Analyzer	4 or 16	Fluorescence
Applied Biosystems	ABI Prism family is the ABI PRISM 3700 DNA Analyzer		Fluorescence
	Comments: Sheath-flow detection, several analyses on 96 samples each may be performed automatically in series, no direct injection of samples from 96-well plates, no possibility of connecting to existing lab robotics, 20-kV max. electrophoresis voltage, dimensions: 135 cm (H), 76 cm (W), 76 cm (D), weight: 232 kg.		
Applied Biosystems	Applied Biosystems 3730 DNA Analyzer and Applied Biosystems 3730xl DNA Analyzer	48 or 96	Fluorescence
	Comments: On-column detection, very sensitive detector (dual-side illumination, backside-thinned CCD), integrated autosampler, integrated stack for sample plates, internal barcode reader, active thermostat (18°C to 70°C), reagents for up to 100 analyses in the equipment.		
Amersham Biosciences	MegaBACE 500, MegaBACE 1000	48, 96	Fluorescence
	Comments: direct injection of samples from 96-well microtiter plates, on-column analysis, no automated operation, no possibility of connecting to existing lab robotics, 488-nm (single mode) or 488 and 532-nm (dual mode) excitation wavelengths, max. 320 V/cm field strength, detection: four channels with two photomultipliers, capillaries in groups of 16 prefitted and ready for use, dimensions: 81 cm (H), 103 cm (W), 88 cm (D), weight: 272 kg.		
Amersham Biosciences	MegaBACE 4000	384	Fluorescence
	Comments: as for MegaBACE 500 and MegaBACE 1000 but 488-nm excitation wavelength, detection: four channels with four photomultipliers.		
SpectruMedix	SCE: Genetic Analysis System and HTS: High Throughput Analysis System	24, 96, 192	Fluorescence
	Comment: direct injection of samples from microtiter plates, on-column analysis, may be connected to existing lab robotics, no moving parts in detection, detector records the visible spectral range of dyes from the major manufacturers, up to 30 colors may be recorded in one process.		
MPI Molekulargenetik	Fully automated DNA-analysis system		Fluorescence
	Comment: fully automatic analysis of up to 40 microtiter plates by using a stack unit and holographic grid enabling additional spectral resolution of the emitted light in addition to the spatial resolution of up to 96 capillaries.		

12.6.1.3   Equipment Systems

There are various lab-on-a-chip measurement equipment and system solutions available from various companies.

Table 12.8 shows a summary of the systems currently available, along with their application areas.

Table 12.8   Lab-on-a-Chip Equipment and System Solutions (a Selection)

Supplier	System	Comments
Agilent	2100 Bioanalyzer	Microfluidic platform for DNA, RNA, protein and cell analysis
		Ideal alternative compared to difficult and time-consuming gel-electrode electrophoresis
		Rapid availability of reproducible, high-quality digital data
		On-chip flow cytometry
		Applications: DNA, RNA, protein analysis, analysis of PCR products; cellular protein expression, GFP expression, gene silencing, apoptosis
Agilent	5100 Auto-mated Lab-on-a-chip (ALP)	Unmatched throughput (several thousand samples per day)
		Size analysis and quantification of DNA and proteins
		Sample handling, electrophoretic analysis and digital data analysis are completely automated
Agilent	HPLC-Chip	First microfluidic system for nanoflow electrospray LC/MS
		With integrated enrichment column, analysis column, hydraulic connections and an electrospray emitter on one single polymer chip
		Advantages: simple handling, sensitivity, productivity and reliability in nanoflow HPLC analysis
Bartels Mikrotechnik	Physiocheck	Complex polymer chip for blood analysis
		Integration of microfluidic structures (200×50 μm size), such as spiral mixers, reaction chambers, micromembranes, reservoir and detection chambers
Caliper Life Science	LabChip 3000 Drug Discovery System	For use in drug screening
		To detect interactions between drug candidates and therapeutic targets
		Assay development, primary screening, selectivity screening, structure-activity relations, screening of therapeutic enzymatic "targets" such as kinase, proteases and phosphatises
		Modular system with a choice of fluorescence excitation and detection wavelengths
		Optional plate-handling automation

Table 12.8   (continued)

Supplier	System	Comments
Caliper Life Science	LabChip 90 System	LabChip electrophoresis
		DNA and protein analysis
		Automated sampling: Transfer plates from a thermal cycler directly into the LabChip 90 System, no sample prep or transfer needed
		Reduces the manual labour associated with gels and frees up valuable resources
		Quantitative sizing and concentration data are automatically reported as each sample is analysed
		Eliminates the need for photo documentation
		DNA and protein analysis can be performed using the same system
		Complete analysis for hundreds of samples in just a few hours; no further processing required
thinXXS	Microfluidic component system	Based on standards typical for laboratories
		Mounting plate the size of a titre plate
		Single chip modules have the size of an object carrier
		Up to four modules (pump, mixer, or chromatography or electrode chip) can be combined into one frame
		Special fluid and line connectors connect up the diagnostics and analysis chips to one another and to the outside world
		System easy to reconfigure

12.6.2   Patch Clamp Technologies

12.6.2.1   Principles of Patch Clamping Technology

Biological membranes have lipids and other proteins, and also specialized proteins that serve ion transport through the membranes. These transport molecules can be classified in a number of ways. Some of these—the "channels"—can be used in patch-clamp measurement. This process, which was developed by Neher and Sakman—for which they received the Nobel Prize in 1991—measures current through a very small area measuring a few μm^2 known as a *patch* in the membrane connected to a certain predefined voltage (*clamp*). A few channels, if more than just one, run through this part of the membrane or patch. Observation of the electrical course reveals abrupt jumps between different levels; in this case, two levels: the channel opens and closes. The current measured (the difference between these two levels) will lie within a few picoampères (pA). Measurement requires a highly sensitive current amplifier [49, 50].

There are three main patch-clamp configurations possible. On-cell configuration is where the pipette is placed on the membrane causing a partial vacuum. Increased suction or a short voltage pulse breaks the membrane; this whole-cell configuration has a direct connection between the cell and pipette [51, 52]. Jerking the pipette away from the cell in the one-cell configuration will cause a piece of the membrane to remain on the pipette; this is referred to as the excised patch configuration.

The patch-clamp technique can be used to measure cell function in the natural environment of the living organism (in vivo).

12.6.2.2   Automating the Patch-Clamping Process

All of the automation approaches involve placing the cell on a micrometer-scale aperture (diameter at least $4\,\mu$m, ideally 8 to $12\,\mu$m) in a planar carrier. The planar automation approach controls this by creating a partial vacuum in the aperture. Increasing this partial vacuum causes a mechanical contact with high electrical sealing resistance. Part of the cell membrane is drawn into the aperture. To optimize the position of the cells, "channel formers" are often added to the electrolyte in the aperture to create pores in the cell membrane, thus enabling access to the inside of the cell and membrane potential clamping. Cytocentering is a solution that avoids the need for channel formers while providing a good yield of high-quality current recording. The apertures necessary for positioning and current recording are separated [53]. The patch aperture is a small central aperture, while a large surrounding opening serves towards positioning the cell. Apart from sufficient suction capacity, a soft partial vacuum keeps the contact aperture clean. In contrast to other processes, the central process of manual patch-clamping remains and is performed in inverse configuration. The hydrodynamic and mechanical forces are exerted on the cell and its membrane, achieving the same current recording quality with a high success rate.

Synchronous control dictates a uniform predetermined experimental period in the microtiter plate approach. Parallel asynchronous operation provides an alternative approach; here, several spatially isolated chips are separately terminated and operated. This means that quality of current measurement, voltage clamp, or cell is individually controlled on each chip regardless of the status of the other chips. Given a sufficient quality level, other tests on the terminated cell are initiated; if the quality is poor, the experiment is halted.

12.6.2.3   Equipment Systems

Table 12.9 shows a summary of automated patch-clamp systems currently available in the market.

12.7   Automation Systems

The use of the high-parallel reader systems presented is only as feasible as the sample prep and injection into the system are suitable for high-throughput applications.

The most important automation factors in high-throughput applications, therefore, include the pipetting and microtiter-plate transport systems.

A typical setup for HTS might also include storing and stacking systems, incubators and harvesters, and detection instrumentation. Instrumentation and equipment suppliers are increasing the compatibility of their equipment with higher-density microplate formats and capabilities in handling nanoliter volumes [1].

Table 12.9   Automated Patch-Clamp Systems

Supplier	Equipment	Comments
Molecular Devices	PatchXpress 7000A	Patch Clamp Automat for individual cells
		SealChip technology: 16 parallel, independent channels
		Automatic single channel pressure control to achieve gigaseal
		Whole-cell recordings
		Integrated pipetting robot with disposable tips
		No continuous perfusion
Molecular Devices	IonWorks Quattro	Primary screening, secondary screening, active agent modification
		Parallel patching of several cells in one array
		PPC: Population Patch Clamp
		Total current in the array cells
		Single-cell measurement not supported, but high success rate and improved data quality (z factors, IC50)
Flyion	Flyscreen 8500	Patch-clamp automat based on OEM pipetting robot
		Robotic recording tip handling
		Automatic cell handling, including creation of gigaseal and whole-cell generation
		Liquid handling including active agent injection and flushing
		pressure and vacuum control, flow control
		Amplifier control and automatic data processing
Cytocentrics	CytoPatch-Chip	Quartz glass material (5 MOhm max. access resistance)
		May be produced in large numbers in existing production facilities (BionChip, NL)
		Chip is surrounded by a silicon housing that enables the microfluidic injection and drainage of nutritional solutions or test substances
		Very rapid solution exchange using closed flow system
		Washing out of substances or use of several concentrations or test substances in one cell possible.

Table 12.9   (continued)

Supplier	Equipment	Comments
		Use of test substances in the μl range comparable to microtitre plates
		Automatic disposal and provision of chips
		Measures up to 200 cells per day
		Up to 50 modules in the automat for increased throughput
		Analysis of voltage-independent ligand-controlled and metabotropic channels
Nanion	Port–A-Patch NPC-1	Semiautomatic system
		Manual liquid handling
		Automatic patch process and sealing
		PC-supported data evaluation
		10-μl sample volumes
Nanion	Patchliner NPC-16	Patch clamp automat based on OEM pipette robot
		Sixteen patch channels in series or parallel according to design
		Proprietary disposable chips
		Rapid liquid transfer
Sophion	QPatch 16	16-channel automat
		Chip-based
		All sixteen chips together on an oplate in microtitre plate format
		Examination of voltage-controlled and ligand-controlled ionic channels possible
Sophion	QPatch HT	48-channel automat
		Not yet commercially available
		Chip-based
		Voltage- and ligand-controlled channels may be analyzed

12.7.1   Dispensing Systems

Automated dispensing of different liquids into microplates can dramatically speed up the biological processes compared to manual pipetting. Typical stand-alone units can deliver liquids into 96- or 384-well plates with very high precision and accuracy. Pipetting units consist of a horizontal stage that moves source and target plates under a dispensing head.

12.7.2 Transportation Systems

Microplate handling devices add great functionality to liquid handling systems since they allow for automated delivery of plates and tips. The most basic functions of microplate handling systems is to feed and remove plates onto and from the deck of liquid handler, plate readers, or plate storage.

12.7.3 Liquid Handling Workstations

Various liquid handling systems are available on the market. The key feature of such systems is the movement of a tip head in the x-y-z-axis across a work surface that may contain plates, tubes, solvents, tips, or other devices. The systems available differ in deck size, number of tips, speed, and pipetting features and capabilities.

The tip head as part of the pipetting system is usually composed of 1, 4, 8, or 96 channels. Typical pipetting tasks performed on liquid handling workstations include the addition of reagents or internal standard solutions, dilutions, and liquid transfer. The workstations are able to wash and rinse fixed tips. Alternatively, they can work with disposable tips. Whereas classical pipetting heads have fixed tips, newer instrumentation includes flexible positioning of up to eight tips (e.g., Span8, Beckman Coulter).

12.8 Future Challenges

Automating biological analyses is currently a wide-ranging area constantly undergoing new development trends. The various development institutions are dependent on the corresponding applications and data to be recorded. The trends in high-content screening and nanodosing technology provide an excellent example of this.

12.8.1 Challenges Facing HCS

Increasing information density from cell-based assays poses increasing demands on data processing and storage. The main challenges in HCS over the next few years will especially involve the development of innovative information technology solutions.

High throughput rates in HCS are only feasible with robotic solutions to reduce the need for time-consuming and cost-intensive manual intervention [1, 54]. This particularly requires integration work on the engineering side, and will also encourage efforts in standardizing device interfaces and data formats.

Connecting complex systems to lab information management systems is another area of focus, enabling uniform planning, control, and management of the systems and the data they produce. The development of modern Web-based technologies based on XML or database technologies may lead to the creation of open uniform solutions for reducing costs while realizing the functionalities required in many different areas.

We may anticipate major advances in 3D cell screening in practical applications, with development work being carried out by both academic and commercial institutions across the world. The use of 3D technologies will improve in vitro modeling, but will also require parallel advances in cell cultivation, assay technologies, and state-of-the-art detection systems.

12.8.2 Nanodosing

The last few years have seen the development of a variety of automation systems for dosing fluids, especially focusing on systems for dosing aqueous compounds in the lower milliliter to lower microliter range. The increasing miniaturization of reaction systems such as the transition towards 1,536-well plates or chip technologies has also created the need for high-precision dosing of ever-decreasing fluid amounts.

There are various processes available for dosing small sample amounts. The most well-known is the ink-jet technique based on the functional principle of ink-jet printers [55].

Apart from that, nanojet processes are used in applications requiring continuous fine jets [56]. There are also processes that combine the advantages of injector pumps and piezoactuators; these include the Active Tip or nQuad technology developed by Tecan [57]. Spot On technology is available for simultaneous realization of dosing and aspiration [58].

Current systems support dosing quantities of less than 100 pl, with standard devotion at less than 7% for volumes down to 50 nl. Standard deviation values of even less than 4% can be reached for higher volumes.

The main drawbacks of the systems available today have so far been the insufficient integration in complex automation systems or their usability with a wide variety of solvents. Also, tracking dosing behavior—the exact determination of volume and number of dosed drops—poses a major challenge on image processing. This not only requires appropriate camera systems, but also specialized image-processing algorithms. This issue becomes particularly acute where high-parallel dosing issues are involved.

12.8.3 Automation and Miniaturization

Regardless of biological analysis process type, there are two major trends: automation and miniaturization. Both trends are based on the requirement for more samples per time interval together with decreased costs per data point.

Automation.   The issue of sequential or parallel sample handling is decisive in the development of automation concepts, as these affect system hardware design and control system and information processing.

Automating manual processes generally poses major issues for engineers and users. On the one hand, manual processes often cannot be implemented on a 1:1 basis, and alterative process designs and modifications are required to reach identical results with the same reproducibility as in manual systems. On the other hand, there is a variety of suppliers for fully automated systems on the market; however, they can only satisfy around 80% to 90% of the requirements that applications pose [59]. The remaining 10% to 20% consists of customer-specific requirements for which there is no standard solution. Additionally, subcomponents in automation systems cannot be freely configured in many cases; cost considerations often require that existing systems be combined or additional subcomponents be inte-grated into the lab automation system. In contrast to PC technology, there are no generally applicable standards for lab devices that would enable plug-and-play tech-nology, and appropriate system integration strategies have yet to be developed.

Miniaturization.   There are two main driving forces behind the increasing minia-turization of reaction systems. First, the tendency is towards processing increasing numbers of samples per unit area and time interval, thus reducing development times and therefore costs. Second, catalysts, proteins, and enzymes are frequently major cost factors in screening examinations—a reduction in the amounts used would also lead to considerable cost savings. At the same time, however, there is also the requirement for the highest possible reproducibility and lowest possible standard deviation in the nanoliter dosing range.

Another necessity to be addressed in increasing miniaturization is upscalability of processes, methods, and techniques. This will avoid additional costs that would otherwise arise from transferring miniaturized processes to the application research lab or production operation level.

References

[1]   HighTech Business Decisions, *High-Throughput Screening 2002: New Strategies and Tech-nologies,* Morgana, CA, May 2002.

[2]   Sannes, L. J., *High Content Screening: Parallel Analysis Fuels Accelerated Discovery and Development*, Report #24, Cambridge Health Institute (CHI), Newton Upper Falls, MA, August 2002.

[3]   Fisler, R., and J. Burke, *High Content Analysis Market Outlook*, Report #36, Cambridge Health Institute (CHI), March 2004.

[4]   Simpson, P. B., "A Model for Efficient Assay Development and Screening at a Small Research Site," *Journal of the Association for Laboratory Automation*, 2006, pp. 100–109.

[5]   Vasudevan, C., et al., "Improving High-Content Screening Assay Performance by Using Division-Arrested Cells," *Assay and Drug Development Technologies*, Vol. 3, No. 5, 2005, pp. 515–523.

[6]   Milligan, G., "High-Content Assays for Ligand Regulation of G-Protein-Coupled Recep-tors," *Drug Discov. Today*, Vol. 8, No. 13, 2003, pp. 579–585.

[7]   Filmore, D., "It's a GPCR World," *Modern Drug Discovery*, Vol. 7, No. 11, 2004, pp. 24–28.

[8]   Felder, R. A., et al., "G Protein-Coupled Receptor Kinase 4 Gene Variants in Human Essen-tial Hypertension," *Proc Natl Acad Sci USA*, Vol. 99, No. 6, March 2002, pp. 3872–3877.

[9]   Borchert, K. M., et al., "High-Content Screening Assay for Activators of the Wnt/Fzd Path-way in Primary Human Cells," *Assay and Drug Development Technologies*, Vol. 3, No. 2, 2005, pp. 133–141.

[10]   Richards, G. R., et al., "A Morphology- and Kinetic Based Cascade for Human Neural Cell High Content Screening," *Assay and Drug Development Technologies*, Vol. 4, No. 2, 2006, pp. 143–152.

[11]   Chan, G. K.Y., et al., "High Content Kinetic Assays of Neuronal Signaling Implemented on BD™ Pathway HT," *Assay and Drug Development Technologies*, Vol. 3, No. 6, 2005, pp. 623–636.

[12]   Price, R. D., et al., "A Simple, Flexible, Nonfluorescent System for the Automated Screening of Neurite Outgrowth," *J. Biomol. Screen*, Vol. 11, No. 2, 2006, pp. 155–164.

[13]   Ramm, P., et al., "Automated Screening of Neurite Outgrowth," *J. Biomol. Screen*, Vol. 8, No. 1, 2003, pp. 7–18.

[14]   Kittler, R., et al., "An Endoribonuclease-Prepared Sirna Screen in Human Cells Identifies Genes Essential for Cell Division," *Nature*, Vol. 432, No. 7020, 2004, pp. 1036–1040.

[15]   Pelkmans, L., et al., "Genome-Wide Analysis of Human Kinases in Clathrin- and Caveolae/Raft-Mediated Endocytosis," *Nature*, Vol. 436, No. 7047, 2005, pp. 78–86.

[16]   Simpson, P. B., and K. A. Wafford, "New Directions in Kinetic High Information Content Assays," *Drug Discovery Technology*, Vol. 11, No. 5/6, 2006, pp. 237–244.

[17]   Lee, M. S., *LC/MS-Applications in Drug Development*, New York: John Wiley & Sons, 2002.

[18]   Lee, M. S., *Integrated Strategies for Drug Discovery Using Mass Spectrometry*, New York: John Wiley & Sons, 2005.

[19]   BioInformatics, LLC, *Mass Spectrometry: Opportunities in the Life Science Market*, Report #05-012, April 2005.

[20]   Hiller, D. L., et al., "Rapid Scanning Technique for the Determination of Optimal Tandem Mass Spectrometric Conditions for Quantitative Analysis," *Rapid Commun. Mass Spectrom.*, Vol. 11, No. 6, 1997, pp. 593–597.

[21]   Zeng, H., J.-T. Wu, and S. E. Unger, "The Investigation and the Use of High Flow Column-Switching LC/MS/MS as a High Throughput Approach for Direct Plasma Sample Analysis of Single and Multiple Components Pharmacokinetic Studies," *J. Pharm. Biomed. Anal.*, Vol. 27, No. 6, 2002, pp. 967–982.

[22]   Dethy, M., et al., "Demonstration of Direct Bioanalysis of Drugs in Plasma Using Nano-electrospray Infusion from a Silicon Chip Coupled with Tandem Mass Spectrometry," *Anal. Chem.*, Vol. 75, No. 4, 2003, pp. 805–811.

[23]   Van Pelt, C. K., S. Zhang, and J. Henion, "Characterization of a Fully Automated Nanoelectrospray System with Mass Spectrometric Detection for Proteomic Analysis," *J. Biomol. Tech.*, Vol. 13, No. 2, 2002, pp. 72–84.

[24]   Guan, F., et al., "Collision-Induced Dissociation Pathways of Anabolic Steroids by Electro-spray Ionization Tandem Mass Spectrometry," *J. Am. Soc. Mass Spectrom.*, Vol. 17, 2006, pp. 477–489.

[25]   Rodgers, R. P., et al., "Jet Fuel Chemical Composition, Weathering, and Identification as a Contaminant at a Remediation Site, Determined by Fourier Transform Ion Cyclotron Resonance Mass Spectrometry," *Anal. Chem.*, Vol. 71, 1999, pp. 5171–5176.

[26]   Kessler, R. W., *Prozessanalytik—Strategien und Fallbeispiele aus der industriellen Praxis*, New York: Wiley-VCH Verlag GmbH & Co KGaA, 2006.

[27]   US005498324 Multiplexed fluorescence detector system for capillary electrophoresis, Edward S. Yeung et al.: Anregung der Detektionszone über Glasfasern (axial oder orthogonal zur Achse der Kapillare).

[28]   CA0002166830A1 Capillary array electrophoresis system, Hideki Kambara et al.: System zur Detektion direkt am Kapillarenausgang bei seitlicher Anregung.

[29]   US0006231739B1 Multi-channel capillary electrophoresis device including sheath-flow cuvette and replaceble capillary array, Eric S. Nordman: Sheat-Flow Zelle mit steckbarem Kapillarenarray.

[30]   US0006120667A Multi-capillary electrophoresis apparatus, Yoshihide Hayashizaki et al.: Anregungs- und Abtastsystem mit konfokaler Optik und Relativbewegung zwischen Array und Abtastsystem (lineare Bewegung).

[31]   US0006054032A Capillary electrophoresis array, Louis C. Haddad et al.: Konstruktion eines flexiblen Kapillarenarrays.

[32]   US0006017765A Fluorescence detection capillary array electrophoresis analyzer, Takashi Yamada et al.: Anordnung mit mehreren Kapillarenarrays, so dass Vorbereitung, Analyse und Gelreplacement an verschiedenen Stationen parallel ablaufen können, Detektion über Sheath-Flow Zelle.

[33] US0005741411A Multiplexed capillary electrophoresis system, Edward S. Yeung et al.: verschiedene Anordnungen mit seitlicher, kohärenter Anregung.

[34] US0005938908A Capillary array electrophoresis system, Takashi Anazawa et al.: Anordnungen zur Erhöhung der Zahl der nutzbaren Kapillaren bei seitlicher Anregung (Fasern als Zylinderlinsen zur Refokussierung, gestufte Anordnung der Kapillaren. Anregung des Arrays von zwei Seiten aus).

[35] US0005833827A Capillary array electrophoresis system, Takashi Anazawa et al.: Anordnungen zur Erhöhung der Zahl der nutzbaren Kapillaren bei seitlicher Anregung (Kapillarenformen, Immersionsmittel).

[36] US0006027627A Automated parallel capillary electrophoretic system, Qingbo Li et al.

[37] US0005916428A Automated system for multi-capillary electrophoresis having a two-dimensional array of capillary ends, Thomas E. Kane et al.

[38] US0005885430A Capillary tube holder for an electrophoretic apparatus, John R. Kernan et al.

[39] WO0200125773A1 Uniform laser excitation and detection in capillary array electrophoresis system and method, Qingbo Li et al.

[40] WO0200116587A1 Automated parallel capillary electrophoresis system with hydrodynamic sample injection Changsheng Liu et al.

[41] WO0009900664A1 Automated parallel capillary electrophoretic system, Thomas E. Kane et al.

[42] WO1997030347A1 (US0005900934A) Capillary Chromatography Detector Apparatus, A. C. Gilby and W. W. Carson.

[43] WO200118528A1 Method of analysing multiple samples by detecting absorption and systems for use in such a method, Edward S. Yeung und Xiaoyi Gong.

[44] Gong, X., and E. S. Yeung, "Novel Absorption Detection Approach for Multiplexed Capillary Electroporesis Using a Linear Photodiode Array," *Analytical Chemistry*, Vol. 71, No. 21, 1999, Patent WO 01/18528 A1.

[45] Manz, A., N. Graber, and H. M. Widmer, "Miniaturized Total Chemical Analysis Systems: A Novel Concept for Chemical Sensing," *Sensors & Actuators*, Vol. 1, Nos. 1–6, 1990, pp. 244–248.

[46] Belder, D., et al., "Poly(Vinyl Alcohol)-Coated Microfluidic Devices for High-Performance Microchip Electrophoresis," *Electrophoresis*, Vol. 23, 2002, pp. 3567–3573.

[47] Schulze, P., et al., "Deep UV Laser-Induced Fluorescence Detection of Unlabeled Drugs and Proteins in Microchip Electrophoresis," *Analytical Chemistry*, Vol. 77, 2005, pp. 1325–1329.

[48] Ludwig, M., and D. Belder, "Coated Microfluidic Devices for Improved Chiral Separations in Microchip Electrophoresis," *Electrophoresis*, Vol. 24, 2003, pp. 2481–2486.

[49] Fertig, N., et al., "Biochip für die Analyse zellulärer Ionenkanäle," *Laborwelt*, Vol. 3, 2002, pp. 16–19.

[49] Owen, D., and A. Silverthorne, "Channelling Drug Discovery: Current Trends in Ion Channel Drug Discovery Research," *Drug Discovery World*, Vol. 3, 2002, pp. 48–61.

[50] Sigworth, F. J., and K. G. Klemic, "Patch Clamp on a Chip," *Biophys. J.*, Vol. 82, 2002, pp. 2831–2832.

[51] Fertig, N., R. H. Blick, and J. C. Behrends, "Whole Cell Patch Clamp Recording Performed on a Planar Glass Chip," *Biophys. J.*, Vol. 82, 2002, pp. 3056–3062.

[52] Knott, T., S. Single, and A. Stett, "Automatisiertes Patch-Clamping: Lösungen und Herausforderungen," *Transkript Laborwelt*, Vol. 4, 2002, pp. 20–22.

[53] Stett, A., et al., "Cytocentering: A Novel Technique Enabling Automated Cell-by-Cell Patch Clamping with the CytoPatch Chip," *Receptors and Channels*, Vol. 9, 2003, pp. 59–66.

[54] Frost & Sullivan, *Analysis of Cell-Based Assay Markets in Europe*, 2004.

[55]   Cooley, P., D. Waace, and B. Antohe, "Application of Ink-Jet Printing Technology to BioMEMS and Microfluidic Systems," *MicroFab Technologies, Inc. SPIE Conference on Microfluidic and BioMEMS*, San Francisco, CA, October 22–25, 2001.

[56]   Firmenschrift, *Mikrodosierung mit Piezo- und Ventil-Technologie*, Microdrop-Gesellschaft für Mikrodosiersysteme mbH, Norderstedt (D), 2003.

[57]   Application Note D-101, *Introduction to Nanoliter Quantitative Aspirate and Dispense*, Cartesian Technologies, U.K., 2001.

[58]   Shevts, S. M., et al., "Spot-On Technology for Low Volume Liquid Handling," *Journal Association Laboratory Automation*, Vol. 7, No. 6, 2002, pp. 125–129.

[59]   Elands, J., "The Evolution of Laboratory Automation," in *Handbook of Drug Screening*, R. Seethala and P. B. Fernandes, (eds.), New York: Marcel Dekker, 2001, pp. 477–492.

CHAPTER 13

In Situ Imaging and Manipulation in Nano Biosystems Using the Atomic Force Microscope

Ning Xi, Guangyong Li, and Donna H. Wang[†]

13.1 Introduction

The daunting challenge that we are facing in the postgenome era is to understand gene and protein function. Tremendous efforts have now been directed towards the development of gene and protein expression profiling, which allows us to look at multiple factors involved in diseases such as essential hypertension, which is known as polygenic disease. However, this global approach merely gives a snapshot of a sequence of events and may not provide cause-and-effect analysis. Understanding the location, structure, and molecular dynamics of biomolecules is of fundamental importance to elucidate their functions. For example, how do proteins behave and function individually? How do they interact with each other? To gain insights into how these biomolecules operate, advanced technologies are required for gaining information at the level of a single molecule. Atomic force microscope (AFM) [1] would be one of the novel tools for this task.

13.1.1 AFM: A Promising Tool for Biological Research

The AFM was initially developed as an instrument mainly used for surface science research. Research efforts in the past few years have indicated that AFM is also a potentially powerful tool for biochemical and biological research [2]. This rapid expansion of AFM applications in biology/biotechnology results from the fact that AFM techniques offer several unique advantages. First, they require little sample preparation, with native biomolecules usually being imaged directly. Second, they are less destructive than other techniques (e.g., electron microscopy) commonly employed in biology. Third, they can work in several environments, including air, vacuum, and liquid, and especially under the condition that the cells are alive [3].

Although membrane proteins are the main drug targets, the study of cell membrane proteins in situ on the molecular level is difficult due to the current technical

† Michigan State University

limitations. Studies on living cells using AFM with high resolution are hampered by cell deformation and tip contamination [4]. Different approaches have been used to obtain high resolution images of soft biological materials. For example, at low temperature, cells stiffen and high resolution imaging becomes feasible [5]. Another solution is to use the tapping mode AFM (TMAFM) in liquid, which gives a substantial improvement in imaging quality and stability over the standard contact mode [6]. Because of the viscoelastic properties of the plasma membrane, the cell may behave like a "hard" material when responding to externally applied high frequency vibration, and it is less susceptible to deformation [4].

13.1.2 Functionalization of AFM Probe: Paving a New Avenue

Recent progress in the spatial resolution of AFM technology has made topographical imaging of a single protein a routine work [7, 8]. The unique capability of the AFM to directly observe single proteins in their native environments provides insights into the interaction of proteins that form functionality assemblies. As far as we know, it is still impossible to recognize specific proteins like receptors only from the topographical information by AFM. The technique to functionalize an AFM tip with certain molecules has opened a promising way to recognize single specific molecules. It has been proven that single receptors can be recognized by an AFM tip functionalized with antibodies through a force mapping technique [9, 10], or directly from a phase image in TMAFM [11, 12]. However, all these results are obtained by imaging well-prepared samples on very flat surfaces, and not in their original locations, that is, cell membranes. In practice, it is very difficult to image the single receptors on cell membranes due to the topographical interference and softness of the membrane. In this chapter, techniques using AFM to probe biomolecules are introduced and reviewed. The state-of-art techniques for characterizing a specific single receptor using the functionalized AFM tip are discussed. An example of studying the angiotensin II type 1 (AT1) receptors expressed in sensory neuronal cells by AFM with a functionalized tip is also given.

13.1.3 From Imaging to Manipulation: A Big Leap

Since the invention of atomic force microscopy, it has become a standard technique in imaging various sample surfaces down to the nanometer scale. Recent development of AFM has rapidly extended its ability from surface imaging to manipulation of nano-objects [13–16]. The main problem of these AFM-based manipulation schemes is the lack of real-time visual feedback. Each operation has to be verified by another new image scan before the next operation. Obviously, this scan-design-manipulation-scan cycle is time-consuming and inefficient. Combining the AFM with virtual reality interface and haptic devices [17, 18] may facilitate the nanomanipulation by simplifying the off-line design, but the operator is still blind because he/she cannot see in real time the environmental changes through the static virtual reality. A new image scan is still necessary after each operation. Thus, any methods which can update the AFM image as close as possible to the real environment in real time will help the operator to perform several operations without the need of a new image scan.

The augmented reality enhanced nanorobotic system developed in [19, 20] aims to provide the operator with both real-time visual feedback and force feedback during manipulation. The real-time visual feedback is a dynamic AFM image of the operating environment, which is locally updated based on real-time force information and system models as well as local scan information. Under the assistance of the augmented reality interface, the operator can perform several operations without the need for a new image scan by AFM. Using the AFM-based manipulation system that is assisted by the augmented reality interface, several experiments have been successfully performed, such as nanoimprints on a soft polycarbonate surface [21], manipulation of latex particles on a glass surface [22], and assembly of nanorods on a polycarbonate surface [23]. However, all the objects in the previous work are rigid and the manipulation is operated under ambient air condition. Most biological entities are soft and have to stay in liquid medium to remain active or alive. In this chapter, manipulation of living neurons under liquid medium is performed by using the AFM-based nanorobotic system. Live cells' images are obtained in their physiological environments using the TMAFM. Single cell surgery can be performed on nanoscale when the cells are alive.

13.2   Reviews of Biomolecular Recognition Using AFM

13.2.1   Surface Functionalization

By immobilizing samples on a very flat surface, such as mica, individual biomolecules can be observed by AFM. For example, the major intrinsic proteins have been immobilized on a freshly cleaved mica surface by incubating with carboxypeptidase at room temperature overnight [8]. The mica surface was silanized in a solution of 2% 3-aminopropy-trithoxysilane in toluene for 2 hours [24] or by exposing it to the vapor of 3-aminopropy-trithoxysilane for several minutes [7].

13.2.2   Functionalization of AFM Probe

Functionalization of AFM tips by chemically and biologically coating with molecules (e.g., biotin-avidin pairs [25, 26] and antigen-antibody pairs [24, 27]) has opened a new research area for studying interactions of molecules on the molecular level. Chemical coating of probes is mainly done by silanization or functionalized with thiols and is often the first step before biological functionalization. Many protocols have been used for attaching proteins to an AFM tip. There are two main ways to functionalize the AFM tip with antibodies. One is to directly coat the antibody on a silanized tip, and the other is to tether the antibody on a tip by a linker (or spacer). The direct coating method is simple and results in high lateral resolution. The tethering method involves much more complicated steps, but it gives better antigen recognition because the interaction between antibody and antigen is highly specific, which involves a high degree of spatial freedom and orientational specificity. The drawback of the tethering method is that the lateral resolution is low. The detailed steps of these two methods are discussed below.

Several direct coating methods are available, and most of them are based on silanizing a solid surface. Here is an example that has been used by the authors to directly attach antibodies to an AFM tip. The silicon nitride tips were treated with 10% nitric acid solution which was left in a bath for 20 minutes at 80°C. This causes the formation of surface hydroxyl groups on the SiN tips. The tips were then thoroughly rinsed with distilled water, placed into 2% 3-aminopropylmethyl diethoxysilane (APrMDEOS) solution in toluene, and kept in a desiccator purged with argon gas for 5 hours. This treatment provides reactive primary amine groups on the nitride surface. The tips were washed thoroughly with phosphate buffer saline (PBS) and placed into a solution of 2-μg/ml Antibody IgG for 10 minutes. The antibody-conjugated tips were then washed thoroughly with PBS and distilled water to remove loosely attached antibodies. These tips should be used immediately before being dried.

Functionalization of the AFM tip with antibody using the tethering method involves many more steps than the direct coating method. It usually needs a spacer to covalently bind the protein in order to orient the protein to expose specific site(s) of the protein. Polyethyleneglycol (PEG) is a common spacer to be used. A terminal thiol group can be first attached to PEG, which then will be attached to a gold-coated silicon nitride tip. An amine group at the other end of the PEG molecule attaches proteins (antibodies for example) via a covalent bond [27]. A detailed protocol to functionalize a silicon nitride AFM tip with AT1 receptor antibody can be found in Section 13.4.4.

13.2.3 Force Interaction Measurement

AFM is capable of measuring forces on the piconewton scale by nature, a property that has been exploited to examine receptor-ligand interactions. An emerging body of literature employing AFM to measure and characterize these interactions is available. A wide range of receptor-ligand pairs has been studied by AFM with a functionalized tip. The first study focuses on very high affinity interactions, such as interactions between biotin and avidin [25, 26]. In the study, a silicon nitride AFM tip was functionalized by avidin through the following steps: The tip was first cleaned and silanized and then incubated in biotinylated bovine serum albumin for 24 hours. After fixed in 1% glutaraldehyde solution for 30 seconds, the avidin was added to bind to the biotin, and the AFM tip was functionalized with avidin. By immobilizing the biotin on an agarose bead, the rupture force between the biotin from the bead and the avidin from the tip was measured using the force modulation mode. When the tip was retracted, it detached from the surface in a series of discrete jumps with each corresponding to breakage of one or more biotin-avidin bindings. The total jump-off force was expected to consist of an integral multiple of single rupture force. Therefore, by constructing a histogram of rupture forces, the single pair unbinding fore was measured. The rupture force has been calibrated from 160 pN for avidin-bioton pair to 260 pN for streptavidin-biotin pair [28]. As understanding and interpretation of these initial studies on the measurements of rupture forces by AFM has been increased, more investigators have reproduced and extended the avidin-biotin findings by determining the bond strength of other examples of receptor-ligand pairs. These studies of AFM binding have been given an extensive overview by Willemsen et al. [29].

13.2.4 Single Receptor Recognition by Force Mapping

Antibody-antigen interaction is of importance in the immune system, which may vary considerably in affinity. Hinterdorfer and coworkers were the first to determine the interaction between individual antibodies and antigens [27]. In their work, they used flexible linkers to couple either the antigen or the antibody to the tip, which provided the antibody and the antigen enough freedom to overcome problems of misorientation due to the high degree of spatial and orientational specificity between antibody and antigen. By coupling the antigens and antibodies to the tip and surface, respectively, via a polyethylene glycol (PEG) spacer (8 nm in length), the binding probability has been significantly improved due to the large mobility provided by the long spacer molecules. When a tip is functionalized at very low antibody density such that only one single antibody at the tip apex has a chance to access an antigen on the surface, the single molecular antibody-antibody complex could be examined. Although high-resolution images can provide some detailed conformal information of molecules, they may not provide information related to any specific protein. Because the interaction between ligands and receptors is highly specific, the technique to functionalize an AFM tip with specific molecules makes investigation of single specific molecule possible. Rupture forces representing biomolecular specific interactions can also be exploited as a contrast parameter to create images in which the individual biomolecules can be recognized. It has been proven that single receptors can be recognized by an AFM tip functionalized with antibody through a force mapping technique [10, 29]. In their study, the functionalized tip was raster-scanned over the surface while a force-distance curve was generated for every pixel. From the force-distance curve, the surface parameters such as stiffness and adhesion force were extracted either by on-line in real time or by off-line analysis. The individual receptors were recognized through this so-called adhesion mode AFM. However, the adhesion AFM image obtained by this method has low lateral resolution and the work is extremely time-consuming.

13.2.5 Single Receptor Recognition by Phase Changing

Another way to recognize specific proteins like receptors is the use of tapping-mode phase imaging. It can differentiate between areas with different properties regardless of their topographical nature [30, 31]. The phase angle is defined as the phase lag of the cantilever oscillation relative to the signal sent to the piezo driving the cantilever. Theoretical simulations and experiments of the cantilever dynamics in air have shown that phase contrast arises from differences in the energy dissipation between the tip and the sample. The phase shift is related analytically to the energy dissipated in the tip sample interaction by the following equation [30, 31]:

$$\sin \psi = \left(\frac{\omega}{\omega_0} \frac{A}{A_0} \right) + \frac{Q E_D}{\pi k A A_0} \tag{13.1}$$

where ψ is the phase angle; ω/ω_0 is the working frequency/resonance frequency; A/A_0 is the set-point amplitude/free amplitude; Q is the quality factor; E_D is the energy dissipation; and k is the cantilever spring constant. The phase shift due to the tip-sample interaction, which involves energy dissipation, is the displacement of the noncontact solution to higher phase shifts and the intermittent contact solution to lower phase shift values. The more dissipative features will appear lighter in the noncontact regime, whereas they will appear darker in the intermittent-contact regime [32].

When scanning the protein immobilized on a mica surface using a tip functionalized with its antibody, the tip-sample interaction force will increase as the tip approaches the receptors. Thus, a significant change of the phase shift will be generated. Since the topographical information is also convoluted to the phase contrast image but with low frequency, a band-pass filter can be used to remove the low frequency topographical information and the high frequency noise. After filtering the phase contrast image, only the receptors' image will be left on the surface. Individual surface receptors have been identified on a mica surface using these techniques [11, 12].

13.3   Techniques for In Situ Probing Membrane Proteins

All the results aforementioned are obtained by imaging well-prepared samples on very flat substrate surfaces, and not by imaging the cells directly. In practice, it is very difficult to image the single receptors on their original locations (cell membranes) due to the topographical interference and softness of the membrane. Therefore, several techniques have to be employed in order to overcome the difficulties. In this section, the critical techniques that are used to overcome the softness of the membrane and the surface convolution will be introduced. An example of studying the AT1 receptors expressed in sensory neurons by AFM with a functionalized tip is also given in the next section.

13.3.1   Cell Fixation

High resolution of cells using AFM has been hampered by cell deformation and tip contamination [4]. TMAFM in liquid gives a substantial improvement in imaging quality and stability over standard contact mode. Due to viscoelastic properties of the plasma membrane, the cell may harden when responding to externally applied high frequency vibration and hence is less susceptible to deformation. Moreover, with this imaging mode, the cantilever oscillates at its resonant frequency and is only in intermittent contact with the cell surface. As a result, the destructive shear force is minimized. Although the TMAFM has significantly improved the resolution, it is still not possible to achieve molecular resolution when imaging the living cells. Therefore, finding an innovative way to fix the cell is necessary. A polymer microgrid has been developed to tackle these challenges [33]. A microgrid can be used to mechanically immobilize the surface of living cells. The microgrid helps to constrain the membrane and to prevent it from deformation caused by the tapping force from the AFM tip. The AFM tip touches the cell membrane through the small openings on the microgrid, provided that the openings are large enough and the thickness is thin enough to allow the AFM tip to probe the cell surface through the hole. The grid design is shown in Figure 13.1. The opening size varies from 2 to 10 μm depending on the cell size. The grid thickness varies from 100 nm to 2 μm.

The fabrication process starts with a blank double-side polished GaAs wafer. A layer of 500-nm Polyimide PI-2556 from HD Micro Systems is deposited on top by a spinning process. It was baked for 60 seconds at 150°C (Hotplate). This layer was then patterned by AZ5214E photoresist. The photoresist and polyimide layers were etched together by AZ300 developer. Then, fully curling of the polyimide was done under vacuum at 200°C. After that, the whole substrate was covered with photoresist and a hole was patterned at the backside of the wafer. H_3PO_4:H_2O_2:H_2O (1:13:12 in volume) was used to back etch GaAs (Gallium Arsenide) substrate. As a result, a suspended polyimide membrane was made.

The microgrid has been fabricated in the authors' laboratory, as shown in Figure 13.2, with a thickness of 500 nm containing 3- and 10-μm holes on it. The force from the gravity of the microgrid may not be strong enough to constrain the cells. In order to generate enough constraint force from the grid to the cell, magnetic beads with diameter of 11 μm are used by depositing them on the top of the microgrid. The constraint force can be adjusted by controlling the number of the magnetic beads as well as the current of the solenoid underneath the sample.

13.3.2   Interleave Lift-Up Scan

Although individual proteins can be recognized efficiently through the tapping-mode phase imaging by a functionalized tip, biomolecules have to be extracted, purified, and attached to a flat and rigid surface. Detecting membrane proteins directly in native environments is a daunting challenge given that the cell membrane surface topographical information will interfere with the antigen-antibody binding. The interference submerges the binding signals and makes the recognition impossible. Fortunately, these problems can be solved by an interleaved lift-up scanning, recently developed, to remove the interference from topography. By scanning the same line twice, the topographical image can be obtained from the first scan, and the phase image can be obtained from the second lift-up scan. The mechanism of interleave lift-up scan is shown in Figure 13.3. During interleave lift-up scan, the

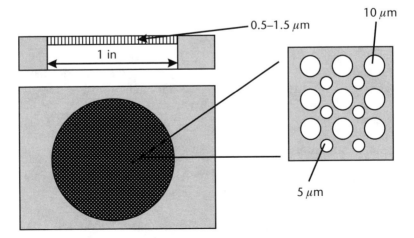

Figure 13.1   The design of polymer microgrid.

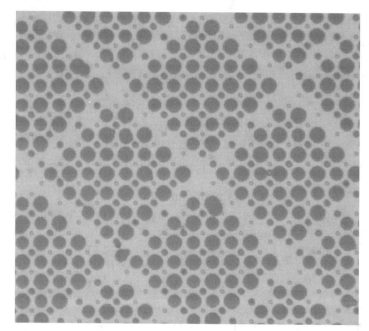

Figure 13.2   Microgrid fabricated with 3- to 10-μm opening.

AFM first scans the surface for one line and records the topographical information of that line, and then the AFM tip is lifted up a certain distance and scans the same line but follows the topography obtained from the previous scan. As long as the lift-up distance is shorter than the length of the PEG linker, the receptor-antibody interaction will not be affected, while the topographical interference caused by the surface can be removed during the second scan. The phase image captured from the second scan only contains the information of receptor-antibody interaction. By integrating the phase image from the second scan with the topographical image captured from the first scan, the location of the receptor can be easily identified and labeled.

13.4   In Situ Probing Angiotensin II Type 1 Receptor

The AT1 receptor expressed in sensory neuronal cell is studied using the developed techniques. Before the experiments, the background of the AT1 receptor and its associated rennin-angiotensin system is introduced.

13.4.1   Renin-Angiotensin System

The renin-angiotensin system plays a major role in the regulation of blood pressure, vasopressin and pituitary hormone release, sodium appetite, and perception of thirst [34, 35]. The cascade pathway of the renin-angiotensin system is illustrated in Figure 13.4. Renin, an enzyme produced primarily by the juxtaglomerular cells of the kidney, converts the liver precursor angiotensinogen to an inactive substance, angiotensin I (AI). AI is in turn converted to the physiologically active peptide,

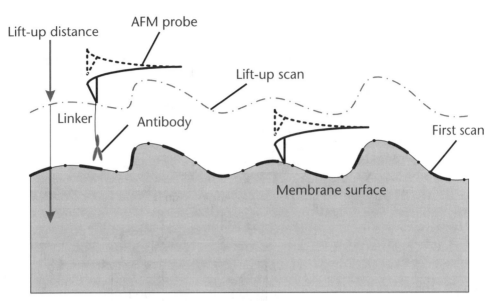

Figure 13.3 The lift-up scan mechnism. The AFM tip scans the first line and records the topographical information, and then the tip is lifted up and scans the same line but follows the topography obtained from the previous scan. In the lift-up scan, only the antigen-antibody interaction has occurred, which can be recorded from the phase image of the lift-up scan.

angiotensin II (AII), by pulmonary angiotensin-converting-enzyme (ACE). AII is thought to be involved in the control of blood pressure, fluid volume, electrolyte balance, and cardiovascular functions by acting on the AT1 receptor [36]. As shown in Figure 13.4, the cascade pathway of the renin-angiotensin system can be blockaded at different steps. Different drugs have been or are in the process of being developed based on the blocking steps. For example, ACE inhibitors or AT1 receptor antagonists have been developed as very effective antihypertensive drugs. Figure 13.4 also shows the major difference in the mode of action of AT1 receptor antagonists as opposed to ACE inhibitors. AT1 receptor antagonists block all of the known actions of angiotensin II through acting on the AT1 receptor directly. In contrast, ACE inhibitors work at the upstream of the renin-angiotensin cascade, inhibiting effects of ACE. ACE acts both on angiotensin I, converting the inactive angiotensin I to the active angiotensin II, and on bradykinin, inactivating this hormone. Since the inactivation of bradykinin is blocked by ACE inhibitors, bradykinin level rises, which results in a side effect: the ACE inhibitor induced cough [37]. That AT1 receptor antagonists are clinically equivalent, superior, or inferior to ACE inhibitors remains unknown because there have been few comparative trials of benefits of AT1 receptor antagonists on morbidity and mortality. Therefore, to develop more AT1 receptor antagonists based on different binding sites of the receptor and bring them to clinical trials may be beneficial in treating hypertensive patients more effectively.

13.4.2 Angiotensin II Type 1 Receptor and Its Antibody

Ang II is an important physiological effector of blood pressure and volume regulation through vasoconstriction, aldosterone release, sodium uptake, and thirst

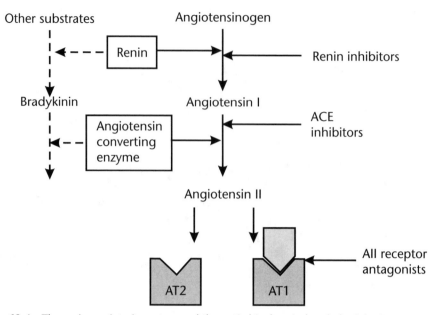

Figure 13.4 The renin-angiotesin system and theoretic biochemical and physiologic consequences of blocking the renin-angiotensin system at different steps in the pathway.

stimulation. The synthesis of novel nonpeptide Ang II receptor antagonists revealed the existence of at least two Ang II receptor types, designated type 1 receptor (AT1) and type 2 receptor (AT2). Although Ang II interacts with two types of cell surface receptors, most of the major cardiovascular effects seem to be mediated through AT1.

The amino acid sequences encoded by the open reading frame of AT1 cDNA and genomic DNA of all mammalian and rodent species consist of 359 amino acid residues [38]. Structural predictions suggest an extracellular NH2-terminus followed by seven helical transmembrane-spanning domains, which are connected by three extracellular and three intracellular loops, linked to the carboxyl-terminus as shown in Figure 13.5. The mechanisms of regulation, activation, and signal transduction of the AT1 receptor have been studied extensively in the decade after its cloning. At present, little is known about the mechanisms regulating intracellular trafficking of the AT1 receptor. However, using confocal laser scanning microscopy, it was demonstrated that the complex of FITC-coupled Ang II and Flag epitope tagged-AT1 receptor was internalized into endosomes. After removal of Ang II, the AT1 receptor was recycled back to the plasma membrane, while Ang II was targeted to the lysosomal degradation pathway [39]. The studies using AFM with tip functionalized by AT1 antibody may provide detailed evidence of receptor intracellular trafficking.

AT1 antibody (Santa Cruz Biotechnology, Inc.) is an affinity purified rabbit polyclonal antibody raised against a peptide mapping near the amino terminus of the AT1 receptor of human origin (identical to a corresponding rat sequence). It was chosen as the end effector functionalization material because it targets against the amino terminus of the AT1 receptor, which is supposed to be outside the cell membrane.

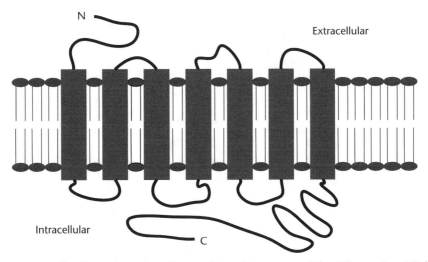

Figure 13.5 Predicted membrane topology and domain structure of the AT1 receptor. AT1 has seven complete transmembrane segments. The amino-terminal region is located outside the plasma membrane, and the carboxyl-terminal region is located inside the plasma membrane.

13.4.3　Living Cell Sample Preparation

Given that the purpose of this research is to study the functionality of the receptors, a native environment for the receptor has to be maintained during the study. In this study, sensory neurons are chosen in which AT1 receptors are expressed [40]. The cell samples are sensory neural cells growing on the glass coverslips with diameter of 15 mm. The cells originate from the dorsal root ganglia (DRG) tissue of male Wistar rats (body weight 125g to 200g). The DRGs from the cervical, thoracic, lumbar, and sacral levels were removed aseptically and collected in F12 medium (Gibco/BRL). The trimmed DRGs were digested in 0.25% collagenase (Boerhinger Mannheim) in F12 medium at 37°C for 90 minutes. Following a 15-minute incubation in PBS containing 0.25% trypsin (Gibco/BRL), the tissues were triturated with a pipette in F12 medium containing DNAse (Sigma, 80 μg/ml), trypsin inhibitor (Sigma, 100 μg/ml) and 10% heat-inactivated horse serum (Hyclone). Cells were then seeded in a 12-well culture plate with polyornithine-coated glass coverslides inside. The cells were cultured in a humid incubator at 37°C with 5% CO_2 and 95% air. The cells are ready for AFM scanning after 7 to 10 days of culture.

13.4.4　Functionalization of AFM Probe with AT1 Antibody

The process is designed to functionalize the tip with AT1 antibody as shown in Figure 13.6, but the protocol can be easily extended to other functional agents. There are several steps needed in the functionalization process:

1. *Modify the AT1 antibody with N-Succinimidyl 3-(Acetylthio)propionate:* Wash PD-10 column with 30 ml of buffer A (100 mM NaCl/50 mM NaH_2PO_4/ 1 mM EDTA, pH 7.5); load 200 μl (4 μg/ml) of antibody in 300 μl of buffer A onto PD-10 column; pass 500 μl of buffer A through the column nine times; collect fractions seven and eight and pool them; add a 10-fold molar excess of N-succinimidyl 3-(acetylthio)propionate (SATP)/

DMSO to antibody solution for 30 to 60 minutes under argon; rinse two PD-10 columns with 30 ml of buffer A; add 500 μl of antibody solution into each PD-10 column, and then pass 9 × 500 μl of buffer A through the columns; collect fractions seven and eight and store them in 100-μl aliquots at −70°C.

2. *Modify AFM tips with aminopropyltriethoxysilane (APTES):* UV-clean cantilevers for 15 minutes; purge a glass desiccator with argon for 2 minutes, and place 30 μl of APTES (99%; Sigma-Aldrich) into it in a small container; place 10 μl of N,N-diisopropylethylamine (99%, distilled; Sigma-Aldrich) into another small container in the desiccator and purge with argon for an additional 2 minutes; place cleaned probes into the desiccator; purge for another 3 minutes, and seal for 0.5 to 2 hours; remove reagents, purge with clean argon, and keep probes stored in this environment.

3. *Tether the cross-linkers (spacer) on tips:* Mix cross-linker and 5 μl of triethylamine in 1 ml of CHCl$_3$; place NH$_2$-modified tips into this solution for 2 to 3 hours; wash tips with CHCl$_3$, and dry with argon.

4. *Link the SATP-labeled antibody to tips:* Incubate the tips in 50 μl of SATP-antibody, 25 μl of NH$_2$OH reagent (500 mM NH$_2$OH.HCl / 25 mM EDTA; pH 7.5), and 50 μl of buffer A for 1 hour; wash tips with buffer A and PBS buffer (150 mM NaCl / 5 mM Na$_2$HPO$_4$; pH 7.5) three times; store tips in PBS buffer at 4°C.

After functionalization, an immunohistochemical-like method has been used to verify the success of the functionalization. By incubating the functionalized tip in PBS, which contains antirabbit IgG (second antibody) labeled with CY3 for 1 hour, the AFM tip can be monitored by confocal fluorescence microscope. A clean AFM tip with functionalization is tested as a control. The fluorescence microscopy results are shown in Figure 13.7. It can be seen that the functionalized tip has significant

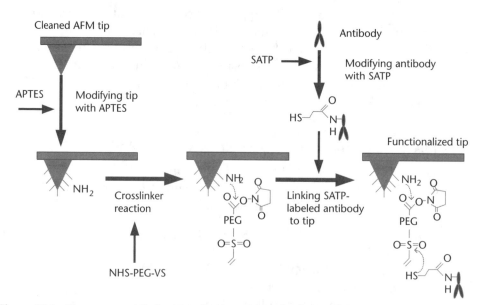

Figure 13.6 The process of tip functionalization with antibody via a linker.

fluorescence signal, but there is no fluorescence signal for the tip without functionalization.

13.4.5   Single AT1 Receptor Recognition

After the cells have been fixed and constrained by the polymer microgrid, a high resolution image of the cell surface is obtainable, as shown in Figure 13.8. However, it is not possible to identify any molecules form the topographical image of the cell. Using the techniques described in the previous section, the living neurons are scanned by the interleave lift-up method through the small opening of the microgrid. During lift-up scan, the tip-sample interaction force will increase as the tip approaches the AT1 receptor, thus a significant change of the phase shift will be generated and recorded in the lift-up phase image, as shown in Figure 13.9 in which the receptors are indicated inside the circles. The experimental result shows that single a biomolecule such as a receptor on the cell membrane can be recognized using the biologically functionalized tip. The result can be further verified by adding enough AT1 antibodies to the medium. The receptors disappear after adding the AT1 antibodies as shown in Figure 13.10. The disappearance of the receptors in the images results from the receptors on the membrane surface being blocked by the antibodies in the medium and so the antibodies on the tip lose their chance to bind the receptors on the membrane.

13.5   Single Cell Surgery by AFM

As in macromanipulation systems, by building up an augmented reality interface based on the modeling of the sample surface within the AFM frame, real-time nanomanipulation is possible with the assistance of a haptic device. Both the tip motion and joystick motion are rendered to the operating interface in real time. The operator can control the tip motion through the joystick and view tip movements

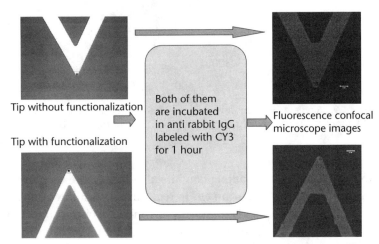

Figure 13.7   The immunohistochemical verification of the the probe functionalization. The functionalized tip has significant fluorescence signal but there is no fluorescence signal for the tip without functionalization, which verifies the success of the tip functionalization.

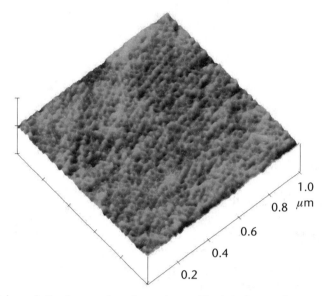

Figure 13.8 High resolution image of a cell membrane. The interference from topography submerges all the molecular images.

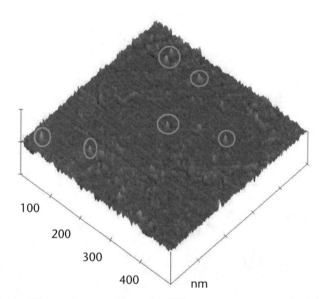

Figure 13.9 Single AT1 receptor reconition. The AT1 receptors are recognized through a lift-up scan by functionalizing the AFM tip with AT1 antibody (e.g., the juts inside the circles).

from the operating environments while also feeling the force from the joystick and seeing the changes of environment in real time. This allows the operator to generate manipulation paths quickly, make rapid control decisions based on force and visual feedback, and also adjust the paths in real time to ensure sufficient contact force without harming the object under manipulation. An augmented reality enhanced nanorobotic system developed in [19] is shown in Figure 13.11. It consists of

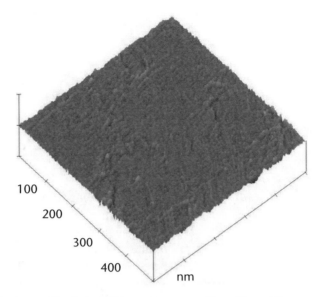

Figure 13.10 Further verification of AT1 receptor recognition. After adding enough AT1 antibodies, all the receptors disppear in the image, which means the receptors on the membrane surface are blocked by the antibodies in the medium and the antibodies on the tip lose their chance to bind with the receptors on the membrane.

real-time force feedback display and real-time visual display. The real-time visual display includes a live image of the cantilever tip from the CCD camera and a graphic image of the operating environment, which is updated based on force information. A haptic device (Phantom, from SensAble Technologies, Inc.) is used for real-time 3D force display.

The AFM based nanorobotic system includes two subsystems: the AFM system and the augmented reality interface, which are shown in Figure 13.11. The AFM

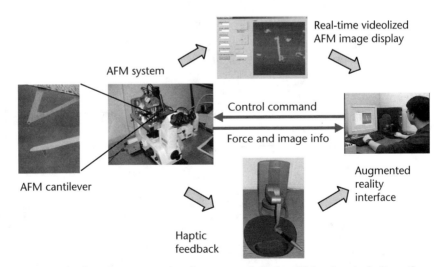

Figure 13.11 AFM based augmented reality system: (left) the AFM system including other accessories for imaging functions; (right) the augmented reality interface providing an interface for an operator to simultaneously control the tip motion through a haptic joystick, view the real-time AFM image, and feel the real-time force during manipulation.

system, called Bioscope (Veeco Instruments), is equipped with a scanner with a maximum xy scan range of 90 μm × 90 μm and a z-range of 5 μm. Peripheral devices include an optical microscope, a charge coupled device (CCD) camera, and a signal access module which can access most real-time signals inside the AFM system. The inverted optical microscope and the CCD camera help the operator to locate the tip, adjust the laser, and search for the interesting areas on the substrate. The augmented reality interface is a computer equipped with a haptic device (Phantom, from Sensable, Co.). Through the signal access module, the deflection signal can go directly into the A/D converter card inside the computer. The augmented reality interface provides enhanced media for the operator to view the real-time AFM image and feel the force feedback during nanomanipulation. The real-time visual display is a dynamic AFM image of the operating environment which is locally updated at video frame rate based on the environment model, tip-subject interaction model, real-time force information, and a local scanning mechanism, as shown in Figure 13.12. The two subsystems are connected through Ethernet.

After cell imaging, manipulation can be performed under the assistance of the augmented reality. The tip can be injected into the cell or can cut the cell membrane at certain locations. The two circles in Figure 13.13 indicate the cuts to a neuron's axon and another neuron's dendrite. The surgery operation is performed in the culture medium and when the cells are still alive.

13.6   Conclusion

The technique using a functionalized tip to measure the interaction force between ligands and receptors by atomic force microscopy has been discussed for more than a decade. The single-molecule recognition on well-prepared samples using a

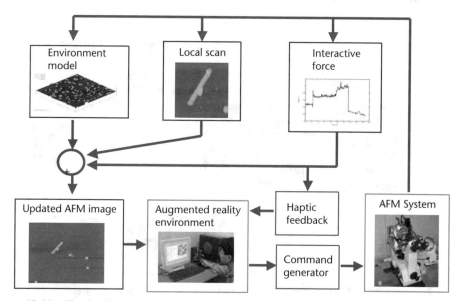

Figure 13.12   The detailed structure of the augmented reality system.

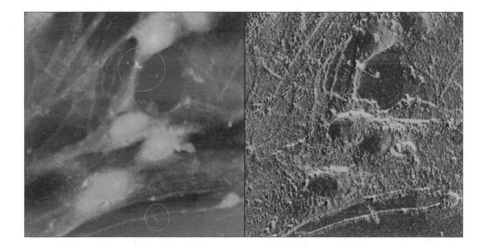

Figure 13.13 The final result of the cutting operation obtained from the AFM image with a scanning range of 90 μm: (left) height image; and (right) phase image.

functionalized tip has also been achieved by recent research advancement. However, none of them can recognize single molecules directly from a cell membrane surface. In this chapter, single AT1 receptor recognition directly from the cell membrane surface is achieved using an AFM tip biologically functionalized with the AT1 antibody. By a lift-up scan and microgrid fixation, single receptor molecules are clearly identified in the processed image. Using this technique, further study of the trafficking behavior of the AT1 receptor is possible. In this chapter, an AFM based nanorobotic system is also introduced. The AFM based nanorobotic system can be operated both in the air and liquid conditions, which makes surgery of a single living cell possible. Since the single receptor molecules can be identified in their physiological condition, it is anticipated that the research that seeks to understand and exploit the interaction forces between nanoprobing mechanisms will provide a leap forward for biomedical research, whose progress is limited by the cumbersome and static multistep methods currently available.

References

[1] Binning, G., C. F. Quate, and C. Gerber, "Atomic Force Microscope," *Phys. Rev. Lett.*, Vol. 56, No. 9, 1986, pp. 930–933.

[2] Henderson, E., "Imaging of Living Cells by Atomic Force Microscopy," *Prog. Surf. Sci.*, 46:39–60, 1994.

[3] You, H. X., and L. Yu, "Atomic Force Microscopy Imaging of Living Cells: Progress, Problems and Propects," *Methods in Cell Science*, Vol. 21, 1999, pp. 1–17.

[4] Putman, C. A. J., et al., "Viscoelasticity of Living Cells Allows High Resolution Imaging by Tapping Mode Atomic Force Microscopy," *Biophysical Journal*, Vol. 67, 1994, pp. 1749–1753.

[5] Prater, C. B., et al., "Atomic Force Microscopy of Biological Samples at Low Temperature," *J. Vac. Sci. Tech. B*, Vol. 9, 1991, pp. 989–991.

[6]   Hansma, P. K., et al., "Tapping Mode Atomic Force Microscopy in Liquids," *Appl. Phys. Lett.*, Vol. 64, 1994, pp. 1738–1740.

[7]   Baker, A. A., et al., "New Insight into Cellulose Structure by Atomic Force Microscopy Shows the i_α Crystal Phase at Near-Atomic Resolution," *Biophysical Journal*, Vol. 79, 2000, pp. 1139–1145.

[8]   Fotiadis, D., et al., "Imaging and Manipulation of Biological Structures with the AFM," *Micron*, Vol. 33, 2002, pp. 385–397.

[9]   Willemsen, O. H., et al., "Simultaneous Height and Adhesion Imaging of Antibody-Antigen Interactions by Atomic Force Microscopy," *Biophysical Journal*, Vol. 75, November 1998, pp. 2220–2228.

[10] Ludwig, M., W. Dettmann, and H. E. Gaub, "AFM Imaging Contrast Based on Molecular Recognition," *Biophysical Journal*, Vol. 72, 1997, pp. 445–448.

[11] Raab, A., et al., "Antibody Recognition Imaging by Force Microscopy," *Nat. Biotechnol.*, Vol. 17, September 1999, pp. 902–905.

[12] Stroh, C., et al., "Single-Molecule Recognition Imaging Microscopy," *Proc. Natl. Acad. Sci. USA*, Vol. 101, No. 34, August 2004, pp. 12503–12507.

[13] Schaefer, D. M., et al., "Fabrication of Two-Dimensional Arrays of Nanometer-Size Clusters with the Atomic Force Microscope," *Appl. Phys. Lett.*, Vol. 66, February 1995, pp. 1012–1014.

[14] Junno, T., et al., "Controlled Manipulation of Nanoparticles with an Atomic Force Microscope," *Appl. Phys. Lett.*, Vol. 66, No. 26, June 1995, pp. 3627–3629.

[15] Hansen, L. T., et al., "A Technique for Positioning Nanoparticles Using an Atomic Force Microscope," *Nanotechnology*, Vol. 9, 1998, pp. 337–342.

[16] Requicha, A. A. G., et al., "Nanorobotic Assembly of Two-Dimensional Structures," *Proc. IEEE Int. Conf. Robotics and Automation*, Leuven, Belgium, May 1998, pp. 3368–3374.

[17] Sitti, M., and H. Hashimoto, "Tele-Nanorobotics Using Atomic Force Microscope," *Proc. IEEE Int. Conf. Intelligent Robots and Systems*, Victoria, B. C., Canada, October 1998, pp. 1739–1746.

[18] Guthold, M., et al., "Controlled Manipulation of Molecular Samples with the Nano-manipulator," *IEEE/ASME Transactions on Mechatronics*, Vol. 5, No. 2, June 2000, pp. 189–198.

[19] Li, G. Y., N. Xi, and M. Yu, "Development of Augmented Reality System for AFM Based Nanomanipulation," *IEEE/ASME Transactions on Mechatronics*, Vol. 9, June 2004, pp. 199–211.

[20] Li, G. Y., et al., "'Videolized' Atomic Force Microscopy for Interactive Nanomanipulation and Nanoassembly," *IEEE Transactions on Nanotechnology*, Vol. 4, No. 5, 2005, pp. 605–615.

[21] Li, G. Y., et al., "Augmented Reality System for Real-Time Nanomanipulation," *Proc. IEEE Int. Conf. Nanotechnology*, San Francisco, CA, August 12–14, 2003.

[22] Li, G. Y., et al., "Modeling of 3-D Interactive Forces in Nanomanipulation," *Proc. IEEE Int. Conf. Intelligent Robots and Systems*, Las Vegas, NV, October 28–30, 2003, pp. 2127–2132.

[23] Li, G. Y., et al., "Assembly of Nanostructure Using AFM Based Nanomanipulation System," *Proc. IEEE Int. Conf. Robotics and Automation*, New Orleans, LA, April 2004, pp. 428–433.

[24] Ros, R., et al., "Antigen Binding Forces of Individually Addressed Single-Chain FV Antibody Molecules," *Proc. Natl. Acad. Sci. USA*, Vol. 95, 1998, pp. 7402–7405.

[25] Florin, E. L., V. T. Moy, and H. E. Gaub, "Adhesion Forces Between Individual Ligand-Receptor Pairs," *Science*, Vol. 264, 1994, pp. 415–417.

[26] Lee, G. U., D. A. Kidwell, and R. J. Colton, "Sensing Discrete Streptavidin-Biotin Interactions with Atomic Force Microscopy," *Langmuir*, Vol. 10, 1994, pp. 354–357.

[27] Hinterdofer, P., et al., "Detection and Localization of Individual Antibody-Antigen Recognition Events by Atomic Force Microscopy," *Proc. Natl. Acad. Sci. USA*, Vol. 93, 1996, pp. 3477–3481.

[28] Moy, V. T., E. L. Florin, and H. E. Gaub, "Intermolecular Forces and Energies Between Ligands and Receptors," *Science,* Vol. 266, 1994, pp. 257–259.

[29] Willemsen, O. H., et al., "Biomolecular Interactions Measured by Atomic Force Microscopy," *Biophysical Journal*, Vol. 79, 2000, pp. 3267–3281.

[30] Cleveland, J. P., et al., "Energy Dissipation in Tapping-Mode Scanning Force Microscopy," *Appl Phys Lett.*, Vol. 72, No. 20, 1998, pp. 2613–2615.

[31] Tamayo, J., and R. Garcia, "Relationship Between Phase Shift and Energy Dissipation in Tapping-Mode Scanning Force Microscopy," *Appl. Phys. Lett.*, Vol. 73, No. 20, 1998, pp. 2926–2928.

[32] James, P. J., et al., "Interpretation of Contrast in Tapping Mode AFM and Shear Force Microscopy: A Study of Nafion," *Langmuir*, Vol. 17, 2001, pp. 349–360.

[33] Li, G., et al., "An AFM Method for In Situ Probing Membrane Proteins Under Physiological Condition," *Proceeding of the 2006 IEEE Nanotechnology Conference*, Cincinnati, OH, July 17–20, 2006.

[34] Bunnemann, B., K. Fuxe, and D. Ganten, "The Rennin-Angiotensin System in the Brain: An Update 1993," *Regul. Pept.*, Vol. 46, No. 3, 1993, pp. 487–509.

[35] Ganong, W. F., "Blood, Pituitary and Brain Rennin-Angiotensin Systems and Regulation of Secretion of Anterior Pituitary Gland," *Front. Neuroendocrinol.*, Vol. 14, No. 3, 1993, pp. 233–249.

[36] Messerli, F. H., M. A. Weber, and H. R. Brunner, "Angiotensin II Receptor Inhibition. A New Therapeutic Principle," *Arch. Intern. Med.*, Vol. 156, No. 17, September 23, 1996, pp. 1957–1965.

[37] Johnston, C. I., and L. M. Burrel, "Evolution of Blockade of the Rennin-Angiotensin System," *J. Hum. Hypertens.*, Vol. 9, 1995, pp. 375–380.

[38] Murphy, T. J., et al., "Isolation of a CDNA Encoding the Vascular Type-1 Angiotensin II Receptor," *Nature,* Vol. 351, 1991, pp. 233–236.

[39] Guo, D. F., et al., "The Angiotensin II Type 1 Receptor and Receptor-Associated Proteins," *Cell Research*, Vol. 11, No. 3, 2001, pp. 165–180.

[40] Thomas, M. A., et al., "Subcellular Identification of Angiotensin I/II- and Angiotensin II (AT1)-Receptor-Immunoreactivity in the Central Nervous System of Rats," *Brain 2,* 2003, pp. 92–104.

CHAPTER 14

Biological Cell and Tissue Manipulation Automation

Yu Sun[†], Xinyu Liu[†], and Bradley J. Nelson[‡]

Automated manipulation, particularly robotic manipulation of biological cells and tissues, is an enabling technology for cellular/developmental biology and robotic-assisted surgery. Over the last decade, many systems and devices have been developed, and tremendous progress has been made in these areas. This chapter provides a review of these technologies for biological cell and tissue manipulation automation.

14.1 Introduction

The ability to analyze individual cells rather than averaged properties over a population is a major step towards understanding the fundamental elements of biological systems. Studies on single cells are a key component in the development of highly selective cell-based sensors, the identification of genes, bacterial synthesis of specific DNA, and certain approaches to gene therapies. Treatments for severe male infertility and the production of transgenic organisms require that individual cells are isolated and individually injected. These recent advances in microbiology, as well as other significant research efforts such as cloning, demonstrate that increasingly complex micromanipulation strategies for manipulating individual biological cells are required.

Instead of focusing on lab automation in which robotic systems or microfluidic devices are employed to automate such processes as culturing and pipetting a large number of cells [1–3], this chapter discusses technologies for single cell manipulation.

Switching from the cellular level to the tissue level, this chapter will continue to discuss robotic systems and tools that have been employed to assist the manipulation of tissues. Although fully autonomous robotic tissue manipulation or surgery has not been realized, many robotic systems have been developed to provide assistance under the surgeon's supervision. Existing applications range from neurosurgery to orthopedic surgery to urology surgery and to cardiac surgery, which will be described later in this chapter.

† Advanced Micro and Nanosystems Lab, University of Toronto
‡ Inst. of Robotics and Intelligent System, ETH-Zürich

14.2 Biological Cell Manipulation

Microrobotics and microsystems technology can play important roles in manipulating cells, a field referred to as biomanipulation. Biomanipulation entails such operations as positioning, grasping, and injecting materials into various locations in cells as shown in Figure 14.1.

Existing biomanipulation techniques can be classified as noncontact manipulation, including laser trapping, dielectrophoresis (DEP) manipulation, and magnetic manipulation; and contact manipulation, referred to as mechanical micro-manipulation (i.e., microrobotic manipulation).

14.2.1 Cell Manipulation Using Laser Trapping

As shown in Figure 14.2, when laser traps (laser tweezers) [4–8] are used for noncontact biomanipulation, a laser beam is focused through a large numerical aperture objective lens, converging to form an optical trap in which the lateral trapping force moves a cell in suspension toward the center of the beam. The longitudinal trapping force moves the cell in the direction of the focal point. The optical trap levitates the cell and holds it in position. The biological applications of laser trapping have been reviewed extensively [9–11]. A combination of laser traps and laser scalpels (laser scissors) was demonstrated to be capable of performing cell surgery [12].

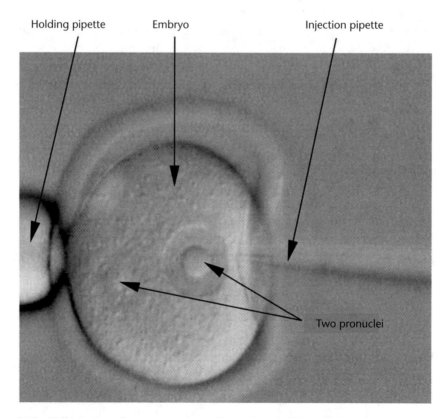

Figure 14.1 Cell injection of a mouse embryo. The embryo is 55 μm in diameter.

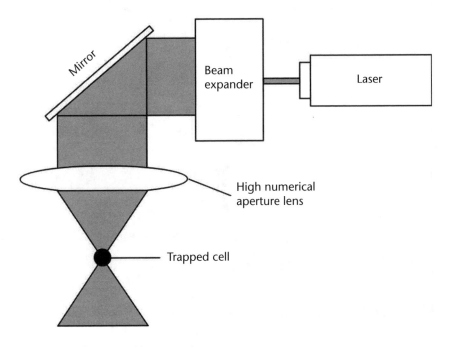

Figure 14.2   Schematic of laser trapping.

Laser traps can work in a well-controlled manner. However, the high dissipation of visible light in aqueous solutions requires the use of high energy light close to the UV spectrum, raising the possibility of damage to the cell and inducing abnormalities in the cells' genetic material, though some researchers claim that such concerns could be overcome by using wavelengths in the near infrared (IR) spectrum [8, 13]. In order to avoid direct irradiation of high power laser to biological cells, microbeads manipulated by laser beams were used to indirectly position biological cells (e.g., yeast cells) [14].

Laser-tweezers based manipulation is difficult to massively parallelize. Most applications to date use a single beam to position cells one at a time. Recent advances, however, have used fiber optics to split a laser beam into many paths, thereby forming an array of optical tweezers [15]. Such a parallelization strategy has the potential to be expanded into array-based manipulation platforms.

14.2.2   Dielectrophoresis Cell Manipulation

One alternative to using laser beams is DEP manipulation. DEP, which is the movement of uncharged particles in a nonuniform electrical field [16, 17], can be used to pattern/position cells. In DEP manipulation, the cells are effectively uncharged particles that can be placed onto predetermined arrays of "traps" in a nonuniform electrical field [18, 19], as shown in Figure 14.3. Once cells are trapped by the dielectric forces, the electrical field can be turned off and cells will attach and spread. This

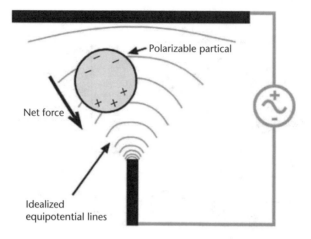

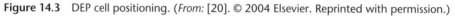

Figure 14.3   DEP cell positioning. (*From:* [20]. © 2004 Elsevier. Reprinted with permission.)

method may be used for examining the biology of single cells, and has also been expanded to allow for the examination of arrays of cells [20].

Using DEP trapping, electric-field induced rotation of cells was demonstrated by Washizu et al. [21, 22], Arnold et al. [23], and Mischel et al. [24]. The use of DEP forces to manipulate cells was also reported by Fuhr et al. [25]. This noncontact cell manipulation technique is based on controlling the phase shift and magnitude of electric fields. These fields, when appropriately applied, produce a torque on the cell. Different system configurations have been established for cell manipulation based on this principle [26, 27].

In the example shown in Figure 14.4, a rotating electric field is generated by arranging the electrical connections to produce a 90° phase difference between adjacent electrodes. A resulting rotational torque is exerted on a cell. Consequently, the cell rotates counter to the clockwise rotating field. Electrorotation techniques can achieve high accuracy in cell positioning. However, this technique lacks a means to

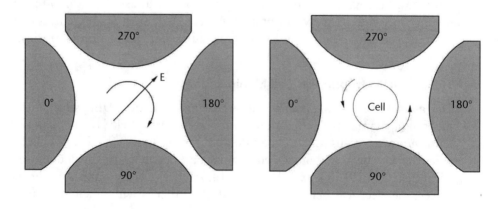

Figure 14.4   Schematic of electrorotation.

hold the cell in place for further manipulation, such as injection, since the magnitude of the electric fields has to be kept low to ensure the viability of cells.

14.2.3 Magnetic Cell Manipulation

As shown in Figure 14.5, magnetic cell positioning utilizes a ferromagnetic particle that acts as a manipulator and follows the motion of a magnet under a glass slide to physically move a cell to the desired locations [28]. Since cells generally are not sensitive to magnetic fields, the cell is manipulated by the ferromagnetic manipulator without being displaced by the magnetic field. In spite of its novelty, this technique is only capable of performing pushing tasks.

For magnet positioning, photolithographic techniques can be used to create arrays of micronscale magnets to generate traps for single cell capture. However, in order for the cells to be patterned onto the traps, they must be magnetized themselves. One method for such magnetization is through the use of magnetic nanowires, which can be absorbed by cells [29, 30].

Cell responses to mechanical forces are different according to which receptor is being stressed. The magnetic bead force application (MBFA) technique permits the application of controlled mechanical forces to specific cell surface receptors using ligand-coated microfabricated ferromagnetic beads. As shown in Figure 14.6, a cell surface-attached magnetic bead is subjected to a high gradient magnetic field generated by the sharpened pole piece of an electromagnet. By controlling the current passing through the electromagnet, the magnetic bead is capable of accurately applying a specified force in the pN scale to specific cell receptors of the cell so that the resulting behavior can be observed. This technique has been applied to mechanical property studies of cells [31] and lipid vesicles [32] as well as neuron studies [33].

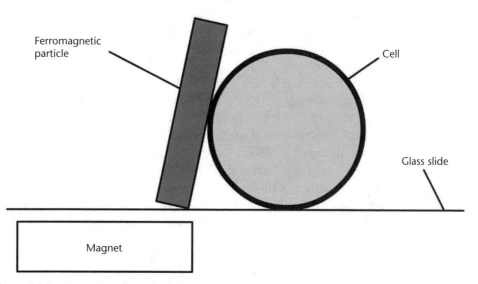

Figure 14.5 Magnetic cell manipulation.

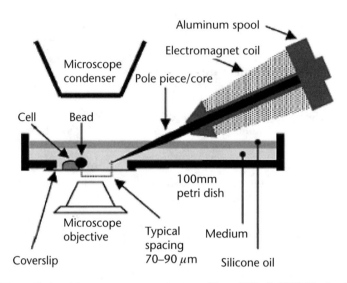

Figure 14.6   Magnetic bead force measurement setup. (*From:* [33]. © 2003 Biophysical Journal. Reprinted with permission.)

14.2.4   Microrobotic Cell Manipulation

The limits of laser trapping, electrorotation, and the limitations inherent to other more complex cell manipulation operations make microrobotic biomanipulation desirable for many cell manipulation tasks. The possible damage caused by laser beams in the laser trapping technique and the lack of a holding mechanism in the electrorotation technique can be overcome by microrobotic mechanical micromanipulation that can achieve more complex biomanipulation tasks.

Cell injection is a procedure used to introduce foreign genetic materials into cells to treat severe male infertility or to create transgenic organisms such as those for cancer studies or to screen target molecules. Besides spermatozoa, molecules such as DNA, morpholinos, and RNA interference (RNAi) need to be injected into mouse embryos, zebra fish embryos, drosophila embryos, or *C. elegans* for mutation screening to identify genes and for sequencing genomes.

In Figure 14.1, a holding pipette holds a mouse embryo and an injection pipette performs the injection task. Conventionally, cell injection is conducted manually. Operators often require at least a year of full-time training to become proficient at the task, and success rates are often disappointingly low. One reason for this is that successful injections are not precisely reproducible. A successful injection is determined greatly by injection speed and trajectory [34]. Automated cell injection can be highly reproducible with precision control of pipette motion. The second reason for the low success rate of conventional cell injection is due to contamination. This also calls for the elimination of direct human involvement. Therefore, the main advantages of automated cell injection are that it reduces the need for extended training and the risk of contamination. Moreover, it is highly reproducible and greatly increases the success rate.

Towards automated cell injection, a joystick-based teleoperative microrobotic system shown in Figure 14.7 was constructed [35]. A human operator controls the motion of an *x-y* motorized stage and two oil-pressure micromanipulators through

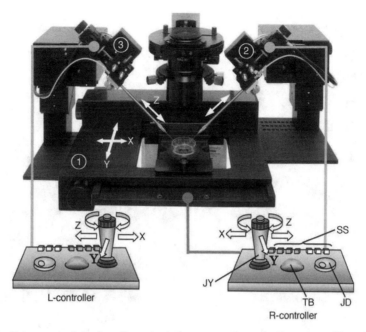

Figure 14.7   Teleoperated single-cell manipulation supporting robotic system. (*From:* [35]. © 2005 Elsevier. Reprinted with permission.)

a pair of joysticks. The system was demonstrated to facilitate the injection process of rice protoplast and mouse embryonic cells, shortening the injection time by 17 times compared to pure human operation.

A visually servoed microrobotic system was demonstrated that is capable of performing automatic pronuclei DNA injection of mouse embryos for the creation of transgenic mice for use in cancer studies [36]. The autonomous microrobotic system shown in Figure 14.8 was developed to conduct autonomous pronuclei DNA injection of mouse embryos. A hybrid control scheme combining visual servoing and position control (Figure 14.9) was developed to fulfill the cell injection task. Cells were brought into focus through the use of an autofocusing technique. The nuclei were detected via pattern recognition. A sum-of-squared-differences optical flow tracking algorithm was adopted for visual servoing. Figure 14.10 illustrates the injection process.

Upon the completion of injection, the DNA-injected embryos were transferred into the ampulla of a pseudo pregnant foster female mouse to reproduce transgenic mice. Results show that the injection success rate was 100%.

From a robotics standpoint, the manipulation of biological cells presents several interesting research issues. Biological cells are highly deformable objects, and the material properties of these objects are not well quantified; therefore, developing strategies for manipulating deformable objects must be addressed. Most biological cells are between 1 and 100 μm in diameter, depending on the cell type; therefore, micromanipulation issues must be explored, including the appropriate use of high resolution, low depth-of-field vision feedback.

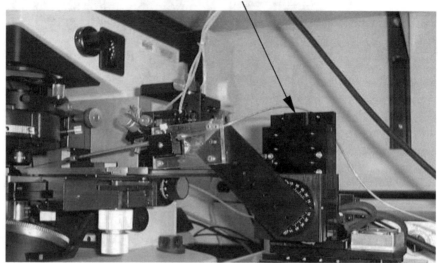

Figure 14.8 3-DOF high precision microrobot with a mounted micropipette.

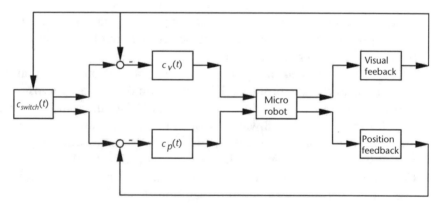

Figure 14.9 Hybrid control scheme for embryo pronuclei DNA injection.

14.2.5   Computer Vision Microscopy

An inverted microscope with a contrast condenser is the best optical platform for microrobotic cell manipulation. Different from the configuration of a conventional upright microscope, the illumination system and condenser of an inverted microscope are above the specimen stage, and the objectives are below the stage. Aided by inverted microscopes, observation of thick specimens is enhanced, such as cultured cells in a petri dish, because the objective lens can get closer to the bottom of the dish where the cells are suspended. Thus, live cells can be observed and manipulated in a relatively larger container such as a petri dish than in a more stressful environment, such as under a cover slip as required for a typical upright microscope. Other

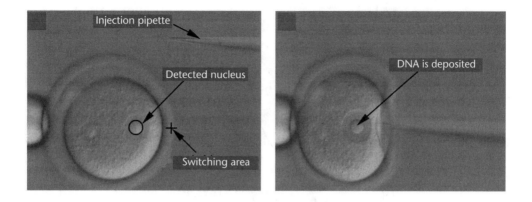

Figure 14.10   Microrobotic embryo pronuclei DNA injection.

advantages of inverted microscopes include higher mechanical stability and optical efficiency, and less sensitivity to vibration compared to upright microscopes [37].

Biological cells are fairly transparent, and bright field illumination does not reveal differences in brightness between structural details (i.e., no contrast is apparent). Structural details can only emerge by staining a live cell, which quite often causes damage to the cell, or via phase difference, which is also called *optical staining*. To produce contrast, edge effects such as diffraction, refraction, and reflection can be utilized, based on which different contrast enhancement condensers are constructed, such as phase contrast, differential interference contrast (DIC), and Hoffman modulation contrast.

In automated cell manipulation, precisely bringing cell samples and manipulation end-effectors (e.g., micropipettes) in focus is a fundamental procedure for the following image processing, pattern recognition, and visual servoing tasks. The autofocusing process comprises the following steps: (1) select a region of interest (ROI) in the image; (2) compute an objective function of a focus algorithm in the ROI; and (3) employ a control algorithm to servo the microscope objectives or samples to the focus position determined by the focus function.

A variety of autofocusing algorithms have been proposed in the literature and implemented in different autofocusing systems [38–43]. A common rationale of focus algorithms is that focused images contain more information and details (e.g., edges) than unfocused images, based on which an objective function indicating the sharpness of images is used to evaluate the degree of focusing. The extremum of the objective function is deemed as the focus position. Figure 14.11 shows two typical focus curves.

As different focus algorithms reveal significant performance differences under a specific experimental microscopy condition, the selection of the optimal focus algorithm for a given application is significant. Several studies have been conducted to investigate the selection of the optimal focus algorithm [44–48]. By comparing the performance of 13 focus algorithms implemented on fluorescence images, it was demonstrated that the AutoCorrelation algorithm [38, 39] is the optimal focus

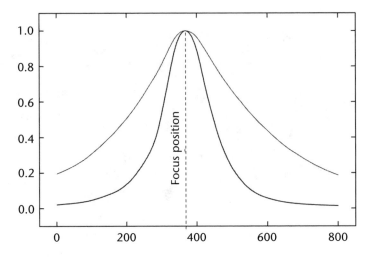

Figure 14.11 Typical focus curves.

algorithm for fluorescence microscopy applications [47]. For nonfluorescence microscopy autofocusing, the Normalized Variance algorithm [44, 46] has been found to provide the best overall performance based on a systematic comparison of 18 commonly used focus algorithms [48].

Based on the selected focus algorithm, the focus position can be located through the use of an algorithm searching for the extremum of the objective function. In previous studies [36, 42, 46, 49], three search algorithms were commonly used for the extreme search.

Exhaustive search [36] is the simplest algorithm to use, which continuously changes the focusing distance and sweeps the objective function values in a certain scope that contains the focus position. The position with the extremum of the objective function is located after the sweeping process. The accuracy of this method depends on the distance between two adjacent steps at which the objective function is computed. Consequently, when a high focusing accuracy is demanded, the sweeping speed is low with a constant image sampling frequency. This low sweeping speed renders the autofocusing process rather time-consuming.

Hill-climbing search [49] utilizes the gradient information of the focus objective function to search the function extremum. This algorithm computes the objective function for two frames of images separated by a step distance and uses the sign difference between the two values to obtain the motor movement direction. The major disadvantage is that the algorithm only determines the movement direction of the motor, requiring adaptive control of the step distance between two successive images.

Fibonacci search [42, 46] has been proven to be the optimal algorithm for autofocusing under the unimodality assumption [42]. The algorithm introduces the Fibonacci series to optimally narrow the initial search range, and only one computation of the objective function is required for all search iterations except the first one. The focusing time mainly depends on the total run time of the focusing motor from its initial position to the final located position, while the computation time of the focus objective function does not noticeably affect the focusing speed.

After the autofocusing step, visual tracking algorithms are required to process well-focused images and extract microscopic features, their spatial relations, and motions, which are essential to autonomous microrobotic cell manipulation. Optical microscopes provide high resolution, low depth-of-field vision feedback. The high deformability of biological cells results in shape changes during manipulation. These unique characteristics require special considerations in designing visual tracking algorithms for microrobotic cell manipulation. As an image in cell manipulation can be typically considered as a two-dimensional problem, the following discussion is limited to two-dimensional computer vision algorithms.

Visual tracking algorithms used in cell manipulation can be classified into two categories: (1) area-based algorithms for temporally correlating image regions in a series of images to obtain motion information; and (2) feature-based algorithms for rapidly detecting edge segments and extracting feature information from detected edges.

In microscopic images with little structural variations, area-based algorithms track a specific pattern by searching for a gray-level feature with the same appearance in terms of its intensity spatial pattern, in which edge detection is not required. The basic assumption for area-based algorithms is *brightness conservation*, the observation that image measurements (e.g., brightness) in a small region remain the same although their location may change. Gradient-based optical flow tracking [50, 51] is a typical area-based algorithm which uses the spatiotemporal image derivatives to estimate flow and is suitable for relatively small motions. Another commonly used area-based algorithm is template matching. This algorithm extracts a local region as a template and seeks to find the region in each image that maximizes the similarity measure with the template. The similarity measure can be sum of absolute values (SAD) or sum of squared differences (SSD). In the microrobotic cell injection system reported in [36], the SSD template matching algorithm was used to successfully track the motion of the injection pipette tip.

Feature-based algorithms typically extract sharp contrast changes (i.e., corners and edges) to detect the presence of target objects. Edge detection algorithms are usually conducted as a precursor step of feature recognition for obtaining edge and corner information. Canny edge detection is the most often used edge detection algorithm [52]. The Canny algorithm detects the zero-crossings of the second directional derivative in a smoothed image. The output is a binary image containing the position of detected edges.

The detected edges containing the geometry and position information of cells and manipulation end-effectors are further analyzed by feature recognition algorithms. Hough transform is a powerful tool that allows recognition of particular shaped features that can be parameterized such as lines, circles, and ellipses within an image [53, 54]. Hough transform locates parameterized features in a parameter space that is quantized into finite accumulator cells corresponding to possible features in the image. Each input (i.e., edge point coordinate) votes for each accumulator cell. The feature with the highest votes represents the final detected feature. In the microrobotic cell injection system reported in [36], Hough transform was used to detect the circular pronucleus of mouse embryos. The major drawbacks of Hough transform include the difficulty of setting an appropriate accumulator cell size and the poor detection performance for noisy images [55].

Hough transform is only capable of extracting nondeformed contour of biological cells. However, microrobotic cell manipulation often produces cell deformations and causes Hough transform to fail. Thus, several deformable model based tracking algorithms were introduced to track deformable biological cells, among which active contour, also known as *snake* [56], is widely used [57–59]. Snakes are energy minimizing splines influenced by external constraint forces and image forces that guide snake points towards features such as lines and edges. Piecewise smoothness constraints are imposed by the internal spline forces, and the image forces are responsible for driving the snakes towards image features. The external constraint forces, which come from user-defined constraints or high-level interpretations, place the snake points near the desired local minimum. The snake energy comprising internal energy, image energy, and external constraint energy is minimized to lock the snakes onto nearby edges.

Active contour tracking was employed for tracking shape changes of human fibroblasts to study the pseudopod kinetics and dynamics of the cell [57]. A "steady-support" criterion based on the topography of the image potential field was proposed to avoid undesired oscillations during optimization. Although satisfactory segmenting and tracking results can be achieved in real time, the snakes are not always capable of finding the desired solution due to the applicability limitation of the steady-support criterion. Based on the parametric active contour model, a segmenting and tracking algorithm for the *E. histolytica* cells was developed to quantitatively analyze cell dynamics in vitro [58]. An edge map based on the average intensity dispersion was employed to detect the cell pseudopods and interfaces between adjacent cells. A repulsive interaction between contours was introduced to allow correct segmentation of cells in contact. Rolling leukocytes were tracked by utilizing parametric active contour for tracking [59]. These active contour based visual tracking algorithms will find important applications in microrobotic cell manipulation.

Another important method for tracking deformable objects is deformable template [60, 61]. Unlike the free-form deformable snakes, the deformable template is a parametric model based on prior geometry information, which can be represented as either a set of parameterized curves or a prototype template under a parametric mapping. In the first scheme, the deformable template consists of several parameterized curves, and the template shape can be changed by varying the parameters of the curves. Similar to active contour model, a potential field based on the salient image features can be constructed, and its energy is minimized by updating the parameters of the deformable template. In the second scheme, the prototype template is derived from the pattern theory [62] and based on the prior knowledge of target objects obtained during sample training. The parametric mapping is selected for describing the possible deformation formats of target objects.

A deformable template matching algorithm was recently integrated into a microrobotic cell manipulation system to track deformable contours of mouse oocytes/embryos for characterizing the biomembrane mechanical properties [63]. Two-dimensional cell contours in the deformable template were assumed to be linearly elastic to properly mimic the cell deformation. The boundary element method [64] was used to model deformed cell contours. A force field was applied to the template for deforming it to match the image. A potential energy based on least-squares errors between the template cell contours and image features is minimized to extract the force field.

14.2.6 Microrobotic Visual Servo Control

Significant progress in the area of visual servo control of robotic manipulators has been made as computation capabilities of control computers are greatly enhanced to be able to process visual feedback at a sufficiently high rate. Because microscopic vision feedback is the major form of feedback available in microrobotic cell injection, microrobotic cell manipulation most often adopts visual servo control. The two major types of visual servo control are position-based visual servoing and image-based visual servoing [65, 66].

Position-based visual servo control [67, 68], as shown in Figure 14.12, extracts feature information from images, which is used for estimating the pose of robotic end-effector according to the known camera model and target geometric model. The estimated pose position together with the input target position generates a position error in the task space, which is computed by the task space controller to provide control signals to the robot. The robot trajectory is directly controlled in the task space. The controller is not responsible for image processing or visual tracking, simplifying controller design, and path planning. However, the feedback error signal is computed from the estimated pose position that is based on the calibrated parameters of the system. Therefore, position-based visual servoing is sensitive to calibration errors [69].

Figure 14.13 depicts a typical structure of a feature-based visual servo control system, in which the position error is directly defined using image feature

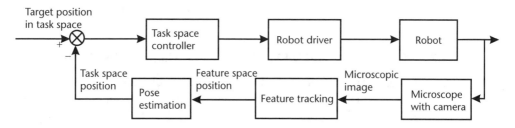

Figure 14.12 Block diagram of position-based microrobotic visual servo control.

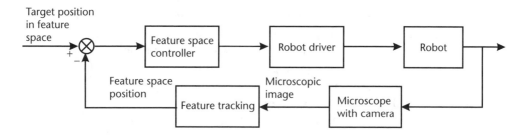

Figure 14.13 Block diagram of image-based visual servo control.

parameters such as image coordinates [70–73], line orientations [74–76], perceived edge length [77], and parameters of an ellipse in the image plane [71]. The controller maps the image feature parameters into task space position changes for controlling the motion of the robot. This mapping is performed through the use of an *image Jacobian* [65, 66], also called *feature sensitivity matrix* [77]. Let x_T, $\dot{x}_T$, represent coordinates and velocity of the robot end-effector in the task space, and x_I, $\dot{x}_I$ coordinates and velocity of the corresponding image feature in the feature space. The image Jacobian, $J_v(x_T)$, is defined as a linear transformation from the tangent space of task space at x_T to the tangent space of feature space at x_I.

$$\dot{x}_I = J_v(x_T)\dot{x}_T \qquad (14.1)$$

where $J_v(x_T) \in \Re^{k \times m}$, and

$$J_v(x_T) = \begin{bmatrix} \dfrac{\partial x_{I1}(x_T)}{\partial x_{T1}} & \cdots & \dfrac{\partial x_{I1}(x_T)}{\partial x_{Tm}} \\ \cdots & \cdots & \cdots \\ \dfrac{\partial x_{Ik}(x_T)}{\partial x_{T1}} & \cdots & \dfrac{\partial x_{Ik}(x_T)}{\partial x_{Tm}} \end{bmatrix} \qquad (14.2)$$

k is the dimension of the image feature space, and m is the dimension of the task space. Because the visual servo task requires a reverse mapping from $\dot{x}_I$, $\dot{x}_T$, the inverse/pseudo-inverse of image Jacobian, $J_v^{-1}(x_T)$, is usually computed.

The main advantage of image-based control over position-based control is that the positioning accuracy is less sensitive to camera calibration errors. However, in some cases, there exist singularities in the feature mapping function which is reflected as unstable points in the inverse Jacobian control law. These instabilities are often less obvious in the equivalent position-based scheme [66].

14.2.7 Microrobots and MEMS/Microfluidic Manipulation Tools

Macroscaled manipulators with a micrometer or submicrometer motion resolution are often used for manipulating micrometer-sized objects including biological cells. These microrobots include stepper motor based [78, 79], dc motor based [80], hydraulic [81, 82], pneumatic [83], and piezoelectric microrobots [84, 85].

Stepper-motor-based microrobots have been employed in biological cell micromanipulation. The microrobots from Sutter, Inc., [78] have three motorized axes, a high resolution of 40 nm, a motion range of 25 mm, and a maximum speed of 2.9 mm/sec along each axis. These microrobots can be controlled by a rotary optical encoder, joystick, or serial port. Eppendorf InjectMan [79] is a programmable, stepper motor based micromanipulator suitable for microinjection of adherent cells and fish larva. It provides many specialized features for biological cell injection such as adjustable injection angle, a pipette breaking avoidance mechanism, and special injection modes for applications in developmental biology. The reported motion resolution of the InjectMan NI2 is 40 nm.

dc motor based microrobots permit complex servo control for micromanipulation, promising better system performance. Siskiyou micromanipulators [80] employ dc motors with encoders that provide position feedback for closed-loop control. Additionally, the preloaded precise lead screw in each axis ensures a drift-free operation.

Hydraulic or pneumatic microrobots [81–83] are driven by hydraulic or pneumatic actuators that can produce a smoother motion than motor-based systems. They are also capable of producing less vibration. However, these systems are susceptible to drift. These two types of microrobots are less often adopted for cell manipulation systems than stepper motor or dc motor based microrobots.

The piezoelectric microrobot provides higher motion resolution down to a few nanometers, which can help further improve the accuracy of biological cell manipulation and help extend microrobotic cell manipulation to probing smaller biological targets such as proteins, DNA molecules, and subcellular structures. PI micro/nanomanipulators [84] use piezoelectric ceramics as actuators and provide up to 6 degree-of-freedom output motion with a nanometer/subnanometer resolution. Capacitive position sensors can also be incorporated into the system for high-resolution closed-loop control. The MINIMAN microrobot [85] integrates positioning and manipulation units and provides a large manipulation motion. The highest motion resolution of MINIMAN was reported to be 10 nm. One drawback of piezoelectric microrobots is that the inherent hysteresis and creeping of piezoelectric actuators require more efforts in the development of sophisticated control algorithms in order to guarantee the positioning accuracy.

While macroscaled robots are employed for cell manipulation, many efforts have also been spent on developing MEMS-based millimeter devices with individual parts of micrometer size that is comparable to biological cells. Most existing microactuators are based on electrostatic, electrothermal, piezoelectric, or magnetic principles. For example, electrostatic and electrothermal microgrippers capable of grasping single cells were demonstrated [86, 87]. Extra care must be taken when using conventional microactuation mechanisms to operate in an aqueous environment for biological cells to survive, which often poses stringent challenges and intricacies in MEMS design, material selection and insulation, and microfabrication.

In order to capture cells in liquids, several micromachined actuators using stimulus-responsive polymeric materials were developed, such as Nafion actuators [88] and conjugated polymer based actuators [89, 90]. In the microrobot design reported in [90], conjugated polymers were used as actuators. As cations diffuse or move out in response to applied voltages, polymer produces a swelling or shrinking motion. Although these microactuators may be capable of capturing cells, none are capable of performing more complex operations such as cell injection and cell surgery.

In order to parallelize cell injection to realize high-throughput operation, arrays of homogenous, sharp needles are desired as manipulation tools. Most existing microneedle arrays were developed for transdermal drug delivery purposes (e.g., [91–93]) that are not applicable to cell injection due to differences in required tip and channel sizes, shapes, and uniformity. An SiO_2 microneedle array was demonstrated for cell injection [94]. However, the flat tips tended to punch cell membranes and damage injected cells. Broken cell membrane pieces caused needle clogging that obstructed liquid flow. Although highly desired, uniform microneedle arrays with beveled tips and flow channels positioned off-center from the needle tips for mini-mizing cell damage and needle clogging do not yet exist.

Development of miniaturized microfluidic cell manipulation devices for sample preparation is a critical step towards micro total analysis systems (μTAS), also known as lab-on-a-chip [95–97]. Many microfluidic cell manipulation devices tar-get cell trapping and sorting, employing mechanisms such as microflitering struc-tures, laser trapping, DEP forces, and magnetic manipulation. We provide a comprehensive review/critique on biological cell sorting in Chapter 16.

14.2.8 Manipulation of Cells with Atomic Force Microscopy

Atomic force microscopy (AFM) renders the capability of imaging living biological materials in their native environments. Meanwhile, many research efforts have also attempted to obtain a better understanding of cell biophysical properties using AFM to manipulate (e.g., push or pull) individual cells. The biological samples that an AFM is capable of manipulating and characterizing range from molecules to cells to tissues. A detailed discussion can be found in [98].

Switching from the tapping mode to contact mode, elasticity measurements are preformed with an AFM by pushing a cantilever tip onto the surface of the sample of interest and monitoring force-distance curves. By fitting the contact mechanics mod-els to the data, the Young's modulus for the sample in the tip contact area is calcu-lated. Using this approach, Weisenhorn et al. [99] quantified the elastic behavior of living lung carcinoma cells from the captured force-distance curves (Young's mod-ulus 0.013 to 0.15 MPa). Radmacher et al. [100] measured the Young's modulus of different organelles of human platelets that ranged from 1 to 50 kPa. Many AFM measurement results on biological samples have been reported, such as those in [101–106]. Besides characterizing biophysical properties of cells, AFM has also been employed to create designed damage to neurons and subsequently image neural response to injury [107].

14.3   Tissue Manipulation: Robotic Surgery

The research area of robotic surgery started in the late 1980s. A Puma industrial robot was used to hold a fixture next to the patient's head to locate a biopsy tool for neurosurgery [108]. Compared with a conventional stereotactic frame, the robotic system demonstrated the capability of automatic and accurate positioning. Also in the late 1980s, IBM developed a robotic system for hip surgery [109] as shown in Figure 14.14, later called "Robodoc" and successfully used in numerous surgeries on human patients in Germany. The robot held a rotating cutter to ream out the proximal femur to take the femoral stem of a prosthetic implant for a total hip replacement. For soft tissue surgery in the transurethral resection of the prostate, a robotic system was developed for benign prostatic hyperplasia at Imperial College in London [110–112].

Current surgical robotic systems do not intend to replace the specialist surgeon; rather, the robot provides assistance under the surgeon's supervision. A robotic sys-tem can be used to produce more precise motion for positioning surgical tools than

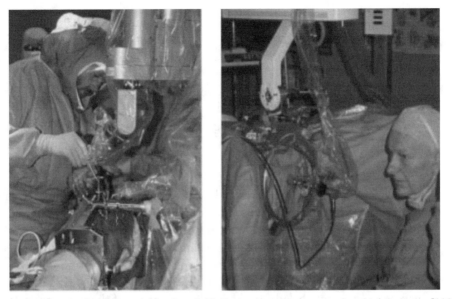

Figure 14.14   Robodoc hip surgery robot (left). Probot prostatectomy robot (right). (*From:* [112]. © 2000 Professional Engineering Publishing. Reprinted with permission.)

the surgeon by scaling large movements made by the surgeon down to micro motion, and the use of a manipulator can help damp out unwanted surgeon tremor. Due to these advantages of robotic surgery, a number of surgical robotic systems have been developed [113]. A recent review of robotic systems used in neurosurgery, orthopedic surgery, urology, and cardiac surgical applications can be found in [114, 115].

Da Vinci Surgical System (Intuitive Surgical, Inc.) and ZEUS (Computer Motion, Inc.) represent two commercial robotic systems employing a master-slave manipulator [113, 116]. The surgeon teleoperates a robot from a console physically separated from the surgical site by using the visual feedback from a camera.

Robots used in surgery can be either powered robots used as passive tool holders or active robots such as those used to manipulate laparoscopes for abdominal minimally invasive surgery [117]. In robotic surgery, particularly in laparoscopic surgery, visual feedback of the internal anatomy is often utilized to track regions of interest and guide the intervention [118, 119].

When manipulating delicate tissues, force feedback or haptic information is important to both the robot and the surgeon for enhanced dexterity and safety. Force feedback enables the surgeon and the robot to percept the stiffness of soft tissues and feel the resistance when the end-effectors contact surrounding tissues during operation. Safe tissue handling requires tissue manipulation that is both secure and nondamaging. Enabling force sensing capability of surgical instruments, enhancing force sensitivity, and developing tactile sensors are among the list of currently and actively pursued research areas [120–126]. Based on measurements of tissue properties, modeling the interaction between surgical tools and deformable, soft tissues has also gained much interest [124, 127, 128].

Recent advances in medical imaging technology [e.g., computer tomography (CT) and magnetic resonance imaging (MRI)] and medical image processing and modeling have provided surgeons with the capability of visualizing anatomic structures in live patients and to use this information in quantitative diagnosis and treatment planning. In robotic surgery, these 3D scans are also essential for the construction of 3D models of the anatomy. The constructed 3D models are further employed to guide the motion of the robot, which greatly strengthens the autonomy of robotic systems.

The development of robotic surgical tools [129] that are endowed with a certain degree of embedded intelligence and autonomy to facilitate teleoperative or semi-autonomous surgery is another active research area. A few examples of robotic surgical tools are described below.

Figure 14.15 shows the steady-hand robot arm and force-instrumented retinal pic developed at Johns Hopkins University [130]. The steady-hand robot arm slows the surgeon's motion and mechanically filters tremor. The force-instrumented retinal pic provides millinewton force sensing resolution at the tip of the retinal pic.

Electroadhesive forceps were developed for manipulating membranes in ophthalmology [131]. In Figure 14.16, the electroforcep lifts and manipulates the soft membrane (arrow labeled). In the left lower corner of the picture is an illumination probe.

Several graspers and tissue stretchers were developed with integrated sensors to acquire tissue force-deflection properties [119, 132] or to enable grasping force control and position sensing [133]. Figure 14.17 shows the Da Vinci wrist for grasping tissues. The device features three-axis motion, mimicking the motion freedom of the human wrist.

Miniaturized devices were also developed, for example, for locomotion and inspection inside the intestine and colon. A miniaturized colonscope with a steerable and telescopic tip and CMOS camera is shown in Figure 14.18 using shape memory alloy [134].

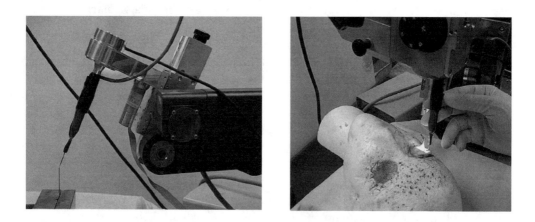

Figure 14.15   JHU's steady-hand robot arm and force instrumented retinal pic. (*From:* [130]. © 2000 IEEE. Reprinted with permission.)

Figure 14.16 Electroadhesive forceps manipulating membranes in ophthalmology. (*From:* [131]. © 2004 SPIE. Reprinted with permission.)

Figure 14.17 Da Vinci wrist for tissue grasping.

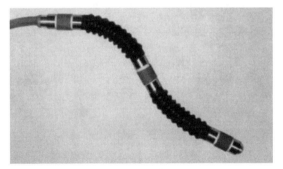

Figure 14.18 Miniaturized integrated robot for colonoscopy. (*From:* [134]. © 2002 Springer. Reprinted with permission.)

14.4 Future Challenges

Robotics has demonstrated the capability of automating the manipulation of biological cells and tissues. Despite the progress that was made over the last decade, existing microrobotic systems are not capable of performing fully autonomous high-throughput single cell manipulation although the requirements of accuracy and the elimination of direct human involvement are satisfied. The ideal approach seen by the majority of the research community is to combine MEMS and microfluidic tools while enhancing the intelligence and robustness of microrobotic systems, which is currently undertaken by many research groups in the robotics and MEMS communities.

On the front of robotic tissue manipulation, efforts are needed in safety enhancement (e.g., controller stability under all operating and fault conditions), medical imaging, CAD-based operation, sensor (e.g., force sensor) development and integration, intelligent instrument development, and man/machine interface, before fully autonomous robotic tissue manipulation becomes a reality in clinical environments.

References

[1]   Kempner, M. E., and R. A. Felder, "A Review of Cell Culture Automation," *Journal of The Association for Laboratory Automation*, Vol. 7, 2002, pp. 56–62.

[2]   Bernard, C. J., et al., "Adjuct Automation to the Cellmate Cell Culture Robot," *Journal of The Association for Laboratory Automation*, Vol. 9, 2004, pp. 209–217.

[3]   "On-Chip Single-Cell Cultivation System," in *Lab-on-Chips for Cellomics: Micro and Nanotechnologies for Life Sciences*, H. Andersson and A. van den Berg, (eds.), Norwell, MA: Kluwer, 2004.

[4]   Ashkin, A., J. M. Dziedzic, and T. Yamane, "Optical Trapping and Manipulation of Single Cells Using Infrared Laser Beams," *Nature*, Vol. 330, No. 6150, 1987, pp. 769–771.

[5]   Ashkin, A., and J. M. Dziedzic, "Optical Trapping and Manipulation of Viruses and Bacteria," *Science*, Vol. 235, No. 4795, 1987, pp. 1517–1520.

[6]   Bruican, T. N., et al., "Automated Single-Cell Manipulation and Sorting by Light Trapping," *Applied Optics*, Vol. 26, No. 24, 1987, pp. 5311–5316.

[7]   Wright, W. H., et al., "Laser Trapping in Cell Biology," *IEEE Journal of Quantum Electronics*, Vol. 26, No. 12, 1990, pp. 2148–2157.

[8]   Conia, J., B. S. Edwards, and S. Voelkel, "The Micro-Robotic Laboratory: Optical Trapping and Scissing for the Biologist," *Journal of Clinical Laboratory Analysis*, Vol. 11, No. 1, 1997, pp. 28–38.

[9]   Svoboda, K., and S. M. Block, "Biological Applications of Optical Forces," *Annual Rev. Biophys. and Biomolec. Structure*, Vol. 23, 1994, pp. 247–285.

[10]   Berns, M. W., W. H. Wright, and R. W. Steubing, "Laser Microbeam as a Tool in Cell Biology," *Int. Rev. Cytology*, 1991, pp. 1–44.

[11]   Nahmias, Y. K., and D. J. Odde, "Analysis of Radiation Forces in Laser Trapping and Laser-Guided Direct Writing Applications," *IEEE Journal of Quantum Electronics*, Vol. 38, No. 2, 2002, pp. 131–141.

[12]   Leitz, G., et al., "The Laser Microbeam Trap as an Optical Tool for Living Cells," *Physiological Chemistry and Physics and Medical NMR*, Vol. 26, 1994, pp. 69–88.

[13]   Ponelies, N., et al., "Laser Micromanipulators for Biotechnology and Genome Research," *J. of Biotechnology*, Vol. 35, 1994, pp. 109–120.

[14]   Arai, F., et al., "Pinpoint Injection of Microtools for Minimally Invasive Micromanipulation of Microbe by Laser Trap," *IEEE/ASME Transactions on Mechatronics*, Vol. 8, No. 1, 2003.

[15]   Biran, I., et al., "Optical Imaging Fiber-Based Live Bacterial Cell Array Biosensor," *Analytical Biochemistry*, Vol. 315, 2003, pp. 106–113.

[16]   Pohl, H. A., *Dielectrophoresis: The Behavior of Neutral Matter in Nonuniform Electric Fields*, Cambridge, U.K.: Cambridge University Press, 1978.

[17] Jones, T. B., *Electromechanics of Particles*, Cambridge, U.K.: Cambridge University Press, 1995.

[18] Pethig, R., "Dielectrophoresis: Using Inhomogeneous AC Electrical Fields to Separate and Manipulate Cells," *Crit. Rev. Biotechnol.*, Vol. 16, 1996, pp. 921–930.

[19] Voldman, J., et al., "Holding Forces of Single-Particle Dielectrophoretic Traps," *Biophysical Journal*, Vol. 80, 2001, pp. 531–541.

[20] Gray, D. S., et al., "Dielectrophoretic Registration of Living Cells to a Microelectrode Array," *Biosensors & Bioelectronics*, Vol. 19, 2004, pp. 1765–1774.

[21] Washizu, M., et al., "Dielectrophoretic Measurement of Bacterial Motor Characteristics," *IEEE Transactions on Industry Applications*, Vol. 29, No. 2, 1993, pp. 286–294.

[22] Washizu, M., and T. B. Jones, "Generalized Multipolar Dielectrophoretic Force and Electrorotational Torque Calculation," *Journal of Electrostatics*, Vol. 38, 1996, pp. 199–211.

[23] Arnold, W. M., and U. Zimmermann, "Electro-Rotation: Development of a Technique for Dielectric Measurements on Individual Cells and Particles," *Journal of Electrostatics*, Vol. 21, No. 2–3, 1988, pp. 151–191.

[24] Mischel, M., A. Voss, and H. A. Pohl, "Cellular Spin Resonance in Rotating Electric Fields," *Journal of Biological Physics*, Vol. 10, No. 4, 1982, pp. 223–226.

[25] Fuhr, G., et al., "Radio Frequency Microtools for Particle and Live Cell Manipulation," *Naturwissenschaften*, Vol. 81, No. 12, 1994, pp. 528–535.

[26] Nishioka, M., et al., "Evaluation of Cell Characteristics by Step-Wise Orientational Rotation Using Optoelectrostatic Micromanipulation," *IEEE Transactions on Industry Applications*, Vol. 33, No. 5, 1997, pp. 1381–1388.

[27] Yamamoto, T., et al., "Molecular Surgery of DNA Based on Electrostatic Micromanipulation," *IEEE Transactions on Industry Applications*, Vol. 36, No. 4, 2000, pp. 1010–1017.

[28] Gauthier, M., and E. Piat, "An Electromagnetic Micromanipulation System for Single-Cell Manipulation," *Journal of Micromechatronics*, Vol. 2, No. 2, 2004, pp. 87–120.

[29] Hultgren, A., et al., "Cell Manipulation Using Magnetic Nanowires," *Journal of Applied Physics*, Vol. 93, 2003, pp. 7554–7556.

[30] Reich, D. H., et al., "Biological Applications of Multifunctional Magnetic Nanowires," *Journal of Applied Physics*, Vol. 93, 2003, pp. 7275–7280.

[31] Alenghat, F. J., et al., "Analysis of Cell Mechanics in Single Vinculin-Deficient Cells Using a Magnetic Tweezer," *Biochem. Res. Commun.*, Vol. 277, 2000, pp. 93–99.

[32] Heinrich, V., and R. E. Waugh, "A PicoNewton Force Transducer and Its Application to Measurement of the Bending Stiffness of Phospholipids Membranes," *Ann. Biomed. Eng.*, Vol. 24, 1996, pp. 595–605.

[33] Fass, J. N., and D. J. Odde, "Tensile Force Dependent Neurite Elicitation Via Anti-B1 Integrin Antibody-Coated Magnetic Beads," *Biophysical Journal*, Vol. 85, 2003, pp. 623–636.

[34] Kimura, Y., and R. Yanagimachi, "Intracytoplasmic Sperm Injection in the Mouse," *Biology of Reproduction*, Vol. 52, No. 4, 1995, pp. 709–720.

[35] Matsuoka, H., et al., "High Throughput Easy Microinjection with a Single-Cell Manipulation Supporting Robot," *Journal of Biotechnology*, Vol. 116, 2005, pp. 185–194.

[36] Sun, Y., and B. J. Nelson, "Biological Cell Injection Using an Autonomous Microrobotic System," *International Journal of Robotics Research*, Vol. 21, No. 10–11, 2002, pp. 861–868.

[37] Inoue, S., and K. R. Spring, *Video Microscopy: The Fundamentals*, 2nd ed., New York: Plenum Press, 1997.

[38] Vollath, D., "Automatic Focusing by Correlative Methods," *J. of Microscopy*, Vol. 147, No. 3, 1987, pp. 279–288.

[39] Vollath, D., "The Influence of the Scene Parameters and of Noise on the Behavior of Automatic Focusing Algorithms," *J. of Microscopy*, Vol. 151, No. 8, 1988, pp. 133–146.

[40] Brenner, J., et al., "An Automated Microscope for Cytologic Research," *J. of Histochem. and Cytochem.*, Vol. 24, No. 1, 1971, pp. 100–111.

[41] Mehdelsohn, M. L., and B. H. Mayall, "Computer-Oriented Analysis of Human Chromosomes-III Focus," *Comput. Biol. Med.*, Vol. 2, 1972, pp. 137–150.

[42] Krotov, E., "Focusing," *International J. of Computer Vision*, Vol. 1, 1987, pp. 223–237.

[43] Nayar, S., and Y. Nakagawa, "Shape from Focus," *IEEE Transactions on Pattern Analysis and Machine Intelligence*, Vol. 16, No. 8, 1994, pp. 824–831.

[44] Groen, I., T. Young, and G. Ligntart, "A Comparison of Different Focus Functions for Use in Autofocus Algorithms," *Cytometry*, Vol. 12, 1985, pp. 81–91.

[45] Firestone, L., et al., "Comparison of Autofocus Methods for Use in Automated Algorithms," *Cytometry*, Vol. 12, 1991, pp. 81–91.

[46] Yeo, S., O. Jayasooriah, and R. Sinniah, "Autofocusing for Tissue Microscopy," *Image and Vis. Comput.*, Vol. 11, 1993, pp. 629–639.

[47] Santos, A., et al., "Evaluation of Autofocus Functions in Molecular Cytogenetic Analysis," *J. of Microscopy*, Vol. 188, 1997, pp. 264–272.

[48] Sun, Y., S. Duthaler, and B. J. Nelson, "Autofocusing in Computer Microscopy: Selecting the Optimal Focus Algorithm," *Microscopy Research and Technique*, Vol. 65, 2004, pp. 139–149.

[49] Jarvis, R. A., "Focus Optimization Criteria for Computer Image Processing," *Microscope*, Vol 24, No. 2, 1976, pp. 163–180.

[50] Horn, B. K. P., and B. G. Schunk, "Determining Optical Flow," *Artificial Intelligence*, Vol. 17, 1981, pp. 185–203.

[51] Lucas, B. D., and T. Kanade, "An Iterative Image Registration Technique with an Application in Stereo Vision," *Proceedings of Seventh International joint Conference on Artificial Intelligence*, Vancouver, 1981, pp. 674–679.

[52] Canny, J., "A Computational Approach to Edge Detection," *IEEE Transactions on Pattern Analysis and Machine Intelligence*, Vol. 8, 1986, pp. 679–698.

[53] Hough, P., "Method and Means for Recognizing Complex Patterns," U.S. Patent No. 3069654, 1962.

[54] Ballard, D. H., "Generalizing the Hough Transform to Detect Arbitrary Shapes," *Pattern Recognition*, Vol. 13, No. 2, 1981, pp. 111–122.

[55] Forsyth, D. A., and J. Ponce, *Computer Vision: A Modern Approach*, Upper Saddle River, NJ: Prentice-Hall, 2002.

[56] Kass, M., A. Witkin, and D. Terzopoulos, "Snakes: Active Contour Models," *Proc. Int. Conf. Comput. Vison*, 1987, pp. 259–268.

[57] Leymarie, F., and M. D. Levine, "Tracking Deformable Objects in the Plane Using an Active Contour Model," *IEEE Transactions on Pattern Analysis and Machine Intelligence*, Vol. 15, No. 6, 1993, pp. 617–634.

[58] Zimmer, C., et al., "Segmentation and Tracking of Migrating Cells in Videomicroscopy with Parametric Active Contours: A Tool for Cell-Based Drug Testing," *IEEE Transactions on Medical Imaging*, Vol. 21, No. 10, 2002, pp. 1212–1221.

[59] Ray, N., S. T. Acton, and K. Ley, "Tracking Leukocytes *In Vivo* with Shape and Size Constrained Active Contours," *IEEE Transactions on Medical Imaging*, Vol. 21, No. 10, 2002, pp. 1222–1235.

[60] Yuille, A. L., P. W. Hallinan, and D. S. Cohen, "Feature Extraction from Faces Using Deformable Template," *International J. of Computer Vision*, Vol. 8, No. 2, 1992, pp. 133–144.

[61] Jain, A. K., Y. Zhong, and S. Lakshmanan, "Object Matching Using Deformable Templates," *IEEE Transactions on Pattern Analysis and Machine Intelligence*, Vol. 18, No. 3, 1996.

[62] Grenander, U., *General Pattern Theory*, Oxford, U.K.: Oxford University Press, 1993.

[63] Sun, Y., M. A. Greminger, and B. J. Nelson, "Investigating Protein Structure Change in the Zona Pellucida with a Microrobotic System," *International Journal of Robotics Research*, Vol. 24, No. 2–3, 2005, pp. 211–218.

[64] Beer, G., *Programming the Boundary Element Method*, New York: John Wiley & Sons, 2001.

[65] Corke, P. I.,"Visual Control of Robot Manipulators—A Review," in *Visual Servoing*, K. Hashimoto, (ed.), Singapore: World Scientific, 1993, pp. 1–31.

[66] Hutchinson, S., G. D. Hager, and P. I. Corke, "A Tutorial on Visual Servo Control," *IEEE Transactions on Robotics and Automation*, Vol. 12, No. 5, 1996, pp. 651–670.

[67] Allen, P., et al., "Automated Tracking and Grasping of a Moving Object with a Robotic Hand-Eye System," *IEEE Transactions on Robotics and Automation*, Vol. 9, No. 2, 1993, pp. 152–165.

[68] Wilson, W., C. W. Hulls, and G. Bell, "Relative End-Effector Control Using Cartesian Position Based Visual Servoing," *IEEE Transactions on Robotics and Automation*, Vol. 12, No. 5, 1996, pp. 684–696.

[69] Huang, T. S., and A. N. Netravali, "Motion and Structure from Feature Correspondences: A Review," *Proceedings of the IEEE*, Vol. 82, No. 24, 1994, pp. 268–252.

[70] Hashimoto, K., et al., "Manipulator Control with Image-Based Visual Servo," *Proceedings of IEEE International Conference on Robotics and Automation*, 1991, pp. 2267–2272.

[71] Espiau, B., F. Chaumette, and P. Rives, "A New Approach to Visual Servoing in Robotics," *IEEE Transactions on Robotics and Automation*, Vol. 8, No. 3, 1992, pp. 313–326.

[72] Castano, A., and S. A. Hutchinson, "Visual Compliance: Task-Directed Visual Servo Control," *IEEE Transactions on Robotics and Automation*, Vol. 10, No. 3, June 1994, pp. 334–342.

[73] Papanikolopoulos, N. P., and P. K. Khosla, "Adaptive Robot Visual Tracking: Theory and Experiments," *IEEE Transactions on Automatic Control*, Vol. 38, No. 3, pp. 429–445, 1993.

[74] Feddema, J. T., and O. Mitchell, "Vision-Guided Servoing with Feature-Based Trajectory Generation," *IEEE Transactions on Robotics and Automation*, Vol. 5, No. 5, 1989, pp. 691–700.

[75] Feddema, J. T., C. S. G. Lee, and O. R. Mitchell, "Weighted Selection of Image Features for Resolved Rate Visual Feedback Control," *IEEE Transactions on Robotics and Automation*, Vol. 7, No. 1, pp. 31–47, 1991.

[76] Mahony, R., and T. Mamel, "Image-Based Visual Servo Control of Aerial Robotic Systems Using Linear Image Features," *IEEE Transactions on Robotics and Automation*, Vol. 21, No. 2, 2005, pp. 227–239.

[77] Sanderson, A. C., L. E. Weiss, and C. P. Neuman, "Dynamic Sensor-Based Control of Robots with Visual Feedback," *IEEE Transactions on Robotics and Automation*, Vol. RA-3, 1987, pp. 404–417.

[78] http://www.sutter.com/products/micromanipulation.html.

[79] http://www.eppendorf.com.

[80] http://www.siskiyou.com.

[81] http://www.somascientific.com/TOChydra.html.

[82] http://www.narishige.co.jp/products/products.htm.

[83] http://www.stoeltingco.com/physio/store/ViewLevel3.asp?keyword3=449.

[84] http://www.physikinstrumente.com/en/products/piezo_motor/index.php?VID=UxHkvuXF2Uo83YRt.

[85] http://wwwipr.ira.uka.de/~microbot/microrobs.html.

[86] Beyeler, F., et al., "Design of a Micro-Gripper and an Ultrasonic Manipulator for Handling Micron Sized Objects," *IEEE/RSJ International Conference on Intelligent Robots and Systems (IROS2006)*, Beijing, China, 2006.

[87]   Chronis, N., and L. P. Lee, "Electrothermally Activated SU-8 Microgripper for Single Cell Manipulation in Solution," *Journal of Microelectromechanical Systems*, Vol. 14, No. 4, 2005, pp. 857–863.

[88]   Chan, H. Y., and W. J. Li, "A Thermally Actuated Polymer Micro Robotic Gripper for Manipulation of Biological Cells," *IEEE International Conference on Robotics and Automation*, Taiwan, September 14–19, 2003, pp. 288–293.

[89]   Smela, E., "Microfabrication of PPy Microactuators and Other Conjugated Polymer Devices," *Journal of Micromech. Microeng.*, Vol. 9, 1999, pp. 1–18.

[90]   Jager, E. W. H., O. Inganas, and I. Lundstrom, "Microrobots for Micrometer-Sized Objects in Aqueous Media: Potential Tools for Single-Cell Manipulation," *Science*, Vol. 288, 2000, pp. 2335–2338.

[91]   Hashmi, S., et al., "Genetic Transformation of Nematodes Using Arrays of Micromechanical Piercing Structures," *BioTechniques*, Vol. 19, 1995, pp. 766–770.

[92]   Henry, S., et al., "Micromachined Needles for the Transdermal Delivery of Drugs," *IEEE Conf. MicroElectroMechanicalSystems*, 1998, pp. 494–498.

[93]   Gardeniers, J. G. E., et al., "Silicon Micromachined Hollow Microneedles for Transdermal Liquid Transport," *J. of Microelectromechanical Systems*, Vol. 12, 2003, pp. 855–862.

[94]   Chun, K., et al., "An Array of Hollow Microcapillaries for the Controlled Injection of Genetic Materials into Animal/Plant Cells," *IEEE Conf. MicroElectroMechanicalSystems*, 1999, pp. 406–411.

[95]   Yi, C., et al., "Microfluidics Technology for Manipulation and Analysis of Biological Cells," *Analytica Chimica Acta*, Vol. 560, 2006, pp. 1–23.

[96]   Andersson, H., and A. van den Berg, "Microfluidic Devices for Cellomics: A Review," *Sensors and Actuators B*, Vol. 92, 2003, pp. 315–325.

[97]   Tai, H. P., and M. L. Shuler, "Integration of Cell Culture and Microfabrication Technology," *Biotechnology Progress*, Vol. 19, 2003, pp. 243–253.

[98]   Morris, V. J., A. R. Kirby, and A. P. Gunning, *Atomic Force Microscopy for Biologists*, London, U.K.: Imperial College Press, 1999.

[99]   Weisenhorn, A. L., et al., "Deformation and Height Anomaly of Soft Surfaces Studied with an AFM," *Nanotechnology*, Vol. 4, 1993, pp. 106–113.

[100]   Radmacher, M., et al., "Measuring the Viscoelastic Properties of Human Platelets with the Atomic Force Microscope," *Biophysical Journal*, Vol. 70, 1996, pp. 556–567.

[101]   Tao, N. J., S. M. Lindsay, and S. Lees, "Measuring the Microelastic Properties of Biological Materials," *Biophys. J.*, Vol. 63, 1992, pp. 1165–1169.

[102]   Rotsch, C., et al., "AFM Imaging and Elasticity Measurements on Living Rat Liver Macrophages," *Cell Biol. Intl.*, Vol. 21, 1997, pp. 685–696.

[103]   Vinckier, A., and G. Semenza, "Measuring Elasticity of Biological Materials by Atomic Force Microscopy," *FEBS Letters*, Vol. 430, 1998, pp. 12–16.

[104]   Sato, M., et al., "Local Mechanical Properties Measured by Atomic Force Microscopy for Cultured Boving Endothelial Cells Exposed to Shear Stress," *J. Biomech.*, Vol. 33, 2000, pp. 127–135.

[105]   Mahay, R. E., et al., "Scanning Probe–Based Frequency–Dependent Microrheology of Polymer Gels and Biological Cells," *Phys. Rev. Lett.*, Vol. 85, 2000, pp. 880–883.

[106]   Dimitriadis, E. K., et al., "Determination of Elastic Moduli of Thin Layers of Soft Materials Using the Atomic Force Microscope," *Biophysical Journal*, Vol. 82, 2002, pp. 2798–2810.

[107]   McNally, H., and R. Borgens, "Three-Dimensional Imaging in Living and Dying Neurons with Atomic Force Microscopy," *Journal of Neurocytology*, Vol. 33, No. 2, 2004, pp. 251–258.

[108]   Kwon, Y. S., et al., "A Robot with Improved Absolute Positioning Accuracy for CT Guided Stereotactic Brain Surgery," *IEEE Transactions on Biomedical Engineering*, Vol. 35, No. 2, 1988, pp. 153–161.

[109] Taylor, R. H., H. A. Paul, and B. Mittelstadt, "Robotic Hip Replacement Surgery in Dogs," *IEEE EMBS International Conference*, 1989, pp. 887–889.

[110] Davis, B. L., et al., "A Surgeon Robot Prostatectomy—A Laboratory Evaluation," *J. Med. Eng. Technol.*, Vol. 13, No. 6, 1989, pp. 273–277.

[111] Ng, W. S., et al., "Robotic Surgery: A Firsthand Experience in Transurethral Resection of the Prostate," *IEEE Engineering in Medicine and Biology*, Vol. 12, 1993, pp. 120–125.

[112] Davis, B., "A Review of Robotics in Surgery," *Proceedings of the Institution of Mechanical Engineers, Part H: Journal of Engineering in Medicine*, Vol. 214, 2000, pp. 129–140.

[113] Low, S. C., and L. Phee, "A Review of Master-Slave Robotic Systems for Surgery," *IEEE Conference on Robotics, Automation and Mechatronics*, Singapore, December 1–3, 2004, pp. 37–42.

[114] Cleary, K., and C. Nguyen, "State of the Art in Surgical Robotics: Clinical Applications and Technology Challenges," *Comput. Aided Surgery*, Vol. 6, No. 6, 2001, pp. 312–328.

[115] Sung, G. T., and I. S. Gill, "Robotic Laparoscopic Surgery: A Comparison of the Da Vinci and Zeus Systems," *Urology*, Vol. 58, No. 6, 2001, pp. 893–898.

[116] Chapman, W. H. H., et al., "Computer-Assisted Laparoscopic Splenectomy with the da Vinci™ Surgical Robot," *Journal of Laparoendoscopic & Advanced Surgical Techniques*, Vol. 12, No. 3, 2002, pp. 155–159.

[117] Sackier, J. M., and Y. Wang, "Robotically Assisted Laparoscopic Surgery: From Concept to Development," in *Computer-Integrated Surgery Technology and Clinical Applications*, R. Taylor, et al., (eds.), Cambridge, MA: MIT Press, 1996.

[118] Casals, A., J. Amat, and E. Laporte, " Automatic Guidance of an Assistant Robot in Laparoscopic Surgery," *IEEE International Conf. on Robotics and Automation*, Minneapolis, MN, 1996, pp. 895–900.

[119] Bicchi, A., et al., "A Sensorized Minimally Invasive Surgery Tool for Detecting Tissue Elastic Properties," *IEEE International Conf. on Robotics and Automation*, Minneapolis, MN, 1996, pp. 884–888.

[120] Eltaib, M. E. H., and J. R. Hewit, "Tactile Sensing Technology for Minimal Access Surgery—A Review," *Mechatronics*, Vol. 13, 2003, pp. 1163–1177.

[121] Duchemin, G., et al., "A Hybrid Position/Force Control Approach for Identification of Deformation Models of Skin and Underlying Tissues," *IEEE Transactions on Biomedical Engineering*, Vol. 52, No. 2, 2005, pp. 160–170.

[122] Peine, W. J., D. A. Kontarinis, and R. D. Howe, "A Tactile Sensing and Display System for Surgical Applications," in *Interactive Technology and the New Paradigm for Health Care*, K. Morgan, et al., (eds.), Washington, D.C.: IOS, 1995.

[123] Fischer, H., B. Neisius, and R. Trapp, "Tactile Feedback for Endoscopic Surgery," in *Interactive Technology and the New Paradigm for Health Care*, K. Morgan, et al., (eds.), Washington, D.C.: IOS, 1995.

[124] Rosen, J., et al., "Force Controlled and Teleoperated Endoscopic Grasper for Minimally Invasive Surgery: Experimental Performance Evaluation," *IEEE Transactions on Biomedical Engineering*, Vol. 46, No. 10, 1999, pp. 1212–1221.

[125] Scott, H. J., and A. Darzi, "Tactile Feedback in Laparoscopic Colonic Surgery," *Br. J. Surg.*, Vol. 84, 1997, pp. 1005–1010.

[126] De Gersem, G., H. Van Brussel, and F. Tendick, "Reliable and Enhanced Stiffness Perception in Soft-Tissue Telemanipulation," *International J. of Robotics Research*, Vol. 24, No. 10, 2005, pp. 805–822.

[127] Okamura, A. M., C. Simone, and M. D. O'Leary, "Force Modelling for Needle Insertion into Soft Tissue," *IEEE Transactions on Biomedical Engineering*, Vol. 51, No. 10, 2004, pp. 1707–1716.

[128] Brett, P. N., A. J. Harrison, and T. A. Thomas, "Schemes for the Identification of Tissue Types and Boundaries at the Tool Point for Surgical Needles," *IEEE Transactions on Information Technology in Biomedicine*, Vol. 4, 2000, pp. 30–36.

[129] Dario, P., B. Hannaford, and A. Menciassi, "Smart Surgical Tools and Augmenting Devices," *IEEE Transactions on Robotics and Automation*, Vol. 19, No. 5, 2003, pp. 782–792.

[130] Kumar, R., et al., "Preliminary Experiments in Cooperative Human/Robot Force Control for Robot Assisted Microsurgical Manipulation," *IEEE International Conf. on Robotics and Automation*, 2000, pp. 610–617.

[131] Vankov, A., et al., "Electro-Adhesive Forceps for Tissue Manipulation," *Ophthalmic Technologies*, SPIE, Vol. 5314, 2004, pp. 270–274.

[132] Ottensmeyer, M. P., and J. K. Salisbury, "In Vivo Data Acquisition Instrument for Solid Organ Mechanical Property Measurement," *Medical Image Computing and Computer-Assisted Intervention*, 2001, pp. 975–982.

[133] Rosen, J., et al., "The blueDRAGON—A System for Measuring the Kinematics and the Dynamics of Minimally Invasive Surgical Tools In Vivo," *IEEE Int. Conf. Robotics and Automation*, 2002, pp. 1876–1881.

[134] Dario, P., et al., "Modeling and Experimental Validation of the Locomotion of Endoscopic Robots in the Colon," *8th Int. Symp. Experimental Robotics*, 2002.

CHAPTER 15

A General Robotic Approach for High-Throughput Automation in the Handling of Biosamples

Yuan F. Zheng,[†] A. Peddi,[†] L. Muthusubramaniam,[†] V. Cherezov,[†] and M. Caffrey[‡]

15.1   Introduction

Research and development activities in life sciences have become a major endeavor in the United States and in many other countries alike in recent years. These activities include DNA sequencing, protein crystallization, cell manipulation, and drug development. Many of these activities involve the screening of a large number of biosamples, in which numerous and repeated experimental trials have to be performed. The purpose is to explore a vast combinatorial space to determine optimal conditions for generating desired results. As a result, automation for high throughput has become a necessity in life science research and has attracted more and more attention in the robotics and automation communities. A considerable number of efforts have been made by researchers in both engineering and life science disciplines including those contained in the special issue of *IEEE Transactions on Automation Science and Engineering* published recently [1] and individually in the literature [2–6].

Take protein crystallization as an example, which is a biochemical process that turns discrete protein molecules into a well-structured crystal. The latter is then used in X-ray crystallography for determining the structure of protein at atomic resolution [7]. The crystallization process involves the mixing of multiple types of solutions, and then delivering them into a large number of containers for incubating at a fixed temperature and for a period of time. The process is tedious and timing consuming using human hands [8], which leads to the desire for automation using robots. It is also important to minimize protein usage per trial to reduce the cost, which can be better achieved by using automated machines.

A number of systems have been developed by research institutions for automated handling of biosamples [9–16]. In [9], a liquid handling robot called Robbins Scientific Tango is a part of a protein crystallization system developed by the

†   The Ohio State University
‡   School of Sciences, University of Limerick, Limerick, Ireland

University of New York at Buffalo. In [10] a fluid sample handling system called ACAPELLA-5K is reported by the University of Washington and is used for automated preparation of DNA samples. In the meantime, a number of commercial products have become available on the market including the Xantus robots [17]. In reviewing the existing systems, however, one can observe that most of the available systems are designed ad hoc for a targeted application which has limited value when the application alternates. Another problem is the lack of theoretical base which governs the development of the automated system as well as the behavior of the biosamples it handles. Scientific and unified solutions for handling of biosamples especially in the micro/nanoliter scales are still nonexistent. Even more challenging is that biosamples to be handled have a wide scope of physical and chemical characteristics. Some are as fluidic as water while others are highly viscous. A general-purpose robotic system which can handle many kinds of biosamples by a program mechanism is clearly attractive and needed.

In this chapter, we present a general robotic approach for the purpose as cited. The philosophy of the approach is to use as many off-the-shelf components as possible and to integrate them together in an optimal way. Such an approach can be duplicated with limited modification for different purposes of biosample handling. The approach to be introduced is based on research activities of the authors for high-throughput crystallization of membrane proteins [18].

In the remainder of this chapter, we will first present a robotic system developed by integrating commercially available components for the purpose of automated handling of biosamples (Section 15.2). That is followed by the procedures for calibrating and optimizing the liquid handling parameters of the system in Section 15.3. In Section 15.4, we describe the newly developed method for the delivery of nanoliter volumes of viscous biosamples. Key factors guaranteeing the successful delivery of the viscous biosamples are discussed. In Section 15.5, a mathematical model describing the behavior of highly viscous materials during dispensing is presented. In each of the above three sections, the performance of the methods in terms of accuracy and reproducibility are tested by using experiments. The chapter is concluded with a conclusions section (Section 15.6).

15.2 The Creation of a Robotic System by Integration

In this section we describe the high-throughput robotic system that can deliver highly viscous biosamples and other various kinds of biosolutions at the nano/microliter scales. As mentioned earlier, the robotic system was developed to automate the crystallization of membrane proteins [18] in which the system is required to dispense nanoliter volumes of a highly viscous biosample call *cubic phase* and microliter volumes of various *precipitant solutions*. The system, however, is not limited to the application of protein crystallization, which will become obvious as our discussion proceeds.

The system consists of two major components: a robot which is built by Xantus (SIAS US, New Castle, Delaware), and a positive displacement microactuator controlled microsyringe pump (UMP II with Micro4 controller, World Precision Instruments, Sarasota, Florida), as shown in Figure 15.1. The standard Xantus robot has a

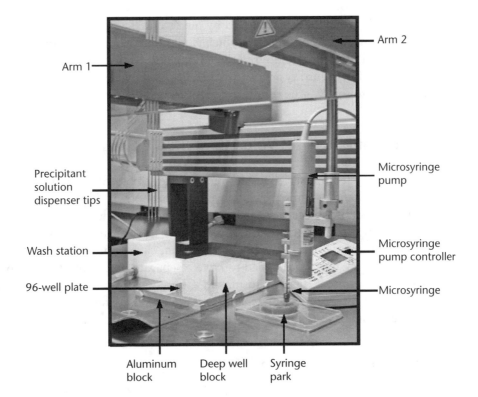

Figure 15.1 The robotic system with the integration of commercial components.

1m × 0.7m deck with two arms, each of which has 3 degrees of freedom (DOF) with linear motion in the *x*-, *y*-, and *z*-directions, respectively. The standard Xantus robot has two arms of which Arm 1 is used to aspirate and dispense solutions and Arm 2 has a plate gripper to move plates. The step size of the *x*-, *y*-, and *z*-motors of both arms is 100 μm.

For crystallization of membrane proteins, the functionality of Arm 1 for handling solutions is retained while the standard tips are replaced with low volume tips to aspirate and dispense microliter volumes of precipitant solutions. There are four low volume liquid handlers held by Arm 1. The functionality of Arm 2 is to deliver nanoliter volumes of highly viscous cubic phase. It is equipped with the microsyringe pump, as mentioned, which is integrated to Xantus through an input/output board. The step size of the Arm 2 *z*-motor was changed from 100 to 12.5 μm as the dispensing distance. The height of the dispensing needle tip above the base of the well which is to contain the sample is a critical factor for successful delivery of viscous biosamples. Reducing the step size of the z motor is for the benefit of fine-tuning the dispensing distance (details of which will be discussed in later sections).

The microsyringe pump can handle syringes from 10 μL to 1 mL. The volume delivered per step for a particular syringe is constant since the step size of the pump is constant, which is 3.189 μm. For the standard Hamilton gas-tight 100-μL syringe with a syringe scale length of 60 mm, the volume per step is thus 5.315 nL. The actual volume dispensed will be the product of volume per step and the number of steps. For a desired volume of 50 nL, for example, 47.83 nL is the volume actually delivered when the plunger is moved by nine steps. The desired volume and rate of delivery is adjusted using Micro4, which is the controller of the microsyringe pump. The standard delivery rate used in the laboratory for dispensing 50-nL cubic phase from a 100-μL standard syringe is 500 nL/sec. As shown in Figure 15.1, this setup is integrated into the Xantus robotic system through an input/output board.

15.3   Delivery of Microliter Volumes of Biosolutions

The robotic system must deliver desired volumes of biosolutions accurately. To do so, the liquid handlers must be calibrated first for both aspiration and dispensing of water to the desired accuracy. Then the parameters, such as aspiration speed and waste volume, are optimized for reproducible and accurate delivery of different solutions. As mentioned in the previous section, each liquid handler (there are four in the current system) is designed to have a separate pump, but the pumps are not equalized in terms of their performance. Consequently, individual liquid handlers have to be calibrated to determine the specific input corresponding to a particular dispense volume.

15.3.1   Calibration Approaches

Here we introduce the calibration approaches. A *gravimetric* technique is used to calibrate the liquid handlers for both aspiration and dispensing of 1, 10, and 100 μL of solution; 1 μL is the smallest volume that the current system can handle, as it is the typical volume of precipitant solutions used for membrane protein crystallization [5]. 10 and 100 μL, on the other hand, are nominal volumes used while preparing precipitant solutions during the optimization procedure. For calibrating the tips, 100 or 10 μL or ten times 1 μL of water is delivered into small lightweight (500 mg) plastic tubes and weighed using a microbalance (AX205, Mettler-Toledo Inc., Columbus, Ohio). Volume is converted to weight using a density of 0.99823g ml^{-1} (293 K, *CRC Handbook of Chemistry and Physics*). Based on the average volume of five deliveries, the calibration factor is adjusted to attain the desired target volume. One should repeat this procedure until the dispensed or aspirated volume is within 5% of nominal volume. A similar procedure is followed for calibrating the liquid handlers to aspirate 1, 10, and 100 μL of water.

Once the tips are calibrated, the liquid handling parameters have to be optimized for delivering different kinds of solutions, on which the accuracy and the reproducibility of the delivery depends. These parameters include aspiration and dispensing speed, system air, transport air, waste volume, and aspiration and dispensing delay, some of which are shown in Figure 15.2. For example, when aspirating viscous solutions there is a tendency for air to be aspirated instead of solutions if speed is too high. So the aspiration speed should be low with a longer postaspiration delay to attain desired accuracy. Similarly, dispensing speed should be lowered to avoid the delivery of air instead of solutions. Furthermore, the set of optimal liquid handling parameters differs for different solutions, as properties such as viscosity and surface tension vary from one solution to another. Thus, the

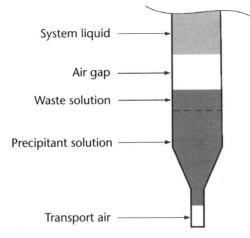

Figure 15.2   Liquid handling parameters of a robotic system tip.

challenging task is to determine a single set of parameters that are optimal for handling a wide range of solutions used in the screening experiments of any biological study.

The liquid handling parameters are determined by delivering 1 μL of water and 25% (w/w) Polyethyleneglycol 4000 (PEG4000) solution in a homemade plate (Figure 15.3). This plate is prepared using a flat glass plate, a 96-hole perforated polymer spacer, and a glass cover slip. The flat glass plate has dimensions of 127.8 × 86.5 mm^2 which conforms to the Society for Biomolecular Sciences (SBS) standard with 1-mm thickness. The spacer (Saunders East, Lombard, Illinois) is a

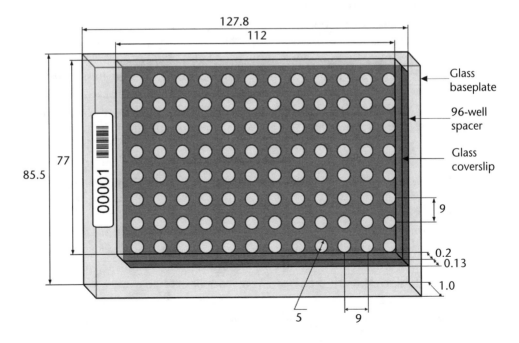

Figure 15.3   Schematic of 96-well glass plate (dimensions in millimeters).

double-stick polyester film that has a foot print of $112 \times 77 \times 0.13$ mm^3 with 96 holes, each 5 mm in diameter and separated by 9 mm. The glass cover slip is made from glass with $112 \times 77 \times 0.2$ mm^3 dimensions. The spacer is placed on the flat glass base plate which creates 96 wells. As the well thickness is 0.13 mm, the 1 μL of dispensed solution in the plate is sandwiched between and in contact with the upper and lower glass surfaces.

An imaging system is used to evaluate the delivery in each well, which was developed in our laboratory and with automated image analysis software. This automated image analysis software is developed by writing plug-ins using *ImageJ* (available free from http://rsb.info.nih.gov/ij/) and *Adobe Photoshop* (Adobe Systems, Inc., San Jose, California). The image system determines the area occupied by the delivered solution, and the volume is calculated as the product of the area and the well spacer thickness (0.13 mm). Water and 25% (w/w) PEG4000 are used to perform the analysis as they cover the range of viscosities likely to be encountered in a typical screen.

From the experiments, it was determined that the aspiration and dispensing speed, waste volume, and system air significantly affect the accuracy of delivery. These parameters were fine tuned to determine the optimum configuration set. For each of these parameters, the delivery is evaluated using the above-mentioned automated image analysis. From the measurements it is determined that the optimal parameters include: dispense speed 50 μL/sec, waster volume 1 μL, system air gap 5 μL, aspiration and dispensing delay 0 second, and transport air 0 μL.

15.3.2 Experiment Using the Calibration Approach

We can use the approaches just described to calibrate the robotic system, which should then perform satisfactorily in delivering various biosolutions. Here we present the evaluation on the accuracy and reproducibility of the system in delivering microliter volumes of solutions. 1 μL of water is dispensed into the first four wells of the 96-well plate by four low volume tips. This is followed by the delivery of 1 μL of 25% PEG4000 solution in the next four wells of the first column. Similarly, all 12 columns of the 96-well crystallization plate are loaded with 1 μL of water and 25% PEG4000. Each plate in this way consists of 12 replicates of a particular solution delivered by a specific tip. The plate is then covered with a cover slip such that the dispensed solution is sandwiched between two glass slides. The plate is scanned immediately using the imaging system mentioned above. The images are analyzed automatically using the image analysis protocol, which determines the area of the solution in each well. The volume of the delivered solution is obtained from the product of the calculated area and the well thickness.

Using the optimal liquid handling parameter configuration described above, the accuracy and reproducibility of the robotic system in delivering water and a 25% PEG4000 solution are presented in Table 15.1. The results are obtained from more than 50 dispenses with each tip. The metrics used to quantify the performance of liquid handlers are the average and coefficient of variance (CV) of replicates. The CV is defined as the ratio of the standard deviation to the sample average. The results indicate that the CV for delivering water and 25% PEG4000 is less than 8%, except for tip 3 which is about 11.3% (slightly higher than the desired 10%). This

Table 15.1 Accuracy and Coefficient of Variance in
Delivering 1 μL of Water and 25% PEG4000

Tip	Water Volume (μL)	25% PEG 4000 CV	Volume (μL)	CV
1	0.98	7.2	0.93	7.9
2	1.08	7.1	1.04	7.2
3	0.95	11.3	0.96	11.1
4	1.03	6.2	1.01	6.5

larger-than-normal CV is due to the higher variation in the output of the pump
associated with tip 3.

15.4 Delivery of Nanovolumes of Viscous Biosamples

15.4.1 Physical and Theoretical Analysis of the Delivering Process

The approach mentioned in the previous section is for dispensing biosolutions of
relatively low viscosity and in microliter volumes. In many biological experiments,
however, it is necessary to deliver highly viscous biosamples such as the cubic phase
used in the crystallization of membrane proteins [8]. Also there is a strong need to
reduce the sample usage per trial to the nanoliter scale. Automatic and accurate
delivery of highly viscous biomaterials is a challenging issue, and there are a number
of factors that impact the dispensing of nanoliter volumes of highly viscous
biomaterials. The approach presented in this chapter can satisfactorily handle
biosamples of viscosity up to 3,000 cP, which is close to that of honey [19].

Automating the delivery of viscous biosamples can be achieved by using the
microsyringe pump and the motors for positioning the needle tip above the base of
the well, both mentioned in Section 15.2. The experimental results indicate that dispensing distance, texture, and volume of material to be delivered, and the diameter
and tip uniformity of the needle, impact the successful delivery of highly viscous
biosamples at the nanoliter scale. Of these, the most important factor is the dispensing distance (i.e., the height of the dispensing needle tip above the base of the well).
The importance of the dispensing distance is clearly demonstrated from the observations, as shown in Figures 15.4.

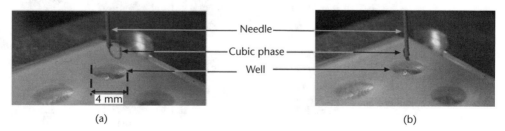

(a)   (b)

Figure 15.4 (a) The cubic phase forms a loop and comes back toward the needle when it is too
far from the bottom of the well during delivery. (b) The cubic phase remains stuck to the needle
when it is too close to the bottom of the well during delivery.

When the dispensing distance is too large, the biosample emerging from the needle forms a continuous tube looping back to the needle and hence it is not delivered to the well [Figure 15.4(a)]. On the other hand, if the needle is too close to the base of well, some of the delivered biosample balls up and sticks to the needle [Figure 15.4(b)] resulting in an unsuccessful delivery. The dispensing distance is clearly too important to be ignored.

To determine the optimal distance, we need to study the delivering process more carefully. Figure 15.5 shows the sequence of events that occur during the nanoliter-scale viscous material dispensing process. In (1), the dispensing needle and residual from the previous trial is at the desired distance above the base of the well. In (2), the desired volume of biosample is expelled. In (3), the needle is retracted back, but there is a small carryover volume at the tip. In (4), one can notice the increase in the carryover volume at the needle tip, which is caused by the relaxation of the viscous material resulting in the increase in the carryover at the needle tip. In (5), the needle tip is shown to have the complete final carryover volume before the next delivery.

The carryover volume affects accurate dispensing, which is determined by two factors. One is where the *break point* of the material is, and the other is how the material relaxes from the needle. These two factors are discussed next.

Figure 15.6 shows how the break point is formed in the sequence of delivery of the viscous biosamples. In (1), the needle is at the adequate distance above the base of the well with the carryover from the previous delivery. In (2), the desired volume of the material is expelled. In (3), the dispenser is retracted back and thus the shape of the material changes because both the needle and well stick to it. In (4), a *break point* is formed which determines the amount of the materials being delivered to the well. In (5), the needle moves back up with the break point carryover volume. In (6), the needle has the complete carryover volume before the next delivery. From (5) to

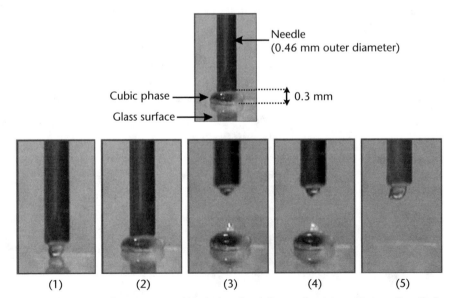

Figure 15.5 Sequence of events occurring during the delivery of a viscous biosample called cubic phase. 100 nL of cubic phase (Monoolein/Methylene blue solution, 63/37 (w/w)) is delivered from 0.3 mm above the base of the well.

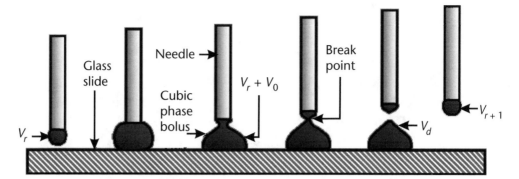

Figure 15.6   The sequence of events during the delivery of viscous biomaterials.

(6) one can see the increase of the carryover volume at the needle tip caused by the relaxation of the materials.

The relaxation of the materials is due to its high viscosity, which strongly resists to flow. Hence, pressure builds up in the needle when the spring is pressed to dispense, which causes the wall of the needle to deflect and the materials to compress. The system relaxes only when the material is expelled from the needle. One has to pay close attention to the time constant of the relaxation and the amount of material expelled during the relaxation. For materials with high viscosity, the relaxation time could be as long as several seconds, which is too long for a high-throughput application. To solve this problem we need to deliver materials with a constant speed and move the needle when the relaxation is taking place. In this way, every delivery will have a uniform carryover, and the total delivery volume to each well will be uniform as well, except for the first few when the time constant is reaching a steady state.

15.4.2   Experiment Verification

As mentioned earlier, the unique capability of the current robotic approach is to deliver nanoliter volumes of highly viscous biosamples. The performance is measured using the same image analysis as described above. The given volume (50 or 20 nL) of the viscous biomaterial cubic phase (Figure 15.5) is delivered in the 96-well plate and then the plate is covered with a cover slip. The entire delivery process is conducted in a humidified atmosphere to ensure that there is no dehydration effect on the cubic phase. The dispensed cubic phase bolus has a relatively uniform disc shape after sandwiching between two glass slides. A sample plate loaded with 100 nL of the cubic phase (Monoolein/Bis-CarboxyEthyl-Carboxy-Fluorescein (BCECF) solution 60/40 (w/w)) and dispensed 23 steps (287.5 μm) above the base of the well is shown in Figure 15.7.

The accuracy and reproducibility of the system in delivering 50 and 20 nL of the cubic phase is shown in Figure 15.8. The results indicate that the reproducibility of the system in delivering 47.83 nL is less than 10% compared to 18% for delivering 21.26 nL. The accuracy is calculated by measuring the volume dispensed after 100 deliveries. The actual volume for 100 deliveries is calculated from the displacement

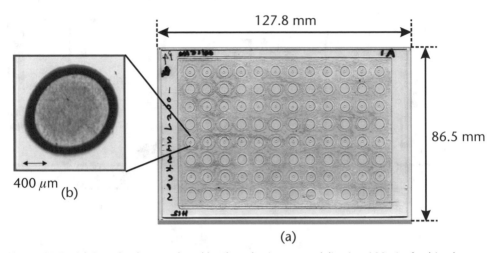

(a)

Figure 15.7 (a) Sample plate produced by the robotic system delivering 100 nL of cubic phase (Monoolein/BCECF solution 60/40 (w/w)) from 23 steps above the base of the well. (b) Zoom-in picture of a particular specific well showing the black outline of the cubic phase bolus.

of the syringe plunger. From the experimental results one can see that the accuracy and reproducibility of the system are much higher when dispensing larger volumes. That is natural because the carryover volume in higher volume delivery is less a factor than in lower volume.

15.5   Mathematical Modeling of the Delivery of Viscous Biosamples

15.5.1   The Modeling Approach

To completely understand the delivery process, we need a mathematical model to determine the actual volume delivered to the well. The mathematical model reveals two parameters of the delivery process: the shape of the delivered material and the formation of the break point.

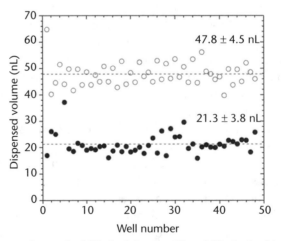

Figure 15.8   Accuracy and reproducibility in delivering 20 and 50 nL of cubic phase.

According to the actual shape of the delivered material, we use an ellipsoidal model as in Figure 15.9 to describe it. The major axis of the ellipsoid represents the diameter of the expelled biosample, and the minor axis represents the height of the needle tip above the base of the well. The volume of the delivered materials can be calculated by the following equation:

$$V = \frac{4}{3}\pi R^2 h \qquad (15.1)$$

where V is the volume of the biosample expelled, R is the radius of the sample at the base of the well, and h is the height of the needle tip above the well base. Using (5.1), it is easy to calculate the radius of dispensed material when the volume and the height are known. The radius will further determine the area of the footprint with which the viscous material contacts the bottom of the well.

We further use a conical model to determine the height above the base of well at which a break point forms, as shown in Figure 15.10. Based on the model, the following equation can be obtained:

$$\frac{R}{r} = \frac{h_1}{h - h_1} \qquad (15.2)$$

where the undefined parameter h_1 is the height of break point from the well base, and r represents the radius of the needle tip.

The radius of the dispensed biosample can be calculated using (15.1). Then the position of the break point can be determined using (15.2), so long as the values of the parameters involved are known.

Based on (15.1) and (15.2), we can develop an equation which governs the volume delivered to a well given the height of delivery, the amount of volume expelled from the syringe, and the radius of the dispensing needle, as follows:

$$V_d = (V_o + V_r)\left(\frac{R^3}{R^3 + r^3}\right) \qquad (15.3)$$

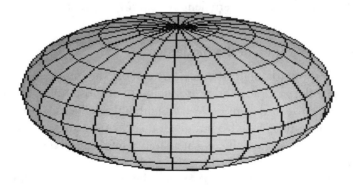

Figure 15.9 An ellipsoid model to describe the shape of the dispensed biomaterials.

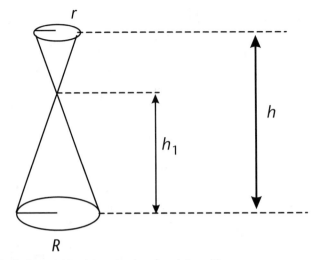

Figure 15.10   Conical model to determine break point position.

where V_d represents the volume delivered to the well, V_o the volume expelled from the syringe, and V_r the break point carryover volume from the previous well. Furthermore, the break point residual volume and the amount of prime volume to be delivered initially are given by the following two equations:

$$V_{r+1} = (V_o + V_r)\left(\frac{r^3}{R^3 + r^3}\right) \qquad (15.4)$$

where V_{r+1} represents the break point carryover volume, and

$$V_p = V_o\left(\frac{r^3}{R^3 + r^3}\right) \qquad (15.5)$$

where V_p is the initial prime volume of the delivery.

We have so far presented a modeling approach for describing the shape and the volume of the dispensed viscous biomaterials. The geometrical models and mathematical equations determine the dispensing process in a quantitative way. These can guide the design of the trajectory of the needle tip and the amount it dispenses. The result is an accurate and efficient system for dispensing viscous biosamples.

15.5.2   Verification of the Model

To verify the mathematical model, we can use the same experiments that are used to determine the accuracy and reproducibility of biosample delivery. A set of 48 wells (alternate rows in the entire 96-well plate) are delivered with a particular composition of the cubic phase from a given dispensing distance. The volume of delivered cubic phase in each bolus is calculated using the direct image analysis described above.

In the experiment, a set of 48 samples with a desired volume of 47.83 nL [V_d in (15.3)] of the cubic phase (Monoolein/BCECF solution 60/40 (w/w)) is delivered.

The cubic phase is delivered from 21 steps (262.5 μm) above the base of the well with a 22-gauge (0.21-mm internal radius) needle. To deliver the desired volume accurately, an initial prime volume serving as V_r is expelled for the first well, as shown in Figure 15.6. Here, V_r is 47.83 nL. Then every well is delivered with a volume of V_o, which is 47.83 nL. From (15.3), one can see that without V_r, V_d will be less than V_o. The result of the experiment is represented in Figure 15.11, where the predicted volume of the cubic phase is calculated using the model proposed and shown as the solid line in the figure. The actual volume of the cubic phase dispensed in the 48 wells is represented by the dashed line in Figure 15.11. From Figure 15.11, one can see that the actual volume of the dispensed cubic phase very much follows the predicted volume. For the 48 deliveries, the average difference between the actual volume delivered and model predicted volume is 3.78 nL (7.9%), while the maximum deviation and minimum deviation of predicted volume from the actual volume in a set of 48 samples is 9.48 nL (19.8%) and 0.11 nL (0.24%), respectively.

The goodness of the model can be further examined by studying the result of the first well. As mentioned above, the initial prime volume of 47.83 nL is expelled out of the needle before the dispensing starts. Using the dispensing height, which is 262.5 μm, and the prime volume, one can calculate the radius of the ellipsoid to be $R = 0.285$ mm by using (15.1). When the cubic phase is delivered to the first well, an additional volume serving as V_o, which is 47.83 nL, is dispensed. Using (15.2), the delivered volume to the first well V_d should be 68.33 nL. The actual volume delivered is 66.54 nL, and the difference between that and the volume predicted by the model is 1.79 nL.

Choosing $V_r = 47.83$ nL is purely experimental; we could choose a smaller value. From the above discussion one can see that after the first dispense, V_r is 27.33 nL, which serves well for the remaining wells. Unfortunately, the initial V_r is difficult to determine considering so many factors being involved. Experimental results

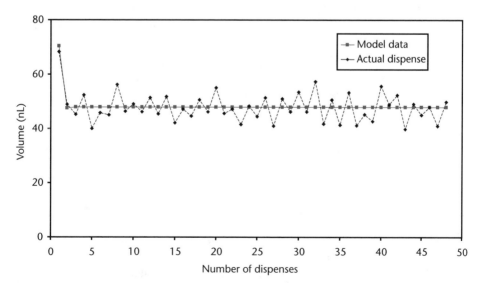

Figure 15.11   Actual volume of the cubic phase delivered in comparison with predicted data when delivering 47.83 nL of the cubic phase (Monoolein/BCECF solution 60/40 (w/w)) from 21 steps (approximately 0.2625 mm) above the base of the well with a 22-gauge (0.42 internal diameter) needle and an initial prime volume of 47.83 nL. Note (15.3) in the figure.

prove that by letting $V_r = V_o$ where V_o is the desired volume of dispense, the actual volume delivered converges to the desired volume quickly. In this way, one can avoid repeated tests to choose an optimal V_r while the performance is still acceptable especially when many wells need to be dispensed.

One may notice that there is an oscillation in the actual volumes of the dispensed biosamples. This is due to the variation of the residual volume V_r carried over from the previous well. Once the dispensed volume is smaller than normal for one particular well, more residual volume is carried over to the next well, which results in a large dispense as indicated in (15.3), and vice versa. The oscillation starts when the variation of the dispensed volume first occurs on a well, which is completely random.

15.6   Conclusions

A general robotic approach for automating the dispensing of biosamples is presented in this chapter. The purpose of the approach is to achieve high-throughput delivery of biosamples in screening operations which are often found in research and development activities of life sciences. The developed approach is based on off-the-shelf robots and devices that can handle many types of biosamples ranging from the highly fluidic (such as water) to highly viscous (such as cubic phase used in the crystallization of membrane proteins). In addition, the approach can dispense biosamples at micro/nanoliter scales.

The approach includes an effective method for scaling and optimizing the liquid handling parameters of the robotic dispensing devices for accurate and reproducible delivery of biosolutions. For dispensing highly viscous biosamples, the dispensing distance is considered as an important factory for successful delivery. Furthermore, a mathematical model to determine the actual volume dispensed and its variation with needle diameter, dispensing distance, and desired volume to be delivered is proposed. All the methods involved in the general approach are validated by performing experiments and quantitative analysis of the biosamples delivered, ranging from fluidic to highly viscous biosamples. The proposed robotic system is inexpensive and can be used for screening operations in life sciences in which automated handling of a large number of biosamples in low and accurate volumes is required.

References

[1]   Zhang, M., et al., "Editorial: Special Issue on Life Science," *IEEE Transactions on Automation Science and Engineering*, Vol. 3, No. 2, April 2005, pp. 137–140.

[2]   Meldrum, D., "Automation for Genomics: Part 1, Preparation for Sequencing," *Genome Research*, Vol. 10, No. 8, August 2000, pp. 1081–1092.

[3]   Meldrum, D., "Automation for Genomics: Part 2, Sequencers, Microarrays, and Future Trends," *Genome Research*, Vol. 10, No. 9, September 2000, pp. 1288–1303.

[4]   Hosfield, D., et al., "A Fully Integrated Protein Crystallization Platform for Small-Molecule Drug Discovery," *J. Struct. Biol.*, 142, 2003, pp. 207–217.

[5]   Cherezov, V., et al., "A Robotic System for Crystallizing Membrane and Soluble Proteins In Lipidic Meso Phases," *Acta. Cryst. D*, Vol. 60, 2004, pp. 1795–1807.

[6]   Sun, Y., and Nelson, "Biological Cell Injection Using an Autonomous Microrobotic System," *Int. J. of Robotics Research*, Vol. 21, 2002, pp. 861–868.

[7]   Michette, A., and S. Pfuntsch, (eds.), *X-Rays: The First Hundred Years*, New York: John Wiley & Sons, 1996.

[8]   Caffrey, M., "Membrane Protein Crystallization," *J. Struct. Biol.*, Vol. 142, 2003, pp. 108–132.

[9]   Luft, J., et al., "A Deliberate Approach to Screening for Initial Crystallization Conditions of Biological Macromolecules," *J. Struct. Biol.*, Vol. 142, 2003, pp. 170–179.

[10] Meldrum, D., et al., "ACAPELLA-5K, a High-Throughput Automated Genome and Chemical Analysis System," *Proc. IEEE/RSJ Int. Conf. on Intelligent Robots and Systems*, Las Vegas, NV, October 2003, pp. 2321–2328.

[11] Stock, D., O. Perisic, and J. Lowe, "Robotic Nano Liter Protein Crystallization at MRC Laboratory of Molecular Biology," *Progress in Biophysics and Molecular Biology*, Vol. 88, 2005, pp. 311–327.

[12] Hannaford, B., et al., "Telerobotic Remote Handling of Protein Crystals," *Proc. Int. Conf. on Robotics and Automation*, Albuquerque, NM, April 1997, pp. 1010–1015.

[13] Stevens, R. C., "High-Throughput Protein Crystallization," *Curr. Opin. Struct. Biol.*, Vol. 10, 2000, pp. 558–563.

[14] Walter, T., et al., "A Procedure for Setting Up High-Throughput Nano Liter Crystallization Experiments," *I. Protocol Design and Validation, Appl. Cryst.*, Vol. 36, 2003, pp. 308–314.

[15] Rupp, B., et al., "The TB Structural Genomics Consortium Crystallization Facility: Towards Automation from Protein to Electron Density," *Acta Cryst. D*, Vol. 58, 2002, pp. 1514–1518.

[16] Heinemann, U., et al., "Facilities and Methods for the High-Throughput Crystal Structural Analysis of Human Proteins," *Acc. Chem. Res.*, Vol. 36, 2003, pp. 157–163.

[17] Xantus Robotic Liquid Handling, http://www.sheepscotmachine.com/viscosity.html.

[18] Muthusubramaniam, L., et al., "Automating Crystallization of Membrane Proteins by Robot with Soft Coordinate Measuring," *Proc. IEEE Int. Conf. on Robotics and Automation*, New Orleans, LA, April 2004, pp. 1450–1455.

[19] http://www.sheepscotmachine.com/viscosity.html.

Biological Cell Sorting Automation

Yu Sun,[†] Deok-Ho Kim,[‡] and Ali Hashemi[†]

This chapter discusses biological cell sorting for potential applications in life science and biotechnology. Fundamental principles and advantages and disadvantages of each cell sorting technique are presented. A set of design considerations are formulated to evaluate each cell sorting technique, improve its applicability, and enable implementation onto lab-on-a-chip devices. Illustrative applications to life science automation are provided. Conclusions and future challenges are presented at the end of the chapter.

16.1 Introduction

In the last two decades, interdisciplinary science and technologies have converged to create exciting challenges and opportunities, involving a new generation of integrated microfabricated devices. These new devices are referred to as lab-on-a-chip or Micro Total Analysis Systems. Their development requires both established and evolving technologies, such as microelectromechanical systems (MEMS) technology, microfluidics, and nanotechnology. As lab-on-a-chip devices become frequently employed, one critical aspect of these devices—cell sorting—plays an important role in the overall utility of the apparatus. Modern methods in molecular biology, drug screening, diagnostics, and cell replacement therapy often require separation and analysis of living cells [1, 2]. For example, it has been estimated that 90% of the cost and 95% of the time needed to obtain molecular diagnostic data today is associated with sample separation, collection, and preparation [3]. The inadequacy of effective sample preparation represents one of the major shortcomings in contemporary molecular analysis. Instruments for separating specific target cells from a mixed cell population are urgently required for biomedical analyses and clinical applications.

Many research efforts have been spent on developing novel biological cell sorting devices using lab-on-a-chip technology. In addition to biochemical interactions, physical phenomena in the molecular and cellular dimensions have been used, which can be implemented in on-chip separation devices, such as mechanical, electrical, magnetic, acoustic, and light origins.

†   Advanced Micro and Nanosystems Lab, University of Toronto
‡   Department of Biomedical Engineering, Johns Hopkins University

Compared with traditional macroscale separation methods, the microfabricated devices have small working volume and subsequently reduced throughput. Crude sample solution containing cells should be diluted to avoid the clogging of the microfluidic chambers, which may result in a greater decrease in separation throughput as well. Despite these disadvantages, on-chip cell separation devices have considerable potential to overcome current limitations because the microfabricated devices can provide unique functions and the capability to separate cells in a sensitive manner. Microfluidic lab-on-a-chip devices can provide an automated, reliable, and efficient system for cell separation.

Here we review recent advances in microfluidic bioanalytical systems for cell separation (Table 16.1), and discuss how these advances can be used to provide appropriate solutions to fundamental problems in the life sciences and biotechnology. We also describe the operating principles, advantages and limitations, and illustrative examples of microanalytical devices developed across many research communities to separate cells. Furthermore, we discuss key opportunities for improved performance of such separation devices as well as future directions and applications.

Table 16.1 Summary of On-Chip Cell Separation Principle and Applications

Separation Principle	Cell Type	First Author (year) [reference number]
μ–FACS	RBC	Wolff (2003) [4]
	E.Coli	Fu (1999, 2001) [5, 6]
μ–MACS	Mouse and rat lymphoid cells	Antoine (1978) [7]
	Human lymphocytes	Zborowski (1995, 1997) [8, 9]
FFF	E. Coli and RBC	Giddings (1993) [10]
	HeLa cell	Caldwell (1984) [11]
	Saccharomyces Cerevisiae	Sanz (2002) [12]
	RBC from nucleated cells	Metreau (1997) [13]
	RBC	Chianea (1999) [14]
	Toxoplasma gondii	Bouamrane (1999) [15]
	Cortical cell	Battu (2001) [16]
	WBC from RBC	Shevkoplyas (2005) [17]
	RBC	Merinodugay (1992) [18]
	RBC	Andreux (1993) [19]
DEP	Human breast cancer MDA-231 cells from blood	Gascoyne (1997) [20]
	Listeria	Li (2002) [21]
	Human T lymphocytes	Pethig (2002) [22]
	Peripheral blood stem cell	Stephens (1996) [23]
	P19 EC cell	Park (2005) [24]
DEP-FFF	Human breast cancer MDA-435 cells from T-lymphocytes and CD34+ hematopoietic stem cells	Wang (2000) [25]
Ultrasonic sorting	E.Coli	Coakley (2000) [26]
	Yeast cell	Hawkes (1996) [27]
	Mouse hybridoma cell	Wang (2004) [28]
	Plasma from blood	Cousins (2000) [29]
	Bacteria and yeast	Hawkes (1997) [30]

16.2 Conventional Macroscale Cell Separation

Cell populations are often heterogeneous, and the cells of interest are suspended in a solution or mixed with different types of chemicals, biomolecules, and cells. Cell separation requires the extraction of one cell type from a larger population. Often, this separation is conducted by exploiting differences in the size, density, or charge of the cells. However, when different cell types that share many of the above-listed physical characteristics are present in a solution, separation is realized by capitalizing on the differences in the number and types of molecules present on cell surfaces. Such a process, termed cell-affinity chromatography, utilizes differences between the adhesion of cells to a ligand-coated surface (where the adhesion is mediated by specific, noncovalent binding between the ligands and the cell surface molecules) to sort cells [31]. These techniques use antibodies which are selected to respond to specific cell surface antigens and are often conjugated to biotin, fluorochromes, or magnetic beads that serve as markers [4, 31]. When the marked cells, located in a larger population, are passed through a sorting mechanism, the target cells are extracted. However, this type of cell separation requires multiple steps, beginning with the staining of target cells with antibodies, followed by cell selection, elution, collection, and finally the release of the conjugated antibodies. Each step in the process reduces the overall yield while increasing the cost, time, and possibility of contamination of the sample [32].

Both fluorescence and magnetic activated cell sorting (FACS and MACS) devices use the same initial steps to enable cell separation. Due to similarities, only the general procedure for MACS is outlined below. Different from MACS, FACS is also capable of sorting cells labeled with fluorescence dyes or tagged with green fluorescence protein (GFP). It should be noted that these two cytometric techniques were first developed on the macroscale and have only recently been adapted for microscale operation [4–6, 31, 33]. In the MACS technique, cells are typically subject to a multistage labeling process. One such process used by Miltenyi et al. involves a three-stage staining and labeling procedure, using biotinylated superparamagnetic beads [31]. In this technique, cells are first stained with a biotinylated antibody, then stained with a fluorescent avidin conjugate and finally, biotinylated superparamagnetic microparticles are bound to remaining free binding sites of the avidin on the cell surface. Since avidin has four biotin-binding sites, the antibody binding cells have both a fluorescent and a magnetic label [31]. Due to the lack of antibodies and known surface markers for some cell types, and the drawback of using animal-derived antibodies, these cell sorting systems are not applicable to many cell lines.

16.3 On-Chip Cell Separation: Principles and Applications

Conventional cell sorting systems such as FACS and MACS frequently require immuno-labeling for separating specific target cells and large sample volumes. Cells for which specific cell-surface antibodies have not been developed are difficult to be sorted by these instruments. Thus, devices that do not require labeling and can be operated easily and safely with small sample and reagent volumes, are desired.

MEMS-based cell separation devices can satisfy these requirements. They have great potential to separate cells, especially valuable for rare cells such as stem cells, since its downscaling results in reduction of cells and reagent consumption as well as time of operation [1, 2, 34]. This decreases costs, reduces waste, and increases the number of assays that can be performed with expensive chemical libraries [35]. Current methods that are commonly used for cell manipulation, concentration, and separation with MEMS techniques employ physical forces from mechanical, hydrodynamic, ultrasonic, optical, electrokinetic, electromagnetic origins.

In this section, the fundamentals and applications of immunological techniques such as micro-FACS (μ-FACS) and micro-MACS (μ-MACS), and nonimmunological techniques such as fractional field flow-based sorting, dielectrophoresis (DEP), and ultrasound field based sorting, are discussed. Each technique has advantages in various aspects: whether it be the speed at which it can perform sorting, the degree of differentiation that can be achieved, or whether they are less intrusive to the cells being sorted. Design considerations are also discussed.

16.3.1 Miniaturized Fluorescence Activated Cell Sorting (μ-FACS)

Fluorescence activated cell sorting involves labeling cells with a particular antibody (which binds differently to each particle in the medium depending on their surface molecular structure), and then employs a laser to sort the cells based on the antibody surface bonding characteristics. The labeling process is multistage and uses different antibodies based on the characteristics of target cells [4]. Once the cells are labeled with a fluorescent marker, they are passed through a microfluidic device that uses a miniaturized laser to sort the cells fluorescently.

FACS is a powerful method for the isolation of subpopulations out of complex cell mixtures. However, its application was often limited by the relatively small separation capacity of approximately 10^7 particles per hour [4]. The isolation of rare cells is more time-consuming and laborious. Processing large numbers of cells (e.g., for molecular analysis or clinical applications) was not possible until the late 1990s [4].

Upon miniaturization, μ-FACS devices have been altered to allow for single-cell or multicell flow [33], thereby greatly increasing their separation capability. The miniaturized fluorescence detection system employs active and passive microoptical components, including semiconductor lasers, ultra bright LED sources, highly sensitive avalanche photodiodes, holographic diffraction gratings, and fiber optics for the transmission and collection of light [33]. μ-FACS devices offer a number of advantages over conventional FACS. A conventional FACS normally applies sorting of droplets in an open system. In contrast, a microsorter structure can be fabricated as a closed system, reducing the risk of infecting the sorted cells and the risk of working with biohazardous materials.

The development of microfabricated devices for cell analysis by flow cytometry [36] or for cell sorting [6, 33, 37] was recently reported. Unfortunately, most of these devices were limited to "proof of concept." No quantitative descriptions were provided on sample throughput or sorting efficiency. Fu et al. were the first to develop a microfabricated elastomeric μ-FACS based on electro-osmotic flow [5]. They demonstrated the sorting of particle and bacteria cells with enrichments of up to 80- to 96-fold over previous attempts. However, the throughput of the system at approximately 10 to 20 cells/s is orders of magnitude lower than what a conventional FACS can offer ($\sim 10^4$ cells/s) [4]. A later pressure-driven μ-FACS from the same group had essentially the same enrichment and a two-fold increase in sample throughput [6]. Concurrently, Wolff et al. designed several integrated functionalities on-chip in the new generation of μ-FACS. The functionalities include a microfluidic structure for sheathing and hydrodynamic focusing of the cell-sample stream, a chip-integrated chamber for holding and culturing of the sorted cells, and integrated optics for the detection of cells [4].

16.3.2   Miniaturized Magnetic Activated Cell Sorting (μ-MACS)

μ-MACS also evolved out of its macro predecessors. Similar to μ-FACS, the operation involves cell labeling; however, in this case, cells are immunologically stained with magnetic particles (diameter $> 0.5\,\mu$m) on the macroscale, while the shift to the microscale brought about the use of particles with diameters much smaller [7, 38].

Advances in the field of microfluidics have allowed for greater refinement of the MACS technique. In order to develop appropriate miniaturization of this technique, the fundamental forces acting on an aqueously suspended, paramagnetically labeled cell should be determined. They are magnetic, buoyant force, gravity, and drag, denoted as F_m, F_{bou}, F_g, and F_d [39]. Assuming a Reynolds number less than 1 (which is valid for microchannels), one can assume that F_d is represented by Stokes drag. Mathematically, these forces are as follows:

$$F_m = A_c\,\alpha\beta F_b \tag{16.1}$$

$$F_g - F_{bou} = \frac{(\rho_c - \rho_f)\pi D_c^3 g}{6} \tag{16.2}$$

$$F_d = -v_c\,3\pi D_c\eta \tag{16.3}$$

where A_c is the surface area of the cell, α is the number of cell surface markers per membrane surface area, β is the number of antibody magnetic bead complexes bound per cell surface marker, F_b is the magnetic force acting on one antibody magnetic bead complex, D_c is the diameter of a cell, g is the acceleration of gravity, v_c is the magnetically induced velocity of a cell, ρ_f is the density of the fluid, ρ_c is the density of a cell, and η is the viscosity of the fluid. One contentious assumption in this derivation is that the binding of immunomagnetic labels does not affect the cell volume or density. It should be noted that while the relationship for F_m appears relatively straightforward, the F_b term, a vector in the direction of the magnetic energy gradient, is highly nonlinear (except under specialized conditions). It can be expressed as

$$F_b = \Delta\chi V_b \nabla B^2 / 2\mu_o \tag{16.4}$$

where μ_o is the magnetic permeability of free space, $\Delta\chi$ is the difference in magnetic susceptibility between the magnetic bead χ_b and the surrounding medium x_f (saline in this case), V_b is the volume of one magnetic bead, and B is the external magnetic field [9].

A change in one key assumption can lead to a fundamentally different solution. If the entire particle is assumed to be paramagnetic, as opposed to (16.1), where the beads attach only to the surface of the cell, then the paramagnetic force on a cell (or particle), which is similar to (16.4), is [40]

$$F_m' = \Delta\chi_c V_c \nabla B^2 / 2\mu_o \qquad (16.5)$$

The difference here is that the volume of a paramagnetic bead, V_b, has been replaced with the volume of the paramagnetically labeled cell (or particle), V_c. Moreover, the difference in the magnetic susceptibility between a paramagnetic bead and the suspending medium, $\Delta\chi_b$, is replaced with a difference between an averaged magnetic susceptibility over the entire cell, or particles, and the suspending medium, $\Delta\chi_c$. With this assumption, applying a force balance on a particle, (16.2) through (16.5) can be solved to obtain

$$\Delta\chi_c = \mu_o \frac{\dfrac{9\eta\nu_c}{2r_c^2} + g\Delta\rho}{\dfrac{1}{2}\left|\nabla B^2\right|} \qquad (16.6)$$

Since most cells are not intrinsically paramagnetic and are instead usually diamagnetic [39], unlike the intrinsic fluorescence of cells observed in FACS systems, greater than three orders of magnitude in difference between magnetically unlabeled and labeled cells are theoretically possible when colloidal or molecular labels are used [9, 39]. If one assumes that a one-to-one correspondence between the cell surface marker number and the number of paramagnetic labels bound to a cell, then it becomes possible to attain a range of at least three orders of magnitude of magnetization to labeled cells (with well-characterized surface markers) compared to unlabeled cells [39]. This means that the selectivity of this process can be exceptionally high.

Several techniques have been developed that either measure or approximate the magnetic susceptibility of small particles or labeled cells. One very early technique from the 1960s used an instrument in which the movement of an aqueous suspended particle could be determined [41]. Using this technique, measurements of the particle velocity of polystyrene latex beads and red blood cells were made through visual, microscopic observations and the use of a stopwatch. In 1995, Zborowski et al. reported on studies in which mean measurements of the magnetic susceptibility of immunomagnetically labeled cells were determined by flowing the cells through a region with a known, high magnetic energy gradient [8]. Under such circumstances, cells that were sufficiently immunomagnetically labeled were deflected and deposited on the bottom surface of a glass substrate in the flow channel. With the knowledge of the magnetic energy gradient and flow condition within the vessel, the mean magnetic susceptibility of the labeled cells can be estimated [39].

16.3.3   Micromachined Thermal Fractional Field Flow (μ-TFFF)

μ-TFFF is an elution separation technique where the separation field is normal to the sample and carrier flow. μ-TFFF utilizes thermal diffusion as the separation field instead of the gel, liquid, or column packing found in other chromatographic separations. This field is accomplished by establishing a temperature gradient across the channel. A schematic of the TFFF system is shown in Figure 16.1 [42].

Separation of the suspended particles is performed in a solvent carrier such as methanol, THF, acetonitrile, or DMSO. It may be used in conjunction with a variety of solvents. Much work has been done to demonstrate how these solvents affect separation characteristics [43]. Water is not typically used as a carrier fluid unless an electrolyte is added. The particles in the solvent react to the temperature gradient by diffusing towards the cold wall [43]. Higher molecular weight particles react more to the thermal gradient and are compacted more tightly against the cold wall than lower molecular weight particles. Because of the laminar velocity profile of the carrier, samples that compact less have a higher average velocity than the samples that compact more. The difference in average velocity results in the spatial and temporal particle separation at the output of the TFFF channel [10]. In TFFF, the resolution requirement in the direction of separation is no longer needed, which means lower field strength, lower power consumption demands, and shorter separation times. TFFF also has the advantage of elution techniques, in that the samples are collected in fractions at given times. As a result, very pure samples can be obtained [10].

To make separation occur in a TFFF channel, the following criteria should be met [44]: (1) difference in molecular weight or diameter of the sample; (2) sample selective perturbation of the samples towards one wall; and (3) laminar velocity profile that results in a different average velocity of each constituent of the sample.

The channel height is a critical parameter in the spatial resolution of two eluting samples. The channel height is inversely proportional to the resolution [44]. Micromachining technologies allow accurate and precise definition of channels of much smaller sizes in μ-TFFF devices (e.g., 25 μm). As a result, higher-resolution separations can be performed [43].

16.3.4   Sedimentation Fractional Field Flow (SdFFF)

SdFFF, commonly referred to as *hyperlayer* [45], utilizes cell size, density, shape, or rigidity as well as channel geometry and flow rate characteristics to separate target cells from the medium. When a constant flow rate and external field strength are

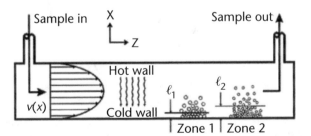

Figure 16.1   Schematic of a thermal field flow channel. (*From:* [42]. © 2002 ACS Publications. Reprinted with permission.)

used, the larger or less dense particles are eluted first [10, 11, 46]. By taking advantage of biophysical properties of cells, SdFFF can rapidly and effectively sort cells in comparison to labeling-dependent methods for specific cell sorting applications, such as stem cell preparation, culture, and transplantation [12, 47, 48]. Figure 16.2 illustrates the principle of this type of fractionation [49].

Cell sorting operation must maintain cell functional integrity (e.g., cell adhesion properties, genomic and proteomic capacity, and metabolism specificity). SdFFF might induce cell death by apoptosis or by necrosis in cell culture or transplantation. The limitation of cell death, apoptosis in particular, depends both on a drastic curtailment of cell-accumulation wall interactions, and on strict application of the cleaning procedure [48]. Additionally, the maturation and differentiation stages of eluted cells should not be altered by the sorting process; that is, the SdFFF elution process should not induce an uncontrolled differentiation of immature or stem cells if the capacity for cell differentiation is to be preserved [48]. Previous results, performed under defined conditions with immature neural cells, have demonstrated that SdFFF elution is a relatively gentle process and does not interfere with the maturation stage for both immature and differentiated cells [48]. Furthermore, as specific labeling is not necessary, SdFFF is particularly useful for applications where labels do not exist, or in which labels might interfere with further cell use (culture, transplantation). Since SdFFF cell sorting is based on the intrinsic biophysical properties, it may be advantageous over FACS or MACS for stem cell sorting.

The most commonly used channel wall material in SdFFF is polycarbonate, while the mobile phase is usually supplemented with a 0.5% to 1% bovine serum albumin (BSA) [13, 14]. However, some specific cell separations cannot be achieved under these conditions, such as rat embryo cortical cells [15]. In order to overcome

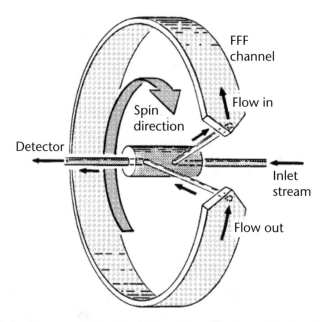

Figure 16.2 Schematic representation of a sedimentation FFF channel that has been wrapped inside a centrifuge basket. The inlet is sealed from the outlet streams by rotary seals. (*From:* [49]. © 1997 Wiley. Reprinted with permission.)

these limitations, replacing the polycarbonate plates with a 2-mm-thick glass wall polystyrene plate, which is highly hydrophobic, leads to effective cell separation with high recovery for all cell types studied thus far, such as nucleated cells, human red blood cells, bacteria, or algae [15, 48]. Moreover, when these polystyrene plates are used, the addition of BSA to the carrier phase is not required, and no modification of retention behavior or separation effectiveness is observed [16].

16.3.5   Gravitational Fractional Field Flow (GrFFF)

As an offshoot of the SdFFF technique, GrFFF utilizes the Earth's gravity field to incite separation among different cells through a microfluidic chamber. In this case, the Earth's gravity is applied perpendicularly to a thin empty channel with a rectangular cross-section. As the carrier fluid flows through the channel, a parabolic velocity profile develops for each particle depending on their size. This technique is particularly effective in sorting particles in the range of 1 to 100 μm.

In contrast to other FFF techniques, GrFFF elutes larger particles first. This reversal is due to the steric exclusion of particles from the channel accumulation wall, which is where the particles would accumulate due to the Earth's gravity field. However, GrFFF differs from real steric elution due to the complex interaction of hydrodynamic forces. Accordingly, these hydrodynamic forces may act on particle motion in the channel by opposing the external gravitational field.

As blood is largely comprised of erythrocytes, platelets, and leukocytes in plasma, with the former dominating the mixture, separating the leukocytes, which comprise less than 1% of the total, is a key objective. GrFFF microfluidic devices were developed to increase the leukocyte concentration by about 34 times [50]. Since erythrocytes are smaller and more deformable than leukocytes, they flow faster in microfluidic conditions. Consequently, the erythrocytes dominate the middle of the flow field, with leukocytes dominating the flow at the edges closest to the wall at the low velocity gradients. By continually bifurcating the flow, through the use of disproportionately sized channels, the proportion of the slower leukocytes may be increased drastically as the faster erythrocytes bypass the small channel [17]. Obviously, this technique will be beneficial in situations where physical differences in particles are marked.

GrFFF has been demonstrated to enable the fractionation of human peripheral red blood cells according to several of their specific characteristic, such as their degree of sphericity [18]. Importantly, the biological activity of the red blood cells is not affected by their migration through the channel. Separation selectivity is attributed to modifications of the physical characteristics of the red blood cells. In another reported study [19], GrFFF was used to examine red blood cell changes during phenylhydrazine-induced hemolytic anemia, separating normal cells from Heinz body-rich cells in a relatively short time.

GrFFF displays high size and density selectivity with respect to the separation of cells, making it possible to correlate direct microscopic observation with cell retention. Moreover, if two cell groups share the same density but have different sizes, or vice versa, GrFFF proves versatile enough to provide effective characterization of the cells in either case [18].

16.3.6 Flow Fractional Field Flow (FlFFF)

As with the other FFF techniques, FlFFF utilizes an applied force to affect the differential migration of particles. FlFFF separates components strictly according to their hydrodynamic size. In addition to a field in the traditional axial dimension, the field is also composed of a carrier fluid flowing across the thin dimension. As the cross-flow moves the target cells to a porous membrane at the accumulation wall, the membrane retains the target cells while allowing the carrier liquid to exit the channel. After an adequate concentration of target cells accumulate at the membrane, they begin to diffuse in the opposite direction against the cross-flow. Figure 16.3 illustrates the concept.

The two opposing motions result in an area where the target cells extend from the accumulation membrane towards the center of the channel. The mean thickness of this area, t, may be calculated if the rate of cross-flow V_c, the channel thickness ω, the geometric volume of the channel V_0, and the target cell's diffusion coefficient D can be determined, according to

$$t = \frac{DV_o}{\omega \dot{V}_c} \tag{16.7}$$

Since each particle in the solution has a different characteristic mean thickness, t, they are consequently transported through the channel at different rates, resulting in the flow characterized in Figure 16.4.

Compared to other chromatography techniques, the open FlFFF channel has a greatly reduced surface area, which reduces the possibility of interactions with the sample. The lack of such interactions has provided a niche for FFF in characterizing materials that are difficult to separate by other methods [52]. FlFFF also has the following advantages: (1) open channel structure; (2) small surface area in contact with the sample material; (3) tunable retention; (4) possibility to concentrate the sample;

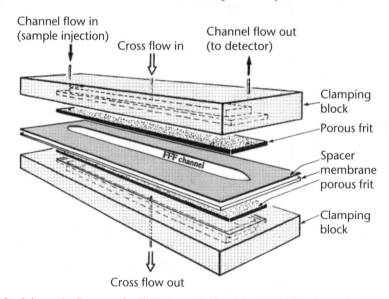

Figure 16.3 Schematic diagram of a FlFFF channel. The clamping blocks are typically PMMA for aqueous operation. For organic solvent use, stainless steel has proven satisfactory. (*From:* [49]. © 1997 Wiley. Reprinted with permission.)

(5) low shear forces; and (6) no need for calibration if diffusion coefficient is determined [53]. One further advantage of FlFFF is that it allows retention times in the channel to be related directly to physicochemical parameters that govern the sample's interaction with the applied field [52]. Finally, the open channel also provides a low-shear environment for the separation of fragile materials minimizing potential damage to cells. Consequently, FlFFF as well as other FFF techniques is often used in characterization of ultrahigh molecular weight polymers, colloids, aggregates, and other materials that tend to break apart in high shear force environments [54].

One modification to FlFFF is the alteration of the top wall such that it becomes impermeable to the solvent. This technique is termed asymmetric FlFFF (AsFlFFF) and results in both cross-flow and channel flow emanating from the inlet flow to the channel. By employing a tapered channel to maintain constant channel flow velocity, AsFlFFF is able to increase separations and has been used for a variety of materials [55]. However, some limitations in versatility occur since less control is possible [49]. FlFFF is a highly selectivity technique, which is usually considered as an advantage. It may, however, lead to high dilution of the sample zone, which may not be desired, especially if the initial sample concentration is low [53].

16.3.7 Dielectrophoresis Sorting

Dielectrophoresis (DEP) utilizes the interactions between the intrinsic dielectric properties of cells and an applied ac electric field, which can separate cells without the use of immunolabeling. The term "dielectrophoresis" was first used in [56]. It involves the introduction of a nonhomogenous alternating electric field to an aqueous dielectric media, in which cells are flowing. DEP forces can affect polarizable particles differently from electrophoresis, in which motion is determined by the net particle charge [57].

When a polarizable particle is placed in an electric field, the positive charge is induced on one side of the particle, and the negative charge is induced on the other

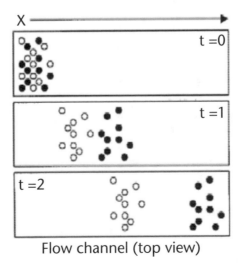

Flow channel (top view)

Figure 16.4 Illustration of the FlFFF principle.

side of the particle. In a uniform electric field, because each side of the induced dipole feels opposite and equal forces, which cancel, the particle does not feel a net force. In a nonuniform electric field, the two sides of the induced dipole experience a different force magnitude and thus, a net force is produced. This force is the origin of DEP sorting, as shown in Figure 16.5.

The magnitude of the dipole moment for a spherical particle suspended in a medium is [58]

$$m = 4\pi r^3 \varepsilon_m \left(\frac{\varepsilon_p^* - \varepsilon_m^*}{\varepsilon_p^* + 2\varepsilon_m^*} \right) E \quad \text{with} \quad \varepsilon^* = \varepsilon - i\frac{\sigma}{\omega} \tag{16.8}$$

where r is the particle radius, ε_m is the permittivity of the medium, ε_p^* and ε_m^* are the complex permittivity of the particle and its suspending medium, and ε is the absolute permittivity, σ is the conductivity, and ω is the radian frequency of electric field. The Clausius-Mossotti factor (f_{CM}) is defined as

$$f_{CM} = \frac{\varepsilon_p^* - \varepsilon_m^*}{\varepsilon_p^* + 2\varepsilon_m^*}$$

Using (16.8) and $f = m\nabla E(x)$, the time-averaged dielectrophoretic force (F_{DEP}) acting on the particle is [34, 58]

$$F_{DEP} = 2\pi\varepsilon_m r^3 \text{Re}(f_{CM})\nabla|E_{rms}|^2 \tag{16.9}$$

where $\text{Re}(f_{CM})$ is the real component of the Clausius-Mossotti factor, and E_{rms} is the root mean square magnitude of the electric field.

Equation (16.9) is the general expression of dielectrophoretic forces. The Clausius-Mossotti factor varies from -0.5 to 1, describing relative polarizability between the particle and a medium. With a positive Clausius-Mossotti factor (i.e., particle is more polarizable than the medium), the force acts in the direction of the

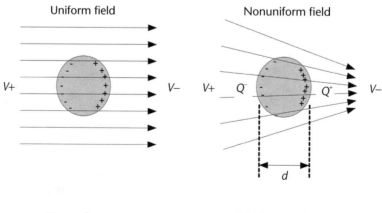

Figure 16.5 The left figure shows the behavior of particle in uniform electric field, while the right figure shows the net force experienced in a nonuniform electric field.

increasing field gradient, and the particle moves to the region of the highest electric field, which is called positive DEP. With a negative Clausius-Mossotti factor, the particle is repelled from regions of a high electric field, which is referred to as negative DEP. Because the force depends on the induced dipole in the particle, it is unaffected by the direction of the electric field and responds only to the field gradient. Capturing cells using positive DEP can only work effectively if there are very large differences among the DEP responses of the cells. Compared to positive DEP, cell handling using negative DEP can avoid the contact between the surface of microelectrode and the cells. Moreover, due to the high permittivity of solutions, most viable biological cells and many artificial particles are not affected by negative DEP, at least at high frequencies [59].

Due to the relative ease of the generation and structuring of an electric field at the microscale, DEP provides flexibility, controllability, and ease of application [60]. As shown in Figure 16.6, DEP is an efficient noninvasive method to separate cells [61], to characterize single cells [62], to detect pathogens [63], and to manipulate viruses [64, 65] and macromolecules [66, 67]. Using DEP, various biological cells have been successfully separated, including bacteria [68, 69], cancer cells [20, 62], stem cells [23, 70], and leukocyte subpopulations [71]. To make DEP separation more adaptable to real cell samples, more effort is required to improve the sensitivity to dielectric properties (e.g., size, morphology, conductivity, and permittivity of cells), sorting efficiency, and throughput. In order to achieve a high sensitivity, proper input conditions such as flow rate and input voltage and the design of microelectrodes and microchannels must be investigated.

From the standpoint of lab-on-a-chip devices for DEP-based cell separation, the reported on-chip DEP separation devices include the dielectric affinity column [20], cell concentrator [21], hyperlayer dielectrophoretic field-flow fraction (DEP-FFF) [72, 73], and 3D microelectrode systems [74, 75]. The dielectric affinity column and the cell concentrator use the same principle—that is, positive DEP traps one cell type while simultaneous negative DEP repels other types. For cell separation, differential DEP affinity experienced by different cell types must be sufficiently large. However, problems in many cell separation applications have been identified, and the small magnitude of the differential forces driving separation is generally associated with low levels of cell separation efficiency [3].

As cells move through a channel of laminar flow, particles strike a balance between their dielectric and gravitational pulls. Different DEP channels can be formed according to DEP migration, affinity/retention, FFF, or traveling wave (TW)-DEP [22, 76]. DEP migration collects different cells in different field regions induced by an ac current. However, once the cell types are separated from the cell mixture according to their dielectric properties, it is difficult to collect the separated cells in the downstream for further processing. DEP affinity or retention captures cells by applying a positive DEP force to the laminar flow, thus attracting and holding the target cells at the electrode edges. A negative DEP force repels unwanted cells away from the electrodes, where they can be swept away by the flow [77]. This approach has the disadvantage that cell sorting cannot be achieved in a continuous fashion.

Hyperlayer DEP-FFF is a very powerful and sensitive method of separating cells [72]. In this method, cells reach equilibrium levitation heights with DEP and

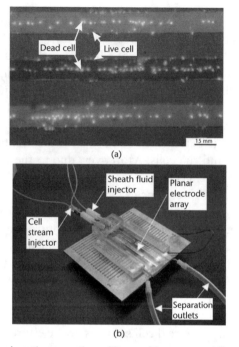

gravitation forces above an electrode array, and are carried through the separation
channel by a hydrodynamic flow profile that causes cell types levitated to different
heights to emerge from the channel outlet at different times. However, it proves difficult to collect the fractionated cell subpopulations [3]. In 3D-microelectrode systems, cell interactions with walls or nonspecific absorption of particles can be
avoided, resulting in easy recovery of the separated cells. In addition, trapping of
particles of interest in a femtoliter-sized volume is also possible for analysis [75].
However, in the conventional 3D-microelectrode system with a fixed gap between
the face-to-face microelectrodes and a constant width of the electrodes, the electric
field is uniform in the microfluidic channel along the transverse direction of flow,
and variation of the dielectrophoretic force generated is small. Thus, the sensitivity
of typical 3D-microelectrode systems is not satisfactory. A 3D-asymmetric microelectrode system was developed to generate a continuously varied electric field and
induce dielectrophoretic forces with more variations [24]. The sensitivity to the
physical properties of cells can therefore be increased by applying different forces to
each distinct cell type.

The DEP-FFF technique achieves separation by using an ac voltage set at a specific frequency (corresponding to the cell types targeted for separation) to drive a set
of planar microelectrodes. All suspended cells experience negative DEP forces. In
such a flow, the velocity of the cells is proportional to their height, as their height
determines the strength of the dielectric force. Consequently, cells of different
heights will require different times to travel through the chamber [12, 78]. A

parabolic flow profile thus emerges, making this type of DEP effective, although it is limited to filtering cells by their dielectric properties and is also not capable of continuous separation [25, 79]. Figure 16.7 illustrates the basic concept of this method [80].

Traveling wave (TW)-DEP utilizes a traveling electric field to affect cell flow velocity. Cells may either travel along with the wave in the same or opposite direction, or they may follow the wave at varying velocities according to velocity profile [81–83].

Recently, a DEP fractionation technique that boasts an improvement in cell separation—essentially further discriminating among the cells caught by the generated DEP fields—has been presented [71]. The device utilizes a nonuniform electric field shaped according to the cell properties. This electric field is created and sustained by microfabricated planar microelectrode arrays which are located on an insulating glass substrate. By individually biasing each microelectrode through a variable frequency ac voltage source, both negative and positive DEP can be used to induce cell separation. The problem of continuous fractionation is partly overcome by establishing a suitable cell stream supported by sheath flow [71].

16.3.8 Ultrasonic Sorting

Ultrasonic manipulation of suspended particles has been reported. In general, it is achieved by moving the particles into one region of their medium and subsequently removing a fraction of the obtained particle from the medium [84]. The same principle can be utilized to separate two different particle types from each other, provided that their acoustic properties are such that the induced acoustic forces on the two particle types have opposite directions. Several particle separation techniques

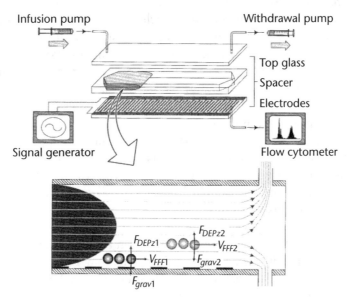

Figure 16.7 DEP-FFF principle and system setup. The DEP-FFF chamber was constructed with microfabricated, interdigitated electrode arrays on its bottom wall. Cells having different dielectric properties are levitated to different heights within a fluid-flow velocity profile inside a thin chamber and are carried with the fluid flow at different velocities. (*From:* [80]. © 2000 Biophysical Journal. Reprinted with permission.)

utilizing ultrasonic waves have been presented. These techniques are based on the fact that suspended particles exposed to an ultrasonic standing wave field are affected by an acoustic radiation force. This force moves the particles towards either the pressure nodes or the pressure antinodes, depending on the density and compressibility of the particles and the medium.

The radiation force is a result of a nonlinear effect in the time-averaged radiation pressure around an object in a sound wave, also known as the time-averaged acoustic Bernoulli pressure. For a propagating sound wave, the radiation force is proportional to the sixth power of the ratio of the object diameter to the sound wavelength. This relationship is valid for an object with a size smaller than the acoustic wavelength. Thus, since this ratio is much less than 1, the radiation force in a propagating wave is too low to have significant effect in most practical cases. However, the effect is much larger in standing waves, where the force is proportional to the third power of this ratio. Thus, most applications of acoustic trapping use standing waves. When the cells are transported from their initial position to a desired position beyond half an acoustic wavelength, phase modulation or position control of the top plate can be conducted. Acoustic energy density must be greater than the critical energy density to keep the cells trapped at nodal positions in the standing wave.

The primary acoustic radiation force is

$$F_r = -\left(\frac{\pi p_o^2 V_c b_w}{2\lambda}\right)\phi(\beta,\rho)\sin(2kx) \quad \text{with} \quad \phi = \frac{5\rho_c - 2\rho_w}{2\rho_c + \rho_w} - \frac{\beta_c}{\beta_w} \qquad (16.10)$$

and this force dominates the motion of suspended particles in the relevant frequency range. The acoustic force traps cells in the closest position of minimum acoustic potential energy, which is separated by half an acoustic wavelength. In (16.10), the densities of the medium and the particles are denoted ρ_w and ρ_c, the corresponding compressibilities are β_w and β_c, p_o is the pressure amplitude, V_c is the volume of the particle, and λ is the ultrasonic wavelength. In the ϕ-factor (i.e., the acoustic contrast factor), k is defined by $2\pi/\lambda$, and x is the distance from a pressure node. The direction of the force is determined by the sign of the ϕ-factor. A positive ϕ-factor results in movement towards a pressure node, and a negative ϕ-factor results in movement towards a pressure antinode.

Ultrasonic sorting employs pairs of ultrasonic waves to pick out cells like tweezers at a faster rate. A microscale ultrasound device creates standing wave traps by providing a minimum of potential energy. A vibrating piezotransducer is used to create a planar ultrasonic field which separates and distributes cells along the chamber. The chamber consists of fluid in between the vibrating glass plate and a rigid glass substrate. As the upper plate vibrates via the transducer, an acoustic wave is emitted into the fluid. This wave is then reflected on the glass substrate creating a standing wave that varies in the longitudial direction along the fluid. With the applied field, cells rearrange themselves according to their potential energy along the lines of the planar field [85].

Thus far, applications of ultrasonic separation have included cell concentration and retention in biosensors or bioreactors [85], 3D manipulation/positioning of microorganisms or particles [86, 87], separation of cells and suspending phase [27,

28] and fractionation or selective retention of cells [88, 89]. Acoustic trapping has been applied to the manipulation of erythrocytes, for example, for the extraction of plasma from whole blood [29, 90] and for autologous blood recovery/wash by separating fat embolies from blood cells [91]. Other microorganism applications include acoustic manipulation of bacteria [30, 92]. Several studies also reported possible damage of cells when they are exposed to ultrasound for standing-wave manipulation applications [85, 93], which must be considered in ultrasonic sorting.

16.4 Design Considerations

In order to determine the optimal mechanism that best fits a biological cell sorting application, it must be noted that little research has been conducted on the categorization and understanding of cell characteristics. Consequently, in order to derive the technique that can capitalize on the inherent properties of biological cells, further research efforts must be spent on determining unique surface markers and physical properties such as the dielectric properties, size, density, and rigidity. Characterized properties will serve as the fundamental rationale for creating lab-on-a-chip devices that quickly and efficiently isolate target cells without causing damage.

Cell sorting subjects target cells to some degree of mechanical force or alteration in cellular function. Since many cells are sensitive to damage (stem cells in particular), it is vital that a lab-on-a-chip device ensures that target cells are not damaged through the sorting process, for example, by estimating the level of shearing or thermal gradient that cells are able to withstand before damage occurs using computational tools or by conducting preliminary experiments on target cells to determine the levels that they can endure.

The DEP method has minimal effects on cells as it only slightly affects their dielectric properties. This method seems to be the most promising for large-scale cell sorting. In contrast, both conventional macroscale sorting techniques that have been adapted to the microscale (μ-FACS and μ-MACS) are relatively invasive and can damage sensitive cells. The adverse effect results from the need of immunologically staining the cells for separation [94].

None of the other separation techniques compares more favorably to the selectivity of FACS or MACS as the degree of selectivity is directly proportional to the characteristics that are highly unique for cells. Relying on differences in densities, height, dielectric properties, and so on, might provide a level of sorting that is not accurate enough in certain applications. While more recent techniques such as TW-DEP can increase the efficiency and selectivity of the sorting process, they are not yet as established as MACS or FACS.

The development of continuous DEP-FFF and centrifugal sorting [95] greatly increases cell sorting efficacy. All of the other FFF techniques have the potential to achieve a similar separation speed since separation is determined by the resultant velocity profile of the cells through the medium. Although conventional FACS machines currently achieve higher throughput and recovery, the other cell sorting techniques provide relatively inexpensive, robust, and flexible ways to sort and manipulate single cells. Moreover, both FACS and MACS techniques can require up to 2 days of manual preparation, testing, and result assessment [96]. Multiplexing cell sorting channels can increase the overall throughput and allow for simultaneous measurements of cells in different compartments.

While each technique has been demonstrated to be feasible for fabrication using established microfabrication techniques, not all of them are amenable to integration in lab-on-a-chip devices. For example, FACS and MACS can be difficult to be incorporated into such devices, as they require multistage staining and labeling of cells before sorting can occur. Another hindrance in the development of highly effective μ-MACS has been the inability to adequately mix the magnetic beads with the target medium in microfluidic chambers due to the exceptionally low Reynolds numbers. Suzuki et al. [97] developed one such mixing method that appears to be able to mix magnetic beads with the target cells in a lab-on-a-chip environment. Two streams, one containing the solution (biological fluid), and the other containing magnetic beads, are introduced to a mixing region. Mixing is achieved by applying a local and time-dependent magnetic field which is generated by microconductors. This field increases the rate of attachment of the beads to the biomolecules.

Besides the techniques discussed in this chapter, more recent development efforts have resulted in new types of devices based on several novel principles including deterministic lateral displacement [98], image dielectrophoresis [99], optical sorting [100], and magnetophoresis [101].

16.5 Future Challenges

The ability to separate and analyze cell populations is required in many biotechnology and biomedical applications, such as cell biology, molecular genetics, clinical diagnostics, and therapeutics. This chapter discussed several automated cell sorting techniques, such as dielectrophoresis, FACS and MACS, ultrasonic, and field flow based approaches. These techniques take advantage of differences in cell density, cell adhesion, electrical charge, or immunological surface markers.

Since the current level of knowledge regarding the properties of cells is still incomplete, future research efforts must be directed towards better understanding cellular properties for the development of novel cell sorting devices. Identification, modeling, and analysis of cellular properties such as biochemical affinity and the mechanical and electrical properties of cells should be investigated to allow and facilitate the development of truly automated lab-on-a-chip devices that will provide tremendous resource savings in the life sciences and biotechnology.

The types of cells that have been separated in microfluidic devices are limited to a few species, such as red blood cells, white blood cells, and *Escherichia coli*. Future challenges include the development of cell sorting systems of higher performance (e.g., separating a target cell line out of many species) and lower cost for emerging applications. As an example, separating stem cells from undifferentiated cells or tumor cells is urgently required in cell replacement therapy [102].

The ability of stem cells to self-renew and give rise to blood cell progeny that retain introduced genes make them an ideal candidate for gene and cell therapy [103]. Characterization and separation of stem cells have created the possibilities of clinical transplantation applications, such as tumor cell purging, T cell depletion, stem cell expansion, and gene therapy [104].

Among the various techniques available for cell separation, the most selective stem cell separation systems by far are FACS sorters and MACS sorters that make use of antibodies specific to the stem cell surface marker. However, these techniques are time-consuming and relatively expensive, and they sometimes induce disturbance to the cell surface. For stem cell separation, it requires much time and effort to seek specific immunological surface markers for differentiating specific stem cells, and techniques using antibodies specific to the stem cell surface marker have not been established yet.

To date, most stem cells and their progenitor populations have been isolated usually using antibodies which capitalize on cell-affinity chromatography methods. Furthermore, it has been a concern that biomarker based and electric-field based separation methods might be invasive to stem cells [105], which is not desired for clinical use. Therefore, a simple biomarker-free analytical device that requires a minimum amount of labor and cost, does not depend on immunological surface markers, and is capable of separating cells with a high throughput, will greatly facilitate stem cell therapy.

Acknowledgments

The authors gratefully thank Dr. P. K. Wong at the University of Arizona and Dr. S. H. Lee at Korea University for a critical reading of the manuscript.

References

[1]   Dittrich, P. S., and A. Manz, "Lab-on-a-Chip: Microfluidics in Drug Discovery," *Nat. Rev. Drug Discov.*, Vol. 5, 2006, pp. 210–218.

[2]   Toner, M., and D. Irimia, "Blood-on-a-Chip," *Annual Review of Biomedical Engineering*, Vol. 7, 2005, pp. 77–103.

[3]   Gascoyne, P., and J. Vykoukal, "Dielectrophoresis-Based Sample Handling in General-Purpose Programmable Diagnostic Instruments," *Proceedings of the IEEE*, Vol. 92, 2004, pp. 22–42.

[4]   Wolff, A., et al., "Integrating Advanced Functionality in a Microfabricated High-Throughput Fluorescent-Activated Cell Sorter," *Lab on a Chip*, Vol. 3, 2003, pp. 22–27.

[5]   Fu, A., et al., "A Microfabricated Fluorescence-Activated Cell Sorter," *Nature Biotechnology*, Vol. 17, 1999, pp. 1109–1111.

[6]   Fu, A., et al., "An Integrated Microfabricated Cell Sorter," *Analytical Chemistry*, Vol. 74, 2002, pp. 2451–2457.

[7]   Antoine, J., et al., "Lymphoid-Cell Fractionation on Magnetic Polyacrylamide-Agarose Beads," *Immunochemistry*, Vol. 15, 1978, pp. 443–452.

[8]   Zborowski, M., et al., "Analytical Magnetapheresis of Ferritin-Labeled Lymphocytes," *Analytical Chemistry*, Vol. 67, 1995, pp. 3702–3712.

[9]   Zborowski, M., et al., "Continuous Fractionation of Human Lymphocytes Using Novel Magnetic Flow-Through Sorters and Immunomagnetic Colloid," *Experimental Hematology*, Vol. 25, 1997, pp. 554–554.

[10]   Giddings, J., "Field-Flow Fractionation—Analysis of Macromolecular, Colloidal, and Particulate Materials," *Science*, Vol. 260, 1993, pp. 1456–1465.

[11]   Caldwell, K., et al., "Separation of Human and Animal-Cells by Steric Field-Flow Fractionation," *Cell Biophysics*, Vol. 6, 1984, pp. 233–251.

[12] Sanz, R., et al., "Steric-Hyperlayer Sedimentation Field Flow Fractionation and Flow Cytometry Analysis Applied to the Study of Saccharomyces Cerevisiae," *Analytical Chemistry*, Vol. 74, 2002, pp. 4496–4504.

[13] J. Metreau, et al., "Sedimentation Field-Flow Fractionation of Cellular Species," *Analytical Biochemistry*, Vol. 251, pp. 178–186, 1997.

[14] Chianea, T., et al., "Field- and Flow-Dependent Trapping of Red Blood Cells on Polycarbonate Accumulation Wall in Sedimentation Field-Flow Fractionation," *Journal of Chromatography B*, Vol. 734, 1999, pp. 91–99.

[15] Bouamrane, F., et al., "Sedimentation Field-Flow Fractionation Application to Toxoplasma Gondii Separation and Purification," *Journal of Pharmaceutical and Biomedical Analysis*, Vol. 20, 1999, pp. 503–512.

[16] Battu, S., et al., "Cortical Cell Elution by Sedimentation Field-Flow Fractionation," *Biochimica et Biophysics Acta*, Vol. 1528, 2001, pp. 89–96.

[17] Shevkoplyas, S., et al., "Biomimetic Autoseparation of Leukocytes from Whole Blood in a Microfluidic Device," *Analytical Chemistry*, Vol. 77, 2005, pp. 933–937.

[18] Merinodugay, A., et al., "Monitoring of an Experimental Red-Blood-Cell Pathology with Gravitational Field Flow Fractionation," *Journal of Chromatography-Biomedical Applications*, Vol. 579, 1992, pp. 73–83.

[19] Andreux, J., et al., "Separation of Red-Blood-Cells by Field-Flow Fractionation," *Experimental Hematology*, Vol. 21, 1993, pp. 326–330.

[20] Gascoyne, P., et al., "Dielectrophoretic Separation of Cancer Cells from Blood," *IEEE Transactions on Industry Applications*, Vol. 33, 1997, pp. 670–678.

[21] Li, H., and R. Bashir, "Dielectrophoretic Separation and Manipulation of Live and Heat-Treated Cells of Listeria on Microfabricated Devices with Interdigitated Electrodes," *Sensors & Actuators B*, Vol. 86, 2002, pp. 215–221.

[22] Pethig, R., et al., "Dielectrophoretic Studies of the Activation of Human T Lymphocytes Using a Newly Developed Cell Profiling System," *Electrophoresis*, Vol. 23, 2002, pp. 2057–2063.

[23] Stephens, M., et al., "The Dielectrophoresis Enrichment of CD34(+) Cells from Peripheral Blood Stem Cell Harvests," *Bone Marrow Transplantation*, Vol. 18, 1996, pp. 777–782.

[24] Park, J., et al., "An Efficient Cell Separation System Using 3D-Asymmetric Microelectrodes," *Lab on a Chip*, Vol. 5, 2005, pp. 1264–1270.

[25] Wang, X., et al., "Cell Separation by Dielectrophoretic Field-Flow-Fractionation," *Analytical Chemistry*, Vol. 72, 2000, pp. 832–839.

[26] Coakley, W., et al., "Analytical Scale Ultrasonic Standing Wave Manipulation of Cells and Microparticles," *Ultrasonics*, Vol. 38, 2000, pp. 638–641.

[27] Hawkes, J., and W. Coakley, "A Continuous Flow Ultrasonic Cell-Filtering Method," *Enzyme and Microbial Technology*, Vol. 19, 1996, pp. 57–62.

[28] Wang, Z., et al., "Retention and Viability Characteristics of Mammalian Cells in an Acoustically Driven Polymer Mesh," *Biotechnology Progress*, Vol. 20, 2004, pp. 384–387.

[29] Cousins, C., et al., "Plasma Preparation from Whole Blood Using Ultrasound," *Ultrasound in Medicine & Biology*, Vol. 26, 2000, pp. 881–888.

[30] Hawkes, J., M. Limaye, and W. Coakley, "Filtration of Bacteria and Yeast by Ultrasound-Enhanced Sedimentation," *Journal of Applied Microbiology*, Vol. 82, 1997, pp. 39–47.

[31] Miltenyi, S., et al., "High-Gradient Magnetic Cell-Separation with MACS," *Cytometry*, Vol. 11, 1990, pp. 231–238, 1990.

[32] Das, C., et al., "Dielectrophoretic Segregation of Different Human Cell Types on Microscope Slides," *Analytical Chemistry*, Vol. 77, 2005, pp. 2708–2719.

[33] Kruger, J., et al., "Development of a Microfluidic Device for Fluorescence Activated Cell Sorting," *Journal of Micromechanics and Microengineering*, Vol. 12, 2002, pp. 486–494.

[34] Muller, T., et al., "The Potential of Dielectrophoresis for Single-Cell Experiments," *IEEE Engineering in Medicine and Biology Magazine*, Vol. 22, 2003, pp. 51–61.

[35] Vetter, D., "Miniaturization for Drug Discovery Applications," *Drug Discovery Today*, Vol. 3, 1998, pp. 404–408.

[36] Schrum, D., et al., "Microchip Flow Cytometry Using Electrokinetic Focusing," *Analytical Chemistry*, Vol. 71, 1999, pp. 4173–4177.

[37] Blankenstein, G., and U. Larsen, "Modular Concept of a Laboratory on a Chip for Chemical And Biochemical Analysis," *Biosensors & Bioelectronics*, Vol. 13, 1998, pp. 427–438.

[38] Kemshead, J., and J. Ugelstad, "Magnetic Separation Techniques—Their Application to Medicine," *Molecular & Cellular Biochemistry*, Vol. 67, 1985, pp. 11–18.

[39] Chalmers, J., et al., "Quantification of Cellular Properties from External Fields and Resulting Induced Velocity: Magnetic Susceptibility," *Biotechnology & Bioengineering*, Vol. 64, 1999, pp. 519–526.

[40] Reddy, S., et al., "Determination of the Magnetic Susceptibility of Labeled Particles by Video Imaging," *Chemical Engineering Science*, Vol. 51, 1996, pp. 947–956.

[41] Gill, S., C. Malone, and M. Downing, "Magnetic Susceptibility Measurements of Signal Small Particles," *Review of Scientific Instruments*, Vol. 31, 1960, pp. 1299–1303.

[42] Edwards, T., B. Gale, and A. Frazier, "A Microfabricated Thermal Field-Flow Fractionation System," *Analytical Chemistry*, Vol. 74, 2002, pp. 1211–1216.

[43] Sisson, R., and J. Giddings, "Effects of Solvent Composition on Polymer Retention in Thermal Field-Flow Fractionation—Retention Enhancement in Binary Solvent Mixtures," *Analytical Chemistry*, Vol. 66, 1994, pp. 4043–4053.

[44] Lou, J., M. Myers, and J. Giddings, "Separation of Polysaccharides by Thermal Field-Flow Fractionation," *Journal of Liquid Chromatography*, Vol. 17, 1994, pp. 3239–3260.

[45] Mozersky, S., et al., "Sedimentation Field Flow Fractionation of Mitochondrial and Microsomal-Membranes from Corn Roots," *Analytical Biochemistry*, Vol. 172, 1988, pp. 113–123.

[46] Schure, M., K. Caldwell, and J. Giddings, "Theory of Sedimentation Hyperlayer Field-Flow Fractionation," *Analytical Chemistry*, Vol. 58, 1986, pp. 1509–1516.

[47] Chmelik, J., "Different Elution Modes and Field Programming in Gravitational Field-Flow Fractionation I. A Theoretical Approach," *Journal of Chromatography A*, Vol. 845, 1999, pp. 285–291.

[48] Battu, S., J. Cook-Moreau, and P. Cardot, "Sedimentation Field-Flow Fractionation: Methodological Basis and Applications for Cell Sorting," *Journal of Liquid Chromatography & Related Technologies*, Vol. 25, 2002, pp. 2193–2210.

[49] Myers, M., "Overview of Field-Flow Fractionation," *Journal of Microcolumn Separations*, Vol. 9, 1997, pp. 151–162.

[50] Williams, P., S. Lee, and J. Giddings, "Characterization of Hydrodynamic Lift Forces by Field-Flow Fractionation—Inertial and Near-Wall Lift Forces," *Chemical Engineering Communications*, Vol. 130, 1994, pp. 143–166.

[51] Moon, M., and J. Giddings, "Size Distribution of Liposomes by Flow Field-Flow Fractionation," *Journal of Pharmaceutical and Biomedical Analysis*, Vol. 11, 1993, pp. 911–920.

[52] Schimpf, M., and K. Wahlund, "Asymmetrical Flow Field-Flow Fractionation as a Method to Study the Behavior of Humic Acids in Solution," *Journal of Microcolumn Separations*, Vol. 9, 1997, pp. 535–543.

[53] Yohannes, G., et al., "Fractionation of Humic Substances by Asymmetrical Flow Field-Flow Fractionation," *Chromatographia*, Vol. 61, 2005, pp. 359–364.

[54] Gao, Y., et al., "Extension of Thermal Field-Flow Fractionation to Ultrahigh (20×106) Molecular-Weight Polystyrenes," *Macromolecules*, Vol. 18, 1985, pp. 1272–1277.

[55] Wahlund, K., and J. Giddings, "Properties of an Asymmetrical Flow Field-Flow Fractionation Channel Having One Permeable Wall," *Analytical Chemistry*, Vol. 59, 1987, pp. 1332–1339.

[56] Pohl, H., *Dielectrophoresis*, Cambridge, U.K.: Cambridge University Press, 1978.

[57]   Pethig, R., and G. Markx, "Applications of Dielectrophoresis in Biotechnology," *Trends in Biotechnology*, Vol. 15, 1997, pp. 426–432.

[58]   Hughes, M. P., *Nanoelectromechanics in Engineering and Biology*, Boca Raton, FL: CRC Press, 2003.

[59]   Schnelle, T., et al., "The Influence of Higher Moments on Particle Behaviour in Dielectrophoretic Field Cages," *Journal of Electrostatics*, Vol. 46, 1999, pp. 13–28.

[60]   Huang, Y., et al., "MEMS-Based Sample Preparation for Molecular Diagnostics," *Analytical and Bioanalytical Chemistry*, Vol. 372, 2002, pp. 49–65.

[61]   Yang, J., et al., "Cell Separation on Microfabricated Electrodes Using Dielectrophoretic/Gravitational Field Flow Fractionation," *Analytical Chemistry*, Vol. 71, 1999, pp. 911–918.

[62]   Cheng, J., et al., "Preparation and Hybridization Analysis of DNA/RNA from E-Coli on Microfabricated Bioelectronic Chips," *Nature Biotechnology*, Vol. 16, 1998, pp. 541–546.

[63]   Green, N., H. Morgan, and J. Milner, "Manipulation and Trapping of Sub-Micron Bioparticles Using Dielectrophoresis," *Journal of Biochemical and Biophysical Methods*, Vol. 35, 1997, pp. 89–102.

[64]   Grom, F., et al., "Accumulation and Trapping of Hepatitis A Virus Particles by Electrohydrodynamic Flow and Dielectrophoresis," *Electrophoresis*, Vol. 27, 2006, pp. 1386–1393.

[65]   Asbury, C., A. Diercks, and G. van den Engh, "Trapping of DNA by Dielectrophoresis," *Electrophoresis*, Vol. 23, 2002, pp. 2658–2666.

[66]   Asbury, C., and G. van den Engh, "Trapping of DNA in Nonuniform Oscillating Electric Fields," *Biophysical Journal*, Vol. 74, 1998, pp. 1024–1030.

[67]   Huang, Y., et al., "Electric Manipulation of Bioparticles and Macromolecules on Microfabricated Electrodes," *Analytical Chemistry*, Vol. 73, 2001, pp. 1549–1559.

[68]   Wang, X., et al., "Seletive Dielectrophoretic Confinement of Bioparticles in Potential-Energy Wells," *Journal of Physics D-Appled Physics*, Vol. 26, 1993, pp. 1278–1285.

[69]   Talary, M., et al., "Dielectrophoretic Separation and Enrichment of CD34+ Cell Subpopulation from Bone-Marrow and Peripheral-Blood Stem-Cells," *Medical & Biological Engineering & Computing*, Vol. 33, 1995, pp. 235–237.

[70]   Yang, J., et al., "Dielectric Properties of Human Leukocyte Subpopulations Determined By Electrorotation as a Cell Separation Criterion," *Biophysical Journal*, Vol. 76, 1999, pp. 3307–3314.

[71]   Li, Y., and K. Kaler, "Dielectrophoretic Fluidic Cell Fractionation System," *Analytica Chimica Acta*, Vol. 507, 2004, pp. 151–161.

[72]   Gascoyne, P., and J. Vykoukal, "Particle Separation by Dielectrophoresis," *Electrophoresis*, Vol. 23, 2002, pp. 1973–1983.

[73]   Muller, T., et al., "Microdevice for Cell and Particle Separation Using Dielectrophoretic Field-Flow Fractionation," *Journal of Liquid Chromatography & Related Technologies*, Vol. 23, 2000, pp. 47–59.

[74]   Muller, T., et al., "A 3-D Microelectrode System for Handling and Caging Single Cells and Particles," *Biosensors & Bioelectronics*, Vol. 14, 1999, pp. 247–256.

[75]   Durr, M., et al., "Microdevices for Manipulation and Accumulation of Micro- and Nanoparticles by Dielectrophoresis," *Electrophoresis*, Vol. 24, 2003, pp. 722–731.

[76]   Chan, K., et al., "Measurements of the Dielectric Properties of Peripheral Blood Mononuclear Cells and Trophoblast Cells Using AC Electrokinetic Techniques," *Biochimica et Biophysica Acta*, Vol. 1500, 2000, pp. 313–322.

[77]   Gascoyne, P., et al., "Dielectrophoretic Separation of Mammalian-Cells Studied by Computerized Image-Analysis," *Measurement Science & Technology*, Vol. 3, 1992, pp. 439–445.

[78]   Hughes, M., "AC Electrokinetics: Applications for Nanotechnology," *Nanotechnology*, Vol. 11, 2000, pp. 124–132.

[79] Wang, X., et al., "Separation of Polystyrene Microbeads Using Dielectrophoretic/Gravitational Field-Flow-Fractionation," *Biophysical Journal*, Vol. 74, 1998, pp. 2689–2701.

[80] Yang, J., et al., "Differential Analysis of Human Leukocytes by Dielectrophoretic Field-Flow-Fractionation," *Biophysical Journal*, Vol. 78, 2000, pp. 2680–2689.

[81] Talary, M., et al., "Electromanipulation and Separation of Cells Using Travelling Electric Fields," *Journal of Physics D*, Vol. 29, 1996, pp. 2198–2203.

[82] Morgan, H., et al., "Large-Area Travelling-Wave Dielectrophoresis Particle Separator," *Journal of Micromechanics & Microengineering*, Vol. 7, 1997, pp. 65–70.

[83] Green, N., et al., "Large Area Multilayered Electrode Arrays for Dielectrophoretic Fractionation," *Microelectronic Engineering*, Vol. 35, 1997, pp. 421–424.

[84] Groschl, M., "Ultrasonic Separation of Suspended Particles—Part I: Fundamentals," *Acustica*, Vol. 84, 1998, pp. 432–447.

[85] Haake, A., et al., "Manipulation of Cells Using an Ultrasonic Pressure Field," *Ultrasound in Medicine & Biology*, Vol. 31, 2005, pp. 857–864.

[86] Hertz, H., "Standing-Wave Acoustic Trap for Nonintrusive Positioning of Microparticles," *Journal of Applied Physics*, Vol. 78, 1995, pp. 4845–4849.

[87] Saito, M., N. Kitamura, and M. Terauchi, "Ultrasonic Manipulation of Locomotive Microorganisms and Evaluation of Their Activity," *Journal of Applied Physics*, Vol. 92, 2002, pp. 7581–7586.

[88] Gaida, T., et al., "Selective Retention of Viable Cells in Ultrasonic Resonance Field Devices," *Biotechnology Progress*, Vol. 12, 1996, pp. 73–76.

[89] Maitz, M., et al., "Use of an Ultrasound Cell Retention System for the Size Fractionation of Somatic Embryos of Woody Species," *Plant Cell Reports*, Vol. 19, 2000, pp. 1057–1063.

[90] Yasuda, K., et al., "Using Acoustic Radiation Force as a Concentration Method for Erythrocytes," *Journal of the Acoustical Society of America*, Vol. 102, 1997, pp. 642–645.

[91] Petersson, F., et al., "Separation of Lipids from Blood Utilizing Ultrasonic Standing Waves in Microfluidic Channels," *Analyst*, Vol. 129, 2004, pp. 938–943.

[92] Limaye, M., and W. Coakley, "Clarification of Small Volume Microbial Suspensions in an Ultrasonic Standing Wave," *Journal of Applied Microbiology*, Vol. 84, 1998, pp. 1035–1042.

[93] Bohm, H., et al., "Ultrasound-Induced Physiological Changes in Cultured Cells of Petunia Hybrida," *Artificial Cells Blood Substitutes & Immobilization Biotechnology*, Vol. 30, 2002, pp. 127–136.

[94] Anderson, G., et al., "MHC Class-II-Positive Epithelium and Mesenchyme Cells Are Both Required for T-Cell Development in the Thymus," *Nature*, Vol. 362, 1993, pp. 70–73.

[95] Castilho, L., and F. Anspach, "CFD-Aided Design of a Dynamic Filter for Mammalian Cell Separation," *Biotechnol. Bioeng.*, Vol. 83, 2003, pp. 514–524.

[96] Mohamed, H., et al., "Development of a Rare Cell Fractionation Device: Application for Cancer Detection," *IEEE Transactions on Nanobioscience*, Vol. 3, 2004, pp. 251–256.

[97] Suzuki, H., C. Ho, and N. Kasagi, "A Chaotic Mixer for Magnetic Bead-Based Micro Cell Sorter," *Journal of Microelectromechanical Systems*, Vol. 13, 2004, pp. 779–790.

[98] Huang, L. R., et al., "Continuous Particle Separation Through Deterministic Lateral Displacement," *Science*, Vol. 304, 2004, pp. 987–990.

[99] Chiou, P. Y., A. T. Ohta, and M. C. Wu, "Massively Parallel Manipulation of Single Cells and Microparticles Using Optical Images," *Nature*, Vol. 436, 2005, pp. 370–372.

[100] Paterson, L., et al., "Light-Induced Cell Separation in a Tailored Optical Landscape," *Applied Physics Letters*, Vol. 87, 2005, pp. 123–901.

[101] Zborowski, M., et al., "Red Blood Cell Magnetophoresis," *Biophysical Journal*, Vol. 84, 2003, pp. 2638–2645.

[102] Bassett, P., *Cell Therapy: Technology, Market and Opportunities,* Global Information, Inc., January 2003.

[103] Kashofer, K., and D. Bonnet, "Gene Therapy Progress and Prospects: Stem Cell Plasticity," *Gene Therapy*, Vol. 12, 2005, pp. 1229–1234.

[104] Zandstra, P., and A. Nagy, "Stem Cell Bioengineering," *Annual Review of Biomedical Engineering*, Vol. 3, 2001, pp. 275–305.

[105] Eppich, H., et al., "Pulsed Electric Fields Function to Size-Select Hematopoietic Cells and Deplete Tumor Cell Contaminants," *Blood*, Vol. 90, 1997, pp. 4326–4326.

PART V

Advanced Life Science Automation

CHAPTER 17

Modeling and Control of Electroporation-Mediated Gene Delivery for Gene Therapy

Ou Ma and Mingjun Zhang

This chapter discusses the modeling and control of a chemo-electro-mechanical process for electroporation-mediated gene delivery [1–25]. Model based feedback control strategies with partial linearization of the process dynamics are introduced. Dynamic simulations are presented to demonstrate the effectiveness of the closed-loop control approach.

17.1   Introduction

Modeling and control of gene delivery (or drug delivery) have attracted many researchers' interests in recent years. It is one of the fastest growing areas in healthcare. The goals of drug delivery are to control the delivery time, speed, and amount of a drug without harming targeted live tissues. This requires consideration of the drug side effects, dynamics, and stability. Recently, the focus of drug delivery research has been moving towards the micro- and nanoscales.

Automation is essential for micro- and nanoscale gene delivery. The goals of automation in gene delivery are to improve accuracy, efficiency, quality, and reliability. Much progress has been made in integrating automation technologies with biological and medical principles to solve practical gene delivery problems, such as MEMS- and NEMS-based systems, distributed control systems, and sensor-network based systems. To further advance the technology for achieving the above-mentioned goals, accurate mathematical modeling of the gene delivery process is expected. A mathematical model of a gene delivery process not only helps us understand the dynamics of the delivery process but also is essential for better and safer control of the delivery process.

This chapter discusses dynamics modeling and control of the electroporation-mediated gene delivery. The electroporation has been of interest to the medical community as a possible means to introduce foreign genes into cells due to several reasons. First, it has less chance of complications and is applicable to a wide range of targets. Second, the method does not need a vector and is relatively efficient.

Third, the cost is low compared with other available approaches. Finally, the gene effect is relatively local, which may lead to fewer side effects. An example of operation is to first introduce delivered genes to a tissue by a needle injection, and then apply electrical pulses to electroporate the tissue cells within a treatment volume [23].

Electroporation often causes a significant increase in the permeability and electrical conductivity of the cell plasma membrane. This is due to creation of transient pores in the cell membrane by an electric field. The electroporation is also regarded as a potential alternative for nonthermal manipulation of cellular functions [10]. The phenomenon is reversible and is highly efficient for introducing foreign genes into cells, especially for mammalian cells [1].

Though fundamentals of the physical or chemical effects of the electroporation process have not been completely understood due to lack of direct measurements [1], efforts have been made to explain the phenomena from various aspects. The process can be theoretically explained as a process of formation of small pores in the plasma membrane. Usually the external electric field needs to surpass the capacitance of the cell membrane. The lipid bilayer membrane can then rearrange to create water-filled pores, which provides a pathway for DNA molecules that normally are impermeable to the membrane [5].

To have genes pass through the cells, the pores need to be created, grow to a size large enough, and last long enough. It has been experimentally observed that, to enable the entry of supercoiled DNA molecules into cells, the pores should be at least 10 nm in diameter and last for 2 to 6 ms [20]. It is known that both the pore opening diameter and the duration are determined by the strength of electrical field and the time duration of the effect. A strong electrical field may create a large pore size. However, excessive exposure of living cells to an electrical field may cause cell damage or even cell death. As seen in experiments, a long and strong pulse may damage a cell. On the other hand, short and small pulse may not create pores with a diameter large enough and a duration long enough for a gene to pass through. One question often arising in the literature is whether a short and high-voltage electrical pulse is more efficient for the gene delivery process [17]. Historically, most studies focused on less than a kilovolt per centimeter electric fields applied in a millisecond range. Recently, studies have been conducted for a process using higher electric fields with a pulse duration of only a few nanoseconds [10, 11]. It is still debatable regarding which one is better. Consequently, a question is raised as how to optimally control the electrical field for gene delivery process.

To develop optimal control for the electroporation-mediated gene delivery, a theoretical model for the gene delivery process is needed. There are ongoing efforts to optimize molecule delivery while preserving cell viability through various technical optimization techniques. New equipment is now commercially available or being developed [7]. Unfortunately, most of these techniques are based on experiences. No feedback control strategies have been proposed due to the lack of an appropriate dynamics model that can accurately describe the dynamics of the process.

To conduct feedback control of a complicated nonlinear dynamic system like the electroporation-mediated gene delivery process, a mathematical model is needed. Some models have been proposed to explain physical and chemical effects of the electroporation process. However, due to a wide variety of membrane compositions and results obtained from different experimental techniques, the models are difficult to compare and often inconsistent with each other. In fact, one obstacle for possible application of the electroporation in clinical practice is the lack of a proper theoretical model [13, 20]. Most current models are less specific for gene therapy. Existing models in the literature are somewhat incorrect for a variety of reasons [10]. Finally, most of these models are computationally complex and thus, not suitable for being integrated into a controller for the gene delivery process. Clearly, a computationally efficient and pro-control model is needed for developing and implementing control strategies.

Efforts have been made recently to model the electroporation-mediated gene delivery. Reference [21] presents the results of molecular dynamics simulations of lipid bilayers under a high transverse electrical field. The applied transmembrane electric fields (0.5 and 1.0 V/nm) induce an electroporation of the peptide bilayer manifested by the formation of water wires and water channels across the membrane. The simulations show that the electroporation process takes place in two stages. First, water molecules organized in single wires penetrate the hydrophobic core of a bilayer. Then, the water wires grow in length and expand into water-filled pores. It is suggested that water wire formation be driven by local field gradients at the water-lipid interface. The electroporation process is independent of the ionic strength of the medium surrounding the membrane.

Electroporation is an electromechanical phenomenon rather than an electrothermal phenomenon [9]. The cell membrane is usually held together by hydrophobic and hydrophylic forces. During the electroporation, the cell membrane undergoes structural changes. In most theoretical analysis, mechanical forces generated by the pulse field and their relationships to the membrane deformation are often overlooked. Neglecting this factor can lead to incorrect analysis [17].

Most models proposed in the literature are derived by the continuum Smoluchowski equation (SE) described as follows:

$$\frac{\partial N(r,t)}{\partial t} - \frac{D}{k_B T} \frac{\partial \left[N(\gamma,\tau) \partial E(\gamma)/\partial \gamma \right]}{\partial \gamma} - D \frac{\partial^2 N(r,t)}{\partial r^2} = F(r) \tag{17.1}$$

where $N(r,t)$ is the pore density; D is the diffusion coefficient of the pore radius; k_B is the Boltzmann constant; T is the absolute temperature; $E(\gamma)$ is the pore energy; and $F(r)$ is the pore formation term. In the model, it is assumed that the hydrophobic pore is created at a rate depending exponentially on the square of the transmembrane potential. A pore with radius greater than an energy barrier for creation of hydrophilic pores will convert spontaneously to long-lived hydrophilic pores. Otherwise, the pore will reseal, revert to hydrophobic configuration, and be destroyed by lipid fluctuations.

Though the above-mentioned SE model describes the creation and evolution of pores, it cannot predict whether pores are large enough to admit plasmid DNA [14, 20]. The postshock growth of pores and the stabilization of their radii at larger diameter cannot be explained by the model. By reexamining the energy cost of the pore creation and expansion, [14, 20] propose a nonlinear extension of the SE models, which can predict the creation of stable large pores. However, the model cannot be used for control purposes due to the extreme constraints on the valid range of variables.

Although some studies have been conducted in control of the electric pulses, most of them are of trial-and-error based on intuitive observations and there is no theoretical foundation for the discussions [20]. We would like to introduce a modified model particularly suitable for closed-loop control of the gene delivery process. Open-loop control is simple for implementation, but it cannot guarantee the desired results in the presence of various uncertainties in the process. To our best knowledge, no closed-loop feedback control strategies for gene delivery have been reported in the open literature.

This chapter extends the dynamics models developed in the open literature and proposes approaches for improvements. The work complements other theoretical studies by providing advanced nonlinear feedback control to the electroporation-mediated gene delivery, and provides a revised model for the process. In the meantime, efforts have been made to reflect time-varying effects of the process. The model is specifically developed for control purposes.

The rest of the chapter is organized as follows: in Section 17.2, a mathematical model describing the chemical, electrical, and mechanical effects of the electroporation process is presented; Section 17.3 presents control strategies for feedback control of the gene delivery process; simulation results and comparisons between open-loop and feedback controls for the gene delivery are provided in Section 17.4; and the chapter is concluded in Section 17.5.

17.2 Dynamics Modeling

The electroporation is a highly nonlinear dynamic phenomenon. All pores are initially created hydrophobically at a rate determined by their energy level [24]. In reality, most of these pores are quickly destroyed by lipid fluctuations. If hydrophobic pores reach a radius of $r \geq r_p$ (where r_p is a radius of energy barrier for creation of hydrophilic pores), the pores convert to long-lived hydrophilic pores.

For the convenience of the model development, it is often assumed that the pores are created uniformly around a cell. In general, the electroporation-mediated gene delivery process can be divided into four phases: pore formation, pore opening, DNA uptake, and pore absence [10, 20]. For the convenience of discussion, we first introduce different parts of the model from the aspects of chemical, electrical, and mechanical effects, respectively. Then, we integrate these different model parts into a system dynamics model.

17.2.1 Chemical Effect

In [4], the pore formulation process under electroporation is formulated as a voltage sensitive chemical reaction. The chemical field is often described by diffusion equations. This transient, permeabilized state can be used to load cells with a variety of molecules, either through simple diffusion in the case of small molecules or through electrophoretically driven processes [7]. For the electroporation-mediated gene delivery, the diffusive and electrophoretic transport is related to the DNA uptake.

The molecular mechanisms of the DNA transfer into cells are not yet clarified. The exact mechanism of DNA uptake is still debatable.

In [16], a DNA uptake model for direct transfer of the plasmid DNA (YEp 351) was proposed. The electrodiffusive transport of DNA into and across the electroporated membrane of the average cell is described by the following Nernst-Planck equation:

$$\frac{1}{s(t)}\frac{dC_i}{dt} = -\frac{D_0}{h}\left[\frac{C_i}{V_{cell}} - \frac{C_o}{V_{sol}}\left(1 - \frac{|z_{eff}|q|V_m(t)|}{k_B T}\right)\right] \tag{17.2}$$

where C_i and C_o are concentrations of the DNA inside and outside the cell, respectively; D_0 is the DNA diffusion coefficient; h is the cell membrane thickness; V_{cell} is the cell volume; z_{eff} is the effective valence of the DNA molecule and q is the elementary charge; $s(t)$ is the electroporated surface area; k_B is the Boltzmann constant; T is the absolute temperature; V_m is the electrical potential difference across the electroporated membrane surface, and V_{sol} is the volume of solution.

In [20], Smith, Neu, and Krassowska modified the above equation by applying it to each individual pore as follows:

$$\frac{dC_i}{dt} = -\frac{D_0}{hV_{cell}}s(t)\left[C_i - C_o\left(1 + \frac{|z_{eff}|q}{kT}\frac{V_m(t)}{1 + R_i/R_p}\right)\right] \tag{17.3}$$

All the variables in (17.3) are identically defined as those in (17.2) except that: $s(t) = \pi r_j^2$ is the area of a single pore; r_j is the radius of the jth pore; $j = 1, 2, ..., n$; and $V_m(t)$ is the actual voltage drop in a cell. As will be discussed in the next section, $R_p = h / \pi\, gr_j^2$ and $R_i = 1/2\, gr_j$ are pore resistance and the input resistance, respectively. Moreover, g is the conductivity of the solution and the term $V_m(t)/(1 + R_j/R_p)$ is corresponding to the fraction of $V_m(t)$, which is the voltage drop across the specific pore.

In addition to the above differences of the two models, [20] used the cell volume to replace the sample volume. The sign of the last term in (17.2) is different from the original equation in [16]. We believe that the sign in [20] is correct.

We agree that it is more appropriate to modify (17.2) for general DNA uptake. Here, we still use the original model from [16] due to the concern that the cell inside concentration is really determined by all pores as opposed to a single pore. Equation (17.12) in [16] is a heuristic equation. To introduce the solution volume into the equation, we define a coefficient ρ and assume the solution volume to be ρV_{cell}. The equation can then reflect concentration inside and outside a single cell with respect to its volume and transmembrane potential drop.

$$\frac{dC_i}{dt} = -\frac{4D_0 N\pi^2 R^2 r_j^2}{hV_{cell}}C_i - \frac{4D_0 N\pi^2 R^2 r_j^2 C_o}{h\rho V_{cell}} + \frac{4D_0 N\pi^2 R^2 r_j^2 C_o}{h\rho V_{cell}}\frac{|z_{eff}|q|V_m(t)|}{k_B T} \tag{17.4}$$

To simplify the model, let

$$\xi_1 = -\frac{4D_0\pi^2 R^2}{hV_{cell}}, \quad \xi_2 = -\frac{\xi_1 C_o}{\rho}, \quad \xi_3 = -\frac{\xi_2 |Z_{eff}|g}{k_B T} \tag{17.5}$$

Equation (17.4) can then be simplified as

$$\frac{dC_i}{dt} = \xi_1 Nr_j^2 C_i + \xi_2 Nr_j^2 + \xi_3 Nr_j^2 V_m(t) \tag{17.6}$$

17.2.2 Electrical Field Effect

The dynamic behavior of cells in electric fields has been widely investigated in [2, 3, 7, 20]. Assume that the cell has a passive resistive-capacitance (RC) membrane. The transmembrane potential induced in a cell by an external field can be described by the following equation [7]:

$$\Delta V_m = fE_{ext}R\cos(\phi) \tag{17.7}$$

where V_m is the transmembrane potential; f is a form factor describing the impact of the cell on the extracellular field distribution; E_{ext} is the applied electric field; R is the cell radius; and ϕ is the polar angle with respect to the external field. Electroporation is achieved when the ΔV_m value superimposed on the resting transmembrane potential is larger than a threshold.

The above equation probably is the earliest one developed for estimating the transmembrane potential drop within a cell under electrical field. Unfortunately, it does not satisfy further understanding of electroporation for a single cell. For the electroporation-mediated gene delivery, electric shocks are applied with controlled strength and duration. Large transmembrane potentials are often induced. Usually the pore's presence affects the transmembrane potential V_m. Based on recent studies in the literature [20], the following revised model is proposed. The circuit representation of the polarized membrane is illustrated in Figure 17.1. The transmembrane potential V_m is governed by the following ODE [20]:

$$C(t)\frac{dV_m}{dt} + \left(\frac{1}{R_0} + \frac{1}{R(t)}\right)V_m + I_t = \frac{V_0}{R_0} \tag{17.8}$$

where $C(t)$ is the cell capacitance. $I_t = \sum_{j=1}^{n} V_m / (R_p + R_j)$ is the combined current through all pores. R_s and $R(t)$ are the series resistance of the experiment cell resistance, respectively. Moreover,

$$C(t) = 4\pi R^2 (1 - NS(t))C_m \tag{17.9}$$

where N is the pore density, which will be defined later. Further, $s(t)$ is the area of one pore and C_m is the surface capacitance of the membrane.

Although $R(t)$ has been assumed to be constant by some researchers in the open literature, we assume it to be time-varying as follows:

$$R(t) = \frac{R_m}{4\pi R^2 (1 - Ns(t))} \tag{17.10}$$

where R_m is the surface resistance of the membrane.

Studies in the literature usually assume uniform polarized membrane. We assume time-varying polarized membrane due to the concerns that: (1) membrane capacitance (related to the membrane surface) is indeed changing significantly with time; and (2) membrane resistance is also changing. In the open literature, cell capacitance is always formulated as a constant. We believe it is more proper to formulate it as a time-varying function affected by pore presence. Define the following parameters:

$$\xi_4 = 4\pi R^2, \quad \xi_5 = 2\pi g \tag{17.11}$$

Then (17.8) can be converted to

$$\frac{dV_m}{dt} = \frac{V_0}{R_0 \left(\xi_4 - \pi\xi_4 Nr_j^2 \right) C_m}$$
$$- \left(\frac{1}{R_0 \left(\xi_4 - \pi\xi_4 Nr_j^2 \right) C_m} + \frac{1}{R_m C_m} \right) V_m - \frac{\xi_5 N V_m r_j^2}{C_m \left(1 - \pi Nr_j^2 \right)\left(2h + \pi r_j \right)} \tag{17.12}$$

A major difference of the proposed model from those in the literature is in the voltage applied to individual cells. In the literature, most models use the voltage U directly applied to the individual cells. Since voltage is from one electrode to another, in reality, this may not be true. There are many pores in between two electrodes. The exact number of pores depends on the cell concentration. Here, it is approximated by assuming that all the cells are connected and parallel to each

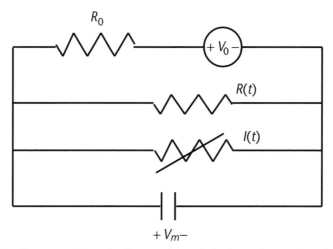

Figure 17.1 Circuit representation of a time-varying polarized membrane. Capacitor $C(t)$ represents the total time-varying capacitance of the membrane. $R(t)$ represents the flow of current through the channel. $I(t)$ is the pore current. R_0 is the resistance of the experimental setup. V_0 is the electrical voltage applied to a single cell.

other. Assume that the distance between the electrodes is L and all the cells are connected and uniformly distributed in parallel. Then, we have

$$V_0 = 2UR / L \qquad (17.13)$$

where U is the applied electrical voltage to the total solution sample.

17.2.3 Mechanical Effect

The pore formation is subject to thermal fluctuations and the existence of threshold energy [12, 22]. The pore dynamics is influenced by the transmembrane potential. Based on continuum Smoluchowski theory, [20] proposed the pore density as the following first-order differential equation:

$$\frac{dN}{dt} = \alpha e^{(V_m / V_{ep})^2} \left(1 - \frac{N}{N_0 e^{(r_m V_m / r_p V_{ep})^2}} \right) \qquad (17.14)$$

where N is the pore density; V_m is the transmembrane potential; V_{ep} is the characteristic voltage of electroporation; α is the creation rate coefficient; N_0 is the pore density; r_m is the minimum energy radius when $V_m = 0$; and r_p is the minimum radius of hydrophilic pores.

To simplify the expression of (17.14), we define

$$\xi_6 = \frac{1}{V_{ep}^2}, \quad \xi_7 = -\frac{\alpha}{N_0}, \quad \xi_8 = 1 - \frac{r_m^2}{r_p^2} \qquad (17.15)$$

then

$$\frac{dN}{dt} = \alpha e^{\xi_6 V_m^2} + \xi_7 N e^{\xi_6 \xi_8 V_m^2} \qquad (17.16)$$

The exponent V_m^2 in the above equation makes the differential equation stiff. Notice that V_m^2 usually works in a very small range for gene delivery. For the convenience of numerical calculation, polynomial functions may be used to approximate (17.16) as follows:

$$\frac{dN}{dt} = \alpha \left(1 + \frac{\xi_6 V_m^2}{n_p} \right)^{n_p} + \xi_7 N \left(1 + \frac{\xi_6 \xi_8 V_m^2}{n_p} \right)^{n_p} \qquad (17.17)$$

where n_p is the order of the polynomial.

The pore presence decreases the bilayer energy by either reducing the area subject to interfacial tension or increasing the energy by introducing line energy of the pore perimeter. Pores change radius to minimize the energy of the entire liquid bilayer. The pores change their sizes following the following equation [20]:

$$\frac{dr_j}{dt} = -\frac{D}{k_B T}\frac{\partial E}{\partial r_j}, \quad j = 1, 2, \ldots, n \tag{17.18}$$

where D is the diffusion coefficient of the pore radius; k_B is the Boltzmann constant; T is the absolute temperature; and n is the total number of pores.

The bilayer energy E is expressed as

$$E = \sum_{j=1}^{n}\left[\beta\left(\frac{r_p}{r_j}\right)^2 + 2\pi\gamma r_j - \pi\sigma_{eff}(S_p)r_j^2 - \int_0^{r_j} F(r_j, V_m)dr\right] \tag{17.19}$$

where β is the steric repulsion energy; γ is the edge energy [20]. The first term of the equation represents the energy due to the steric repulsion of the lipid heads. This term creates an energy barrier at r_p and slows down the resealing of pores. The second term is for line energy of the pore perimeter. The third term contains the effective tension of the membrane. Moreover, σ_{eff} represents the tension of a membrane without pores [14, 20]. $S = 4\pi R^2(1 - Ns(t))$ is the total area of the lipid bilayer. $S_p = \sum_{j=1}^{n}\pi r_j^2$ is the combined area of pores. We revise this term with the result in [20] as

$$\sigma_{eff}(r_j) = \frac{\sigma_0(1 - N\pi r_j^2)^2 - 2\sigma'(2 - 3N\pi r_j^2)N\pi r_j^2}{(1 - 2N\pi r_j^2)^2} \tag{17.20}$$

where σ' is the energy per area of the hydrocarbon-water interface and σ_0 is the tension of the intact bilayers.

For large and highly conductive pores, it is necessary to consider energy induced by transmembrane potential. The fourth term in (17.19) was given in [15] derived from the Maxwell stress tensor. It implies that the electrical force expanding the pore asymptotes to a constant value for large pores. In [20], the contribution of the transmembrane potential V_m to the bilayer energy is described as follows:

$$F(r_j, V_m) = \frac{F_{max}}{1 + \dfrac{r_h}{r_j + r_t}}V_m^2 \tag{17.21}$$

where F_{max}, r_h, and r_t are constants. The equation predicts that the force F approaches a constant value F_{max} as the pore radius increases. In fact, the above equation is only a numerical approximation of the force [15]. Here, we propose the following equation to approximate the force:

$$F(r_j, V_m) = k_f \ln_n(k_o \gamma_o V_m + e)V_m^2 \tag{17.22}$$

where k_f and k_o are coefficient constants.

Assume all pores have the same radii at any fixed time, and let

$$\xi_9 = \beta r_p^4, \quad \xi_{10} = 2\pi\gamma, \quad \xi_{11} = r_t + r_h, \quad \xi_{12} = -\frac{D}{kT} \tag{17.23}$$

Then, (17.19) can be expressed as

$$E = \sum_{j=1}^{n} \left[\frac{\xi_9}{r_j^4} + \xi_{10} r_j - \pi\sigma_{eff}(S_p) r_j^2 - \int_0^{r_j} F(r_j, V_m) dr \right]$$

$$= \sum_{j=1}^{n} \left\{ \frac{\xi_9}{r_j^4} + \xi_{10} r_j - \pi r_j^2 \frac{\sigma_0 (1 - N\pi r_j^2)^2 - 2\sigma'(2 - 3N\pi r_j^2) N\pi r_j^2}{(1 - 2N\pi r_j^2)^2} \right.$$

$$\left. - \frac{k_f}{k_0 V_m} \left[(k_0 r_j V_m + e) \ln(k_0 r_j V_m + e) - e \right] \right\}$$

(17.24)

Equation (17.18) can then be expressed as

$$\frac{dr_j}{dt} = \xi_{12} \frac{\partial E}{\partial r_j} = \xi_{12} \left[-\frac{4\xi_9}{r_j^5} + \xi_{10} - 4\pi\sigma' r_j + \right.$$

$$\left. \frac{2\pi(2\sigma' - \sigma_0) r_j (1 - \pi N r_j^2)(1 - \pi N r_j^2 + 2\pi^2 N^2 r_j^4)}{(1 - 2N\pi r_j^2)^3} - k_f \ln(k_0 r_j V_m + e) \right]$$

(17.25)

where $j = 1, 2, \cdots, n$.

17.2.4 Membrane Resealing

When the electric pulse is turned off, the membrane discharges rapidly through the existing pores, V_m drops to zero. The pores shrink to near the minimum energy radius r_m and the pore reseal. The cell's discharging of potential is usually faster than the cell's prolonged recovery period due to the slow rate of pore resealing. The time constant of pore resealing can be evaluated as follows [2]:

$$\tau_N = \frac{N_0}{\alpha} e^{(\eta-1)\left(\frac{V_m}{V_{ep}}\right)^2}$$

(17.26)

For example, if $V_m = 0$ mV and $\tau_N = 1.5$ seconds, the pore density will decrease exponentially and requires about 20 seconds to decrease to its preshock distribution.

17.2.5 Integrated Chemo-Electro-Mechanical Model

The above-discussed electroporation process can be divided into four steps:

Step 1: Pore formation. During this step, the pores are created as hydrophobic/nonconducting due to an externally applied electrical field. If the radius of a pore is greater than a threshold (r_p), the pore transforms into a conducting hydrophilic pore. Otherwise, it is destroyed.

Step 2: Pore opening. Due to fluctuation and energy change, all pores go through evolution process.

Step 3: DNA uptake. Due to chemical diffusion, the DNA molecules pass through the pores and enter the cell.

Step 4: Pore absence. After turning off the electrical field, the pores reduce sizes, merge, and eventually disappear.

Based upon the discussions in the previous sections, the above four-step electroporation process can be described by the following system of differential equations:

$$
\begin{bmatrix} \dfrac{dV_m}{dt} \\ \dfrac{dC_i}{dt} \\ \dfrac{dN}{dt} \\ \dfrac{dr_j}{dt} \end{bmatrix} = \begin{bmatrix} \dfrac{V_0}{R_0 C_m \xi_4 \left(1-\pi Nr_j^2\right)} - \dfrac{V_m}{R_0 C_m \xi_4 \left(1-\pi Nr_j^2\right)} - \dfrac{V_m}{R_m C_m} - \dfrac{\xi_5 V_m Nr_j^2}{C_m\left(1-\pi Nr_j^2\right)\left(2b+\pi r_j\right)} \\ \dfrac{\left(\xi_1 C_i + \xi_2 - \xi_3 V_m\right)Nr_j^2}{\alpha e^{\xi_6 V_m^2} + \xi_7 Ne^{\xi_6 \xi_8 V_m^2}} \\ \dfrac{2\pi\xi_{12}\left(2\sigma'-\sigma_0\right)r_j\left(1-\pi Nr_j^2\right)\left[1-\pi Nr_j^2 + 2\left(\pi Nr_j^2\right)^2\right]}{\left(1-2\pi Nr_j^2\right)^3} + \\ \xi_{10}\xi_{12} - \dfrac{4\xi_9\xi_{12}}{r_j^5} - 4\pi\xi_{12}\sigma'r_j - \xi_{12}k_f\ln\left(k_0 V_m r_j + e\right) \end{bmatrix}
$$

$$(17.27)$$

where ξ_1, ξ_2, ξ_3 are intermediate variables defined in (17.5); ξ_4, ξ_5 are intermediate variables defined in (17.11); ξ_6, ξ_7, ξ_8 are intermediate variables defined in (17.15); $\xi_9, \xi_{10}, \xi_{11}, \xi_{12}$ are intermediate variables defined in (17.23). All other parameters and variables are defined earlier.

This is a highly nonlinear dynamical model. It has only one input V_0, which can be used to control the process, and hence, the process is very difficult to control. In the open literature, open-loop pulse control methods have been proposed for such a system. As well-known, however, open-loop control strategies cannot always achieve a desired control goal. The reason is that the controller may keep the pore open to a proper radius in a time duration. However, the setting and various changes (disturbances) in the system components during the process will disturb the diffusion process. In addition, a predefined open-loop control input cannot guarantee to work. Feedback control strategies would be the best choice.

The concentration of DNA molecules inside the cell is a key controlled goal for the electroporation-mediated gene delivery. It is acceptable to assume that pore radius, duration, and potential drop along a cell are the only necessary conditions. In addition, the potential drops are expected not to cause any damage to the cells. Ideally, pore radius, potential drop, and DNA concentration inside the cells are three quantities that need to be controlled in the process. Unfortunately, the desired values of these three quantities are not those in the equilibriums of the dynamic system and the system is very complicated. Thus, the desired values cannot be easily and precisely reached without an active control. We will address the control problem by proposing feedback control strategies in the next section.

Note that the dynamic system described by (17.27) is not only difficult to control, but also difficult to simulate. This is because the differential equations are not only nonlinearly coupled, but also very stiff. For example, the third equation has two exponential terms, one with a large positive exponent and the other with a negative exponent. Such a dynamic system is very difficult to integrate numerically. So, a special integration method for solving stiff differential equations must be used in simulation.

17.3   Control Strategies

To develop an appropriate controller, the dynamics model presented in (17.27) is rewritten into the canonical form of nonlinear control system as follows:

$$\dot{\mathbf{x}} = \mathbf{f}(\mathbf{x}) + \mathbf{g}(\mathbf{x})u$$
$$\mathbf{y} = \mathbf{h}(\mathbf{x})$$

(17.28)

where all the bold-faced terms are vectors. In the above model, the state variables are defined as

$$\mathbf{x} = \begin{bmatrix} x_1 \\ x_2 \\ x_3 \\ x_4 \end{bmatrix} \equiv \begin{bmatrix} V_m \\ C_i \\ N \\ r_j \end{bmatrix}$$

(17.29)

and the other terms of the model are defined as follows:

$$\mathbf{f}(\mathbf{x}) = \begin{bmatrix} f_1(\mathbf{x}) \\ f_2(\mathbf{x}) \\ f_3(\mathbf{x}) \\ f_4(\mathbf{x}) \end{bmatrix} \equiv \begin{bmatrix} -\dfrac{x_1}{R_0 C_m \xi_4 \left(1 - \pi x_3 x_4^2\right)} - \dfrac{x_1}{R_m C_m} - \dfrac{\xi_5 x_1 x_3 x_4^2}{C_m \left(1 - \pi x_3 x_4^2\right)\left(2h + \pi x_4\right)} \\ \dfrac{\left(\xi_1 x_2 + \xi_2 - \xi_3 x_1\right) x_3 x_4^2}{\alpha e^{\xi_6 x_1^2} + \xi_7 x_3 e^{\xi_6 \xi_8 x_1^2}} \\ \dfrac{2\pi\xi_{12}\left(2\sigma' - \sigma_0\right)x_4\left(1 - \pi x_3 x_4^2\right)\left[1 - \pi x_3 x_4^2 + 2\left(\pi x_3 x_4^2\right)^2\right]}{\left(1 - 2\pi x_3 x_4^2\right)^3} + \\ \xi_{10}\xi_{12} - \dfrac{4\xi_9\xi_{12}}{x_4^5} - 4\pi\xi_{12}\sigma' x_4 - \xi_{12} k_f \ln(k_0 x_1 x_4 + e) \end{bmatrix}$$

(17.30)

$$\mathbf{g}(\mathbf{x}) = \begin{bmatrix} g_1(\mathbf{x}) \\ g_2(\mathbf{x}) \\ g_3(\mathbf{x}) \\ g_4(\mathbf{x}) \end{bmatrix} \equiv \begin{bmatrix} \dfrac{1}{R_0 C_m \xi_4 \left(1 - \pi x_3 x_4^2\right)} \\ 0 \\ 0 \\ 0 \end{bmatrix}$$

(17.31)

$$u = V_0$$

(17.32)

Scalar variable u is the control input which can be designed for the system based on desired control goal and dynamic performance. Moreover, the output function $\mathbf{h}(\mathbf{x})$ varies depending on the specific output quantities of interest. All the other parameters of the above equations are constants defined in the earlier sections.

This is a highly nonlinear, multidimensional dynamic system with only one control input, which makes it very difficult to simultaneously control all the variables. As mentioned earlier, the desired values of the state variables may not be those of the equilibrium state of the dynamic system. It is almost impossible to control all four state variables simultaneously using just a single-input u. Adding additional control inputs means a redesign of the gene delivery system and thus is beyond the scope of this chapter. In the remaining parts of this section, we will investigate the possibilities of controlling one state variable while keeping the rest of the state variables within a working range. The basic idea is to use the known dynamics model of the system to partially linearize the dynamics of the system, so that well-developed linear control techniques can be used to control the resulting system.

Among the four state variables, the most interesting variable for controlling the gene delivery process would be the radius of the pore, namely, x_4. In this case, the output function of (17.28) becomes

$$y = \mathbf{h}(\mathbf{x}) = x_4 \tag{17.33}$$

Since $g_4(\mathbf{x}) = 0$, the last differential equation of (17.28) reduces to

$$\dot{x}_4 = f_4(\mathbf{x}) + g_4(\mathbf{x})u = f_4(\mathbf{x}) \tag{17.34}$$

By theory, it is impossible for us to design a controller $u(\mathbf{x})$ based on the known dynamics model such that the nonlinear term $f_4(\mathbf{x})$ in (17.28) can be directly cancelled because the equation does not involve $u(\mathbf{x})$ at all. An alternative approach is to apply the input-output linearization technique [25]. This can be done by differentiating the output equation (17.33) with respect to time twice (the second order Lie derivatives), so that the control input u will explicitly appear in the resulting second derivative. Following this approach, we have

$$\ddot{y} = \ddot{x}_4 = \frac{\partial f_4(\mathbf{x})}{\partial \mathbf{x}} \dot{\mathbf{x}} = \frac{\partial f_4(\mathbf{x})}{\partial \mathbf{x}} \left[\mathbf{f}(\mathbf{x}) + \mathbf{g}(\mathbf{x})u \right] \tag{17.35}$$

Because $\partial f_4(\mathbf{x})/\partial x_1 \neq 0$, a particular state-feedback control law $u(\mathbf{x})$ can be constructed. The controller is expected to cancel all nonlinear terms in (17.34) and thus linearize the closed-loop equation. We can design the following controller:

$$u = \left[\frac{\partial f_4(\mathbf{x})}{\partial x_1} g_1(\mathbf{x}) \right]^{-1} \left[-\frac{\partial f_4(\mathbf{x})}{\partial \mathbf{x}} \mathbf{f}(\mathbf{x}) + \ddot{x}_{4d} + K_v \left(\dot{x}_{4d} - \dot{x}_4 \right) + K_p \left(x_{4d} - x_4 \right) \right] \tag{17.36}$$

This controller consists of two parts. The first part is the known dynamics model of the system which is used to cancel the corresponding nonlinear term of the plant dynamics, and the second part is a usual linear controller. Parameters K_v and K_p are gains of the linear controller and $x_{4d} = y_d$ is the desired output (reference output). Substituting the above control law into (17.34), we can show that the dynamics equation reduces to the following closed-loop error equation, which is a standard second order linear system:

$$\left(\ddot{x}_{4d} - \ddot{x}_4 \right) + K_v \left(\dot{x}_{4d} - \dot{x}_4 \right) + K_p \left(x_{4d} - x_4 \right) = 0 \tag{17.37a}$$

or

$$\ddot{e} + K_v \dot{e} + K_p e = 0 \tag{17.37b}$$

where $e = x_{4d} - x_4$ is the error between the desired output and the actual output of the dynamic system. This is an easily controllable second order linear system. The block diagram of the corresponding control system is illustrated in Figure 17.2. In the diagram, the plant represents the gene delivery process, which is assumed to have the model described by (17.27) or (17.28). As one can see, the control system consists of two state feedback loops. One loop (on the top portion of the diagram) is to calculate the two nonlinear terms $[f_4(\mathbf{x})/\partial\mathbf{x}]\mathbf{f(x)}$ and $\{[\partial f_4(x)/\partial x_1]/g_1(\mathbf{x})\}^{-1}$ of the control input and then feed the calculated terms into the plant in order to partially cancel the nonlinearity of the plant. The other loop (on the bottom portion of the diagram) is a usual PD servo loop for reducing the errors between the current state and the desired state. If this control strategy works, the control of the pore radius, as shown in (17.37a) and (17.38b), becomes relatively simple. However, our simulation results suggest that this approach does not work well. The best explanation is that, in the process of controlling x_4, one or more of the other three uncontrolled state variables are pushed out of range and thus destabilize the system. For this dynamic system, partial linearization and control of the state variables may not be sufficient to make the entire system even loosely working.

In fact, control of x_1 will be less difficult than the control of x_4 because $g_1(\mathbf{x}) \neq 0$ and thus, a properly designed input function $u(\mathbf{x})$ can directly cancel the nonlinearity of the first differential equation of (17.28). Hence, we turned our control goal to the first state variable (i.e., x_1), which is the transmembrane potential V_m. In this case, the output equation of (17.28) becomes

$$\mathbf{y} = \mathbf{h(x)} = x_1 \tag{17.38}$$

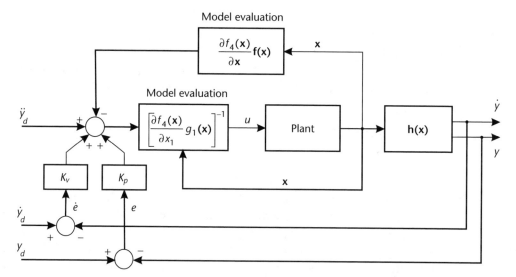

Figure 17.2 The closed-loop control system using input-output linearization technique.

Assuming that the dynamics model is known, we can partially linearize the plant dynamics (17.28) by proposing the following control law:

$$u = [g_1(\mathbf{x})]^{-1}[-f_1(\mathbf{x}) + K_p(x_{1d} - x_1) + K_v(\dot{x}_{1d} - \dot{x}_1)] \qquad (17.39)$$

where $x_{1d} = y_d$ is the desired value of the output variable y or state variable x_1. Again, the controller consists of two parts. The first part is derived from the known dynamics model of the system which is used to partially cancel the nonlinearity of the plant dynamics and the second part is a linear servo controller. Parameters K_p and K_v are control gains to be determined based on required control performance specifications. Substituting the control law into (17.28), we obtain

$$\mathbf{x} = \mathbf{f}(\mathbf{x}) + \mathbf{g}(\mathbf{x})[g_1(\mathbf{x})]^{-1}[-f_1(\mathbf{x}) + K_p(x_{1d} - x_1) + K_v(\dot{x}_{1d} - \dot{x}_1)] = \bar{\mathbf{f}}(\mathbf{x}) \qquad (17.40)$$

where

$$\bar{\mathbf{f}}(\mathbf{x}) = \begin{bmatrix} \bar{f}_1(\mathbf{x}) \\ f_2(\mathbf{x}) \\ f_3(\mathbf{x}) \\ f_4(\mathbf{x}) \end{bmatrix} \equiv \begin{bmatrix} K_p(x_{1d} - x_1) + K_v(\dot{x}_{1d} - \dot{x}_1) \\ (\xi_1 x_2 + \xi_2 - \xi_3 x_1)x_3 x_4^2 \\ \alpha e^{\xi_6 x_1^2} + \xi_7 x_3 e^{\xi_6 \xi_8 x_1^2} \\ \dfrac{2\pi\xi_{12}(2\sigma' - \sigma_0)x_4(1 - \pi x_3 x_4^2)[1 - \pi x_3 x_4^2 + 2(\pi x_3 x_4^2)^2]}{(1 - 2\pi x_3 x_4^2)^3} + \\ \xi_{10}\xi_{12} - \dfrac{4\xi_9\xi_{12}}{x_4^5} - 4\pi\xi_{12}\sigma' x_4 - \xi_{12}k_f \ln(k_0 x_1 x_4 + e) \end{bmatrix} \qquad (17.41)$$

Apparently, the first equation in (17.41) is a simple linear equation, while the others remain unchanged as they originally appear in (17.30). With this controller, x_1 can be precisely controlled to match a given value. The corresponding block diagram of this feedback control system is shown in Figure 17.3. In the diagram, the plant represents the gene deliver process, whose dynamics can be described by (17.27) or (17.28). Similarly, as the one shown in Figure 17.2, this control system also consists of two state feedback loops. One loop (on the top portion of the diagram) is to calculate the nonlinear terms $f_1(\mathbf{x})$ and $1/g_1(\mathbf{x})$ of the control input and then feed the two calculated terms into the plant in order to partially cancel the nonlinearity of the plant. The other loop (on the bottom portion of the diagram) is a usual PD servo loop for reducing the error between the current state and the desired state. Again, whether this approach works or not is dependent on whether or not the other three left-alone state variables can behave stably during the course of the control. We performed simulation tests for various scenarios and the simulation results indicated that all the three remaining state variables vary within acceptable range when we actively control only the first state variable. As shown in simulation Case Two described in the next section, their resulting values are not harmful to the cells. This finding tells us that we can at least control one state variable of the system while having the other state variables stably vary in an acceptable range. Although such an approach is not our ultimate goal (i.e., to robustly control all the state

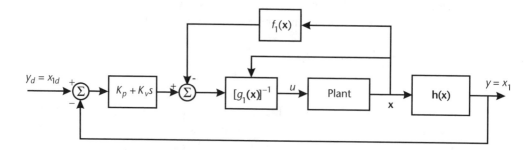

Figure 17.3 Block diagram of the strategy for controlling x_1.

variables), it is a very encouraging first step toward the eventual solution of this very difficult nonlinear control problem.

It should be noted that the above-discussed nonlinear control strategies require state feedback. Measurement or estimation of these state variables can be a very challenging task. To simplify the study at this stage, it is assumed that the values of all the state variables are available as needed. However, practical implementation of the control technology will have to include state estimation, which is another open research topic.

17.4   Simulation Results

The foregoing dynamics model and control strategies have been implemented in a simulation model using MATLAB. MATLAB's ODE solver "ode23tb" and a self-coded second order fixed-step integrator are used for numerical integration. Function ode23tb is a low order ODE solver for solving stiff differential equations. The fixed-step integrator is convenient for debugging the code because of its simple algorithm and open code.

Two simulation examples are presented in this section to illustrate the effectiveness of the above-mentioned control strategies. The first example uses an open-loop control strategy and the second uses the nonlinear feedback control as described in the previous section. All the values of the model parameters used in our simulation study are taken from published references and they are listed in Table 17.1 for quick reference. Simulation results of the two examples are compared and discussed in order to demonstrate the benefits of the control strategy.

17.4.1   Simulation Case One

In this case, the gene delivery process is under open-loop control with the following constant voltage input:

$$u(t) = 0.0137V \tag{17.42}$$

Table 17.1 Physical Variables, Values, and Descriptions

Parameter	Value	Description
D_0	2×10^{-12} m^2/s	DNA diffusion coefficient
R	$10 \, \mu$m	The cell radius
h	5×10^{-9}m	Thickness of cell membrane
V_{cell}	3.14×10^{-16} m^3	Cell volume
C_o	2×10^{-6} mol/m^3	Outside DNA concentrations
z_{eff}	-27	Effective valence of the DNA [20]
q	1.602×10^{-19}C	Elementary charge
k_B	1.381×10^{-23} J/K	The Boltzmann constant
T	310K	The absolute body temperature
g	2 S/m	Conductivity of the solution [15]
f	1.5	Form factor
ΔV_s	1V	Potential threshold
R_0	100Ω	External resistance
R_m	0.523 Ωm^2	Surface resistance
C_m	9.5×10^{-3} F/m^2	Membrane surface capacitance
F_{max}	0.7×10^{-9} N/V^2	Maximum electric force for $V_m = 1$V
r_h	0.97×10^{-9}m	Constant for equation (refforce)
r_t	0.31×10^{-9}m	Constant for equation (refforce)
α	10^9 1/(m^{2}s)	Creation rate coefficient
V_{ep}	0.258V	Electroporation characteristic voltage
N_0	1.5×10^9 1/m^2	Equilibrium pore density at $V_m = 0$
r_p	0.51×10^{-9}m	Minimum radius of hydrophilic pores
r_m	0.8×10^{-11} J/m	Minimum energy radius at $V_m = 0$
D	5×10^{-14} m^2/s	Pore radius diffusion coefficient [10]
β	1.4×10^{-19} J	Steric repulsion energy
γ	1.8×10^{-11} J/m	Edge energy
σ_0	1×10^{-3} J/m^2	Tension of the intact bilayer [6]
σ'	2×10^{-2} J/m^2	Interfacial energy per area of the hydrocarbon-water interface [8]
η	2.46	Electroporation constant
k_f	2×10^{-9}	Coefficient for (17.22)
k_o	1×10^{-10}	Coefficient for (17.22)

Source: [13, 14, 20].

This particular input voltage has been regarded as the most appropriate electrical field applied to cells in the open literature. The initial conditions of the dynamic system for the simulation are assumed to be

$$\mathbf{x}(0) = \begin{bmatrix} V_m(0) \\ C_i(0) \\ N(0) \\ r_j(0) \end{bmatrix} = \begin{bmatrix} 1.4 \times 10^{-3} \text{ V} \\ 1.0 \times 10^{-9} \text{ mol} / \text{m}^2 \\ 1.5 \times 10^{9} \text{ 1} / \text{m}^2 \\ 0.6 \times 10^{-9} \text{ m} \end{bmatrix} \quad (17.43)$$

Simulation output results are plotted in Figures 17.4 through 17.10. Figure 17.4 shows the control input, which is apparently a constant, as defined by (17.42). Figures 17.5 through 17.8 plot the time histories of the four state variables defined in (17.29), respectively. As we can see, the transmembrane potential V_m dropped to about 0.008V from its initial value in about 1 μs. The DNA concentration inside the cell C_i continuously dropped in the process. The number of pores per cell quickly increased to 10^5 in the very beginning of the process and then was kept in that density throughout the remaining time period of the process. The pore quickly opened to the maximum value of about 19 nm within about 1 μs and then kept that size throughout the remaining period of the process. Figures 17.9 and 17.10 plot energy versus pore radius and the time history of the pore electrical force, respectively.

It should be pointed out that, although the simulation demonstrated that the system can satisfactorily respond to an open-loop control signal (i.e., the control input u), the result cannot be guaranteed when disturbances present in the system because no feedback control is applied. In other words, the dynamic response seen in the plots is very sensitive to various disturbances to the system and thus cannot be guaranteed if any condition of the system changes.

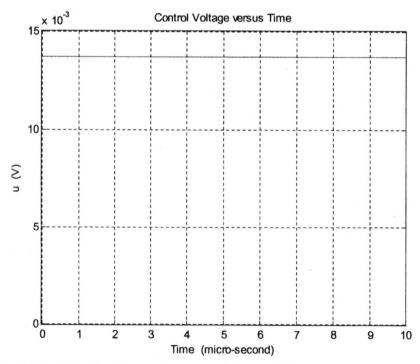

Figure 17.4 Control input of the open-loop controlled gene delivery process.

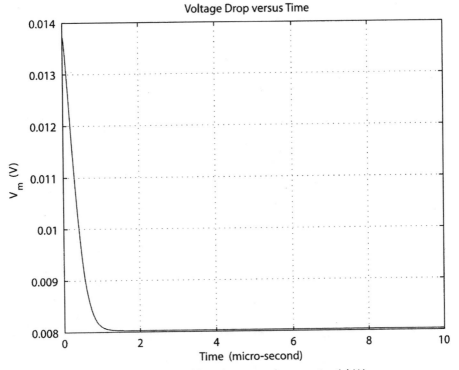

Figure 17.5 Time response of state variable x_1 (transmembrane potential V_m).

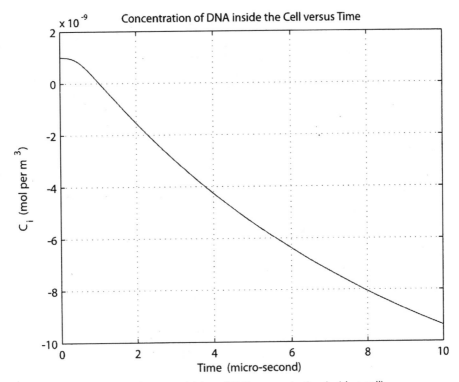

Figure 17.6 Time response of state variable x_2 (DNA concentration inside a cell).

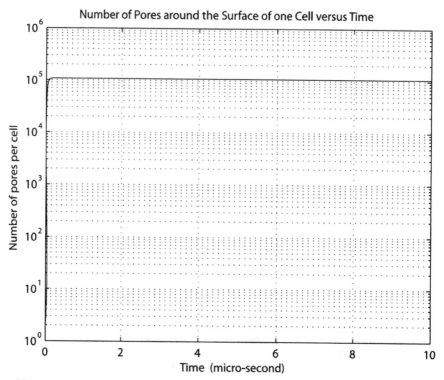

Figure 17.7   Time response of state variable x_3 (number of pores around suffice of one cell).

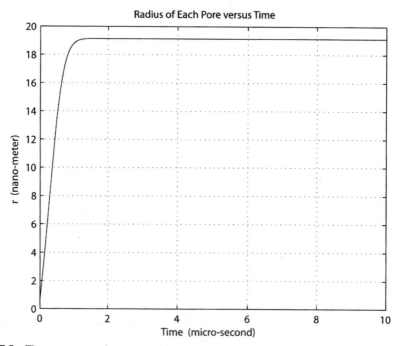

Figure 17.8   Time response of state variable x_4 (radius of a pore).

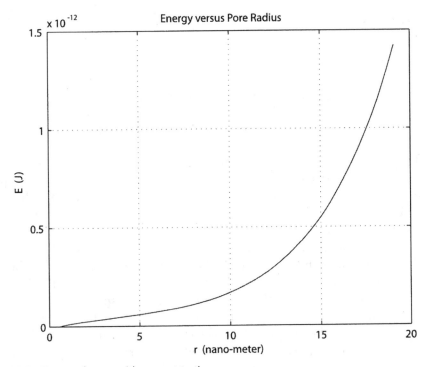

Figure 17.9   Energy change with respect to time.

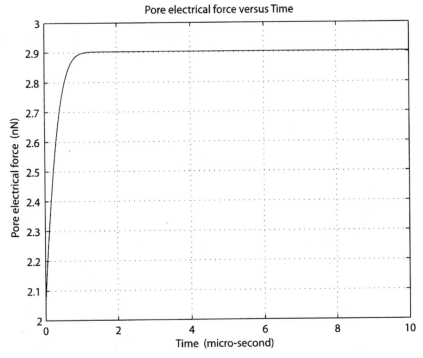

Figure 17.10   Electrical force with respect to time.

17.4 2   Simulation Case Two

In this case, the state feedback control law proposed in (17.39) is applied to the electroporation process. The control goal is to regulate the first state variable x_1 (i.e., V_m) while having other state variables stably settled to acceptable values. The desired output of the control system is set to

$$y_d = x_{1d} = 0.008\text{V},\ \dot{y}_d = \dot{x}_{1d} = 0 \text{ V}/\text{s} \tag{17.44}$$

The initial conditions of the system are exactly the same as those for Case One, given by (17.43). The two control gains are set to $K_p = 5 \times 10^6$ and $K_v = 0.6$, respectively. In this case, the criterion for tuning the two control gains is to make the three uncontrolled state variables as close to those in Case One as possible while keeping the control system stable.

Simulation results are plotted in Figures 17.11 through 17.17. Figure 17.11 shows the control input which was high in the beginning and then became nearly constant after the system had settled to its steady state. Figures 17.12 through 17.15 plot the time histories of the four state variables, respectively. As one can see, the transmembrane potential V_m monotonically dropped to the desired value of 0.008V from its initial value in about $2\,\mu$s. The DNA concentration inside the cell C_i continuously dropped in the process. The number of pores per cell quickly reached about 10^5 in the beginning and then kept that density level throughout the rest of the process. The radius of the pore opened to the maximum value of about 19 nm within nearly $2\,\mu$s and then kept that size to the end of the process. Figures 17.16 and 17.17

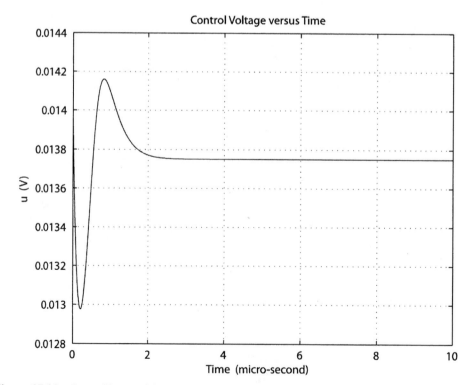

Figure 17.11   Control input of the open-loop controlled gene delivery process.

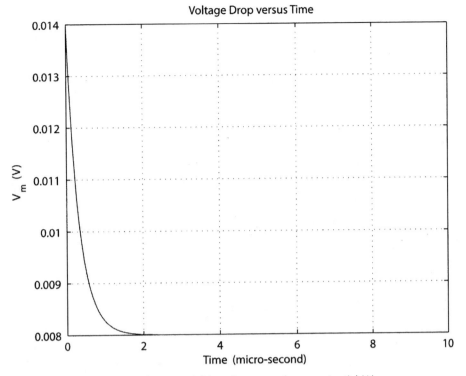

Figure 17.12   Time response of state variable x_1 (transmembrane potential V_m).

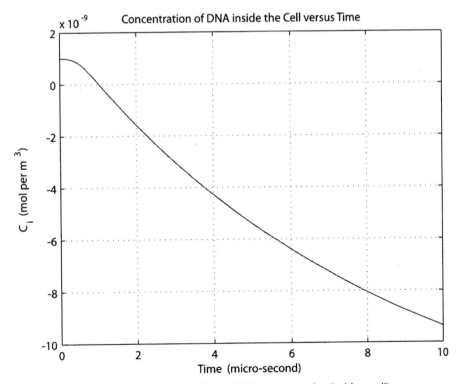

Figure 17.13   Time response of state variable x_2 (DNA concentration inside a cell).

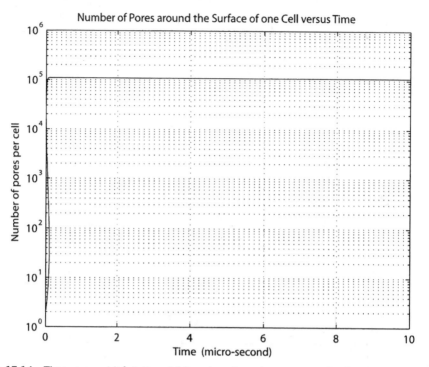

Figure 17.14   Time response of state variable x_3 (number of pores around suffice of one cell).

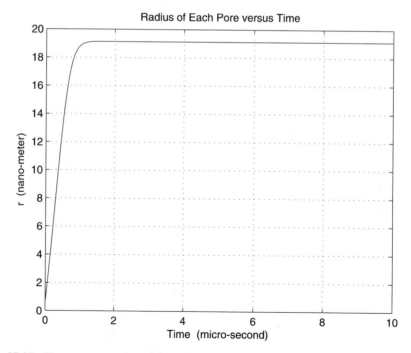

Figure 17.15   Time response of state variable x_4 (radius of a pore).

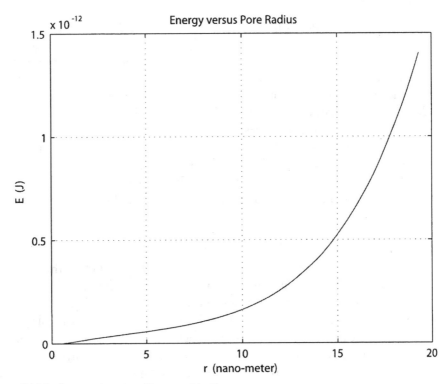

Figure 17.16   Energy change with respect to time.

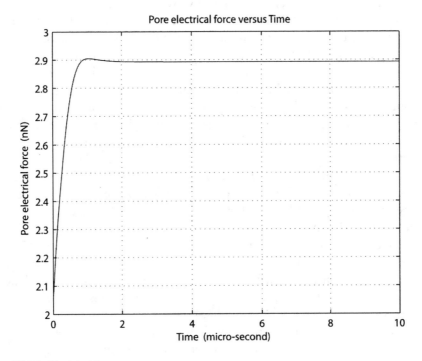

Figure 17.17   Electrical force.

plot energy versus pore radius and the time history of the pore electrical force. Note that, since the closed-loop system for V_m has become a second order linear system after the linearization, its response can be easily tuned to show certain required characteristics. For example, an overshoot can be eliminated by increasing the derivative gain K_v and the settling time can be shortened by increasing the proportional gain K_p, but higher proportional gain will reduce stability margins as this is a feedback control system. It is very interesting to notice that, although only the first state variable is actively controlled, all the other state variables also behave stably within acceptable ranges (close to the values in Case One). The steady-state values of these state variables can also be tuned to other desired levels by adjusting the two control gains while still keeping the first state variable locked in the same desired value.

17.4.3   Discussion of the Two Case Studies

The immediate difference between Case One and Case Two is that the former is an open-loop control system and thus we cannot predefine a desired output by any means. If we need to achieve a desired output, a trial-and-error method has to be used to adjust the input voltage u. Case Two is a closed-loop control system and thus we can define a desired output value, which is x_1 in this case, and guarantee that the controlled output will reach and stay in that value. In addition, the two control gains can also be used to adjust the values of the other uncontrolled state variables as long as the closed-loop system is kept stable. This is why we can see that the simulation outputs of the two cases presented in the previous two subsections are close because we have adjusted the control target and control gains to make the dynamics response of Case Two close to that of Case One. This suggests that we can achieve certain control goals even when we actively control only the first state variable as opposed to controlling all four state variables. In other words, although the proposed nonlinear control strategy only partially controls the state variables, it still can provide a good means for an approximate "control" of the entire system.

Another difference between Case One and Case Two is that, as an open-loop control, the output of the Case One is very sensitive to any disturbances (such as initial conditions, parameter variations, unmodeled factors), while the output of the Case Two is more robust to uncertainties. For instance, if we add a 5% error to the initial conditions given in (17.43), all four state variables will settle in different steady-state values in Case One (see Figures 17.18 and 17.19, for example), but only the three uncontrolled states will settle in different steady-state values in Case Two (see Figures 17.20 and 17.21, for example). If we add 10% error in the initial conditions, Case One becomes unstable but Case Two is still stable (although the three uncontrolled state variables further deviate), as shown in Figures 17.20 and 17.21. Especially, the controlled output, x_1, is almost unaffected by the fluctuations in the initial conditions. This demonstrated that the application of feedback control is greatly beneficial in improving the stability and robustness of the gene delivery process.

Of course, the output of Case Two is certainly not as robust as if we actively control all four state variables in the process. Unfortunately, as discussed earlier, the problem of full control of such a complicated nonlinear process is a very difficult and unsolved problem. Nevertheless, the approach proposed in this chapter

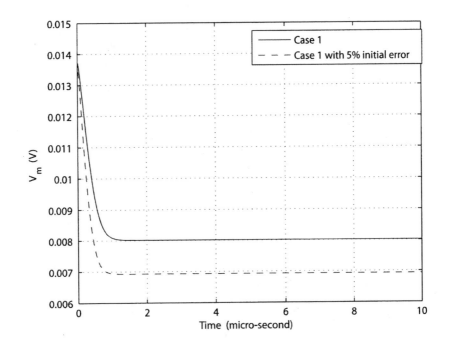

Figure 17.18 Variations of x_1 in Case One after a 5% change in the initial conditions. (The system becomes unstable with a 10% change of initial conditions and is thus not plotted here.)

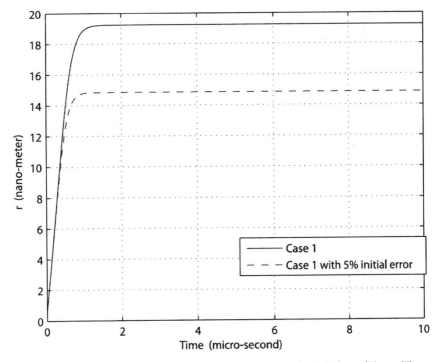

Figure 17.19 Variations of x_4 in Case One after a 5% change in the initial conditions. (The system becomes unstable with a 10% change of initial conditions and is thus not plotted here.)

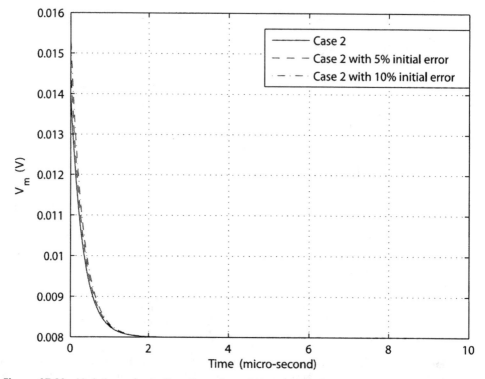

Figure 17.20   Variations of x_1 in Case Two after a 5% and 10% changes in all initial conditions.

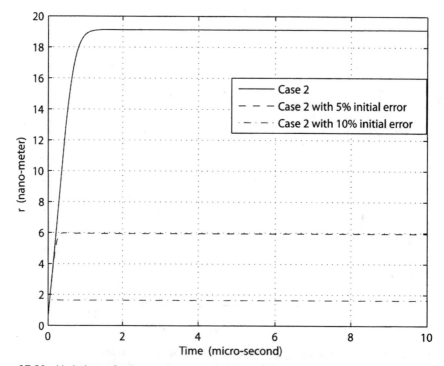

Figure 17.21   Variations of x_4 in Case Two after a 5% and 10% changes in all initial conditions.

provides a good starting point for a complete solution to this difficult nonlinear control problem.

17.5 Discussions and Conclusions

This paper proposes an integrated chemo-electro-mechanical dynamics model for the electroporation-mediated gene delivery process and investigated feedback control of the gene delivery process. The model is specific in three aspects. First, it is control enabled. By that, we mean the model is presented in the canonical control form and is in a form that is computationally more efficient (a property required by implementing model-based control strategies). The modified model can be readily used for designing and testing different control strategies for gene delivery process. Second, the model has been revised to take consideration of limitations for different models in the open literature, especially time-varying effects, which makes the model more general in nature. The electrical field is applied explicitly to a single cell rather than the whole solution. We believe that the model opens some interesting discussions regarding the dynamics model of gene delivery. Finally, some points still missing in the current open literature are addressed in this chapter, including the modification of voltage applied to a single cell, the change of sample solution volume, and the approximation of exponential functions with polynomial.

The second part of this chapter presented an investigation of the control of the gene delivery process. Model based nonlinear feedback control strategies are discussed. It was shown that, due to the high nonlinearity of the system plus the special structure of the dynamics model, a complete control of all the state variables is extremely difficult. Our main idea is to use the proposed dynamics model to partially linearize the nonlinear plant dynamics and then apply servo control techniques to control the linearized system. Simulation study has demonstrated that the proposed strategy can satisfyingly control one state variable while having all other state variables stably settled in acceptable ranges. Since the proposed control strategy relies on state feedback, it has been assumed that the system is observable.

Our analysis and simulations indicate that the electroporation-mediated process is a highly nonlinear dynamic process. The dynamics is quite complicated and sensitive to electrical field. It is difficult to control both the radius and concentration at the same time. This is due to the natural structure of the dynamic system. We proposed an idea to actively control the transmemberance potential through which the radius of pores can be indirectly controlled. Simulation results demonstrated the effectiveness of this control strategy. This is an encouraging first step toward a complete control of such a difficult nonlinear process.

References

[1]   Cima, L. F., and L. M. Mir, "Macroscopic Characterization of Cell Electroporation in Biological Tissue Based on Electrical Measurements," *Applied Physics Letters*, Vol. 85, No. 19, 2004, pp. 4520–4522.

[2]   DeBruin, K. A., and W. Krassowska, "Modeling Electroporation in a Single Cell I: Effects of Fields Strength and Rest Potential," *Biophysical Journal*, Vol. 77, September 1999, pp. 1213–1224.

[3]   DeBruin, K. A., and W. Krassowska, "Modeling Electroporation in a Single Cell II: Effects of Ionic Concentration," *Biophysical Journal*, Vol. 77, September 1999, pp. 1225–1233.

[4]   Bier, M., et al., "Electroporation of a Lipid Bilayer as a Chemical Reaction," *Bioelectromagnetics*, Vol. 25, 2004, pp. 634–637.

[5]   Diaz-Rivera, R. E., and B. Rubinsky, "Electrical and Thermal Characterization of Nanochannels Between a Cell and a Silicon Based Micro-Pore," *Biomed. Microdevices*, Vol. 8, 2006, pp. 25–34.

[6]   Freeman, S. A., M. A. Wang and J. C. Weaver, "Theory of Electroporation of Planar Bilayer Membranes: Predictions of the Aqueous Area, Change in Capacitance, and Pore-Pore Separation," *Biophys. J.*, Vol. 67, 1994, pp. 42–56.

[7]   Gehl, J., "Electroporation: Theory and Methods, Perspectives for Drug Delivery, Gene Therapy And Research," *Acta Physical Scand*, Vol. 177, 2003, pp. 437–447.

[8]   Israelachvili, J., *Intermolecular and Surface Forces*, New York: Academic Press, 1985.

[9]   Jayaram, S. H., and S. A. Boggs, "Optimization of Electroporation Waveforms for Cell Sterilization," *IEEE Transactions on Industrial Applications*, Vol. 40, No. 6, 2004, pp. 1489–1497.

[10]   Joshi, R. P., Q. Hu and K. H. Schoenbach, "Dynamical Modeling of Cellular Response to Short-Duration, High-Intensity Electrical Effects," *IEEE Transactions on Dielectrics and Electrical Insulation*, Vol. 10, No. 5, October 2003, pp. 778–787.

[11]   Joshi, R. P., et al., "Self-Consisting Simulations of Electroporation Dynamics in Biological Cells Subjected to Ultrafast Electrical Pulses," *Phys. Review E*, Vol. 64, 2001, pp. 11913-01–11913-13.

[12]   Lister, J. D., "Stability of Lipid Bilayers and Red Blood Cell Membranes," *Phys. Lett. A*, Vol. 53, 1975, pp. 193–194.

[13]   Neu, J. C., and W. Krassiwska, "Asymptotic Model of Electroporation," *Phys. Rev. E.*, Vol. 59, 1999, pp. 3471–3482.

[14]   Neu, J. C., and W. Krassiwska, "Modeling Postshock Evolution of Large Electropores," *Phys. Rev. E.*, Vol. 67, No. 021915, 2003.

[15]   Neu, J. C., K. C. Smith and W. Krassiwska, "Electrical Energy Required to Form Large Conducting Pores," *Bioelectrochemistry*, Vol. 60, 2003, pp. 107–114.

[16]   Neumann, E., et al., "Calcium-Mediated DNA Adsorption to Yeast Cells and Kinetics of Cell Transformation by Electroporation," *Biophysical Journal*, Vol. 71, 1996, pp. 868–877.

[17]   Pawlowski, P., et al., "Electrorheological Modeling of the Permeabilization of the Stratum Corneum: Theory and Experiment," *Biophysical Journal*, Vol. 75, December 1998, pp. 2721–2731.

[18]   Rols, M. P., and J. Teissie, "Electropermeabilization of Mammalian Cells to Macromolecules: Control by Pulse Duration," *Biophyical Journal*, Vol. 74, 1998, pp. 1415–1423.

[19]   Rybenkov, V. V., A. V. Vologodskii and N. R. Cozzarelli, "The Effect of Ionic Conditions on the Conformations of Supercoiled DNA I: Sedimentation Analysis," *J. Mol. Biol.*, Vol. 267, 1997, pp. 299–311.

[20]   Smith, K. C., J. C. Neu and W. Krassowska, "Model of Creation and Evolution of State Electropores for DNA Delivery," *Biophysical Journal*, Vol. 86, May 2004, pp. 2813–2826.

[21]   Tarek, M., "Membrane Electroporation: A Molecular Dynamics Simulation," *Biophysical Journal*, Vol. 88, June 2005, pp. 4045–4053.

[22]   Taupin, C., M. Dvolaitzky and C. Sauterey, "Osmotic Pressure Induced Pores in Phospholipid Vesicles," *Biochem.*, Vol. 14, 1975, pp. 4771–4775.

[23]   Weaver, J. C., "Electroporation of Cells and Tissues," *IEEE Transactions on Plasma Science*, Vol. 28, No. 1, 2000, pp. 24–33.

[24] Weaver, J. C., and Y. A. Chizmadzhev, "Theory of Electroporation: A Review," *Bioelectrochem. Bioenerg.*, Vol. 41, 1999, pp. 135–160.

[25] Sontag, E. D., *Mathematical Control Theory*, 2nd ed., New York: Springer, 1998.

CHAPTER 18

Nanotechnology for Advanced Life Science Automation

Lixin Dong and Bradley J. Nelson[†]

This chapter overviews the state of the art of the advancements of nanoimaging, nanofabrication, nanomanipulation, and nanoelectromechanical systems (NEMS), and focuses on their applications and potential applications in advanced life science automation.

18.1   Introduction

As the twenty-first century unfolds, the impact of nanotechnology on the health, wealth, and security of humankind is expected to be at least as significant as the combined influences in the twentieth century of antibiotics, the integrated circuit, and human-made polymers. N. Lane stated in 1998, "If I were asked for an area of science and engineering that will most likely produce the breakthroughs of tomorrow, I would point to nanoscale science and engineering" [1]. The great scientific and technological opportunities that nanotechnology present have stimulated extensive exploration of the "nano" world and initiated an exciting worldwide competition. The advancements of nanoinstrumentation, nanofabrication, nanomanipulation, and NEMS will enable many new possibilities in advanced life science automation.

By the early 1980s, scanning tunneling microscopes (STMs) [2] radically changed the way in which we interacted with and even regarded single atoms and molecules. The very nature of proximal probe methods encourages the exploration of the nanoworld beyond conventional microscopic imaging. Scanning probes now allow us to perform "engineering" operations on single molecules, atoms, and bonds, thereby providing a tool that operates at the ultimate limits of fabrication. They have also enabled exploration of molecular properties on an individual nonstatistical basis.

Furthermore, a variety of new nanomaterials and structures have been synthesized or fabricated. Quantum dots (QDs) [3], fullerenes, carbon nanotubes (CNTs) [4], nanowires [5], nanobelts [6], and helical nanostructures provide a new family of building blocks for many potential applications, especially in nanoelectronics [7],

†   Institute of Robotics and Intelligent Systems, ETH Zurich, Switzerland

NEMS, and other nanodevices [8]. These devices will in turn enable new exploring tools for life science. QDs are semiconductor nanocrystals with a diameter in the range of a few nanometers, and as a result, their energy levels can no longer be treated as continuous but become quantized with values directly related to the size of the QD. This leads to broadband absorption and to emission of light in a fixed and very narrow spectrum with intensities and stability unmatched by other fluorophores. The well-defined geometries, exceptional mechanical properties, and extraordinary electric characteristics, among other outstanding physical properties, of CNTs [4] qualify them for many such applications. For NEMS, some of the most important characteristics of nanotubes include their nanometer diameter, large aspect ratio (10 to 1,000), TPa scale Young's modulus [9], excellent elasticity [10], ultra-small interlayer friction, excellent capability for field emission, various electric conductivities, high thermal conductivity, high current carrying capability with essentially no heating, sensitivity of conductance to various physical or chemical changes, and charge-induced bond-length change. Helical 3D nanostructures have been synthesized from different materials including helical carbon nanotubes [11] and zinc oxide nanobelts [6]. A new method of creating structures with nanometer-scale dimensions has recently been presented [12] and can be fabricated in a controllable way [13]. The structures are created through a top-down fabrication process in which a strained nanometer thick heteroepitaxial bilayer curls up to form 3D structures with nanoscale features [14, 15]. Helical geometries and tubes with diameters between 10 nm and 10 μm have been achieved. Because of their interesting morphology, mechanical, electrical, and electromagnetic properties, potential applications of these nanostructures in NEMS include nanosprings, electromechanical sensors, magnetic field detectors, chemical or biological sensors, generators of magnetic beams, inductors, actuators, and high-performance electromagnetic wave absorbers.

Progress in robotics over the past years has dramatically extended our ability to explore the world from perception, cognition, and manipulation perspectives at a variety of scales, extending from the edges of the solar system, to the bottom of the sea, down to individual atoms. At the lower end of this scale, technology has been moving toward greater control of the structure of matter, suggesting the feasibility of achieving thorough control of the molecular structure of matter atom by atom, as Richard Feynman first proposed in 1959 in his prophetic article on miniaturization [16]: "What I want to talk about is the problem of manipulating and controlling things on a small scale...I am not afraid to consider the final question as to whether, ultimately—in the great future—we can arrange the atoms the way we want: the very atoms, all the way down!" He asserted that "At the atomic level, we have new kinds of forces and new kinds of possibilities, new kinds of effects. The problems of manufacture and reproduction of materials will be quite different. The principles of physics, as far as I can see, do not speak against the possibility of maneuvering things atom by atom." The "great future" of Feynman began to be realized in the 1980s. Some of the capabilities he dreamed of have been demonstrated, while others are being actively pursued.

Nanorobotics represents the next stage in miniaturization for maneuvering nanoscale objects. Nanorobotics is the study of robotics at the nanometer scale, and includes robots that are nanoscale in size (i.e., nanorobots) and large robots capable of manipulating objects that have nanometer dimensions with nanometer resolution (i.e., nanorobotic manipulators). The field of nanorobotics brings together several disciplines, including nanofabrication processes used for producing nanoscale robots, nanoactuators, nanosensors, and physical modeling at nanoscales. Nanorobotic manipulation technologies, including the assembly of nanometer-sized parts, the manipulation of biological cells or molecules, and the types of robots used to perform these types of tasks, also form a component of nanorobotics. Nanorobotics will play a significant role as an enabling nanotechnology and could ultimately be a core part of nanotechnology if Drexler's machine-phase nanosystems based on self-replicative molecular assemblers via mechanosynthesis can be realized [17].

18.2 Nanoimaging in an Aqueous Environment

New imaging tools and approaches used in the research area of life science have allowed for new applications that go beyond conventional optical microscopy, electron microscopy, and scanning probe microscopy. Examples include fluorescence, scanning near-field optical microscopy (SNOM), environmental scanning electron microscopy (ESEM), and atomic force microscopy (AFM). Here we show the enhancement of fluorescence microscopy by using quantum dots.

The environment in a biological cell or the human body is aqueous and as such has requirements quite different from the high-vacuum conditions required by electron microscopy. Real-time observation must take place while avoiding the damaging side effects of the electron beam.

The work presented here is part of our effort to make our fabrication methods and observation strategies more compatible with the conditions to be expected in real-world environments, such as for biomedical applications. We require methods for interfacing nanodevices with other synthetic building blocks as well as with natural biomolecules such as biomolecular motors in order to build advanced hybrid transducers and NEMS.

The attachment of QDs onto CNTs will create the possibility of assessing the position and orientation of individual decorated CNTs in a liquid environment as well as the tracking of motion [e.g., for drug delivery and during dielectrophobic (DEP) assembly]. The morphological changes of transducers such as telescoping multiwalled carbon nanotubes (MWNTs) could be visualized in real time without needing high-vacuum electron microscopy. At the same time, the high selectivity of functionalized QDs to chemically active sites can be used as a means to indicate the active sites (such as the defect sites on CNTs) and to facilitate electron microscopy with the identification of covalent bonds. Furthermore, the knowledge of the formation and the selective exploitation of attachment sites enable the assembly of more complex devices as well as the "debugging" of the intermediate stages in the process.

Covalent attachment of QDs to CNTs has been previously achieved. This prior work has been concerned mainly with the tuning of the electronic properties of carbon nanotubes in order to exploit them as heterojunctions in combination with semiconducting materials. Haremza et al. reported the conjugation of single amine-terminated QDs to the sidewalls of single-walled carbon nanotubes (SWNTs) and only rarely to the ends of shorter SWNTs (i.e., with lengths < 200 nm) [18]. The bulk material was characterized with IR and UV-visual spectroscopy and the individual tubes with AFM and transmission electron microscopy (TEM). No optical microscopy was used to observe or characterize the single tubes. Banerjee and Wong conjugated amine terminated QDs to SWNTs to form heterostructures [19]. TEM and IR spectroscopy were used to characterize single tubes and bulk material, respectively. Ravindran et al. reported the attachment of QDs to MWNTs, using SEM and TEM to characterize individual tubes [20]. However, it seems that no characterization using optical microscopy has been used.

Our work confirms and continues previous efforts and offers further evidence to support assumptions and claims regarding covalent attachment of functionalized QDs. In addition, we show high resolution transmission electron microscope (HRTEM) images of clusters of QDs on MWNTs, and we used optical and confocal laser scanning microscopy (CLSM) to prove the feasibility of using QDs for locating and tracking particular sites on nanodevices and building blocks for fabrication processes and applications other than nanoelectronics. Furthermore, we are interested in gentle, practical, and low-cost methods to work with small quantities when prototyping new devices.

QDs are commercially available with various types of coating to achieve even higher chemical stability, nontoxicity, and water solubility. Functional end groups allow for specific chemical attachment to other objects, such as CNTs.

As opposed to common fluorophores used in life sciences, core-shell QDs (such as the CdSe/ZnS QDs used in this study) are very stable and do not seem to bleach after hours of constant illumination [21]. They have been successfully tracked with a CLSM for up to a few hours [22]. Long fluorescence lifetimes enable the use of time-gated detection and suppression of background fluorescence and noise [23]. Another phenomenon which can be exploited for the identification of single QDs is *blinking*: surface defects in the crystal structure act as temporary "traps" for electron or hole preventing their radiative recombination. Alternation of trapping and untrapping events results in intermittent fluorescence. As a result, single QDs have been successfully observed with a lateral resolution of about 40 nm and their position has been determined with a time resolution of up to 10 ms [24], effectively beating the best optical resolution of about 250 nm by an order of magnitude. Optical diffraction pattern and tracking algorithms for tracking single fluorescent particles can be used to theoretically locate the center of a point source to within approximately 5 nm [25]. Dahan et al. claimed to have tracked single dots for at least 20 minutes with a signal-to-noise ratio (SNR) of around 50 (at 75-ms integration time) and with an even higher resolution of 5 to 10 nm [26]. When using two colors (i.e., when different types of QDs for different sites are possible), the resolution can be even higher and approach a few nanometers.

As in previous studies, the attachment of functionalized QDs to MWNTs has been successfully achieved and reproduced in multiple experiments with variations of the low-tech and low-cost method described. The QDs indeed form distinct and dense clusters as expected. Hence, this suggests that the clusters tend to form around defect sites on the tube walls and at the opened caps (Figure 18.1), which is desired for selective functionalization of telescoping structures.

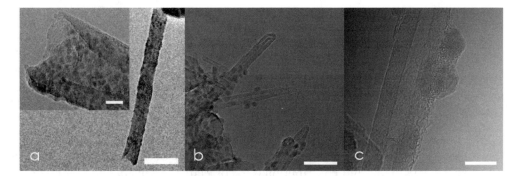

Figure 18.1   (a) TEM images showing an MWNT with an open cap and a characteristic inward cone. The inset shows the tip in more detail, particularly the same sized and roughly spherical CdSe/ZnS QDs with an average crystal diameter of 5.2 nm (Evident Technologies, New York). Scale bars are 200 and 20 nm. We used gold wires as sample substrates for TEM observation. For the attachment of the sample powder we used conductive silver paint or carbon cement. The complete samples were then suspended for about 25 minutes in a solution of amine-functionalized QDs. Dicyclohexyl carbodiimide (DCC) was added as a reagent to aid in the formation of covalent amid bonds between functional amine groups on the QD surface and the carboxylic acid end groups on the tube wall surface. The attachment phase was followed by a gentle but extensive washing phase. The samples were suspended in about 20 ml of constantly stirred deionized water with a small amount of surfactant in order to remove loose particles and unbonded QDs and side products. (b, c) HRTEM images of MWNTs (MER Corp., Tucson, Arizona), with QDs 8 × 15 nm CdSe/ZnS crystals (Qdot, California) on the tube wall. (b) Tubes protruding from a mass of entangled tubes (scale bar is 50 nm). Note the closed caps. (c) Close-up at atomic resolution with tube walls and crystal lattices clearly visible (scale bars are 10 nm).

QDs with water soluble PEG coating are highly unlikely to resist the concentration gradient during washing and remain on the predominantly nonpolar surface of carbon nanotubes. It is likely that most of these QDs form covalent bonds as opposed to attaching only due to van der Waals interaction. The nanocrystals have been confidently identified with atomic resolution TEM imaging showing the characteristic crystal lattice (Figure 18.1).

Densely decorated tubes are highly fluorescent. Decoration density depends on the purity and quality of the bulk material, the subsequent acid treatment, and the ratio of QDs to CNTs during conjugation. Untreated tubes remain mostly undecorated and nonfluorescent while the QDs either stay in suspension or attach to the more reactive amorphous particles remaining in the sample. Experiments of various levels of decoration density depending on the intensity of acid treatment and as a result of the degree of oxidization are in further support of this assumption (images not shown). The QDs used in this study are photostable under intense laser illumination during hours of observation and provided constant and strong emission.

The presence of conjugated QDs has been demonstrated in earlier studies with HRTEM. Here we show TEM images, but for the first time combine these images with optical microscopy to correlate QD presence with single-tube fluorescence. Multimode imaging of the identical tube has been a greater challenge than anticipated. General difficulties during preparation included the boundary conditions for sample preparation, as mentioned in the materials and methods section. Specimens have been repeatedly rendered nonfluorescent when exposed to the high-vacuum required for electron microscopy.

Chemical and physical decay, such as micelle degradation and bleaching effects, are likely, but they are further complicated by the possibly destructive nature of the electron beam and electron beam induced deposition (EBID). The darkening was particularly true for regions where the beam had been focused, which further supports the explanation that the destruction was most likely the result of a thin coating induced by the electron beam. The coating seems to effectively shield the QDs.

Recently, we have achieved the multimode imaging of individual tubes [27]. In our work we present concepts and solutions that focus on applications using gentle, low-cost protocols for selective treatment and the observation of the result by optical means. Thus, our contribution is to correlate the density of quantum dot decoration on individual multiwalled carbon nanotubes with their actual fluorescence for the first time using optical and high resolution electron microscopy. The protocols for preparation and separation of bulk material are simpler than those previously reported, lower cost, and more practical for device prototyping using small quantities and single particles.

18.3   Nanofabrication

The design and fabrication of NEMS is an emerging area being pursued by an increasing number of researchers. Two approaches to nanofabrication—top-down and bottom-up—have been identified by the nanotechnology research community and are being independently investigated by various researchers. Top-down approaches are based on microfabrication and include technologies such as nanolithography, nanoimprinting, and chemical etching. Presently, these are 2D fabrication processes with relatively low resolution. Bottom-up strategies are assembly-based techniques. Currently these strategies include techniques such as self-assembly, dip-pen lithography, and directed self-assembly. These techniques can generate regular nanopatterns at large scales.

In this section, we will introduce the applications of two major nanofabrication techniques: one is e-beam lithography, and the other is epitaxy and strain engineering for creating nanostructures.

E-beam lithography is an attractive alternative technique for fabricating nanostructures [28]. It uses an electron beam to expose an electron-sensitive resist such as polymethyl methacrylate (PMMA) dissolved in trichlorobenzene (positive) or poly chloromethylstyrene (negative). E-beam lithography has been used to define Ni catalyst dots [Figure 18.2(a)] and nanoelectrodes [Figure 18.2(b)] for direct growth of CNTs and dielectrophoretic assembly of CNTs enabling the integration of CNTs in either vertical or horizontal direction [29].

Atomic precision deposition techniques such as molecular beam epitaxy (MBE) and ultrahigh-vacuum chemical vapor deposition (UHV-CVD) have proven to be effective tools in fabricating a variety of quantum confinement structures and devices.

Three-dimensional helical structures are created through these top-down fabrication processes in which a strained nanometer-thick heteroepitaxial bilayer curls up [Figure 18.3(a)] to form 3D structures with nanoscale features such as SiGe/Si tubes [Figure 18.3(b); diameters between 10 nm and 10 μm], Si/Cr ring [Figure

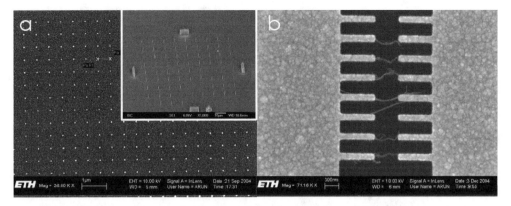

Figure 18.2 (a) Ni catalyst nanodots (50 nm) defined using e-beam lithography. Inset shows arrays of individual MWNT grown from them. (b) Nanoelectrodes fabricated with e-beam lithography for dielectrophoretic assembly of nanotubes. Results from composite ac-dc field electrophoresis were carried out with arrays of 100-nm nanoelectrodes located on larger electrodes. It can be seen that nanotubes are placed only across the nanoelectrode gaps (100-nm-wide electrode array with 500-nm spacing; deposition time: 4 minutes).

18.3(c)], SiGe/Si coils [13] [Figure 18.3(d)], InGaAs/GaAs coils [14] [Figure 18.3(e)], small-pitch InGaAs/GaAs coil [Figure 18.3(f)], Si/Cr claws [Figure 18.3(g)], Si/Cr spirals [30] [Figure 18.3(h)], and small-pitch SiGe/Si/Cr coils [15] [Figure 18.3(i)].

18.4   Nanorobotic Manipulation

Nanorobotics represents the next stage in miniaturization for maneuvering nanoscale objects. Nanorobotics is the study of robotics at the nanometer scale, and this includes robots that are nanoscale in size (i.e., nanorobots) and large robots capable of manipulating objects that have dimensions in the nanoscale range with nanometer resolution (i.e., nanorobotic manipulators). Robotic manipulation at nanometer scales is a promising technology for structuring, characterizing, and assembling nano-building blocks into NEMS. Combined with recently developed nanofabrication processes, a hybrid approach is realized for building NEMS and other nanorobotic devices from individual carbon nanotubes and SiGe/Si nanocoils. Material science, biotechnology, electronics, and mechanical sensing and actuation will benefit from advances in nanorobotics.

Nanomanipulation, or positional and/or force control at the nanometer scale, is a key enabling technology for nanotechnology. It fills the gap between top-down and bottom-up strategies, and may lead to the appearance of replication-based molecular assemblers [17]. These types of assemblers have been proposed as general purpose manufacturing devices for building a wide range of useful products as well as copies of themselves (self-replication).

Presently, nanomanipulation can be applied to the scientific exploration of mesoscopic physical phenomena, biology, and the construction of prototype nanodevices. It is a fundamental technology for property characterization of

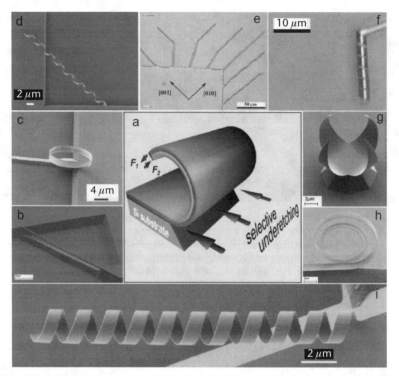

Figure 18.3 3D helical nanostructures. (a) Schematic diagram of rolled-up helical structures. (b) SiGe/Si tube. (*Source:* [13].) (c) Si/Cr ring. (d) SiGe/Si coil. (*Source:* [13].) (e) InGaAs/GaAs coils. (*Source:* [14].) (f) Small-pitch InGaAs/GaAs coil. (g) Si/Cr claws. (h) Si/Cr spiral. (*Source:* [30].) (i) Small-pitch SiGe/Si/Cr coil. (*Source:* [15].)

nanomaterials, structures, and mechanisms, for the preparation of nano-building blocks, and for the assembly of nanodevices such as NEMS.

Nanomanipulation was enabled by the inventions of STMs [2], AFMs [31], and other types of SPMs. Besides these, optical tweezers (laser trapping) [32] and magnetic tweezers [33] are also possible nanomanipulators. Nanorobotic manipulators (NRMs) [34, 35] are characterized by the capability of 3D positioning, orientation control, independently actuated multiple end-effectors, and independent real-time observation systems, and they can be integrated with scanning probe microscopes. NRMs largely extend the complexity of nanomanipulation.

An STM can be applied to particles as small as atoms with atomic resolution. However, limited by its 2D positioning and available strategies for manipulations, standard STMs are ill-suited for complex manipulation and cannot be used in 3D space. An AFM is another important type of nanomanipulator. There are three imaging modes for AFMs: contact mode, tapping mode (periodic contact mode), and noncontact mode. The later two are also called dynamic modes and can attain higher imaging resolution than contact mode. Atomic resolution is obtainable with noncontact mode. Manipulation with an AFM can be done in either contact or dynamic mode. Generally, manipulation with an AFM involves moving an object by touching it with a tip. A typical manipulation can occur as follows: image a particle first in noncontact mode, then remove the tip oscillation voltage and sweep the tip across the particle in contact with the surface and with the feedback disabled. Mechanical pushing can exert larger forces on objects and, hence, can be applied for the manipulation of relatively larger objects. Nanosized objects can be manipulated on a 2D substrate. However, the manipulation of individual atoms with an AFM remains a challenge. By separating the imaging and manipulation functions, nanorobotic manipulators can have many more degrees-of-freedom, including rotation for orientation control, and, hence, can be used for the manipulation of 0D (symmetric spheres) to 3D objects in 3D free space. Limited by the relative lower resolution of electron microscopes, NRMs are difficult to use for the manipulation of atoms. However, their general robotic capabilities, including 3D positioning, orientation control, independently actuated multiple end-effectors, separate real-time observation system, and integrations with SPMs inside, make NRMs quite promising for complex nanomanipulation.

The first nanomanipulation experiment was performed by Eigler and Schweizer in 1990 [36]. They used an STM and materials at low temperatures (4K) to position individual xenon atoms on a single-crystal nickel surface with atomic precision. The manipulation enabled them to fabricate rudimentary structures of their own design, atom by atom. The result is the famous set of images showing how 35 atoms were moved to form the three-letter logo "IBM," demonstrating that matter could indeed be maneuvered atom by atom as Feynman suggested [16].

A nanomanipulation system generally includes nanomanipulators as the "arms," microscopes as the "eyes," various end-effectors including probes and tweezers among others as its "fingers," and types of sensors (force, displacement, tactile, and strain) to facilitate the manipulation and/or to determine the properties of the objects. Key technologies for nanomanipulation include observation, actuation, measurement, system design and fabrication, calibration and control, communication, and human-machine interface.

Strategies for nanomanipulation are basically determined by the environment—air, liquid, or vacuum—which is further decided by the properties and size of the objects and observation methods. In order to observe manipulated objects, STMs can provide subangstrom imaging resolution, whereas AFMs can provide atomic resolutions. Both can obtain 3D surface topology. Because AFMs can be used in an ambient environment, they provide a powerful tool for biomanipulation that may require a liquid environment. The resolution of SEM is limited to about 1 nm, whereas field-emission SEM (FESEM) can achieve higher resolutions. SEM/FESEM can be used for 2D real-time observation for both the objects and end-effectors of manipulators, and large UHV sample chambers provide enough space to contain an NRM with many DOFs for 3D nanomanipulation. However, the 2D nature of the observation makes positioning along the electron-beam direction difficult. High-resolution transmission electron microscopes (HRTEM) can provide atomic resolution. However, the narrow UHV specimen chamber makes it difficult to incorporate large manipulators. In principle, optical microscopes (OMs) cannot be used for nanometer scale (smaller than the wavelength of visible lights) observation because of diffraction limits. SNOMs break this limitation and are promising as a real-time observation device for nanomanipulation, especially for ambient environments. SNOMs can be combined with AFMs, and potentially with NRMs for nanoscale biomanipulation.

Nanomanipulation processes can be broadly classified into three types: (1) lateral noncontact, (2) lateral contact, and (3) vertical manipulation. Generally, lateral noncontact nanomanipulation is mainly applied for atoms and molecules in UHV with an STM or bio-object in liquid using optical or magnetic tweezers. Contact nanomanipulation can be used in almost any environment, generally with an AFM, but is difficult for atomic manipulation. Vertical manipulation can be performed by NRMs.

Nanorobotic manipulators are the core components of nanorobotic manipulation systems. The basic requirements for a nanorobotic manipulation system for 3D manipulation include nanoscale positioning resolution, a relative large working space, enough DOFs including rotational ones for 3D positioning and orientation control of the end-effectors, and usually multiple end-effectors for complex operations.

A commercially available nanomanipulator (MM3A from Kleindiek) installed inside a SEM (Carl Zeiss DSM962) is shown in Figure 18.4(a). The manipulator has 3 DOFs, and nanometer to subnanometer scale resolution. Figure 18.4(b) shows a nanorobotic manipulation system that has 16 DOFs in total and can be equipped with three to four AFM cantilevers as end-effectors for both manipulation and measurement. The functions of the nanorobotic manipulation system include nanomanipulation, nanoinstrumentation, nanofabrication, and nanoassembly. The positioning resolution is subnanometer order and strokes are centimeter-scale. For the construction of MWNT based nanostructures, manipulators position and orient nanotubes for the fabrication of nanotube probes and emitters, for performing nanosoldering with EBID [37], for the property characterization of single nanotubes for selection purposes, and for characterizing junctions to test connection strength.

Nanomanipulation is a promising strategy for nanoassembly. Key techniques for nanoassembly include the structuring and characterization of nano-building blocks, the positioning and orientation control of the building blocks with nanometer scale resolution, and effective connection techniques. Nanorobotic manipulation, which is characterized by multiple DOFs with both position and orientation controls, independently actuated multiprobes, and a real-time observation system, have been shown effective for assembling nanotube-based devices in 3D space.

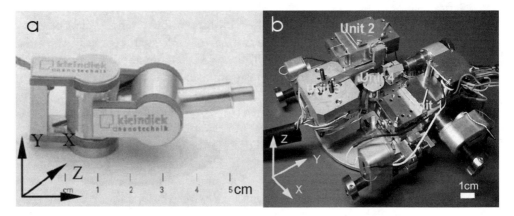

Figure 18.4 Nanorobotic manipulators. (a) A commercially available manipulator (MM3A from Kleindiek) for an SEM. (b) A 4-probe, 16-DOF nanorobotic manipulation system.

NEMS based on individual carbon nanotubes and nanocoils are of increasing interest, indicating that capabilities for incorporating these individual building blocks at specific locations on a device must be developed. Random spreading [38], direct growth [39], self-assembly [40], dielectrophoretic assembly [29], and nano-manipulation [41] have been demonstrated for positioning as-grown nanotubes on electrodes for the construction of these devices. However, for nanotube-based structures, nanorobotic assembly is still the only technique capable of in situ structuring, characterization, and assembly. Because the as-fabricated nanocoils are not free-standing from their substrate, nanorobotic assembly is virtually the only way to incorporate them into devices at present.

Nanotube manipulation in two dimensions on a surface was first performed with an AFM by contact pushing on a substrate. The first demonstration was given by Lieber and coworkers for measuring the mechanical properties of a nanotube [42] by pushing one end of it and fixing the other end. The same strategy was used for the investigation of the behaviors of nanotubes under large strain [43]. Avouris and coworkers combined this technique with an inverse process, namely straightening by pushing along a bent tube, and realized the translation of the tube to another location and between two electrodes to measure the conductivity [44]. This technique was also used to place a tube on another tube to form a SET with a cross-junction of nanotubes [45].

Manipulation of CNTs in 3D space is important for assembling CNTs into structures and devices. Basic techniques for the nanorobotic manipulation of carbon nanotubes are shown in Figure 18.5 [46]. These serve as the basis for handling, structuring, characterizing, and assembling NEMS.

The basic procedure is to pick up a single tube from nanotube soot [Figure 18.5(a)]. This has been shown first by using dielectrophoresis [35] through nanorobotic manipulation [Figure 18.5(b)]. By applying a bias between a sharp tip and a plane substrate, a nonuniform electric field can be generated between the tip and the substrate with the strongest field near the tip. This field can cause a tube to orient along the field or further "jump" to the tip by electrophoresis or dielectrophoresis (determined by the conductivity of objective tubes). Removing the bias, the tube can be placed at other locations at will. This method can be used for free-standing tubes on nanotube soot or on a rough surface on which surface van der Waals forces are generally weak. A tube strongly rooted in CNT soot or lying on a flat surface cannot be picked up in this way. The interaction between a tube and the atomic flat surface of AFM cantilever tip has been shown to be strong enough for picking up a tube onto the tip [Figure 18.5(c)] [47]. By using EBID, it is possible to pick up and fix a nanotube onto a probe (Figure 18.5(d)) [48]. For handling a tube, a weak connection between the tube and the probe is desired.

Bending and buckling a CNT, as shown in Figure 18.5(e, f), are important for in situ property characterization of a nanotube [49], which is a simple way to get the Young's modulus of a nanotube without damaging the tube (if performed within its elastic range) and, hence, can be used for the selection of a tube with desired properties.

Stretching a nanotube between two probes or a probe and a substrate has generated several interesting results [Figure 18.5(g)]. The first demonstration of 3D nanomanipulation of nanotubes took this as an example to show the breaking

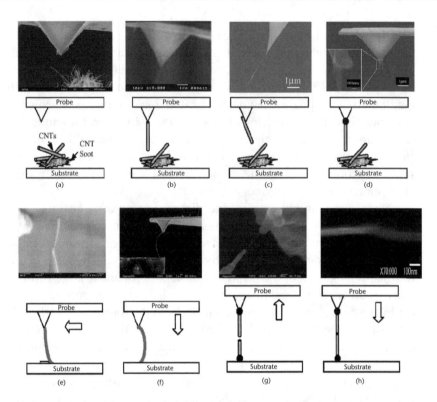

Figure 18.5 Nanorobotic manipulation of CNTs. The basic technique is to pick up an individual tube from CNT soot, as in (a), or from an oriented array; (b) shows a free-standing nanotube picked up by dielectrophoresis generated by a nonuniform electric field between the probe and substrate; and (c) and (d) show the same manipulation by contacting a tube with the probe surface or fixing (e.g., with EBID) a tube to the tip (inset shows the EBID deposit). Vertical manipulation of nanotubes includes (e) bending, (f) buckling, (g) stretching/breaking, and (h) connecting/bonding. All examples, with the exception of (c), are from the authors' work.

mechanism of a MWNT [34], and to measure the tensile strength of CNTs [50]. By breaking a MWNT in a controlled manner, interesting nanodevices have been fabricated. This technique—destructive fabrication—has been presented to get sharpened and layered structures of nanotubes, and to improve the length control of a nanotube [51]. Typically, a layered and a sharpened structure can be obtained from this process, similar to that achieved from electric pulses [52]. Bearing motion has been observed in an incompletely broken MWNT, and the interlayer friction has been predicted to be very small [53].

The reverse process, namely the connection of broken tubes [Figure 18.5(h)], has been demonstrated recently, and the mechanism is revealed as rebonding of unclosed dangling bonds at the ends of broken tubes. Based on this interesting phenomenon, mechanochemical nanorobotic assembly has been performed [54]. Assembly of nanotubes is a fundamental technology for enabling nanodevices. The most important tasks include the connection of nanotubes and placing of nanotubes onto electrodes. Pure nanotube circuits [55] created by interconnecting nanotubes of different diameters and chirality could lead to further size reductions in devices. Nanotube intermolecular and intramolecular junctions are basic elements for such systems. An intramolecular kink junction behaving like a rectifying diode has been reported [56]. With a cross-junction of two SWNTs (semiconducting/metallic), three- and four-terminal electronic devices have been made [57]. A suspended cross-junction can function as electromechanical nonvolatile memory [40].

Although some kinds of junctions have been synthesized with chemical methods, there is no evidence yet showing that a self-assembly based approach can provide more complex structures. SPMs were also used to fabricate junctions, but they are limited to a 2D plane. We have presented 3D nanorobotic manipulation based nanoassembly, which is a promising strategy, both for the fabrication of nanotube junctions and for the construction of more complex nanodevices with such junctions as well.

Nanotube junctions can be classified into different types by (1) the kinds of components—SWNTs or MWNTs; (2) geometric configurations—V- (kink), I-, X- (cross), T-, Y- (branch), and 3D junctions; (3) conductivity—metallic or semiconducting; and (4) connection methods—intermolecular (connected with van der Waals force, EBID) or intramolecular (connected with chemical bonds) junctions.

3D nanorobotic manipulation has opened a new route for the structuring and assembly of nanotubes into nanodevices. However, at present, nanomanipulation is still performed in a serial manner with master-slave control, which is not a large-scale production oriented technique. Nevertheless, with advances in the exploration of mesoscopic physics, better control on the synthesis of nanotubes, more accurate actuators, and effective tools for manipulation, high-speed and automatic nanoassembly will be possible. Another approach might be parallel assembly by positioning building blocks with an array of probes [58] and joining them together simultaneously, for example, with the parallel EBID [48] approach we presented. Further steps might progress towards automatic assembly, and in the far future to self-replicating assembly [17].

Focusing on the unique aspects of manipulating nanocoils due to their helical geometry, high elasticity, single-end fixation, and strong adhesion of the coils to the substrate from wet etching, a series of new processes is presented using a manipulator (MM3A, Kleindiek) installed in an SEM (Zeiss DSM962). Special tools have been fabricated including a nanohook prepared by controlled "tip-crashing" of a commercially available tungsten sharp probe (Picoprobe T-4-10-1mm and T-4-10) onto a substrate, and a "sticky" probe prepared by "tip–dipping" into a double-sided SEM silver conductive tape (Ted Pella, Inc.). As shown in Figure 18.6, experiments demonstrate that nanocoils can be released from a chip by lateral pushing, picked up with a nanohook or a "sticky" probe, and placed between the probe/hook and another probe or an AFM cantilever (Nano-probe, NP-S). Axial pulling/pushing, radial compressing/releasing, and bending/buckling have also been demonstrated. These processes have shown the effectiveness of manipulation for the characterization of coil-shaped nanostructures and their assembly for NEMS, which have been otherwise unavailable.

From fabrication and characterization results, the helical nanostructures appear to be suitable to function as inductors. They would allow further miniaturization compared to state-of-the-art microinductors. For this purpose, higher doping of the bilayer and an additional metal layer would result in the required conductance. Conductance, inductance, and quality factor can be further improved if, after curling up, additional metal is electroplated onto the helical structures. Moreover, a

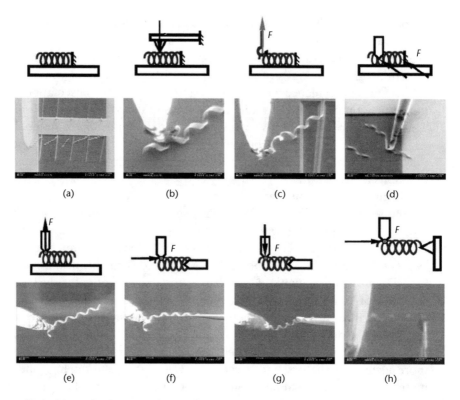

Figure 18.6 Nanorobotic manipulation of nanocoils: (a) original state; (b) compressing/releasing; (c) hooking; (d) lateral pushing/breaking; (e) picking; (f) placing/inserting; (g) bending; and (h) pushing and pulling. The coils being manipulated are approximately 20 nm thick and form a coil approximately 1.2 μm in diameter.

semiconductive helical structure, when functionalized with binding molecules, can be used for chemical sensing under the same principle as demonstrated with other types of nanostructures. With bilayers in the range of a few monolayers, the resulting structures would exhibit very high surface-to-volume ratio with the whole surface exposed to an incoming analyst.

18.5 NEMS

NEMS involve tools, sensors, and actuators at the nanometer scale. Shrinking device size makes it possible to manipulate nanosized objects with nanosized tools, measure mass in femtogram ranges, sense force at piconewton scales, and induce gigahertz motion, among other amazing advancements.

Top-down and bottom-up strategies for manufacturing such nanodevices have been independently investigated by a variety of researchers. Top-down strategies are based on nanofabrication and include technologies such as nanolithography, nanoimprinting, and chemical etching. Presently, these are 2D fabrication processes with relatively low resolution. Bottom-up strategies are assembly-based techniques. At present, these strategies include such techniques as self-assembly, dip-pen lithography, and directed self-assembly. These techniques can generate regular nanopatterns at large scales. With the ability to position and orient nanometer scale objects, nanorobotic manipulation is an enabling technology for structuring, characterizing, and assembling many types of nanosystems [46]. By combining bottom-up and top-down processes, a hybrid nanorobotic approach (as shown in Figure 18.7) based on nanorobotic manipulation provides a third way to fabricate NEMS by structuring as-grown nanomaterials or nanostructures. This new nanomanufacturing technique can be used to create complex 3D nanodevices with such building blocks. Nanomaterial science, bionanotechnology, and nanoelectronics will also benefit from advances in nanorobotic assembly.

The configurations of nanotools, sensors, and actuators based on individual nanotubes that have been experimentally demonstrated are summarized, as shown in Figure 18.8.

For detecting deep and narrow features on a surface, cantilevered nanotubes [Figure 18.8(a)] [48] have been demonstrated as probe tips for an AFM [59], an STM, and other types of SPMs. Nanotubes provide ultra-small diameters, ultra-large aspect ratios, and excellent mechanical properties. Manual assembly, direct growth [60], and nanoassembly have proven effective for their construction. Cantilevered nanotubes have also been demonstrated as probes for the measurement of ultra-small physical quantities, such as femtogram mass [61], and piconewton-order force sensors on the basis of their static deflections or change of resonant frequencies detected within an electron microscope. Deflections cannot be measured from micrographs in real time, which limits the application of this kind of sensor. Interelectrode distance changes cause emission current variation of a nanotube emitter and may serve as a candidate to replace microscope images.

Bridged individual nanotubes [Figure 18.8(b)] [29] have been the basis for electric characterization. Opened nanotubes [Figure 18.8(c)] [62] can serve as an

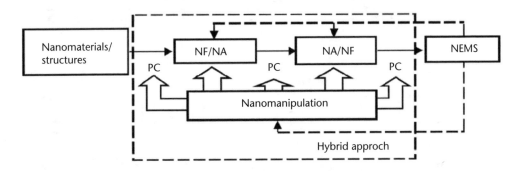

Figure 18.7 Hybrid approach to NEMS. (PC: property characterization; NF: nanofabrication; NA: nanoassembly). In this system, nanofabrication based top-down processes and nanoassembly based bottom-up processes can be performed in an arbitrary order. Consider nanofabrication processes in which nanomaterials or nanostructures can be fabricated into nano-building blocks by removing unwanted parts. These building blocks can then be assembled into NEMS. Conversely, nanoassembly can be performed first and nanomaterials or nanostructures can be assembled into higher level (i.e., more complex, 3D, arrays, and so forth) structures, and then the high-level structures can be further modified into NEMS by nanofabrication. NEMS manufactured from this system can be "feedback" (as indicated with dashed lines) to the system to shrink the sizes of the system itself.

atomic or molecular container. A thermometer based on this structure has been shown by monitoring the height of the gallium inside the nanotube using TEM [63].

A new family of nanotube actuators can be constructed by taking advantage of the ultra-low interlayer friction of a multiwalled nanotube. Linear bearings based on telescoping nanotubes have been demonstrated [53]. A microactuator with a nanotube as a rotation bearing was demonstrated in 2003 [64]. A preliminary experiment on a promising nanotube linear motor with field emission current serving as position feedback has been shown with nanorobotic manipulation [Figure 18.8(d)] [62].

Cantilevered dual nanotubes have been demonstrated as nanotweezers [65] and nanoscissors [Figure 18.8(e)] [41] by manual and nanorobotic assembly, respectively. Based on electric resistance change under different temperatures, nanotube thermal probes [Figure 18.8(f)] have been demonstrated for measuring the temperature at precise locations. These thermal probes are more advantageous than nanotube based thermometers because the thermometers require TEM imaging. Gas sensors and hot-wire based mass/flow sensors can also be constructed in this configuration rather than a bridged one. The integration of the above-mentioned devices can be realized using the configurations shown in Figure 18.8(g, h) [29]. The arrays

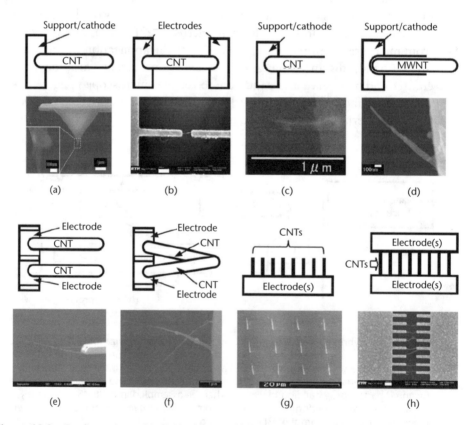

Figure 18.8 Configurations of individual nanotube-based NEMS. Scale bars: (a) 1 μm (inset: 100 nm); (b) 200 nm; (c) 1 μm; (d) 100 nm; (e, f) 1 μm; (g) 20 μm; and (h) 300 nm. All examples are from the authors' work.

of individual nanotubes can also be used to fabricate nanosensors, such as position encoders [66].

Configurations of NEMS based on 3D helical nanostructures are shown in Figure 18.9. The cantilevered structures shown in Figure 18.9 (a tubes, d rings, g coils, and j spirals) can serve as nanosprings using their elasticity in axial (tubes and coils), radial (rings), and tangential/rotary (spirals) directions. Nanoelectromagnets, chemical sensors, nanoinductors and capacitors involve building blocks bridged between two electrodes (two or four for rings) as shown in Figure 18.9 (b tubes, e rings, h coils, and k spirals). Electromechanical sensors can use a similar configuration but with one end connected to a moveable electrode as shown in Figure 18.9 (c tubes, f rings, i coils, and l spirals). Mechanical stiffness and electrical conductivity are fundamental properties for these devices which must be further investigated. Electron microscopy imaging or their intrinsic electromechanical coupling property can serve as readout mechanisms.

The construction of NEMS using 3D helical nanostructures involves the assembly of as-fabricated building blocks, which is a significant challenge from a fabrication standpoint. Processes are developed for the manipulation of as-fabricated 3D helical nanostructures. As shown in Figure 18.10, experiments demonstrate that the as-fabricated nanostructures can be released from a chip by being picked up with a "sticky" probe from their free ends [Figure 18.10(a), tubes], fixed ends [Figure 18.10(d), coils], external surfaces [Figure 18.10(g), rings], or internal surfaces [Figure 18.10(j), spirals], and bridged between the probe and another probe [Figure 18.10(k)] or an AFM cantilever [Figure 18.10(b, e, h)], showing a promising approach for robotic assembly of these structures into complex systems. Axial pulling [Figure 18.10(f1-4)]/pushing, radial compressing [Figure 18.10(i1-5)]/releasing, bending/buckling [Figure 18.10(c1-4)], and unrolling [Figure 18.10(l1-5), spirals; and Figure 18.10(n1-8), claws] have also been demonstrated for property characterization. The stiffness of the tube, the coil, and the ring has been measured from the SEM images by extracting the AFM tip displacement and the deformation of the structures. The stiffness of the tube, the ring, and the coil springs was estimated to be ~10 N/m, 0.137 N/m, and 0.003 N/m (calibrated AFM cantilever stiffness: 0.038 N/m), showing a large range for selection. The linear elastic region of the small pitch coils reaches up to 90%. Unrolling experiments show these structures have an excellent ability to "remember" their original shapes.

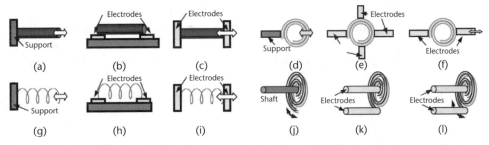

Figure 18.9 Configuration of 3D helical nanostructures based NEMS. (a–c) Tubes. (d–f) Rings. (g–i) Coils. (j–l) Spirals. (a, d, g, j) Cantilevered. (b, e, h, k) Bridged (fixed). (c, f, i, l) Bridged (moveable).

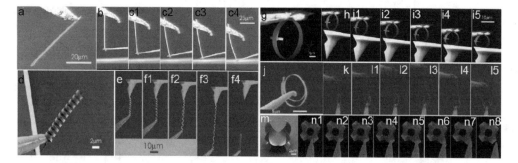

Figure 18.10 Nanorobotic manipulation of 3D helical structures. Pick up a tube (a), bridge it between a probe and an AFM cantilever (b), and buckle it (c1-4) for electromechanical property characterization for force measuring. Pick up a small pitch coil (d), bridge it between a probe and an AFM cantilever (e), and pull it for mechanical property characterization for building a "spring balance" (f1-4). Pick up a ring (external diameter: 12.56 μm; strip width: 1.2 μm; number of turns: 2.5; thickness: Si/Cr 35 nm/10 nm) (g), bridge it between a probe and an AFM cantilever (h), and compress it for mechanical property characterization for understanding its stiffness (i1-5). Pick up a spiral (Si/Cr layer thickness: 35/10 nm) (j), bridge it between a probe and another probe (k), and unroll it for mechanical property characterization for understanding its interlayer interaction (taken from a video clip) (l1-5). Unroll a leaf of claws (m) for mechanical property characterization for understanding its "shape memory" (taken from a video clip) (n1-8).

These processes demonstrate the effectiveness of manipulation for the characterization of the 3D helical nanostructures and their assembly for NEMS, which have otherwise been unavailable.

The excellent elasticity of nanocoils suggests that they can be used to sense ultra-small forces by monitoring the deformation of the spring as a "spring balance" [Figure 18.10(f1-4)]. If working in an SEM, suppose an imaging resolution of 1 nm can be obtained (the best commercially available FESEM can provide such a resolution in an ideal environment), a spring balance constructed with the calibrated coil (10 turns, 0.003 N/m) can provide a 3-pN/nm resolution for force measurement. With smaller stripe widths or more turns, nanocoils can potentially provide fN resolution. In the SEM used in these experiments, the available imaging resolution is 10 nm, which provides a 30-pN/10-nm resolution. Figure 18.10(f1-4) shows a way to use such a coil to measure the adhesive force between a coil and adhesive silver tape. Comparing the length difference, the extension of the spring can be found and converted to force according to the calibrated spring constant. For Figure 18.10(f1-3), the relevant forces are determined to be 15.31±0.03 nN, 91.84±0.03 nN, (intermediate steps), and 333.67±0.03 nN (maximum holding/releasing force). It can be seen from Figure 18.10(f4) that the coil recovered its shape after releasing.

Electrical properties can be characterized by placing a coil between two probes or electrodes [67]. An interesting phenomena found in the measurements is that the SiGe/Si nanocoils with Cr layers can shrink further by passing current through them or by placing a charged probe on them. A 5-turn as-fabricated coil was observed to become an 11-turn coil, showing the possibility of structuring them.

Despite the claims of many futurists, such as Issac Asimov's legendary submarine Proteus inside the human body [68] and Robert A. Freitas's nanomedical robots [69], the form nanorobots of the future will take and what tasks they will actually perform remain unclear. It is clear, however, that nanotechnology is progressing towards the construction of intelligent sensors, actuators, and systems that are smaller than 100 nm. These NEMS will serve as both the tools to be used for fabricating future nanorobots as well as the components from which these nanorobots may be developed. Shrinking device size to these dimensions presents many fascinating opportunities such as manipulating nanoobjects with nanotools, measuring mass in femtogram ranges, sensing forces at piconewton scales, and inducing gigahertz motion, among other new possibilities waiting to be discovered. These capabilities will, of course, drive the tasks that future nanorobots constructed by and with NEMS will perform.

References

[1]   Roco, M. C., R. S. Williams, and P. Alivisatos, *Nanotechnology Research Directions: Interagency Working Group on Nanoscience, Engineering and Technology (IWGN) Workshop Report (Vision for Nanotechnology R&D in the Next Decade),* Dordrecht: Kluwer Academic Publishers, 2000.

[2]   Binnig, G., et al., "Surface Studies by Scanning Tunneling Microscopy," *Phys. Rev. Lett.,* Vol. 49, 1982, pp. 57–61.

[3]   Alivisatos, A. P., "Semiconductor Clusters, Nanocrystals, and Quantum Dots," *Science,* Vol. 271, No. 5251, February 1996, pp. 933–937.

[4]   Iijima, S., "Helical Microtubules of Graphitic Carbon," *Nature,* Vol. 354, 1991, pp. 56–58.

[5]   Huang, Y., et al., "Logic Gates and Computation from Assembled Nanowire Building Blocks," *Science,* Vol. 294, 2001, pp. 1313–1317.

[6]   Kong, X. Y., and Z. L. Wang, "Spontaneous Polarization-Induced Nanohelixes, Nanosprings, and Nanorings of Piezoelectric Nanobelts," *Nano Lett.,* Vol. 3, 2003, pp. 1625–1631.

[7]   Tans, S. J., A. R. M. Verchueren, and C. Dekker, "Room-Temperature Transistor Based on a Single Carbon Nanotube," *Nature,* Vol. 393, 1998, pp. 49–52.

[8]   Baughman, R. H., A. A. Zakhidov, and W. A. d. Heer, "Carbon Nanotubes—The Route Toward Applications," *Science,* Vol. 297, 2002, pp. 787–792.

[9]   Treacy, M. J., T. W. Ebbesen, and J. M. Gibson, "Exceptionally High Young's Modulus Observed For Individual Carbon Nanotubes," *Nature,* Vol. 381, 1996, pp. 678–680.

[10] Yu, M. F., et al., "Strength and Breaking Mechanism of Multiwalled Carbon Nanotubes Under Tensile Load," *Science,* Vol. 287, 2000, pp. 637–640.

[11] Zhang, X. B., et al., "The Texture of Catalytically Grown Coil-Shaped Carbon Nanotubules," *Europhys. Lett.,* Vol. 27, 1994, pp. 141–146.

[12] Golod, S. V., et al., "Fabrication of Conducting GeSi/Si Micro- and Nanotubes and Helical Microcoils," *Semicond. Sci. Technol.,* Vol. 16, 2001, pp. 181–185.

[13] Zhang, L., et al., "Controllable Fabrication of SiGe/Si and SiGe/Si/Cr Helical Nanobelts," *Nanotechnol.,* Vol. 16, 2005, pp. 655–663.

[14] Bell, D. J., et al., "Fabrication and Characterization of Three-Dimensional InGaAs/GaAs Nanosprings," *Nano Lett.,* Vol. 6, No. 4, April 2006, pp. 725–729.

[15] Zhang, L., et al., "Anomalous Coiling of SiGe/Si and SiGe/Si/Cr Helical Nanobelts," *Nano Lett.,* Vol. 6, No. 7, 2006, pp. 1311–1317.

[16] Feynman, R. P., "There's Plenty of Room at the Bottom," *Caltech's Eng. Sci.,* Vol. 23, 1960, pp. 22–36.

[17] Drexler, K., *Nanosystems: Molecular Machinery, Manufacturing and Computation,* New York: Wiley-Interscience, 1992.

[18] Haremza, J. M., et al., "Attachment of Single CdSe Nanocrystals to Individual Single-Walled Carbon Nanotubes," *Nano Lett.*, Vol. 2, No. 11, 2002, pp. 1253–1258.

[19] Banerjee, S., and S. S. Wong, "Synthesis and Characterization of Carbon Nanotube-Nanocrystal Heterostructures," *Nano Lett.*, Vol. 2, No. 3, 2002, pp. 195–200.

[20] Ravindran, S., and C. S. Ozkan, "Self-Assembly of ZnO Nanoparticles to Electrostatic Coordination Sites of Functionalized Carbon Nanotubes," *Nanotechnol.*, Vol. 16, 2005, pp. 1130–1136.

[21] Dubertret, B., et al., "In Vivo Imaging of Quantum Dots Encapsulated in Phospholipid Micelles," *Science*, Vol. 298, No. 5599, November 2002, pp. 1759–1762.

[22] Lacoste, T. D., et al., "Ultrahigh-Resolution Multicolor Colocalization of Single Fluorescent Probes," *Proc. Natl. Acad. Sci. USA*, Vol. 97, No. 17, August 2000, pp. 9461–9466.

[23] Dahan, M., et al., "Time-Gated Biological Imaging by Use of Colloidal Quantum Dots," *Optics Lett.*, Vol. 26, No. 11, June 2001, pp. 825–827.

[24] Bausch, A. R., and D. A. Weitz, "Tracking the Dynamics of Single Quantum Dots: Beating the Optical Resolution Twice," *J. Nanopart. Res.*, Vol. 4, No. 6, December 2002, pp. 477–481.

[25] Cheezum, M. K., W. F. Walker, and W. H. Guilford, "Quantitative Comparison of Algorithms for Tracking Single Fluorescent Particles," *Biophysical Journal*, Vol. 81, No. 4, October 2001, pp. 2378–2388.

[26] Dahan, M., et al., "Diffusion Dynamics of Glycine Receptors Revealed by Single-Quantum Dot Tracking," *Science*, Vol. 302, No. 5644, October 2003, pp. 442–445.

[27] Frutiger, D. R., L. X. Dong, and B. J. Nelson, "Optical Tracking of Multi-Walled Carbon Nanotubes by Attaching Functionalized Quantum Dots," *Proc. of 2006 IEEE Int. Conf. on Nano/Micro Engineered and Molecular Systems (IEEE-NEMS2006)*, Zhuhai, China, 2006, pp. 1181–1186.

[28] Rai-Choudhury, P., (ed.), *Handbook of Microlithography, Micromachining, and Microfabrication*, Bellingham, WA: SPIE, 1997.

[29] Subramanian, A., et al., "Micro and Nanorobotic Assembly Using Dielectrophoresis," in *Robotics: Science and Systems I*, S. Thrun, et al., (eds.), Cambridge, MA: MIT Press, 2005, pp. 327–334.

[30] Zhang, L., et al., "Fabrication and Characterization of Freestanding Si/Cr Micro- and Nano-spirals," *Microelectron. Eng.*, Vol. 83, 2006, pp. 1237–1240.

[31] Binnig, G., C. F. Quate, and C. Gerber, "Atomic Force Microscope," *Phys. Rev. Lett.*, Vol. 56, 1986, pp. 93–96.

[32] Ashkin, A., and J. M. Dziedzic, "Optical Trapping and Manipulation of Viruses and Bacteria," *Science*, Vol. 235, 1987, pp. 1517–1520.

[33] Crick, F. H. C., and A. F. W. Hughes, "The Physical Properties of Cytoplasm: A Study by Means of the Magnetic Particle Method, Part I. Experimental," *Exp. Cell Res.*, Vol. 1, 1950, pp. 37–80.

[34] Yu, M. F., et al., "Three-Dimensional Manipulation of Carbon Nanotubes Under a Scanning Electron Microscope," *Nanotechnol.*, Vol. 10, 1999, pp. 244–252.

[35] Dong, L. X., F. Arai, and T. Fukuda, "3D Nanorobotic Manipulation of Nano-Order Objects Inside SEM," *Proc. of 2000 Int. Symp. on Micromechatronics and Human Science (MHS2000)*, Nagoya, 2000, pp. 151–156.

[36] Eigler, D. M., and E. K. Schweizer, "Positioning Single Atoms with a Scanning Tunneling Microscope," *Nature* Vol. 344, 1990, pp. 524–526.

[37] Koops, H. W. P., et al., "Characterization and Application of Materials Grown by Electron-Beam-Induced Deposition," *Jpn. J. Appl. Phys. Part 1*, Vol. 33, 1994, pp. 7099–7107.

[38] Martel, R., et al., "Single- and Multi-Wall Carbon Nanotube Field-Effect Transistors," *Appl. Phys. Lett.*, Vol. 73, 1998, pp. 2447–2449.

[39] Franklin, N. R., et al., "Patterned Growth of Single-Walled Carbon Nanotubes on Full 4-Inch Wafers," *Appl. Phys. Lett.,* Vol. 79, 2001, pp. 4571–4573.

[40] Rueckes, T., et al., "Carbon Nanotube-Based Nonvolatile Random Access Memory for Molecular Computing Science," *Science,* Vol. 289, 2000, pp. 94–97.

[41] Fukuda, T., F. Arai, and L. X. Dong, "Assembly of Nanodevices with Carbon Nanotubes Through Nanorobotic Manipulations," *Proc. IEEE,* Vol. 91, No. 11, November 2003, pp. 1803–1818.

[42] Wong, E. W., P. E. Sheehan, and C. M. Lieber, "Nanobeam Mechanics: Elasticity, Strength, and Toughness of Nanorods and Nanotubes," *Science,* Vol. 277, 1997, pp. 1971–1975.

[43] Falvo, M. R., et al., "Bending and Buckling of Carbon Nanotubes Under Large Strain," *Nature,* Vol. 389, 1997, pp. 582–584.

[44] Avouris, P., et al., "Carbon Nanotubes: Nanomechanics, Manipulation, and Electronic Devices," *Appl. Surf. Sci.,* Vol. 141, 1999, pp. 201–209.

[45] Ahlskog, M., et al., "Single-Electron Transistor Made of Two Crossing Multiwalled Carbon Nanotubes and Its Noise Properties," *Appl. Phys. Lett.,* Vol. 77, 2000, pp. 4037–4039.

[46] Dong, L. X., "Nanorobotic Manipulations of Carbon Nanotubes," Ph.D. thesis, Nagoya University, March 2003.

[47] Hafner, J. H., et al., "High-Yield Assembly of Individual Single-Walled Carbon Nanotube Tips for Scanning Probe Microscopies," *J. Phys. Chem. B,* Vol. 105, 2001, pp. 743–746.

[48] Dong, L. X., F. Arai, and T. Fukuda, "Electron-Beam-Induced Deposition with Carbon Nanotube Emitters," *Appl. Phys. Lett.,* Vol. 81, No. 10, September 2002, pp. 1919–1921.

[49] Dong, L. X., F. Arai, and T. Fukuda, "3D Nanorobotic Manipulations of Multi-Walled Carbon Nanotubes," *Proc. of 2001 IEEE Int. Conf. on Robotics and Automation (ICRA2001),* Seoul, 2001, pp. 632–637.

[50] Yu, M. F., et al., "Tensile Loading of Ropes of Single Wall Carbon Nanotubes and Their Mechanical Properties," *Phys. Rev. Lett.,* Vol. 84, 2000, pp. 5552–5555.

[51] Dong, L. X., F. Arai, and T. Fukuda, "Destructive Constructions of Nanostructures with Carbon Nanotubes Through Nanorobotic Manipulation," *IEEE-ASME Transactions on Mechatron.,* Vol. 9, No. 2, June 2004, pp. 350–357.

[52] Cumings, J., P. G. Collins, and A. Zettl, "Peeling and Sharpening Multiwall Nanotubes," *Nature,* Vol. 406, 2000, p. 58.

[53] Cumings, J., and A. Zettl, "Low-Friction Nanoscale Linear Bearing Realized from Multiwall Carbon Nanotubes," *Science,* Vol. 289, 2000, pp. 602–604.

[54] Dong, L. X., F. Arai, and T. Fukuda, "Nanoassembly of Carbon Nanotubes Through Mechanochemical Nanorobotic Manipulations," *Jpn. J. Appl. Phys. Part 1,* Vol. 42, No. 1, January 2003, pp. 295–298.

[55] Chico, L., et al., "Pure Carbon Nanoscale Devices: Nanotube Heterojunctions," *Phys. Rev. Lett.,* Vol. 76, No. 6, February 1996, pp. 971–974.

[56] Yao, Z., et al., "Carbon Nanotube Intramolecular Junctions," *Nature,* Vol. 402, 1999, pp. 273–276.

[57] Fuhrer, M. S., et al., "Crossed Nanotube Junctions," *Science,* Vol. 288, 2000, pp. 494–497.

[58] Minne, S. C., et al., "Automated Parallel High-Speed Atomic Force Microscopy," *Appl. Phys. Lett.,* Vol. 72, 1998, pp. 2340–2342.

[59] Dai, H. J., et al., "Nanotubes as Nanoprobes in Scanning Probe Microscopy," *Nature,* Vol. 384, 1996, pp. 147–150.

[60] Hafner, J. H., C. L. Cheung, and C. M. Lieber, "Growth of Nanotubes for Probe Microscopy Tips," *Nature,* Vol. 398, 1999, pp. 761–762.

[61] Poncharal, P., et al., "Electrostatic Deflections and Electromechanical Resonances of Carbon Nanotubes," *Science,* Vol. 283, 1999, pp. 1513–1516.

[62] Dong, L. X., et al., "Towards Linear Nano Servomotors," *IEEE Transactions on Automation Science and Engineering,* Vol. 3, No. 3, 2006, pp. 228–235.

[63] Gao, Y. H., and Y. Bando, "Carbon Nanothermometer Containing Gallium—Gallium's Macroscopic Properties Are Retained on a Miniature Scale in This Nanodevice," *Nature*, Vol. 415, No. 6872, February 2002, p. 599.

[64] Fennimore, A. M., et al., "Rotational Actuators Based on Carbon Nanotubes," *Nature*, Vol. 424, 2003, pp. 408–410.

[65] Kim, P., and C. M. Lieber, "Nanotube Nanotweezers," *Science*, Vol. 286, 1999, pp. 2148–2150.

[66] Dong, L. X., et al., "Nano Encoders Based on Vertical Arrays of Individual Carbon Nanotubes," *Adv. Robotics*, Vol. 20, No. 11, 2006, pp. 1281–1301.

[67] Bell, D. J., et al., "Three-Dimensional Nanosprings for Electromechanical Sensors," *Sens. Actuator A-Phys.*, Vol. 130–131, August 2006, pp. 54–61.

[68] Asimov, I., *Fantastic Voyage*, New York: Bantam Books, 1966.

[69] Freitas, R. A., *Nanomedicine, Volume I: Basic Capabilities*: Landes Bioscience, 1999.

About the Editors

Mingjun Zhang is an R&D engineer for Agilent Technologies in Palo Alto, California. At Agilent his work is focused on quantitative and automation approaches to life sciences and includes DNA gene-chip and protein microarray fabrication, modeling and control of drug delivery systems, and molecular diagnostics. Dr. Zhang is an associate editor of the *IEEE Transactions on Automated Science and Engineering* and a member of the *Nanomedicine* editorial board. He was awarded the first Early Career Award for industrial/government professionals by the IEEE Robotics and Automation Society in 2003. Dr. Zhang received a D.Sc. from Washington University in St. Louis, Missouri, and a Ph.D. from Zhejiang University, China.

Bradley Nelson is a professor of robotics and intelligent systems at ETH Zurich, Switzerland, and the founder of the Institute of Robotics and Intelligent Systems at ETH Zurich. Professor Nelson serves on, or has been a member of, the editorial boards of the *IEEE Transactions on Robotics, IEEE Transactions on Nanotechnology, Journal of Micromechatronics, Journal of Optomechatronics,* and the *IEEE Robotics and Automation Magazine.* He has also chaired several international workshops and conferences. In 2005 he was named to the *Scientific American 50 Award,* the magazine's annual list recognizing leadership in science and technology. He received a Ph.D. in robotics from Carnegie Mellon University.

Robin Felder is the director of the Medical Automation Research Center and a professor of pathology at the University of Virginia. Dr. Felder founded the Association for Laboratory Automation and was the founding editor of the *Journal of the Association for Laboratory Automation,* a peer-reviewed journal. He has co-authored more than 290 papers and has been awarded 11 patents. Professor Felder also serves on the board of seven medical and biotechnology companies. He received a Ph.D. in biochemistry at Georgetown University and did his postdoctoral work at the National Institute of Mental Health.

Index

Recent Related Artech House Titles

Advanced Methods and Tools for ECG Data Analysis, Gari D. Clifford, Francisco Azuaje, and Patrick E. McSharry, editors

Biomolecular Computation for Bionanotechnology, Jian-Qin Liu and Katsunori Shimohara

Electrotherapeutic Devices: Principles, Design, and Applications, George D. O'Clock

Intelligent Systems Modeling and Decision Support in Bioengineering, Mahdi Mahfouf

Matching Pursuit and Unification in EEG Analysis, Piotr Durka

Microfluidics for Biotechnology, Jean Berthier and Pascal Silberzan

Recent Advances in Diagnostic and Therapeutic 3-D Ultrasound Imaging for Medical Applications, Jasjit S. Suri, Ruey-Feng Chang, Chirinjeev Kathuria, and Aaron Fenster

Systems Bioinformatics: An Engineering Case-Based Approach, Gil Alterovitz and Marco F. Ramoni, editors

Text Mining for Biology and Biomedicine, Sophia Ananiadou and John McNaught

For further information on these and other Artech House titles, including previously considered out-of-print books now available through our In-Print-Forever® (IPF®) program, contact:

Artech House Publishers	Artech House Books
685 Canton Street	46 Gillingham Street
Norwood, MA 02062	London SW1V 1AH UK
Phone: 781-769-9750	Phone: +44 (0)20 7596 8750
Fax: 781-769-6334	Fax: +44 (0)20 7630 0166
e-mail: artech@artechhouse.com	e-mail: artech-uk@artechhouse.com

Find us on the World Wide Web at: www.artechhouse.com